NEURAL MECHANISMS OF CARDIOVASCULAR REGULATION

NEURAL MECHANISMS OF CARDIOVASCULAR REGULATION

edited by

Nae J. Dun, Ph.D.
*Temple University School of Medicine
Philadelphia, PA, USA*

Benedito H. Machado, Ph.D.
*School of Medicine of Ribeirão Preto
University of São Paulo, Brazil*

Paul M. Pilowsky, B.M.B.S., Ph.D.
*Faculty of Medicine
University of Sydney, Australia*

KLUWER ACADEMIC PUBLISHERS
Boston / Dordrecht / London

Distributors for North, Central and South America:
Kluwer Academic Publishers
101 Philip Drive
Assinippi Park
Norwell, Massachusetts 02061 USA
Telephone (781) 871-6600
Fax (781) 681-9045
E-Mail <kluwer@wkap.com>

Distributors for all other countries:
Kluwer Academic Publishers Group
Post Office Box 322
3300 AH Dordrecht, THE NETHERLANDS
Telephone 31 786 576 000
Fax 31 786 576 254
E-Mail <services@wkap.nl>

 Electronic Services <http://www.wkap.nl>

Library of Congress Cataloging-in-Publication Data

Neural mechanisms of cardiovascular regulation / edtied by Nae J. Dun, Benedito H. Machado, Paul M. Pilowsky
 p. cm.
 Includes bibliographical references and index.
 ISBN 1-4020-7711-4 (alk. paper)
  1. Neurons. 2. Neurophysiology. 3. Nervous system. 4. Cardiovascular system. I. Dun, Nae J. II. Machado, Benedito Honório. III. Pilowsky, P. M. (Paul M.)

QP361.N4623 2004
612.1—dc22   2003064005

Copyright © 2004 by Kluwer Academic Publishers.

All rights reserved. No part of this work may be reproduced, stored in a retrieval system, or transmitted in any form or by any means, electronic, mechanical, photocopying, microfilming, recording, or otherwise, without the written permission from the Publisher, with the exception of any material supplied specifically for the purpose of being entered and executed on a computer system, for exclusive use by the purchaser of the work

Permission for books published in Europe: permissions@wkap.nl
Permissions for books published in the United States of America: permissions@wkap.com

Printed on acid-free paper.
Printed in the United States of America

The Publisher offers discounts on this book for course use and bulk purchases.
For further information, send email to <melissa.ramondetta@wkap.com>.

Table of Contents

Preface .... *vii*

Chapter 1   **The Baroreceptor Reflex: Novel Methods and Mechanisms**
Mark W. Chapleau and Francois M. Abboud 1

Chapter 2   **Chemoreflex and Sympathoexcitation**
Benedito H. Machado 31

Chapter 3   **Cardiovascular Integration in the Nucleus of the Solitary Tract**
Michael C. Andresen 59

Chapter 4   **Neurotransmitters in the Nucleus Tractus Solitarius Mediating Cardiovascular Function**
Hreday N. Sapru 81

Chapter 5   **Cardiovascular Pathways Revealed With Functional Neuroanatomy**
Teresa L. Krukoff 99

Chapter 6   **The Hypothalamus and Cardiovascular Regulation**
John H. Coote 117

Chapter 7   **Cellular Properties of Autonomic-Related Neurons in the Paraventricular Nucleus of the Hypothalamus**
Javier Stern 147

Chapter 8   **The Anterior Hypothalamus and Salt-Sensitive Hypertension**
Suzanne Oparil and J. Michael Wyss 163

Chapter 9   **The Presympathetic Cells of the Rostral Ventrolateral Medulla (RVLM): Anatomy, Physiology and Role in the Control of Circulation**
Patrice Guyenet and Ruth L. Stornetta 187

Chapter 10	Serotonin Neurons in the Brainstem and Spinal Cord: Diverse Projections and Multiple Functions Paul M. Pilowsky ....…..….219
Chapter 11	Medullary Raphe Neurons in Autonomic Regulation Shaun F. Morrison ……………………………….…... 245
Chapter 12	Interneuronal Inputs to Sympathetic Preganglionic Neurons: Evidence from Transected Spinal Cord Ida J. Llewellyn-Smith[1] and Lynne C. Weaver[2] ….…..265
Chapter 13	Sympathetic Preganglionic Neurons: Electrical Properties and Response to Neurotransmitters G. Cristina Brailoiu and Nae J. Dun ………..….………285
Chapter 14	Neurochemical Heterogeneity in Sympathetic Ganglia and Its Implications for Cardiovascular Regulation Miguel A. Morales[1], John C. Hancock[2], and Donald B. Hoover[2] ……………………………..…………………….303
Chapter 15	Mammalian Cardiac Ganglia as Local Integration Centers: Histochemical and Electrophysiological Evidence Rodney L. Parsons ……………………………….…....335
Chapter 16	Parasympathetic Influences on Cerebral Circulation: A Link to Arterial Baroreflexes William T. Talman ……………………………...…….357
Chapter 17	Brainstem Premotor Cardiac Vagal Neurons David Mendelowitz ……………………………..…….371
Chapter 18	Genes Regulating Cardiovascular Function as Revealed Using Viral Vectors Julian F. R. Paton, Hidefumi Waki, Mohan Raizada, and Sergey Kasparov ……………………………..……...399

Index ……………………………….……….....................…..411

PREFACE:

The works here, represent a truly multidisciplinary and multinational approach to our understanding of some of the current questions about how neurons in the central and peripheral nervous systems regulate the cardiovascular system. The Editors make no apologies for the eclectic and diverse nature of the reviews presented here. In fact, it was never our aim to provide a complete, comprehensive, examination of every aspect of neural control mechanisms. Rather, we aimed to generate a series of thoughtful reviews that would provoke and illuminate, with the intention of revealing some of the ideas that current practitioners in the field of cardiovascular research are using to generate their current studies. If these prove in any way controversial and lead to more study then so much the better. We, therefore, apologize in advance if any reader considers that their favorite field is underrepresented.

Despite any apparent failings, which are all due to the Editors, we feel that the readers are treated here to a series of detailed expositions that range from the input side of cardiovascular control in terms of chemo- and baro- reflexes all the way to outputs that include local integration in vagal motoneurons, cardiac ganglia and sympathetic postganglionic neurons. The role of the nervous system in the genesis of hypertension is also addressed in the chapter on salt-sensitive hypertension.

There are a number of chapters that address the roles played within the brain by key centres and key populations of neurons including those in the nucleus tractus solitarius, ventrolateral medulla and hypothalamus. The role of new approaches such as viral vectors and modulation of gene function is also addressed. The importance of neurotransmitters and receptors is dealt with also, especially in terms of the ventrolateral medulla and spinal sympathetic nuclei.

It is our hope that this work will stimulate further discussion, constructive criticism and above all new experimentation. It is only through the questioning of accepted wisdoms that new knowledge may emerge.

The editors are extremely grateful to the many people who have made this work a reality. A multi-authored text is a difficult beast to control. In this case, the authors of the chapters delivered their works with good humor and in a timely fashion. The criticisms that came from the full peer review process were dealt with in good part, good humor and were speedily resolved. All of the authors thank the granting agencies that support the research that takes place in their laboratories and the Editors, in particular, would like to thank Ms Natalie Costin for her careful proof-reading of each Chapter and Mrs. Lottie Winters for her superb work in proof-reading and coordinating the production of this volume.

N.J. DUN	B.H. MACHADO	P.M. PILOWSKY
East Tennessee State University	University of Sao Paulo	University of Sydney
USA	BRAZIL	AUSTRALIA
dunnae@mail.etsu.edu	bhmachad@fmrp.usp.br	pilowsky@med.usyd.edu.au

Chapter 1

THE BARORECEPTOR REFLEX: NOVEL METHODS AND MECHANISMS

Mark W. Chapleau[1] and Francois M. Abboud[2]
The Cardiovascular Center,[1,2] the Departments of Internal Medicine,[1,2] and Physiology and Biophysics,[1,2] The University of Iowa, Iowa City, IA 52242, and the Veterans Affairs Medical Center,[1] Iowa City, IA 52246, USA

Abstract: The baroreceptor reflex is a key blood pressure regulatory mechanism. This chapter provides an overview of recent advances focusing on novel experimental approaches and mechanisms that determine baroreflex sensitivity. Gene discovery and technological advances now enable molecular mechanisms essential to baroreflex function to be defined. Recent discoveries include: mechanosensitive DEG/ENaC ion channels that mediate baroreceptor mechanoelectrical transduction, the complementary roles of voltage-gated ion channels and autocrine/paracrine factors in modulation of baroreceptor sensitivity, and novel molecular determinants of parasympathetic efferent control of heart rate. Mechanisms contributing to decreased baroreflex sensitivity in pathological states associated with endothelial dysfunction, oxidative stress, platelet activation, and neurohumoral activation are discussed, and future directions for research are suggested.

Key words: pressoreceptors, mechanosensitive channels, parasympathetic nerve activity, sympathetic nerve activity, oxidative stress, platelet activation, gene transfer

INTRODUCTION

Changes in arterial blood pressure (BP) are "sensed" by baroreceptor nerve endings located in carotid sinuses, aortic arch, and the origin of the right subclavian artery (Kirchheim, 1976). Baroreceptor activity is transmitted to the brainstem where the signals are integrated and relayed through a network of central neurons that determine parasympathetic nerve activity (paraSNA) and sympathetic nerve activity (SNA) to effector organs including the heart, vasculature, and kidneys (Abboud et al., 1976; Kirchheim, 1976; Abboud and Thames, 1983) (Fig. 1).

Changes in baroreceptor activity trigger reflex adjustments that buffer or oppose the change in BP (Abboud et al., 1976; Kirchheim, 1976; Abboud and Thames, 1983). A rise in BP increases baroreceptor activity leading to reflex inhibition of SNA, activation of paraSNA, and subsequent decreases in vascular resistance and heart rate (HR). Conversely, a fall in BP decreases baroreceptor activity eliciting a reflex increase in SNA, inhibition of paraSNA, and increases in vascular resistance and HR. In addition, circulating levels of norepinephrine, epinephrine, renin, and vasopressin are modulated by the baroreflex.

ParaSNA and SNA also influence electrical properties of the heart (Podrid et al., 1990). Related to this action, decreased baroreflex sensitivity predicts susceptibility to arrhythmias and sudden cardiac death in pathological states including myocardial infarction and heart failure (Kaye and Esler, 1995; LaRovere et al., 1998).

Thus, the baroreflex provides a powerful moment-to-moment negative feedback regulation of BP thereby reducing BP lability and its adverse consequences. In addition, the reflex may protect the heart from arrhythmias by providing appropriate and rapid modulation of autonomic tone.

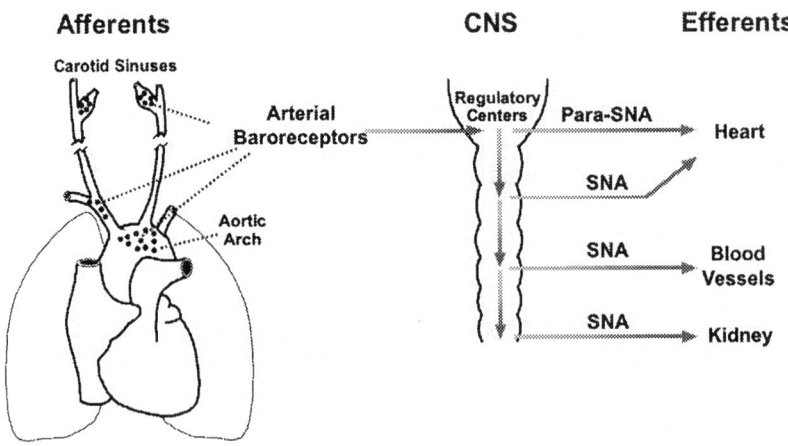

Figure 1. Afferent, central, and efferent baroreflex pathways. Shown are sites of baroreceptor innervation (filled circles) and efferent paraSNA and SNA directed to heart, blood vessels, and kidneys. Adapted and reprinted from Chapleau (2003) with permission.

NOVEL EXPERIMENTAL APPROACHES

Integrative Baroreflex Function

Traditional methods of assessing baroreflex function require interventions to change transmural arterial BP (e.g., administration of vasoconstrictor and dilator agents, neck suction/pressure) in order to evoke baroreflex-mediated changes in HR, SNA, and/or vascular resistance (Kirchheim, 1976; Mancia and Mark, 1983). These methods provide control of the BP stimulus to the baroreceptors and can assess the regional selectivity of the reflex, but do *not* measure the effectiveness of baroreflex buffering of BP changes. New methods of "baroreflex testing" summarized below may provide useful, unique information and are applicable to humans.

Baroreflex Buffering of Blood Pressure Changes

Baroreflex BP buffering capacity can be assessed in humans by measuring the potentiation of the BP response to the vasoconstrictor phenylephrine by ganglionic blockade (Jordan et al., 2000, 2002; Christou et al., 2003; Jones et al., 2003). The main determinants of the increase in BP are the vascular responsiveness to α-adrenergic receptor stimulation and the extent of baroreflex buffering of the rise in BP. Consequently, interruption of the baroreflex by ganglionic blockade enhances the pressor response in proportion to the baroreflex buffering capacity. Baroreflex buffering capacity is a major determinant of the BP response to vasoactive drugs in humans with important clinical implications (Jordan et al., 2002). Furthermore, buffering capacity is impaired in patients with multiple system atrophy, in subsets of hypertensive patients, and with aging (Jordan et al., 2000, 2002; Jones et al., 2003).

Spontaneous Baroreflex Sensitivity

The laboratory-based methods discussed above require specialized equipment and invasive procedures and are not suited for screening large numbers of patients. Furthermore, vasoactive drugs may influence baroreflex sensitivity by changing the compliance of the carotid sinuses and aortic arch and/or by actions in the central nervous system (CNS). To avoid these limitations, methods have been developed to calculate *spontaneous baroreflex sensitivity* from spontaneous fluctuations in BP and HR without the need for mechanical or pharmacological interventions. The techniques enable measurement of baroreflex sensitivity during natural behaviors, assessment of environmental or behavioral influences, and screening of large numbers of subjects.

Two general methods have been used. Calculation of the transfer function or α coefficient between fluctuations in HR (R-R interval) and BP at defined frequencies using *spectral analysis* provides a measure of baroreflex sensitivity (Robbe et al., 1987; Parati et al., 2000). The *"sequence method"* involves detection of sequences of three or more consecutive BP pulses where changes in systolic BP and R-R interval change in the same direction and are positively correlated (Parati et al., 2000). The average slope of the BP–R-R interval relationships is indicative of baroreflex sensitivity. The relative number of baroreflex sequences (baroreflex sequences/total BP sequences) provides an additional index of baroreflex function (Di Rienzo et al., 2001).

Assessment of "Mechanical" and "Neural" Components of Baroreflex

The baroreflex involves sensory transduction, central mediation of the reflex, and efferent neurocardiac and neurovascular transmission (Fig. 1). Sensory transduction involves a "mechanical component" related to arterial compliance and ionic mechanisms that generate nerve activity in the baroreceptor endings. Decreased arterial compliance, as occurs with aging (Abboud and Huston, 1961; Andresen, 1984), is often assumed to be the sole cause of decreased baroreflex sensitivity. The "neural component" of the reflex must also be considered (Hajduczok et al., 1991). Determining which component is altered under various states can be problematic in humans.

Simultaneous measurements of carotid diameter and R-R interval enable separate assessment of the "neural component" of the reflex (carotid diameter–HR relation) along with the "mechanical component" of sensory transduction (BP–carotid diameter relation) (Hunt et al., 2001a; Kornet et al., 2002). Age-related impairment of both components of the reflex has been demonstrated in humans using this approach (Hunt et al., 2001b; Kornet et al., 2002).

Application of Genetic and Molecular Approaches

Transgenic Animals and "Knockout" Mice

Gene discovery and the ability to create transgenic animals and "knockout" mice provide new opportunities to define molecular mechanisms essential to baroreflex function (Smithies, 1997; Gassmann and Hennet, 1998; Bockamp et al., 2002). Methods of analyzing baroreflex function have been successfully implemented in mice including analysis of spontaneous baroreflex sensitivity and afferent, central, and efferent components of the reflex (Stauss et al., 1999; Whiteis et al., 2000; Ma et al., 2002; Gross et al., 2002). Baroreceptor activity in the aortic depressor nerve (ADN) and SNA can be recorded during drug-induced changes in BP (Ling et al., 1998; Ma et al., 2001a, 2002) (Fig. 2). Measurement of reflex responses to electrical stimulation of the ADN provides an assessment of central/efferent mediation of the reflex while responses to stimulation of the peripheral end of the crushed right vagus nerve reflect efferent neurocardiac transmission (Ma et al., 2002).

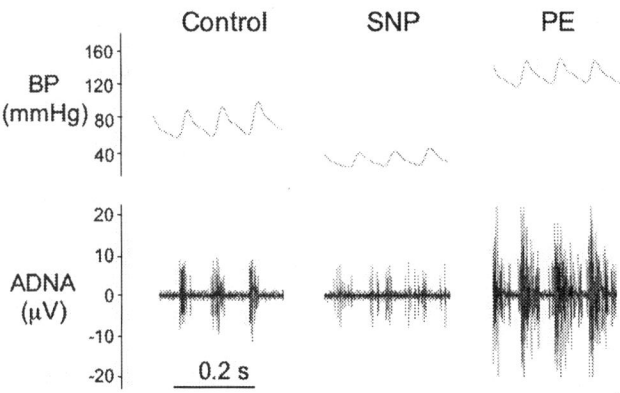

Figure 2. Shown are original recordings of BP and baroreceptor activity in ADN under baseline conditions and after administration of the vasodilator sodium nitroprusside (SNP) and the vasoconstrictor phenylephrine (PE) in an anesthetized mouse. Reprinted from Ma et al. (2002) with permission.

Measurement of reflex responses to bilateral carotid artery occlusion (BCO) provides an additional assay of baroreflex function (Krieger, 1963). Quantitation of the BCO reflex in mice during room air and again during 100% oxygen ventilation enables the relative contribution of the baroreflex and chemoreflex to the pressor response to be estimated (Alcayaga et al., 1986; Sun et al., 2000; Sun et al., 2002). Baroreflex function has been evaluated in a variety of genetically-modified mice (Table 1).

Table 1. Studies of baroreflex in genetically modified mice.

Model	HR	SNA	BP	Reference
Knockout Mice				
Angiotensin II AT_2 receptor	x			Gross et al., 2002
Neurokinin-1 receptor	x			Butcher et al., 1998
Bradykinin B_2 receptor	x			Madeddu et al., 1999
$\alpha 7$ nicotinic acetylcholine receptor	x			Franceschini et al., 2000
$\alpha 1 a/c$-adrenergic receptor	x			Rokosh & Simpson, 2002
G protein-gated K^+ channel GIRK4	x			Wickman et al., 1998
Type I neuronal NOS	x			Jumrussirikul et al., 1998
Type III endothelial NOS	x			Stauss et al., 1999
Endothelin-1	x	x		Ling et al., 1998
Atrial natriuretic peptide	x			Ackermann & Deliva, 2001
G protein-coupled receptor kinase-3	x			Walker et al., 1999
Dystrophin	x			Chu et al., 2002
Synapsin I and II		x		Zhang et al., 2000
Rab3A (GTP binding protein)		x		Zhang et al., 2000
Apolipoprotein E (apoE)		x		Ma et al., 2000
			x	Meyrelles & Chapleau, 2000
			x	Sun et al., 2001
Transgenic Mice				
β_1 adrenergic receptor in atria	x			Mansier et al., 1996
GTP-binding protein $G_{s\alpha}$ in heart	x			Uechi et al., 1998
Renin and angiotensinogen ($R^+ A^+$)	x			Merrill et al., 1996
		x		Ma et al., 1999
	x			Sakai et al., 2003

Gene Transfer

Gene transfer provides an additional approach to investigate molecular mechanisms and provide novel therapies for disease (Lafont et al., 1996; Slack and Miller, 1996). Local injections of viral vectors enable site-specific increases or decreases in gene expression (Meyrelles et al., 1997; Sinnayah et al., 2002; Stec et al., 2002). Gene transfer to carotid sinus, parasympathetic neurons, and key CNS cardiovascular regulatory centers has revealed novel mechanisms of cardiovascular regulation (Sakai et al., 2000;

1. THE BARORECEPTOR REFLEX: NOVEL METHODS AND MECHANISMS

Kishi et al., 2001; Paton et al., 2001; Li et al., 2002; Mohan et al., 2002; Zimmerman et al., 2002; Meyrelles et al., 2003).

Isolated Baroreceptor Neurons in Culture

Studies of baroreceptor afferents generally rely on measurement of action potential discharge in fibers distant from the site of mechanoelectrical transduction in the nerve endings, thereby limiting investigation of sensory transduction. Furthermore, the majority of baroreceptor afferents (85-90%) are small diameter unmyelinated C-fibers while ~10-15% are large diameter myelinated A-fibers (Kirchheim, 1976). Difficulty in recording activity from C-fibers has resulted in limited information on their properties.

Studies on isolated baroreceptor neurons in culture overcome some of these limitations (Cunningham et al., 1995; Li et al., 1997a, 1988; Sullivan et al., 1997; Drummond et al., 1998; Kreske et al., 1998; Snitsarev et al., 2002a). Aortic baroreceptor neurons can be labeled *in vivo* by application of the fluorescent dye DiI to the aortic arch and the baroreceptor neurons isolated and studied 1-3 weeks later (Fig. 3). Differential expression of ion channels and neurochemical markers in myelinated and unmyelinated afferents enables A-type neurons (myelinated) and C-type neurons (unmyelinated) to be distinguished in culture (Belmonte and Gallego, 1983; Undem and Weinreich, 1993; Schild and Kunze, 1997; Doan and Kunze, 1999; Drew et al., 2002). Ligand receptors and ion channels present on the nerve endings are also present on the soma of cultured nodose neurons (Fowler et al., 1985a; Stansfield et al., 1986; Christian et al., 1989; Undem and Weinreich, 1993). Differences in functional properties (e.g., mechanosensitivity and spike frequency adaptation) between different sensory nerve terminals are also evident at the soma (Harper, 1991; Drew et al., 2002). Cultured baroreceptor neurons are mechanosensitive (Cunningham et al., 1995; Sullivan et al., 1997; Kraske et al., 1998; Drummond et al., 1998; Snitsarev et al., 2002a).

Thus, although differences in expression and regulation of molecules in sensory terminals vs. isolated neuron are expected, the isolated baroreceptor neuron appears to be a valid model to investigate sensory mechanisms.

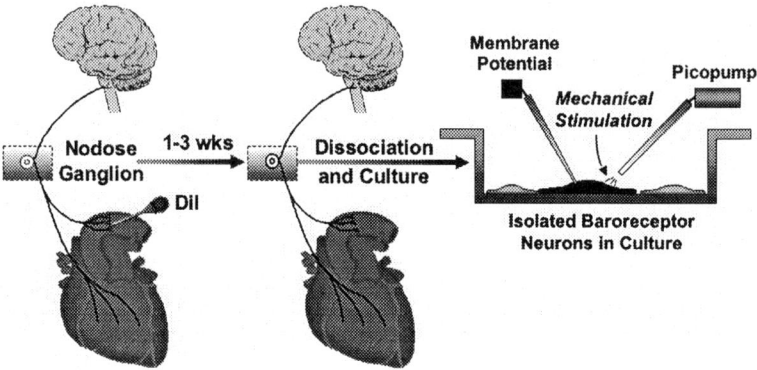

Figure 3. Identification of isolated baroreceptor neurons in culture. The schematic shows the approach used to label baroreceptor neurons with the fluorescent dye DiI in vivo, enabling later identification and study of the labeled neurons in culture. Adapted and reprinted from Chapleau et al. (2001) with permission.

BARORECEPTOR SENSORY TRANSDUCTION

Baroreceptor sensory transduction involves three major processes: vascular distension and transmission of mechanical force to the baroreceptor endings, mechanically-induced depolarization of the endings (mechano-electrical transduction), and encoding of depolarization into action potential discharge at the spike initiating zone (SIZ) (Katz, 1950) (Fig. 4).

The compliance of carotid sinus and aortic arch influences the magnitude of vascular distension and is an important determinant of baroreceptor sensitivity (Kirchheim, 1976; Andresen, 1984; Hunt et al., 2001b). The neural component of sensory transduction is reviewed below.

1. THE BARORECEPTOR REFLEX: NOVEL METHODS AND MECHANISMS

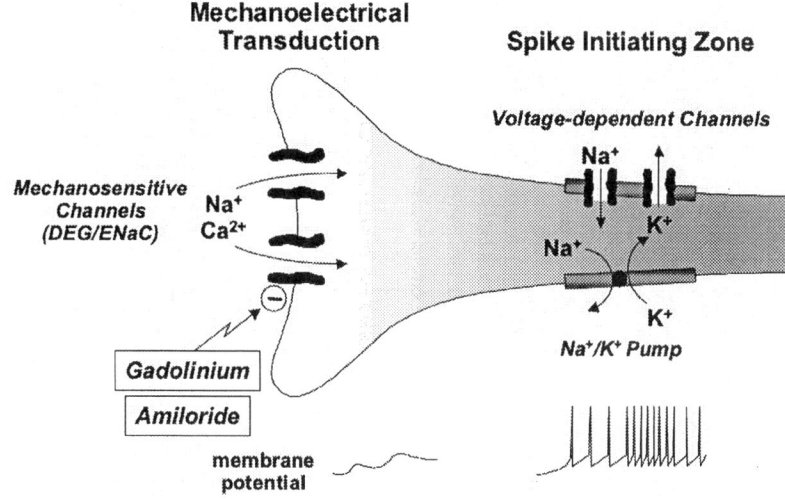

Figure 4. Mechanisms of baroreceptor activation during mechanical stimulation. Mechanoelectrical transduction involves opening of mechanosensitive channels leading to membrane depolarization. Depolarization of the spike initiating zone triggers action potential discharge by opening voltage-dependent channels. Adapted and reprinted from Chapleau et al., 2001 with permission.

Mechanosensitive Ion Channels

Mechanosensitive ion channels have been implicated as mechanosensors in a variety of tissues and cells (Welsh et al., 2002). Gadolinium, a mechanosensitive ion channel blocker (Yang and Sachs, 1989), attenuates pressure-induced increases in baroreceptor activity in rabbits (Hajduczok et al., 1994) and blocks mechanically-induced inward current and increases in cytosolic Ca^{2+} in isolated baroreceptor neurons (Cunningham et al., 1995; Sullivan et al., 1997).

Accumulating evidence suggests that members of the Degenerin/Epithelial Na^+ Channel (DEG/ENaC) family are components of mechanosensitive channels in mechanoreceptors (Tavernarakis and Driscoll, 2001; Welsh et al., 2002). Amiloride or its analog benzamil (DEG/ENaC channel blockers) inhibit pressure-induced baroreceptor activity in rabbits and mechanically-induced depolarization and increases in cytosolic Ca^{2+} in isolated baroreceptor neurons (Drummond et al., 1998; Snitsarev et al.,

2002a). The mechanically-induced depolarization is abolished by amiloride at concentrations (1 μM) that do not attenuate action potential discharge evoked by depolarizing current injection (Snitsarev et al., 2002a) (Fig. 5). Furthermore, mRNA for β and γ subunits of ENaC and Acid Sensing Ion Channel 2 (ASIC2, also referred to as BNC1) are present in nodose ganglia and γENaC protein has been localized in DiI-labeled baroreceptor sensory terminals (Drummond et al., 1998; Ma et al., 2001b).

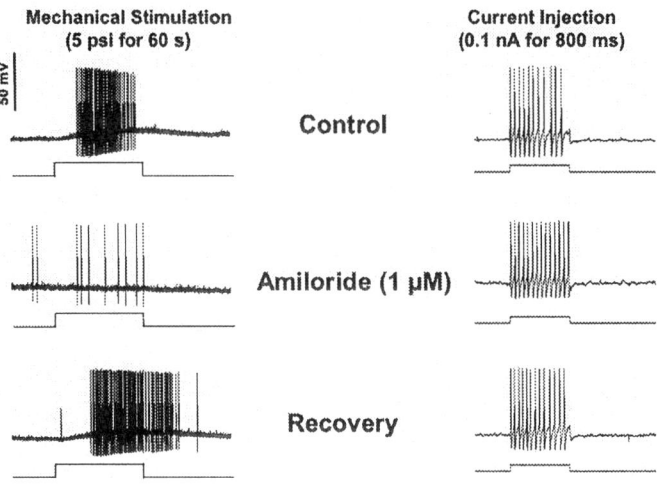

Figure 5. Shown are membrane potential and action potential responses to mechanical stimulation (left) and depolarizing current injection (right) in an isolated nodose sensory neuron. Amiloride (1 □M) abolished the mechanically-induced depolarization without inhibiting current-evoked action potential firing. The neuron was mechanically stimulated by buffer ejected from a micropipette under pressure. Reprinted from Snitsarev et al. (2002a).

The functional role of ASIC2 (BNC1) was investigated *in vivo* by recording baroreceptor activity in ASIC2-deficient and littermate control mice (Ma et al., 2001b). The ability to maintain increased baroreceptor activity over time during a phenylephrine-induced increase in BP was impaired in ASIC2-deficient mice consistent with a role of ASIC2 in mechanoelectrical transduction.

These results, taken together, strongly support the hypothesis that DEG/ENaC subunits are components of the mechanosensitive ion channel complex in baroreceptors.

Voltage-gated Ion Channels and Na$^+$/K$^+$ Pump

Mechanically-induced depolarization is localized to sensory terminals and rapidly decays with distance from the site of stimulation (Katz, 1950). Consequently, action potential generation and propagation are essential for signaling the CNS of a rise in BP. Depolarization is encoded into action potential discharge when membrane potential reaches and exceeds the "threshold" for opening of voltage-dependent Na$^+$ and K$^+$ channels (Katz, 1950) (Fig. 4). The frequency of action potential discharge increases with further depolarization and is critically dependent on the expression and properties of voltage-dependent channels and pumps at the SIZ (Fig. 4).

Opening of tetrodotoxin-sensitive (TTX-S) and TTX-resistant (TTX-R) *Na$^+$ channels* mediate the upstroke of the action potential. TTX-S and TTX-R Na$^+$ channels are differentially expressed in A-type and C-type baroreceptor neurons and the relative expression of these channels influences neuronal excitability (Schild and Kunze, 1997).

Opening of *K$^+$ channels* mediate membrane hyperpolarization, repolarization of the action potential, spike after-hyperpolarization, and interspike interval. K$^+$ channels susceptible to inhibition by low concentrations of 4-aminopyridine (4-AP) and α-dendrotoxin are important determinants of neuronal excitability (Stansfeld et al., 1986; Glazebrook et al., 2002). Large and small conductance Ca^{2+}–activated K$^+$ channels mediate brief and prolonged spike after-hyperpolarizations in nodose neurons (Fowler et al., 1985b).

Hyperpolarization-activated inward current (I$_H$) is highly expressed in A-type nodose neurons with less expression in C-type neurons. I$_H$ contributes to the resting membrane potential and is an important determinant of membrane excitability (Doan and Kunze, 1999).

Depolarization and Na$^+$ influx during action potential discharge activates an electrogenic *Na$^+$/K$^+$ pump* that exerts a hyperpolarizing influence on membrane potential.

In addition to encoding membrane depolarization into action potential discharge, voltage-dependent channels/pumps are key mediators of two forms of *dynamic baroreceptor modulation* discussed below.

Baroreceptor Adaptation and Resetting

During a *sustained* increase in BP, baroreceptor activity increases initially but declines *(adapts)* over time as the elevated pressure is maintained (Coleridge et al., 1984; Chapleau et al., 1993). Furthermore, baroreceptor activity is inhibited (*post-excitatory depression*), the pressure threshold is increased, and the BP-activity function curve is *reset* to higher levels of BP after a period of acute hypertension (Saum et al., 1976; Heesch et al., 1984; Andresen, 1984).

Baroreceptor adaptation is markedly attenuated by the K^+ channel blocker 4-AP (Chapleau et al., 1993). The inhibitory effect of 4-AP is selective; it does *not* influence peak nerve activity during the initial rise in pressure, the slope of the pressure-activity relation, vascular compliance, or the post-excitatory depression and baroreceptor resetting that occur after the period of elevated pressure (van Brederode et al., 1990; Chapleau et al., 1993; Drummond and Seagard, 1994). 4-AP also attenuates spike frequency adaptation in isolated sensory neurons during sustained injections of depolarizing current (Stansfeld et al., 1986).

Conversely, inhibitors of the Na^+ pump (e.g. ouabain, low K^+) significantly attenuate post-excitatory depression and baroreceptor resetting without influencing spike frequency adaptation (Saum et al., 1976; Heesch et al., 1984; Chapleau et al., 1993).

The results suggest that opening of 4-AP-sensitive K^+ channels mediates baroreceptor adaptation during acute increases in BP while Na^+ pump activation mediates post-excitatory depression and acute baroreceptor resetting.

Chemical Sensitivity of Baroreceptors

Vasoactive hormones may affect baroreceptor activity indirectly by altering vascular diameter and compliance (Kirchheim, 1976). In addition, a variety of circulating and locally-produced chemical factors modulate baroreceptor sensitivity through direct actions on the nerve endings. Voltage-gated ion channels represent key molecular targets of chemical factors.

Inhibition of *"leak" K^+ channels* and/or *Ca^{2+}– activated K^+ channels* mediate the *excitatory actions* of several factors on nodose sensory neurons (Fowler et al., 1985a; Christian et al., 1989; Undem and Weinreich, 1993; Weinreich et al., 1995). *Prostacyclin* (PGI_2) depolarizes and increases excitability of isolated baroreceptor neurons through inhibition of these channels (Li et al., 1997a; Snitsarev et al., 2001). Enhancement of *TTX-R Na^+ current* also may contribute to activation of C-type neurons by prostanoids (Gold et al., 1996).

Nitric oxide (NO) and *reactive oxygen species* (ROS) inhibit baroreceptor activity (Matsuda et al., 1995; Li et al., 1996), apparently through direct

1. THE BARORECEPTOR REFLEX: NOVEL METHODS AND MECHANISMS

interaction with Na^+ and/or K^+ channels. NO donors applied to isolated baroreceptor neurons inhibit voltage-gated Na^+ currents through a cyclic GMP-independent, nitrosylation-dependent mechanism (Li et al., 1998; Bielefeldt et al., 1999). The baroreceptor Na^+ currents are also inhibited by oxidizing agents (Li et al., 1997b). NO and ROS enhance the activity of specific voltage-gated and Ca^{2+}-activated K^+ channels via interactions at nitrosylation consensus sequences and cysteine and methionine residues (Ruppersberg et al., 1991; Ciorba et al., 1997; Ciorba et al., 1999; Tang et al., 2001). NO contributes to spike after-hyperpolarizations mediated by Ca^{2+}-activated K^+ channels in C-type nodose neurons (Cohen et al., 1994).

MODULATION OF BAROREFLEX SENSITIVITY

A wide variety of mechanisms acting at sensory, CNS, and efferent sites modulate baroreflex sensitivity. A comprehensive review of the mechanisms is beyond the scope of this chapter. Examples of recently-discovered mechanisms of potential clinical significance are summarized.

Paracrine Modulation of Baroreceptor Sensitivity

The importance of paracrine modulation of vascular tone and platelet function, e.g. by endothelium-derived NO and PGI_2, is widely appreciated (Mombouli and Vanhoutte, 1999). Furthermore, endothelial dysfunction, oxidative stress, and platelet activation are major underlying causes of cardiovascular disease (Mombouli and Vanhoutte, 1999). Increasing evidence indicates that paracrine mechanisms modulate the sensitivity of baroreceptor afferents with important clinical implications.

Prostacyclin and Nitric Oxide

While *PGI_2* and *NO* both exert vasodilator and anti-platelet actions, these factors exert opposing actions on baroreceptor sensitivity. PGI_2 injected into isolated carotid sinuses of rabbits *increases* baroreceptor sensitivity (McDowell et al., 1989; Chen et al., 1990), while NO and NO donors *decrease* sensitivity (Matsuda et al., 1995). The effects of PGI_2 and NO are *not* related to their effects on vascular tone and appear to be mediated through direct interaction with K^+ and Na^+ channels on the baroreceptor endings (Nuyt et al., 1995; Li et al., 1997a, 1998).

Exposure of the isolated carotid sinus to the cyclooxygenase inhibitor indomethacin decreases baroreceptor activity suggesting that endogenous PGI_2 contributes to activation of baroreceptors during increases in BP (Chen et al., 1990; Xie et al., 1990; Wang et al., 1993). Indomethacin fails to decrease baroreceptor activity in hypertensive or atherosclerotic rabbits suggesting that impaired production of PGI_2 in these states may contribute to decreased baroreceptor sensitivity (Xie et al., 1988, 1990).

Modification of paracrine influences may provide an approach to *chronically* modulate baroreceptor function. Gene transfer of eNOS to carotid sinus adventitia produces a sustained resetting of the baroreceptor pressure-activity curve to higher pressures (Meyrelles et al., 2003) (Fig. 6).

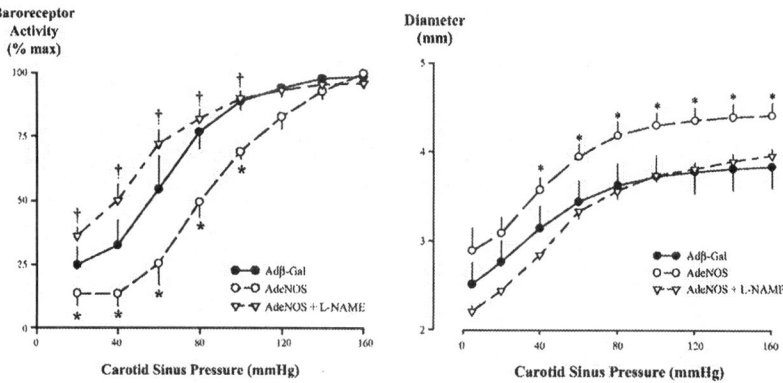

Figure 6. Baroreceptor resetting by gene transfer of eNOS. Adenoviral vectors encoding the reporter gene β-Gal (control, n=5) or the eNOS gene (n=5) were applied topically to carotid sinuses of rabbits 45 days before isolation of the sinus and recording of baroreceptor activity (left) and carotid artery diameter (right). The NOS inhibitor L-NAME was applied together with AdeNOS in a third group of rabbits (n=5). NOS gene transfer inhibited baroreceptor activity, reset the pressure-activity curve to higher pressures, and increased carotid diameter. Responses were blocked by L-NAME. Adapted and reprinted from Meyrelles et al. (2003) with permission.

Oxidative Stress

The hypothesis that oxidative stress contributes to decreased baroreceptor sensitivity in atherosclerosis has been evaluated in two animal models. Brief exposure (10-15 min) of the isolated carotid sinus to the ROS scavengers superoxide dismutase (SOD) and catalase significantly increases baroreceptor activity in cholesterol-fed atherosclerotic rabbits, but fails to increase activity in normal healthy rabbits (Li et al., 1996). Chemical

1. THE BARORECEPTOR REFLEX: NOVEL METHODS AND MECHANISMS

generation of ROS in the isolated carotid sinus with xanthine and xanthine oxidase decreased baroreceptor activity, confirming that ROS impair baroreceptor sensitivity (Li et al., 1996).

Apolipoprotein E (apoE) deficient mice with atherosclerosis exhibit impaired baroreflex control of renal SNA (Ma et al., 2000) and a decrease in the baroreflex component of the pressor response to carotid occlusion (Meyrelles and Chapleau, 2000; Sun et al., 2001). Gene transfer of SOD and catalase to carotid sinus adventitia restores the carotid occlusion reflex in the apoE knockout mice (Meyrelles and Chapleau, 2000).

The concept that oxidative stress decreases baroreflex sensitivity is supported by the recent finding that acute intravenous infusion of vitamin C partially restores baroreflex sensitivity in patients with heart failure (Nightingale et al., 2003; Piccirillo et al., 2003).

Platelet Activation

Carotid sinuses are common sites of platelet aggregation in patients with cardiovascular disease (Isaka et al., 1989). We hypothesized that factors released from activated platelets modulate baroreceptor sensitivity. Injection of thrombin-activated platelets into the isolated carotid sinus, while maintaining carotid sinus pressure constant, rapidly *increases* the activity of a subpopulation of baroreceptor afferents and reflexly abolishes renal SNA resulting in profound hypotension (Chapleau et al., 1995; Mao et al., 1996) (Fig. 7). The rapid platelet-induced inhibition of SNA and hypotension are transient and are mediated by platelet-derived serotonin (5-HT) acting on 5-HT_3 and 5-HT_2 receptors (Mao et al., 1996).

After sustained exposure (10-15 min) to activated platelets, baroreceptor sensitivity to changes in carotid sinus pressure is markedly *decreased* (Li et al., 1992, 1995). Cell-free releasate collected from activated platelets also decreases baroreceptor sensitivity indicating that the inhibition is mediated by a stable diffusible factor (Li and Chapleau, 1995). The platelet-induced decrease in baroreceptor sensitivity is *not* inhibited, and is actually enhanced after exposure of the sinus to indomethacin indicating that it is *not* mediated by thromboxane or other cyclooxygenase metabolites (Li et al., 1995). Interestingly, human subjects with higher platelet counts (that may correspond to increased platelet activation) tend to exhibit lower baroreflex sensitivity (Yasumasu et al., 2003).

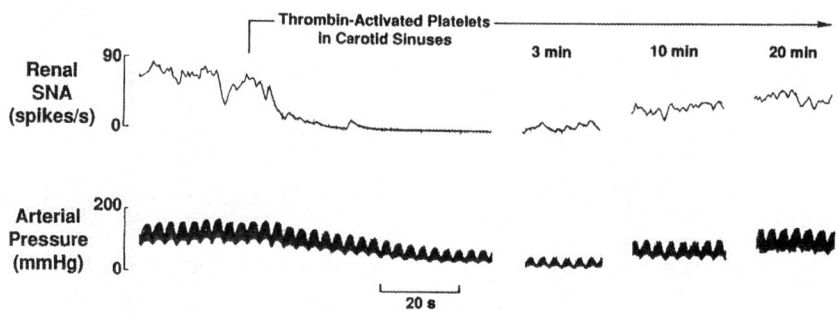

Figure 7. Inhibition of renal SNA and hypotension triggered by platelet activation in carotid sinuses. Thrombin-activated rabbit platelets were injected into the isolated carotid sinuses of anesthetized rabbits while maintaining carotid sinus pressure constant at 80 mmHg. Reprinted from Mao et al. (1996) with permission.

Putative Autocrine Factors

Mechanical or electrical stimulation of sensory neurons is known to release chemical mediators including substance P, calcitonin gene related peptide, and adenosine triphosphate (ATP). These "autocrine factors" may serve diverse functions, e.g., vasodilation to increase local blood flow, modulation of sensory nerve activity, and/or participation as a central neurotransmitter. Recent studies suggest that brain derived neurotrophic factor (BDNF), NO, and ROS may function as autocrine factors in baroreceptor neurons.

BDNF is expressed in carotid sinuses and aortic arch of the fetus and is essential for development of baroreceptor innervation at these sites (Brady et al., 1999). After birth, BDNF is expressed in baroreceptor neurons, is released from nodose/petrosal ganglion neuronal cultures during electrical stimulation, and modulates neurotransmission in the NTS (Brady et al., 1999; Balkowiec and Katz, 2000; Balkowiec et al., 2000).

Neuronal NOS (nNOS) is present in carotid sinus afferent nerve fibers and expressed in a subpopulation of isolated baroreceptor neurons (Hohler et al., 1994; Li et al., 1998). NO scavengers or NOS inhibitors increase voltage-gated Na^+ currents in isolated baroreceptor neurons, suggesting that NO may function as an autocrine regulator of membrane excitability (Li et al., 1998). Preliminary results suggest that NO and ROS mediate activity-dependent inhibition of membrane excitability in isolated baroreceptor neurons (Snitsarev et al., 2002b, 2002c).

Baroreflex Dysfunction Caused by Neurohumoral Activation

Decreased baroreflex sensitivity may contribute to excessive neurohumoral drive in pathological states such as hypertension and heart failure. Less appreciated is the potential for neurohumoral activation to exacerbate the impairment in baroreflex sensitivity.

Excessive Sympathetic Nerve Activity

Chemical sympathectomy prevents impairment of baroreflex-mediated bradycardia in rats with myocardial infarction-induced heart failure, suggesting that *elevated SNA* contributes to the decreased baroreflex sensitivity (Mircoli et al., 2002). Excessive SNA and increased levels of norepinephrine may restrain baroreflex–mediated bradycardia by presynaptic inhibition of acetylcholine release from cardiac vagal efferent nerves, antagonism of acetylcholine action on the sinoatrial node, and/or decreasing large artery compliance and baroreceptor sensitivity (Vizi, 1974; Mangoni et al., 1997). Enhanced sympathoexcitatory reflexes in heart failure also may contribute to decreased baroreflex sensitivity (Wang and Zucker, 1996; Ponikowski et al., 1997).

Activation of Renin-Angiotensin-Aldosterone System

Activation of the renin-angiotensin-aldosterone system contributes to decreased baroreflex sensitivity in heart failure and hypertension. Central administration of either the angiotensin AT_1 receptor antagonist losartan or the aldosterone receptor antagonist spironolactone restores baroreflex sensitivity in heart failure (DiBona et al., 1995; Francis et al., 2001). Multiple actions of *angiotensin* may influence baroreflex function (Reid, 1992; Brooks, 1997; Paton et al., 2001). In addition to its central action, *aldosterone* infused *chronically* decreases the sensitivity of *baroreceptor afferents* (Wang, 1994). Administration of ouabain restores baroreceptor sensitivity in both aldosterone-infused dogs and dogs with heart failure, suggesting that the inhibitory effect is mediated by increased Na^+ pump activity (Wang et al., 1990; Wang, 1994). Interestingly, aldosterone administered into the isolated carotid sinus decreases baroreceptor sensitivity *acutely* (Wang et al., 1992). This rapid inhibition is *not* reversed by ouabain but is abolished by endothelial denudation, suggesting that it is mediated by an "inhibitory factor" released from endothelium.

Molecular Determinants of Vagal Efferent Control of HR

ParaSNA plays a major role in baroreflex control of HR and generation of HR variability in mice as well as in humans (Jumrussirikul et al., 1998; Wickman et al., 1998; Just et al., 2000). Recent studies utilizing knockout mice and gene transfer have revealed important roles of *nNOS* and *G-protein-gated K^+ channels* in mediating the vagal efferent limb of baroreflex control of HR (Jumrussirikul et al., 1998; Wickman et al., 1998; Choate et al., 2001; Mohan et al., 2002). These molecules are attractive targets for therapy designed to reduce risk of arrhythmias and sudden death in pathological states.

Neuronal Nitric Oxide Synthase

Conscious nNOS-deficient mice exhibit increased resting HR and decreased HR variability due primarily to loss of parasympathetic control (Jumrussirikul et al., 1998). NO produced by nNOS in cardiac vagal efferent nerve terminals acts presynaptically to facilitate acetylcholine release during vagal efferent stimulation resulting in an enhanced decrease in HR (Choate et al., 2001; Mohan et al., 2002).

G protein-gated K^+ Channels

The postsynaptic binding of acetylcholine to cardiac muscarinic receptors decreases HR through multiple signaling pathways including activation of G protein-gated K^+ channels (I_{Kach}) that consist of GIRK subunits. Bradycardic responses to efferent vagal stimulation and to baroreceptor stimulation are each attenuated by ~50 % and low-frequency HR variability is essentially abolished, both at rest and during baroreceptor stimulation, in conscious GIRK4-deficient mice (Wickman et al., 1998).

FUTURE DIRECTIONS FOR RESEARCH

Despite the voluminous literature on the baroreceptor reflex, many questions remain to be answered. Fundamental mechanisms essential to afferent, central, and efferent components of the reflex are not understood at the molecular level. Very little is known regarding effects of development, aging, and chronic disease on gene expression in baroreflex pathways or the effect of altered gene expression on reflex function.

While the baroreflex is universally accepted as being of major importance in mediating rapid circulatory adjustments to acute changes in BP, its role in long-term regulation of BP has been questioned (Cowley et al., 1973). Recent studies, however, have observed sustained baroreflex-mediated changes in SNA and BP over a period of days to weeks (Thrasher, 2002; Lohmeier et al., 2001; Barrett et al., 2003), encouraging further research into the role of the reflex in longer-term BP regulation.

Of clinical importance is the development of "therapies" to reverse or prevent baroreflex dysfunction. A variety of interventions including exercise training, antioxidant administration, and implementation of bionic computer-based feedback regulation have shown potential for restoring baroreflex sensitivity in animals and humans (Monahan et al., 2000; Hunt et al., 2001b; LaRovere et al., 2002; Liu et al., 2002; Sato et al., 2002; Piccirillo et al., 2003; Nightingale et al., 2003). Additional studies in this area are encouraged.

How can the areas discussed above be addressed experimentally? The availability of genome sequences and model systems amenable to genetic analysis (e.g., *C. elegans* and *Drosophila*) enables discovery of novel molecules and identification of mammalian homologues (Tavernarakis and Driscoll, 2001; Welsh et al., 2002). The ability to measure and experimentally manipulate gene expression should greatly facilitate study of effects of development, aging and disease on baroreflex function as well as the role of the baroreflex in long-term BP regulation. Novel animal models of disease can be created that represent a single genetic defect, or the interaction of multiple defects as so often occurs in patients. The enormous power of this technology is just beginning to be realized. Particularly exciting are techniques that enable spatial targeting of gene deletion (or overexpression) to specific cell types or anatomical regions and the creation of conditional or inducible knockouts and transgenics where gene expression can be altered reversibly (Bockamp et al., 2002).

The knowledge gained from sequencing the human genome should be applied to human studies. The influence of polymorphisms in candidate genes on baroreflex sensitivity can be evaluated. For example, polymorphisms in a Ca^{2+}–activated K^+ channel and a mutation in the norepinephrine transporter gene have recently been associated with baroreflex sensitivity and orthostatic intolerance, respectively (Shannon et al., 2000; Gollasch et al., 2002). Patients with a rare type of autosomal dominant genetic hypertension demonstrate severe impairment of baroreflex BP buffering capacity with only mild inhibition of baroreflex control of HR (Jordan et al., 2000), suggesting a selective central impairment of the reflex.

Thus, as in most biomedical research, the future for baroreflex research is in a multidisciplinary approach including molecular and genetic techniques, functional assays *in vitro*, and studies of integrative physiology in intact animals and humans in health and disease.

The challenge is great, but so are the opportunities.

ACKNOWLEDGMENTS

The authors acknowledge the contributions of Drs. Xiuying Ma, Vladislav Snitsarev, Zhi Li, Silvana Meyrelles, Heather Drummond, Xin Su, Huizhen Mao, Wei Sun, Rubens Fazan Jr., and Klaus Bielefeldt to work from our laboratory included in this review. This publication was made possible by a VA Merit Review Award to MWC from the Department of Veterans Affairs, an NIH Program Project Grant (HL14388), and funds from the Iowa Affiliate of the American Heart Association. Its contents are solely the responsibility of the authors and do not necessarily represent the official views of the NIH, the Department of Veterans Affairs, or the AHA.

REFERENCES

Abboud, F.M., Heistad, D.D., Mark, A.L., Schmid, P.G., 1976. Reflex control of the peripheral circulation. Prog. Cardiovasc. Dis. 18, 371-403.

Abboud, F.M., Huston, J.H., 1961. The effects of aging and degenerative vascular disease on the measurement of arterial rigidity in man. J. Clin. Invest. 40, 933-939.

Abboud, F.M., Thames, M.D., 1983. Interaction of cardiovascular reflexes in circulatory control. In: Shepherd, J.T., Abboud, F.M. (Eds.), Handbook of Physiology, Section 2, The Cardiovascular System, Vol. III, Peripheral Circulation and Organ Blood Flow, American Physiological Society, Bethesda, MD, pp. 675-753.

Ackermann, U., Deliva, R.D., 2001. Reduced baroreceptor sensitivity during hypotension in ANP-knockout mice. Can. J. Physiol. Pharmacol. 79, 201-205.

Alcayaga, J., Iturriaga, R., Zapata, P., 1986. Carotid body chemoreceptor excitation produced by carotid occlusion. Acta Physiol. Pharmacol. Latinoam. 36, 199-215.

Andresen, M.C., 1984. Short- and long-term determinants of baroreceptor function in aged normotensive and spontaneously hypertensive rats. Circ. Res. 54, 750-759.

Balkowiec, A., Kunze, D.L., Katz, D.M., 2000. Brain-derived neurotrophic factor acutely inhibits AMPA-mediated currents in developing sensory relay neurons. J. Neurosci. 20, 1904-1911.

Balkowiec, A., Katz, D.M., 2000. Activity-dependent release of endogenous brain-derived neurotrophic factor from primary sensory neurons detected by ELISA in situ. J. Neurosci. 20, 7417-7423.

Barrett, C.J., Ramchandra, R., Guild, S.-J., Lala, A., Budgett, D.M., Malpas, S.C., 2003. What sets the long-term level of renal sympathetic nerve activity: a role for angiotensin II and baroreflexes? Circ. Res. 92, 1330-1336.

Belmonte, C., Gallego, R., 1983. Membrane properties of cat sensory neurones with chemoreceptor and baroreceptor endings. J. Physiol. 342, 603-614.

Bielefeldt, K., Whiteis, C.A., Chapleau, M.W., Abboud, F.M., 1999. Nitric oxide enhances slow inactivation of voltage-dependent sodium currents in rat nodose neurons. Neurosci. Lett. 271(3), 159-162.

Bockamp, E., Maringer, M., Spangenberg, C., Fees, S., Fraser, S., Eshkind, L., Oesch, F., Zabel, B., 2002. Of mice and models: improved animal models for biomedical research. Physiol. Genomics 11, 115-132.

Brady, R., Zaidi, S.I., Mayer, C., Katz, D.M., 1999. BDNF is a target-derived survival factor for arterial baroreceptor and chemoafferent primary sensory neurons. J. Neurosci. 19, 2131-2142.

Brooks, V.L., 1997. Interactions between angiotensin II and the sympathetic nervous system in the long-term control of arterial pressure. Clin. Exp. Pharmacol. Physiol. 24, 83-90.

Butcher, J.W., De Felipe, C., Smith, A.J.H., Hunt, S.P., Paton, J.F.R., 1998. Comparison of cardiorespiratory reflexes in NK_1 receptor knockout, heterozygous and wild-type mice in vivo. J. Auton. Nerv. Syst. 69, 89-95.

Chapleau, M.W., 2003. Arterial baroreflexes. In: Izzo, J.L., Black, H.R. (Eds.), Hypertension Primer, 3rd Edition. American Heart Association. Dallas, TX, pp. 103-106.

Chapleau, M.W., Li, Z., Meyrelles, S.S., Ma, X., Abboud, F.M., 2001. Mechanisms determining sensitivity of baroreceptor afferents in health and disease. Ann. N.Y. Acad. Sci. 940, 1-19.

Chapleau, M.W., Lu, J., Hajduczok, G., Abboud, F.M., 1993. Mechanism of baroreceptor adaptation in dogs: Attenuation of adaptation by the K^+ channel blocker 4-aminopyridine. J. Physiol. (Lond.) 462, 291-306.

Chapleau, M.W., Su, X., Li, Z., 1995. Platelets aggregating in carotid sinus selectively modulate activity of baroreceptor fiber types (abstract). FASEB J. 9(3), A7.

Chen, H.I., Chapleau, M.W., McDowell, T.S. Abboud, F.M., 1990. Prostaglandins contribute to activation of baroreceptors in rabbits: possible paracrine influence of endothelium. Circ. Res. 67, 1394-1404.

Choate, J.K., Danson, E.J., Morris, J.F., Paterson, D.J., 2001. Peripheral vagal control of heart rate is impaired in neuronal NOS knockout mice. Am. J. Physiol. Heart Circ. Physiol. 281, H2310-H2317.

Christian, E.P., Taylor, G.E., Weinreich, D., 1989. Serotonin increases excitability of rabbit C-fiber neurons by two distinct mechanisms. J. Appl. Physiol. 67, 584-591.

Christou, D.D., Jones, P.P., Seals, D.R., 2003. Baroreflex buffering in sedentary and endurance exercise-trained healthy men. Hypertension 41, 1219-1222.

Chu, V., Otero, J.M., Lopez, O., Sullivan, M.F., Morgan, J.P., Amende, I., Hampton, T.G., 2002. Electrocardiographic findings in mdx mice: a cardiac phenotype of Duchenne muscular dystrophy. Muscle Nerve 26, 513-519.

Ciorba, M.A., Heinemann, S.H., Weissbach, H., Brot, N., Hoshi, T., 1997. Modulation of potassium channel function by methionine oxidation and reduction. Proc. Natl. Acad. Sci. USA 94, 9932-9937.

Ciorba, M.A., Heinemann, S.H., Weissbach, H., Brot, N., Hoshi, T., 1999. Regulation of voltage-dependent K^+ channels by methionine oxidation: effect of nitric oxide and vitamin C. FEBS Lett. 442, 48-52.

Cohen, A.S., Weinreich, D., Kao, J.P.Y., 1994. Nitric oxide regulates spike frequency accommodation in nodose neurons of the rabbit. Neurosci. Lett. 173, 17-20.

Coleridge, H.M., Coleridge, J.C.G., Poore, E.R., Roberts, A.M., Schultz, H.D., 1984. Aortic wall properties and baroreceptor behavior at normal arterial pressure and in acute hypertensive resetting in dogs. J. Physiol. (Lond) 350, 309-326.

Cowley, A.W., Liard, J.F., Guyton, A.C., 1973. Role of baroreceptor reflex in daily control of arterial blood pressure and other variables in dogs. Circ. Res. 32, 564-576.

Cunningham, J.T., Wachtel, R.E., Abboud, F.M., 1995. Mechanosensitive currents in putative aortic baroreceptor neurons in vitro. J. Neurophysiol. 73, 2094-2098.

DiBona, G.F., Jones, S.Y., Brooks, V.L., 1995. ANG II receptor blockade and arterial baroreflex regulation of renal nerve activity in cardiac failure. Am. J. Physiol. Regulatory Integr. Comp. Physiol. 269(38), R1189-R1196.

Di Rienzo, M., Parati, G., Castiglioni, P., Tordi, R., Mancia, G., Pedotti, A., 2001. Baroreflex effectiveness index: an additional measure of baroreflex control of heart rate in daily life. Am. J. Physiol. Regulatory Integr. Comp. Physiol. 280, R744-R751.

Doan, T.N., Kunze, D.L., 1999. Contribution of the hyperpolarization-activated current to the resting membrane potential of rat nodose sensory neurons. J. Physiol. 514(1), 125-138.

Drew, L.J., Wood, J.N., Cesare, P., 2002. Distinct mechanosensitive properties of capsaicin-sensitive and –insensitive sensory neurons. J. Neurosci. 22, RC228 (1-5).

Drummond, H.A., Seagard, J.L., 1994. Lack of effect of 4-aminopyridine on acute resetting of the type I carotid baroreceptor. Neurosci. Lett. 173, 45-49.

Drummond, H.A., Price, M.P., Welsh, M.J., Abboud, F.M., 1998. A molecular component of the arterial baroreceptor mechanotransducer. Neuron 21, 1435-1441.

Fowler, J.C., Greene, R., Weinreich, D., 1985b. Two calcium-sensitive spike after-hyperpolarizations in visceral sensory neurones of the rabbit. J. Physiol. 365, 59-75.

Fowler, J.C., Wonderlin, W.F., Weinreich, D., 1985a. Prostaglandins block a Ca^{2+}–dependent slow spike after hyperpolarization independent of effects on Ca^{2+} influx in visceral afferent neurons. Brain Res. 345, 345-349.

Franceschini, D., Orr-Urtreger, A., Yu, W., Mackey, L.Y., Bond, R.A., Armstrong, D., Patrick, J.W., Beaudet, A.L., DeBiasi, M., 2000. Altered baroreflex responses in α7 deficient mice. Behav. Brain Res. 113, 3-10.

Francis, J., Weiss, R.M., Wei, S.-G., Johnson, A.K., Beltz, T.G., Zimmerman, K., Felder, R.B., 2001. Central mineralcorticoid receptor blockade improves volume regulation and reduces sympathetic drive in heart failure. Am. J. Physiol. Heart Circ. Physiol. 281, H2241-H2251.

Gassmann, M., Hennet, T., 1998. From genetically altered mice to integrative physiology. News Physiol. Sci. 13: 53-57.

Glazebrook, P.A., Ramirez, A.N., Schild, J.H., Shieh, C.-C., Doan, T., Wible, B.A., Kunze, D.L., 2002. Potassium channels Kv1.1, Kv1.2 and Kv1.6 influence excitability of rat visceral sensory neurons. J. Physiol. 541(2), 467-482.

Gold, M.S., Reichling, D.B., Shuster, M.J., Levine, J.D., 1996. Hyperalgesic agents increase a tetrodotoxin-resistant Na^+ current in nociceptors. Proc. Natl. Acad. Sci. USA. 93, 1108-1112.

Gollasch, M., Tank, J., Luft, F.C., Jordan, J., Maass, P., Krasko, C., Sharma, A.M., Busjahn, A., Bahring, S., 2002. The BK channel β1 subunit gene is associated with human baroreflex and blood pressure regulation. J. Hypertension 20, 927-933.

Gross, V., Plehm, R., Tank, J., Jordan, J., Diedrich, A., Obst, M., Luft, F.C., 2002. Heart rate variability and baroreflex function in AT_2 receptor-disrupted mice. Hypertension 40, 207-213.

Hajduczok, G., Chapleau, M.W., Abboud, F.M., 1991. Rapid adaptation of central pathways explains the suppressed baroreflex with aging. Neurobiol. Aging 12, 601-604.

Hajduczok, G., Chapleau, M.W., Ferlic, R.J., Mao, H.Z., Abboud, F.M., 1994. Gadolinium inhibits mechanoelectrical transduction in rabbit carotid baroreceptors: Implication of stretch-activated channels. J. Clin. Invest. 94, 2392-2396.

Harper, A.A., 1991. Similarities between some properties of the soma and sensory receptors of primary afferent neurones. Exp. Physiol. 76, 369-377.

Heesch, C.M., Abboud, F.M., Thames, M.D., 1984. Acute resetting of carotid sinus baroreceptors, II: Possible involvement of electrogenic Na^+ pump. Am. J. Physiol. 247, H833-H839.

Hohler, B., Mayer, B., Kummer, W., 1994. Nitric oxide synthase in the rat carotid body and carotid sinus. Cell Tissue Res. 276, 559-564.

Hunt, B.E., Fahy, L., Farquhar, W.B., Taylor, J.A., 2001a. Quantification of mechanical and neural components of vagal baroreflex in humans. Hypertension 37, 1362-1368.

Hunt, B.E., Farquhar, W.B., Taylor, J.A., 2001b. Does reduced vascular stiffening fully explain preserved cardiovagal baroreflex function in older, physically active men? Circ. 103, 2424-2427.

Isaka, Y., Kimura, K., Uehara, A., Hashikawa, K., Mieno, M., Matsumoto, M., Handa, N., Nakabayashi, S., Imaizumi, M., Kamada, T., 1989. Platelet aggregability and in vivo platelet deposition in patients with ischemic cardiovascular disease--evaluation by indium-111 platelet scintigraphy. Thromb. Res. 56, 739-749.

Jones, P.P., Christou, D.D., Jordan, J., Seals, D.R., 2003. Baroreflex buffering is reduced with age in healthy men. Circ. 107, 1770-1774.

Jordan, J., Tank, J., Shannon, J.R., Diedrich, A., Lipp, A., Schroder, C, Arnold, G., Sharma, A.M., Biaggioni, I., Robertson, D., Luft, F.C., 2002. Baroreflex buffering and susceptibility to vasoactive drugs. Circ. 105, 1459-1464.

Jordan, J., Toka, H.R., Heusser, K., Toka, O., Shannon, J.R., Tank, J., Diedrich, A., Stabroth, C., Stoffels, M., Naraghi, R., Oelkers, W., Schuster, H., Schobel, H.P., Haller, H., Luft, F.C., 2000. Severely impaired baroreflex-buffering in patients with monogenic hypertension and neurovascular contact. Circ. 102, 2611-2618.

Jumrussirikul, P., Dinerman, J., Dawson, T.M., Dawson, V.L., Ekelund, U., Georgakopoulos, D., Schramm, L.P., Calkins, H., Snyder, S.H., Hare, J.M., Berger, R.D., 1998. Interaction between neuronal nitric oxide synthase and inhibitory G protein activity in heart rate regulation in conscious mice. J. Clin. Invest. 102, 1279-1285.

Just, A., Faulhaber, J., Ehmke, H., 2000. Autonomic cardiovascular control in conscious mice. Am. J. Physiol. Reg. Integr. Comp. Physiol. 279, R2214-R2221.

Katz, B., 1950. Depolarization of sensory terminals and the initiation of impulses in the muscle spindle. J. Physiol. (Lond) 111, 261-282.

Kaye, D.M., Esler, M.D., 1995. Abnormalities of the autonomic nervous system in heart failure. Cardiol. Rev. 3(4), 184-195.

Kirchheim, H.R., 1976. Systemic arterial baroreceptor reflexes. Physiol. Rev. 56, 100-176.

Kishi, T., Hirooka, Y., Sakai, K., Shigematsu, H., Shimokawa, H., Takeshita, A., 2001. Overexpression of eNOS in the RVLM causes hypotension and bradycardia via GABA release. Hypertension 38, 896-901.

Kornet, L., Hoeks, A.P., Janssen, B.J., Willigers, J.M., Reneman, R.S., 2002. Carotid diameter variations as a non-invasive tool to examine carotid baroreceptor sensitivity. J. Hypertension 20, 1165-1173.

Kraske, S., Cunningham, J.T., Hajduczok, G., Chapleau, M.W., Abboud, F.M., Wachtel, R.E., 1998. Mechanosensitive ion channels in putative aortic baroreceptor neurons. Am. J. Physiol. 275 (Heart Circ. Physiol. 44), H1497-H1501.

Krieger, E.M., 1963. Carotid occlusion in the rat: circulatory and respiratory effects. Acta Physiol. Latinam. 13, 350-357.

Lafont, A., Guerot, C., Lemarchand, P., 1996. Prospects for gene therapy in cardiovascular disease. Eur. Heart J. 17, 1312-1317.

LaRovere, M.T., Bersano, C., Gnemmi, M., Specchia, G., Schwartz, P.J., 2002. Exercise-induced increase in baroreflex sensitivity predicts improved prognosis after myocardial infarction. Circ. 106, 945-949.

LaRovere, M., Bigger, Jr. J.T., Marcus, F.I., Mortara, A., Schwartz, P.J., 1998. Baroreflex sensitivity and heart-rate variability in prediction of total mortality after myocardial infarction. Lancet 351, 478-484.

Li, Y.-F., Roy, S.K., Channon, K.M., Zucker, I.H., Patel, K.P., 2002. Effect of in vivo gene transfer of nNOS in the PVN on renal nerve discharge in rats. Am. J. Physiol. Heart Circ. Physiol. 282, H594-H601.

Li, Z., Bates, J.N., Lee, H.-C., Abboud, F.M., Chapleau M.W., 1997b. Oxidation selectively inhibits tetrodotoxin-sensitive sodium current in isolated baroreceptor neurons (abstract). Soc. Neurosci. Abstr. 23 (part 2), 1519.

Li, Z., Chapleau, M.W., 1995. Platelet-induced suppression of baroreceptor activity is mediated by a stable diffusible factor. J. Auton. Nerv. Syst. 51, 59-65.

Li, Z., Abboud, F.M., Chapleau, M.W., 1992. Aggregating human platelets in carotid sinus of rabbits decrease sensitivity of baroreceptors. Circ. Res. 70, 644-650.

Li, Z., Chapleau, M.W., Bates, J.N., Bielefeldt, K., Lee, H.-C., Abboud, F.M., 1998. Nitric oxide as an autocrine regulator of sodium currents in baroreceptor neurons. Neuron 20, 1039-1049.

Li, Z., Lee, H.C., Bielefeldt, K., Chapleau, M.W., Abboud, F.M., 1997a. The prostacyclin analogue carbacyclin inhibits Ca^{2+}-activated K^+ current in aortic baroreceptor neurones of rats. J. Physiol. (Lond) 501(2), 275-287.

Li, Z., Mao, H., Abboud, F.M., Chapleau, M.W., 1996. Oxygen derived free radicals contribute to baroreceptor dysfunction in atherosclerotic rabbits. Circ. Res. 79, 802-811.

Li, Z., Su, X., Chapleau, M.W., 1995. Role of cyclooxygenase metabolites in platelet-induced baroreceptor dysfunction. Am. J. Physiol. 269 (Heart Circ. Physiol. 38), H599-H608.

Ling, G.-Y., Cao, W.-H., Onodera, M., Ju, K.-H., Kurihara, H., Kurihara, Y., Yazaki, Y., Kumada, M., Fukuda, Y., Kuwaki, T., 1998. Renal sympathetic nerve activity in mice: comparison between mice and rats and between normal and endothelin-1 deficient mice. Brain Res. 808, 238-249.

Liu, J.-L., Kulakofsky, J., Zucker, I.H., 2002. Exercise training enhances baroreflex control of heart rate by a vagal mechanism in rabbits with heart failure. J. Appl. Physiol. 92, 2403-2408.

Lohmeier, T.E., Lohmeier, J.R., Reckelhoff, J.F., Hildebrandt, D.A., 2001. Sustained influence of the renal nerves to attenuate sodium retention in angiotensin hypertension. Am. J. Physiol. Regulatory Integr. Comp. Physiol. 281, R434-R443.

Ma, X., Abboud, F.M., Chapleau, M.W., 2002. Analysis of afferent, central, and efferent components of the baroreflex in mice. Am. J. Physiol. Regulatory Integr. Comp. Physiol. 283, R1033-R1040.

Ma, X., Price, M.P., Drummond, H.A., Welsh, M.J., Chapleau, M.W., Abboud, F.M., 2001b. The DEG/ENaC ion channel family member BNC1 mediates mechanical transduction of arterial baroreceptor nerve activity in vivo (abstract). FASEB J. 15(5), A1146.

Ma, X.Y., Abboud, F.M., Chapleau, M.W., 2001a. A novel effect of angiotensin on renal sympathetic nerve activity in mice. J. Hypertension 19, 609-618.

Ma, X.Y., Abboud, F.M., Chapleau, M.W., 2000. Altered baroreflex control of renal sympathetic nerve activity in normotensive apo-E deficient atherosclerotic mice (abstract). The Physiologist 43(4), 275.

Ma, X.Y., Shaffer, R.A., Sigmund, C.D., Abboud, F.M., Chapleau, M.W., 1999. Mechanisms of impaired baroreflex control of sympathetic nerve activity in hypertensive renin-angiotensinogen double transgenic mice (abstract). FASEB J. 13(5) Pt II, A775.

Madeddu, P., Salis, M.B., Emanueli, C., 1999. Altered baroreflex control of heart rate in bradykinin B_2-receptor knockout mice. Immunopharmacology. 45, 21-27.

Mancia, G., Mark, A.L., 1983. Arterial baroreflexes in humans. In: Handbook of Physiology. Section 2, vol. III, part 2, Bethesda, MD: American Physiological Society; 755-793.

Mangoni, A.A., Mircoli, L., Giannattasio, C., Mancia, G., Ferrari, A.U., 1997. Effect of sympathectomy on mechanical properties of common carotid and femoral arteries. Hypertension 30, 1085-1088.

Mansier, P., Medigue, C., Charlotte, N., Vermeiren, C., Coraboeuf, E., Deroubai, E., Ratner, E., Chevalier, B., Clairambault, J., Carre, F., Dahkli, T., Bertin, B., Briand, P., Strosberg, D., Swynghedauw, B., 1996. Decreased heart rate variability in transgenic mice overexpressing atrial β_1-adrenoceptors. Am. J. Physiol. Heart Circ. Physiol. 271, H1465-H1472.

Mao, H.Z., Li, Z., Chapleau, M.W., 1996. Platelet activation in carotid sinuses triggers reflex sympathoinhibition and hypotension. Hypertension 27(Pt 2), 584-590.

Matsuda, T., Bates, J.N., Lewis, S.J., Abboud, F.M., Chapleau, M.W., 1995. Modulation of baroreceptor activity by nitric oxide and S-nitrosocysteine. Circ. Res. 76, 426-433.

McDowell, T.S., Axtelle, T.S., Chapleau, M.W., Abboud, F.M., 1989. Prostaglandins in carotid sinus enhance baroreflex in rabbits. Am. J. Physiol. Reg. Integr. Comp. Physiol. 257(26), R445-R450.

Merrill, D.C., Thompson, M.W., Carney, C.L., Granwehr, B.P., Schlager, G., Robillard, J.E., Sigmund, C.D., 1996. Chronic hypertension and altered baroreflex responses in transgenic mice containing the human renin and human angiotensinogen genes. J. Clin. Invest. 97, 1047-1055.

Meyrelles, S., Chapleau, M., 2000. Impaired carotid occlusion reflex in apoE deficient mice and its reversal by gene transfer to carotid sinus (abstract). J. Hypertension 18 (supp 4), S202.

Meyrelles, S.S., Mao, H.Z., Heistad, D.D., Chapleau, M.W., 1997. Gene transfer to carotid sinus in vivo: A novel approach to investigation of baroreceptors. Hypertension 30 [Pt 2], 708-713.

Meyrelles, S.S., Sharma, R.V., Mao, H.Z., Abboud, F.M., Chapleau, M.W., 2003. Modulation of baroreceptor activity by gene transfer of nitric oxide synthase to carotid sinus adventitia. Am. J. Physiol. Reg. Integr. Comp. Physiol. 284, R1190-R1198.

Mircoli, L., Fedele, L., Benetti, M., Bolla, G.B., Radaelli, A., Perlini, S., Ferrari, A.U., 2002. Preservation of the baroreceptor heart rate reflex by chemical sympathectomy in experimental heart failure. Circ. 106, 866-872.

Mohan, R.M., Heaton, D.A., Danson, E.J.F., Krishnan, S.P.R., Cai, S., Channon, K.M., Paterson, D.J., 2002. Neuronal nitric oxide synthase gene transfer promotes cardiac vagal gain of function. Circ. Res. 91, 1089-1091.

Mombouli, J.-V., Vanhoutte, P.M., 1999. Endothelial dysfunction: from physiology to therapy. J. Mol. Cell. Cardiol. 31, 61-74.

Monahan, K.D., Dinenno, F.A., Tanaka, H., Clevenger, C.M., DeSouza, C.A., Seals, D.R., 2000. Regular aerobic exercise modulates age-associated declines in cardiovagal baroreflex sensitivity in healthy men. J. Physiol. 529(1), 263-271.

Nightingale, A.K., Blackman, D.J., Field, R., Glover, N.J., Pegge, N., Mumford, C., Schmitt, M., Ellis G.R., Morris-Thurgood, J.A., Frenneaux, M.P., 2003. Role of nitric oxide and oxidative stress in baroreceptor dysfunction in patients with chronic heart failure. Clin. Sci. 104, 529-535.

Nuyt, A.M., Mao, H.Z., Abboud, F.M., Chapleau, M.W., 1995. Nitric oxide modulates arterial baroreceptor activity via Ca^{2+}-activated K^+ channels in vivo (abstract). Pediatric Res. 37, 31A.

Parati, G., Di Rienzo, M., Mancia, G., 2000. How to measure baroreflex sensitivity: from the cardiovascular laboratory to daily life. J. Hypertension 18, 7-19.

Paton, J.F.R., Deuchars, J., Ahmad, Z., Wong, L.-F., Murphy, D., Kasparov, S., 2001. Adenoviral vector demonstrates that angiotensin II-induced depression of the cardiac baroreflex is mediated by endothelial nitric oxide synthase in the nucleus tractus solitarii of the rat. J. Physiol. 531, 445-458.

Piccirillo, G., Nocco, M., Moise, A., Lionetti, M., Naso, C., Di Carlo, S., Marigliano, V., 2003. Influence of vitamin C on baroreflex sensitivity in chronic heart failure. Hypertension 41, 1240-1245.

Podrid, P.J., Fuchs, T., Candinas, R., 1990. Role of the sympathetic nervous system in the genesis of ventricular arrhythmia. Circ. 82 (suppl I), I-103-I-113.

Ponikowski, P., Chua, T.P., Piepoli, M., Ondusova, D., Webb-Peploe, K., Harrington, D., Anker, S.D., Volterrani, M., Colombo, R., Mazzuero, G., Giordano, A., Coats, A.J.S.,

1. THE BARORECEPTOR REFLEX: NOVEL METHODS AND MECHANISMS

1997. Augmented peripheral chemosensitivity as a potential input to baroreflex impairment and autonomic imbalance in chronic heart failure. Circ. 96, 2586-2594.

Reid, I., 1992. Interactions between ANG II, sympathetic nervous system, and baroreceptor reflexes in regulation of blood pressure. Am. J. Physiol. 262 (Endocrinol. Metab. 25), E763-E778.

Robbe, H.W.J., Mulder, L.J.M., Ruddel, H., Langewitz, W.A., Veldeman, J.B.P., Mulder, G., 1987. Assessment of baroreceptor reflex sensitivity by means of spectral analysis. Hypertension 10, 538-543.

Rokosh, D.G., Simpson, P.C., 2002. Knockout of the α1A/C-adrenergic receptor subtype: The α1A/C is expressed in resistance arteries and is required to maintain arterial blood pressure. Proc. Natl. Acad. Sci. USA 99, 9474-9479.

Ruppersberg, J.P., Stocker, M., Pongs, O., Heinemann, S.H., Frank, R., Koenen, M., 1991. Regulation of fast inactivation of cloned mammalian $I_K(A)$ channels by cysteine oxidation. Nature 352, 711-714.

Sakai, K., Hirooka, Y., Matsuo, I., Kenichi, E., Shigematsu, H., Shimokawa, H., Takeshita, A., 2000. Overexpression of eNOS in NTS causes hypotension and bradycardia in vivo. Hypertension 36, 1023-1028.

Sakai, K., Chapleau, M.W., Sigmund, C.D., 2003. Neuron-derived angiotensin II causes cardiac-sympathetic baroreflex resetting in the absence of hypertension (abstract). Hypertension (In press).

Sato, T., Kawada, T., Sugimachi, M., Sunagawa, K., 2002. Bionic technology revitalizes native baroreflex function in rats with baroreflex failure. Circ. 106, 730-734.

Saum, W.R., Brown, A.M., Tuley, F.H., 1976. An electrogenic sodium pump and baroreceptor function in normotensive and spontaneously hypertensive rats. Circ. Res. 39, 497-505.

Schild, J.H., Kunze, D.L., 1997. Experimental and modeling study of Na^+ current heterogeneity in rat nodose neurons and its impact on neuronal discharge. J. Neurophysiol. 78, 3198-3209.

Shannon, J.R., Flattem, N.L., Jordan, J., Jacob, G., Black, B.K., Biaggioni, I., Blakely, R.D., Robertson, D., 2000. Clues to the origin of orthostatic intolerance: a genetic defect in the cocaine- and antidepressant-sensitive norepinephrine transporter. New Engl. J. Med. 342, 541-549.

Sinnayah, P., Lindley, T.E., Staber, P.D., Cassell, M.D., Davidson, B.L., Davisson, R.L., 2002. Selective gene transfer to key cardiovascular regions of the brain: comparison of two viral vector systems. Hypertension 39, 603-608.

Slack, R.S., Miller, F.D., 1996. Viral vectors for modulating gene expression in neurons. Curr. Opin. Neurobiol. 6, 576-583.

Smithies, O., 1997. A mouse view of hypertension. Hypertension 30, 1318-1324.

Snitsarev, V., Whiteis, C.A., Abboud, F.M., Chapleau, M.W., 2002a. Mechanosensory transduction of vagal and baroreceptor afferents revealed by study of isolated nodose neurons in culture. Auton. Neurosci. 98, 59-63.

Snitsarev, V., Whiteis, C.A., Abboud, F.M., Chapleau, M.W., 2002b. Nitric oxide functions as an autocrine negative-feedback inhibitor of membrane excitability in nodose sensory neurons (abstract). Program No. 46.24. 2002 Abstract Viewer/Itinerary Planner. Washington, DC: Society for Neuroscience. CD-ROM.

Snitsarev, V., Yermolaieva, O., Whiteis, C.A., Abboud, F.M., Heinemann, S.H., Hoshi, T., Chapleau, M.W., 2002c. Reactive oxygen species generated during action potential discharge mediate "activity-dependent resetting" of baroreceptor and vagal afferent neurons in culture (abstract). Circ. 106(19), II-66.

Snitsarev, V., Whiteis, C.A., Abboud, F.M., Chapleau, M.W., 2001. Effect of prostacyclin analog on mechanosensitive vs. voltage-gated ion channels in nodose neurons (abstract). Soc. Neurosci. Abstr. 27, 1812.

Stansfeld, C.E., Marsh, S.J., Halliwell, J.V., Brown, D.A., 1986. 4-Aminopyridine- and dendrotoxin-induced repetitive firing in rat visceral sensory neurons by blocking a slowly inactivating outward current. Neurosci. Lett. 64, 299-304.

Stauss, H.M., Godecke, A., Mrowka, R., Schrader, J., Persson, P.B., 1999. Enhanced blood pressure variability in eNOS knockout mice. Hypertension 33, 1359-1363.

Stec, D.E., Keen, H.L., Sigmund, C.D., 2002. Lower blood pressure in floxed angiotensinogen mice after adenoviral delivery of cre-recombinase. Hypertension 39, 629-633.

Sullivan, M.J., Sharma, R.V., Wachtel, R.E., Chapleau, M.W., Waite, L.J., Bhalla, R.C., Abboud, F.M., 1997. Non-voltage-gated calcium influx through mechanosensitive ion channels in aortic baroreceptor neurons. Circ. Res. 80, 861-867.

Sun, W., Ma, X., Abboud, F.M., Chapleau, M.W., 2002. Differential effects of aging on baro- and chemoreflex regulation of arterial pressure: selective impairment of baroreflex in senescent mice (abstract). Clin. Auton. Res. 12, 323.

Sun, W., Abboud, F.M., Chapleau, M.W., 2001. Altered baro- and chemoreflex sensitivity revealed by carotid artery occlusion reflex in apo-E knockout mice (abstract). FASEB J. 15(5) Pt. II, A1146.

Sun, W., Abboud, F.M., Chapleau, M.W., 2000. Evaluation of baroreflex and chemoreflex by carotid artery occlusion in mice: A method for phenotypic analysis of deletion of candidate sensory molecules (abstract). Circ. 102(18), II-700.

Tang, X.D., Daggett, H., Hanner, M., Garcia, M.L., McManus, O.B., Brot, N., Weissbach, H., Heinemann, S.H., Hoshi, T., 2001. Oxidative regulation of large conductance calcium-activated potassium channels. J. Gen. Physiol. 117, 253-273.

Tavernarakis, N., Driscoll, M.A., 2001. Degenerins: At the core of the Metazoan mechanotransducer? Ann. NY Acad. Sci. 940, 28-41.

Thrasher, T.N., 2002. Unloading arterial baroreceptors causes neurogenic hypertension. Am. J. Physiol. 282, R1044-R1055.

Uechi, M., Asai, K., Osaka, M., Smith, A., Sato, N., Wagner, T.E., Ishikawa, Y., Hayakawa, H., Vatner D.E., Shannon, R.P., Homcy, C.J., Vatner, S.F., 1998. Depressed heart rate variability and arterial baroreflex in conscious transgenic mice with overexpression of cardiac $G_{s\alpha}$. Circ. Res. 82, 416-423.

Undem, B.J., Weinreich, D., 1993. Electrophysiological properties and chemosensitivity of guinea pig nodose neurons in vitro. J. Auton. Nerv. Syst. 44, 17-34.

van Brederode, J.F.M., Seagard, J.L., Dean, C., Hopp, F.A., Kampine, J.P., 1990. An experimental and modeling study of the excitability of carotid sinus baroreceptors. Circ. Res. 66, 1510-1525.

Vizi, E.S., 1974. Interaction between adrenergic and cholinergic system: presynaptic inhibitory effect of noradrenaline on acetylcholine release. J. Neural Trans. Suppl. 11, 61-78.

Walker, J.K., Peppel, K., Lefkowitz, R.J., Caron, M.G., Fisher, J.T., 1999. Altered airway and cardiac responses in mice lacking G protein-coupled receptor kinase 3. Am. J. Physiol. Regulatory Integr. Comp. Physiol. 276, R1214-R1221.

Wang, W., Brandle, M., Zucker, I.H., 1993. Indomethacin reduces acute baroreceptor resetting in the dog. J. Physiol. 469, 139-151.

1. THE BARORECEPTOR REFLEX: NOVEL METHODS AND MECHANISMS

Wang, W., Chen, J.S., Zucker, I.H., 1990. Carotid sinus baroreceptor sensitivity in experimental heart failure. Circ. 81, 1959-1966.

Wang, W., McClain, J.M., Zucker, I.H., 1992. Aldosterone reduces baroreceptor discharge in the dog. Hypertension 19, 270-277.

Wang, W., Zucker, I.H., 1996. Cardiac sympathetic afferent reflex in dogs with congestive heart failure. Am. J. Physiol. Regulatory Integr. Comp. Physiol. 271, R751-R756.

Wang, W., 1994. Chronic administration of aldosterone depresses baroreceptor reflex function in the dog. Hypertension 24, 571-575.

Weinreich, D., Koschorke, G.M., Undem, B.J., Taylor, G.E., 1995. Prevention of the excitatory actions of bradykinin by inhibition of PGI_2 formation in nodose neurones of the guinea-pig. J. Physiol. 483(3), 735-746.

Welsh, M.J., Price, M.P., Xie, J., 2002. Biochemical basis of touch perception: mechanosensory function of degenerin/epithelial Na^+ channels. J. Biol. Chem. 277, 2369-2372.

Whiteis, C.A., Fazan, Jr. R., Abboud, F.M., Chapleau, M.W., 2000. Analysis of baroreflex sensitivity in conscious mice using spectral analysis (abstract). The Physiologist 43, 275.

Wickman, K., Nemec, J., Gendler, S.J., Clapham, D.E., 1998. Abnormal heart rate regulation in GIRK4 knockout mice. Neuron 20, 103-114.

Xie, P., McDowell, T.S., Hajduczok, G., Chapleau, M.W., Abboud, F.M., 1988. Contribution of cyclooxygenase metabolites to baroreceptor activation in hypercholesterolemic (HC) rabbits (abstract). Circ. 78, II-177.

Xie, P., Chapleau, M.W., McDowell, T.S., Hajduczok, G., Abboud, F.M., 1990. Mechanism of decreased baroreceptor activity in chronic hypertensive rabbits. Role of endogenous prostanoids. J. Clin. Invest. 86, 625-630.

Yang, X.-C., Sachs, F., 1989. Block of stretch-activated ion channels in Xenopus oocytes by gadolinium and calcium ions. Science 243, 1068-1071.

Yasumasu, T., Takahara, K., Sadayasu, T., Date, H., Isozumi, K., Kouzuma, R., Nakashima, Y., 2003. Influence of coronary artery disease risk factors on baroreflex sensitivity in the elderly. Clin. Auton. Res. (In press).

Zhang, W., Li, J.-L., Hosaka, M., Janz, R., Shelton, J.M., Albright, G.M., Richardson, J.A., Sudhof, T.C., Victor, R.G., 2000. Cyclosporine A-induced hypertension involves synapsin in renal sensory nerve endings. Proc. Natl. Acad. Sci. USA 97, 9765-9770.

Zimmerman, M.C., Lazartigues, E., Lang, J.A., Sinnayah, P., Ahmad, I.M., Spitz, D.R., Davisson R.L., 2002. Superoxide mediates the actions of angiotensin II in the central nervous system. Circ. Res. 91, 1038-1045.

Chapter 2

CHEMOREFLEX AND SYMPATHOEXCITATION

Benedito H. Machado
Department of Physiology, School of Medicine of Ribeirao Preto, University of Sao Paulo, Ribeirao Preto, SP, 14049-900, Brazil

Abstract: Activation of the peripheral chemoreflex in awake rats produces a complex pattern of cardiovascular, respiratory and behavioral responses. The focus of the present chapter is the different aspects involved in the neurotransmission of sympathoexcitatory component of the chemoreflex, which is responsible for a large increase in arterial pressure. Experiments performed in awake and anesthetized rats reveal important differences relative to the pattern of response to chemoreflex activation as well as the effects of blockade of neurotransmission of this reflex at the nucleus tractus solitarius (NTS) level. The main aspect of this chapter is related to the fact that the sympathoexcitatory component of the chemoreflex was not blocked by microinjection of different antagonists of the excitatory amino acids (EAA) receptors into the NTS. Several experiments described in this chapter strongly suggest that l-glutamate and EAA receptors are not directly involved in the neurotransmission of the sympathoexcitatory component of the chemoreflex and also indicate that the processing of this component of the chemoreflex is not modulated by inhibitory amino acids such as GABA and glycine. In addition, experiments suggesting a potential role for ATP and purinergic receptors in the processing of the sympathoexcitatory component of the chemoreflex at the NTS level are presented. The involvement of other areas of the brain in the neural pathways of the chemoreflex, including the paraventricular nucleus of hypothalamus and the parabrachial nucleus, is emphasized. Lastly, the possibility that the complex neural mechanisms involved in the generation of the pressor response to chemoreflex activation may be part of the pathophysiological substrate for autonomic dysfunction such as arterial hypertension is discussed.

Key words: arterial pressure, hypoxia, nucleus tractus solitarius, paraventricular nucleus of hypothalamus, excitatory amino acid receptors, parabrachial nucleus, l-glutamate, ATP, neurotransmission, neuromodulation.

INTRODUCTION

One of the most important topics related to the central neural control of circulation is the generation and modulation of sympathetic activity. Several experimental models can be used to produce sympathoexcitation, including chemoreflex activation, which, in addition to the respiratory and behavioral adjustments, produces an acute and remarkable increase in mean arterial pressure, essentially dependent on the increase in sympathetic activity.

In this chapter different aspects of the processing of sympathoexcitatory component of the chemoreflex in the central nervous system will be considered. The main aspect to be explored is related to the neurotransmission of the chemoreflex afferents at the nucleui tracti solitarii (NTS) level with focus on the sympathoexcitatory component of the response. In the NTS the neurotransmitters and neuromodulators potentially involved in the processing of this sympathoexcitatory component will be analyzed. Different areas of the brain such as the paraventricular nucleus of the hypothalamus (PVN) and parabrachial nucleus (PBN) that take part in the complex processing of the autonomic, respiratory and behavioral responses to chemoreflex activation, particularly in awake rats, are also reviewed. Moreover, the current knowledge and the perspectives concerning the mechanisms involved in the generation and modulation of the sympathoexcitatory component of the chemoreflex in normal as well as in pathophysiological conditions such as hypertension are considered in this chapter.

HISTORICAL ASPECTS RELATED TO THE STUDY OF THE ARTERIAL CHEMOREFLEX

A histological study by de Castro (1928) suggested that the carotid body might be the sensory apparatus which detects changes in the chemical composition of arterial blood (cf. Daly, 1997). Classical studies by Corneille Heymans and co-workers in the first half of the last century (1931a, b) showed that cyanide produced a cytotoxic stimulus in the aortic and carotid bodies which responded with an increase in ventilation and bradycardia. They also documented that these responses were dependent on the carotid sinus nerve and were not related to a central effect of cyanide and clearly illustrated in dogs the functional role of the carotid chemoreceptors in the respiratory and cardiovascular adjustments in response to changes in the arterial blood gases. In addition, the carotid chemoreceptors were also stimulated by arterial hypoxia, hypercapnia, acidemia and arterial hypotension. Studies by Heymans et al. (1932, 1933, 1934) also documented that the activation of arterial chemoreceptors produced an important increase in baseline mean arterial pressure (MAP), which they suggested to be related to the activation of vasomotor centers in the central nervous system.

The main cardiovascular responses to chemoreflex activation are bradycardia and vasoconstriction but the most important pattern of neurovegetative response to chemoreflex activation is the increase in respiratory ventilation. Therefore, a precise interaction of respiratory and sympathetic adjustments in the central nervous system is also a relevant aspect of the complex pattern of responses to chemoreflex activation (Daly, 1997).

CHEMOREFLEX IN ANESTHETIZED ANIMALS

Anesthetics may modify the integrative autonomic, respiratory and cardiovascular responses to chemoreflex activation when compared to the responses of unanesthetized animals (Daly, 1997). The sites in the central nervous system and the mechanisms of action of anesthetics are not completely known but are intentionally used to reduce the brain function related to neural integration. Studies by Franchini and Krieger (1993) documented that pentobarbital, more than chloralose, almost abolished the bradycardic response to chemoreflex activation but these 2 anesthetics similarly blocked the pressor response component. Therefore, studies related to the central processing of the chemoreflex, due to its multiple integrative aspects, should be preferably performed in unanesthetized rats.

Studies by Vardhan et al. (1993) in anesthetized rats documented that a mild pressor response to chemoreflex activation, in the range of 10 to 15 mmHg, was blocked by combined microinjection of AP-7 (a NMDA receptor antagonist) and DNQX (a non-NMDA receptor antagonist) into the NTS. Zhang and Mifflin (1993) also verified that the pressor response to electrical stimulation of the carotid sinus nerve was significantly reduced by microinjection of kynurenic acid, a non-selective antagonist of EAA receptors, into the NTS of anesthetized rats. Although the magnitude of the pressor response to chemoreflex activation in anesthetized rats is on average significantly smaller than the pressor response in awake rats (Haibara et al., 1995, 1999), studies by Vardhan et al. (1993) and Zhang and Mifflin (1993), performed under anesthesia, showed that the pressor response was blocked using excitatory amino acid (EAA) receptor antagonists. As it will be emphasized in this chapter, the pressor response in the absence of anesthesia was not blocked by different antagonists of EAA receptors microinjected into different sub-regions of the NTS, indicating that a direct influence of the anesthetics may alter the processing of this reflex.

NEURAL PATHWAYS OF THE CHEMOREFLEX IN THE BRAINSTEM

Is the NTS really the first synaptic station of the chemoreflex afferents in the central nervous system? Different studies have shown that the first synapse of the carotid chemoreceptor afferents occurs in the lateral (Finley and Katz, 1992; Mifflin, 1992) and caudal commissural NTS (Chitravanshi et al., 1994; Chitravanshi and Sapru, 1995). The latter location is also supported by recent morphological data by Paton et al. (2001) revealing cell bodies of some chemoreceptive NTS neurons within this region. There is evidence that the rostral ventrolateral medulla (RVLM) neurons are the final relay in the brainstem for the sympathetic outflow. Sun and Spyer (1991) suggested that the excitatory influence of the chemoreflex activation on the RVLM neurons may be mediated via the hypothalamus or a direct projection from the NTS to the RVLM. There is functional (Urbanski and Sapru, 1988) and anatomical evidence (Aicher et al., 1996) of the existence of direct projections from the NTS to RVLM. Studies by Koshiya et al. (1993) performed on anesthetized rats have suggested that the projections of sympathoexcitatory component of the chemoreflex in the medulla are independent of the baroreflex projections. Neurons of the RVLM that receive excitatory influence from chemoreflex activation are not known but they seem to be C1 adrenergic neurons that receive direct projections from the NTS (Ross et al., 1985). In accordance with studies by Koshiya and Guyenet (1994), the A5 area in the caudal ventrolateral pons containing noradrenergic neurons is also part of the chemoreflex pathways.

Studies by Mifflin (1992) have documented that the integration of chemoreceptor afferents is not homogeneous in the neurons of the NTS and there is also evidence that peripheral chemoreceptors activate NTS neurons that are intermingled with neurons mediating other cardiovascular and non-cardiovascular inputs (Paton and Kasparov, 2000). These facts may explain the wide diversity of autonomic (sympathetic and parasympathetic), respiratory and behavioral integrative responses to the chemoreflex activation at the NTS level. In this case we may suggest that each of these "specific reflexes" in the NTS is mediated by different neurotransmitters and influenced by different neuromodulators, which may act on interneurons or on highest order neurons projecting to different areas of the brain involved in the complex pattern of the responses to chemoreflex activation. The approach to the study of the chemoreflex in awake animals is even more important if we consider that both cardiovascular and respiratory responses to chemoreflex activation with potassium cyanide (KCN) were almost abolished by urethane or chloralose anesthesia (Franchini and Krieger, 1993).

Although microinjection of l-glutamate into the NTS of awake rats (Machado and Bonagamba, 1992; Colombari et al., 1994) and activation of the chemoreflex produce similar pressor responses, studies from our laboratory (Mauad and Machado, 1998) indicate that the activation of a sympatho-excitatory pathway in the NTS by microinjection of l-glutamate in this area apparently is not the same as that used for the activation of the chemoreceptor afferents in the NTS. Mauad and Machado (1998) showed that the pressor response to microinjection of l-glutamate into the NTS was blocked by microinjection of kynurenic acid into the ipsilateral RVLM. However, microinjection of kynurenic acid into the RVLM ipsilateral to the carotid chemoreceptor activated by KCN was not effective in blocking the pressor response (Mauad and Machado, 2001). Therefore, considering that chemoreceptor afferents have their first synapse at the NTS level, we suggest that from this site there are also probably projections to the contralateral RVLM, which may explain the maintenance of the pressor response. Alternatively, we may also consider projections from the NTS to several other nuclei such as the PBN (Loewy, 1990) and PVN (Ricardo and Koh, 1978), which project to the RVLM (Chamberlin and Saper, 1992; Yang and Coote, 1998). The question that remains to be answered is related to the functional role of the sympatho-excitatory projections from NTS to RVLM considering that there is functional (Urbanski and Sapru, 1988; Mauad and Machado, 1998) and anatomical evidence (Aicher et al., 1996) indicating that it exists. Electrophysiological studies by Koshiya et al. (1993) performed on anesthetized rats documented that RVLM neurons play an important role in the sympathetic nerve discharge in response to chemoreflex activation, and Koshiya and Guyenet, (1996) documented that chemosensitive neurons of the NTS involved in the respiratory modulation project to the RVLM and arborize in this region.

The findings in awake rats reported by Mauad and Machado (2001) differ from those reported by Amano et al. (1994), who observed that the pressor response to unilateral chemoreflex activation was significantly reduced by ipsilateral microinjection of kynurenic acid into the RVLM, suggesting that the sympatho-excitatory response of the chemoreflex is mediated only by ipsilateral neurons of the RVLM. The difference in this case seems to be related to the fact that our experiments were peformed on awake rats while the study by Amano et al. (1994) was performed on anesthetized rats. Considering that the cardiovascular responses to chemoreflex activation (Franchini and Krieger, 1993) as well as the cardiovascular responses to microinjection of l-glutamate into the NTS are deeply affected by anesthesia (Machado and Bonagamba, 1992), the findings by Mauad and Machado (2001) and those by Amano et al. (1994) cannot be easily compared, but we may suggest that under anesthesia the direct projection from NTS to RVLM may play a more important role than in awake animals, in which more rostral areas of the brain play a relevant role in the integration of ventilatory, autonomic and behavioral responses to chemoreflex activation.

NEUROTRANSMISSION OF THE CHEMOREFLEX AFFERENTS IN THE NTS

The complete understanding of the neural integration that occurs at the NTS level is not a simple task due to a wide diversity of neurotransmitters/neuromodulators, receptors and interneurons found in this area. The topographic division of the NTS apparently is not enough to permit a better understanding of the complexity of this area, because multiple integrative systems are processed side-by-side and the activation of projections related to the autonomic control of the circulation, for example, may affect different neuro-vegetative responses.

To study the chemoreflex in awake rats we use intravenous injection of KCN, initially described in the classical studies by Heymans to activate the chemoreflex in dogs and more recently adapted to rats by Franchini and Krieger (1993). KCN (iv) produces a tachypneic response, increased arterial pressure, bradycardia and an important behavioral response when injected to awake rats. The cytotoxic hypoxia produced by KCN in the dose range used (from 80 to 320 µg/kg), is restricted to the chemosensitive cells of the carotid bodies because ligation of the carotid body arteries abolished all the response to chemoreflex activation (Haibara et al., 1995). Studies by Barros et al. (2002) verified that the activation of chemoreflex by a hypoxic stimulus, such as that occurring when the rats placed inside a closed chamber and submitted for a short period of time to a hypoxia 7-5% O_2, produced a pattern of pressor and bradycardic responses similar to the responses to chemoreflex activation with KCN. In addition, these cardiovascular responses to 7-5% O_2 were abolished by ligation of the carotid body arteries (Fig. 1).

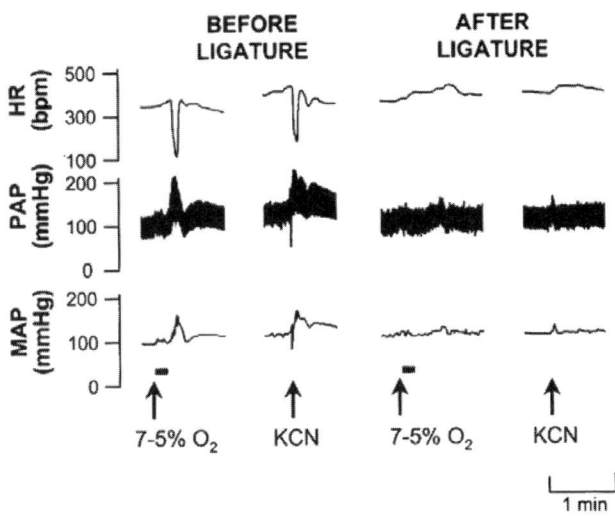

Figure 1. Typical tracings of one rat showing the changes in heart rate (HR), pulsatile arterial pressure (PAP) and mean arterial pressure (MAP) in response to chemoreflex activation with hypoxic-hypoxia (7-5% O_2 induced by 100% N_2 in the pletysmographic chamber) or cytotoxic-hypoxia with potassium cyanide (KCN, intravenous) before (BEFORE LIGATURE) and after bilateral ligature of the carotid body arteries (AFTER LIGATURE). Note that the cardiovascular responses to 7-5% O_2 or KCN were similar in the control and were abolished after ligature. Data reprinted from Barros et al. (2002) with permission.

Therefore, the cardiovascular responses produced by KCN or 100% N_2 are similar, indicating that KCN is an appropriate tool for the activation of chemoreflex, with the advantage that this experimental procedure by intravenous injection is easier to perform than with the rat inside a closed chamber, especially considering the manipulations for microinjections into the brainstem areas. The bradycardic response to chemoreflex activation is not a consequence of the baroreflex activation because Haibara et al. (1995) verified that the blockade of the pressor response with prazosin, an alpha adrenergic antagonist, produced no effect on the bradycardic response (Fig. 2), which was abolished only after methyl-atropine. Therefore, the chemoreflex activation in awake rats produces two independent cardiovascular responses, i.e., an increase in arterial pressure (sympathoexcitation) and bradycardia (parasympathoexcitation) (Franchini and Krieger, 1993; Haibara et al., 1995). In a series of experiments reported here, we tried to determine the role of ionotropic and metabotropic l-glutamate receptors involved in the processing of these two autonomic components of the chemoreflex at the NTS level.

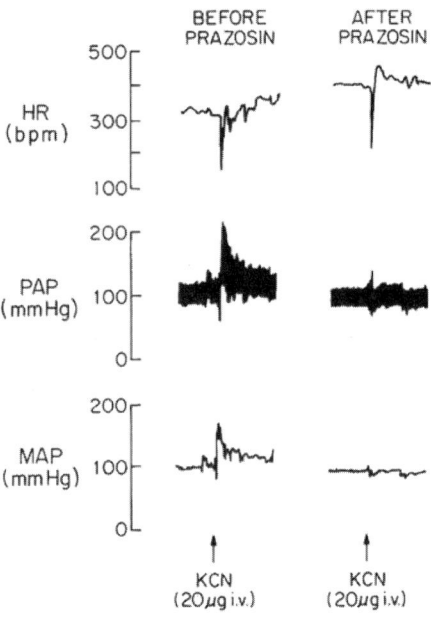

Figure 2. Typical tracings of one rat showing the changes in heart rate (HR), pulsatile arterial pressure (PAP) and mean arterial pressure (MAP) in response to chemoreflex activation with with potassium cyanide (KCN, iv) before and after prazosin (1 mg.kg^{-1}, iv), a α-1 adrenoceptor antagonist.

To evaluate the role of NMDA receptors in the processing of the chemoreflex at the NTS level, the chemoreflex was activated before and after bilateral microinjection of AP-5, a selective NMDA receptor antagonist, into the NTS (Haibara et al., 1995). Increasing doses of AP-5 produced a dose-dependent blockade of the bradycardic response and no changes in the pressor response to chemoreflex activation (Fig. 3). These data indicate that NMDA receptors are involved in the bradycardic response but not in the neurotransmission of the sympathoexcitatory component (pressor response) of the chemoreflex activation in the NTS. The possible involvement of NMDA receptors in the parasympathetic component of the chemoreflex is also supported by findings of another study from our laboratory (Haibara et al., 1999), in which the blockade of the bradycardic response to chemoreflex activation was effective when kynurenic acid was microinjected into the rostral aspect of the commissural NTS. In that study, the microinjection of kynurenic acid into the caudal aspect of the commissural NTS produced no effect on the bradycardic response,

2. CHEMOREFLEX AND SYMPATHOEXCITATION

indicating that the processing of the parasympathetic component of the chemoreflex is restricted to the rostral aspect of the commissural NTS.

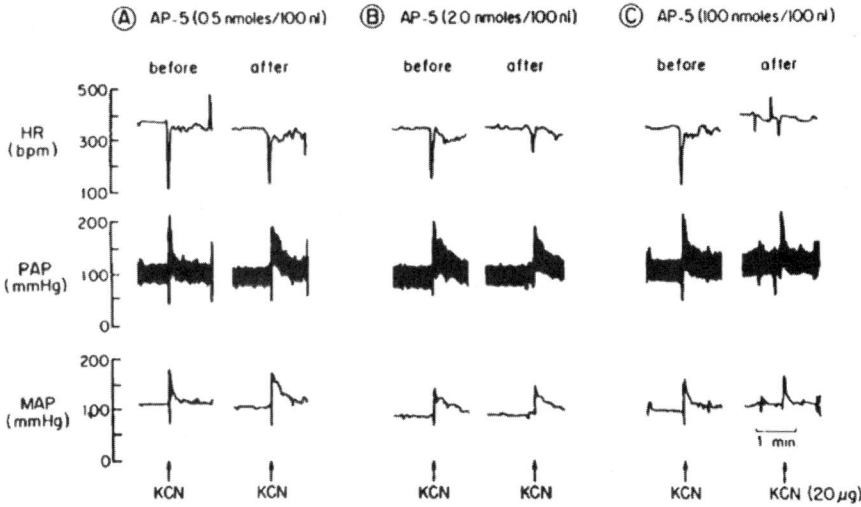

Figure 3. Tracings of three rats representative of their respective groups showing the changes in heart rate (HR), pulsatile arterial pressure (PAP), and mean arterial pressure (MAP) in response to injection of KCN (20 µg.rat^{-1}0.1ml^{-1} iv) before and after bilateral microinjection of phosponovaleric acid (AP-5) at 0.5 (A), 2.0 (B), and 10.0 nmol/100 nL (C) into the nuclei tracti solitarii. Data reprinted from Haibara et al. (1995) with permission.

These data also suggest that NMDA receptors were not involved in the processing of the sympathoexcitatory component of the chemoreflex in the NTS because the pressor response was not affected by bilateral microinjection of AP-5. As a consequence of these previous findings, the involvement of non-NMDA receptors in the processing of the sympathoexcitatory component of the chemoreflex reported in the study by Haibara et al. (1999) was explored in another experimental protocol, as shown in Fig. 4. In this case, we activated the chemoreflex before and after bilateral microinjection of increasing doses of DNQX, a selective non-NMDA receptor antagonist, into the NTS. It is important to note that bilateral microinjection of DNQX into the NTS produced a large increase in baseline MAP. This increase in baseline MAP is related to the blockade of non-NMDA receptors located in the neurons of the NTS probably involved in the sympathoinhibitory projections from the NTS to the RVLM, via the caudal ventrolateral medulla (CVLM), which is integral to the baroreflex pathways. The blockade of this sympathoinhibitory projection by DNQX may increase the activity of RVLM neurons and consequently increase baseline MAP. The reduction in the pressor response to chemoreflex activation after DNQX was probably due to the large increase in baseline MAP but, nevertheless, DNQX in a range of doses selective for non-NMDA receptors was ineffective in blocking the pressor response to chemoreflex.

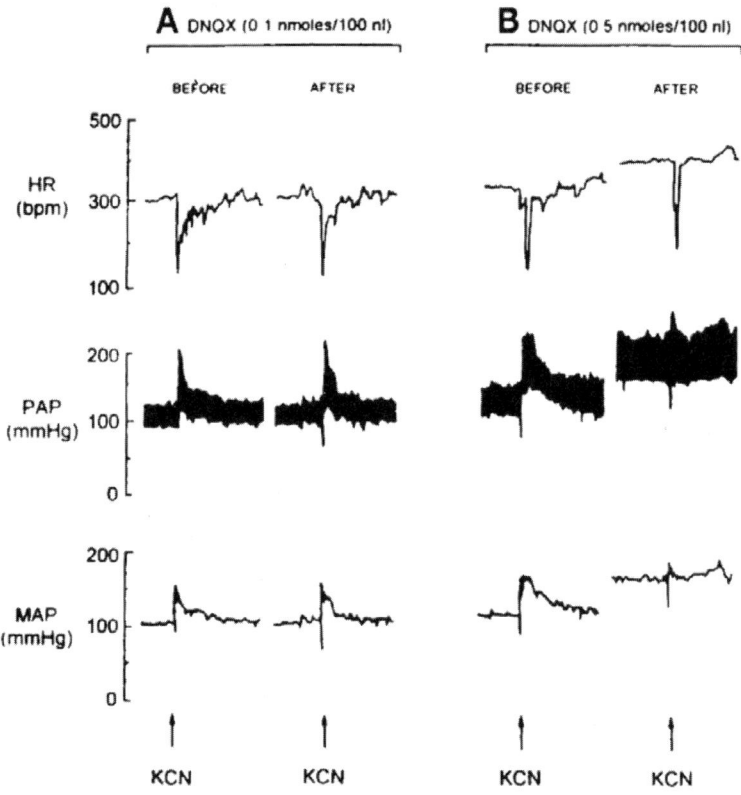

Figure 4. Tracings of two rats representative of their respective groups showing the changes in heart rate (HR), pulsatile arterial pressure (PAP), and mean arterial pressure (MAP) in response to KCN (20 µg.rat^{-1}.0.1moles^{-1} iv) before and 10 min after bilateral microinjection of 6,7-dinitroquinoxaline-2, 3-dione (DNQX) [0.1 (A), and 0.5 nmoles/100 nL (B)] in a range of doses selective to non-NMDA receptors into the rostral aspect of the commissural nuclei tracti solitarii. Data reprinted from Haibara et al. (1999) with permission.

For a better evaluation of the role of EAA receptors in the processing of the sympathoexcitatory component of the chemoreflex, Haibara et al. (1999) evaluated the effect of simultaneous triple microinjections of kynurenic acid, a non-selective EAA receptor antagonist, into the rostral (bilaterally) and caudal (medial) aspects of the commissural NTS, considering that several experimental lines of evidence suggest that the processing of the chemoreflex afferents occurs mainly at the caudal aspect of the commissural NTS (Vardhan et al., 1993; Zhang and Mifflin, 1993; Ciriello et al., 1994). Therefore, in the next step kynurenic acid was microinjected into the rostral commissural NTS or into the midline portion of the caudal commissural NTS or simultaneously at three sites into the NTS (rostral and caudal aspect of the commissural NTS) in order to block all ionotropic receptors in these areas. In spite of these procedures, the data summarized in Fig. 5 show that kynurenic acid microinjected into different subregions of the NTS produced only a reduction of approximately 50% of the pressor response of the chemoreflex. In this case it is also important to note that kynurenic acid produced a large increase in baseline MAP similar to the effect of DNQX. These findings have several implications: 1) l-glutamate ionotropic receptors seem to be integral to the sympathoinhibitory pathways of the baroreflex; 2) the reduction in the magnitude of the pressor response to chemoreflex activation was secondary to the large increase in baseline MAP, and 3) l-glutamate ionotropic receptors apparently are not involved in the neurotransmission of the sympathoexcitatory component of the chemoreflex. These different aspects strongly support the hypothesis that the neurotransmission of the sympathoexcitatory component of the chemoreflex at the NTS level is mediated by a neurotransmitter other than l-glutamate. The involvement of l-glutamate metabotropic receptors in the processing of the sympathoexcitatory component of the chemoreflex in the NTS was also evaluated by Haibara et al. (1999) using MCPG, a metabotropic receptor antagonist. These authors verified that bilateral microinjection of MCPG into the NTS produced no effect on the pressor or bradycardic responses to chemoreflex activation, indicating that metabotropic receptors also play no major role in the neurotransmission of the sympathoexcitatory component of the chemoreflex at the NTS level. In a series of recent studies from our laboratory, we verified that the magnitude of the pressor response to chemoreflex activation, when the increase in the baseline MAP due to microinjection of kynurenic acid into the NTS was normalized by sodium nitroprusside infusion, was similar to control values. This evidence gives additional support to the hypothesis that l-glutamate and ionotropic and metabotropic EAA receptors are not involved in the neurotransmission of the sympathoexcitatory component of the chemoreflex at the NTS level and suggests that other neurotransmitters or co-transmitters may participate in this processing.

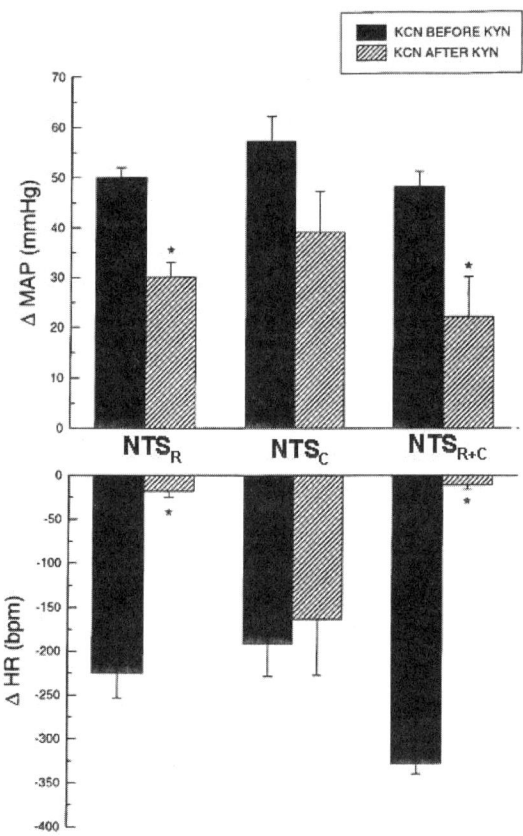

Figure 5. Changes in mean arterial pressure (Δ MAP) and heart rate (Δ HR) in response to KCN (20 µg.0.1 ml^{-1}.rat^{-1}, iv) before and 10 min after bilateral microinjection of kynurenic acid (10 nmol/100 nL) into the rostral aspect of the commissural NTS (NTS$_R$, n=6), midline aspect of the caudal commissural NTS (NTS$_C$, n=6) or simultaneously into the rostral and caudal commissural NTS (NTS$_{R+C}$, n=6) of 3 different groups of rats. Data reprinted from Haibara et al. (1999) with permission.

The possible role of substance P in the neurotransmission of the sympathoexcitatory component of the chemoreflex in the NTS of awake rats was evaluated by Zhang et al. (2000), considering that previous studies have shown an increase of substance P in the NTS during hypoxia. Bilateral microinjection of an NK-1 receptor antagonist (WIN) into the NTS of awake rats produced no changes in the cardiovascular responses to chemoreflex activation. Studies by de Paula and Machado (2001) evaluated the possible

2. CHEMOREFLEX AND SYMPATHOEXCITATION

role of adenosine and adenosine receptors in this neurotransmission. For this purpose, the chemoreflex was activated before and after bilateral microinjection of DPCPX, a non-selective A1 and A2 receptor antagonist, at a dose that was effective in blocking the cardiovascular responses to microinjection of adenosine into the NTS. The data showed that the cardiovascular responses to chemoreflex activation were not affected by DPCPX, indicating that adenosine and adenosine receptors are also not involved in this neurotransmission in the NTS.

Several recent studies have indicated that ATP and adenosine can act as neurotransmitters or neuromodulators in the central nervous system (CNS), including areas involved in cardiovascular regulation such as the NTS (Mosqueda-Garcia et al., 1989; Barraco et al., 1991; Mosqueda-Garcia et al., 1991; Tao and Abdel-Rahman, 1993; Ergene et al., 1994; Barraco et al., 1996; Lawrence and Jarrott, 1996; Phillis et al., 1997; Scislo et al., 1997; Scislo and O'Leary, 1998; Scislo et al., 1998; de Paula and Machado, 2001). Studies on anesthetized rats have documented that microinjection of ATP into the NTS produced a fall in blood pressure and bradycardia, which were blocked by suramin, a non-selective antagonist of ATP receptors (Ergene et al., 1994). These cardiovascular responses to ATP exhibited a typical pattern of fast-transmitter similar to the responses to microinjection of l-glutamate into the NTS of anesthetized rats, which produces a fall in arterial pressure and bradycardia (Talman et al., 1980; Leone and Gordon, 1989). However, microinjection of increasing doses of ATP into the commissural NTS of awake rats produced a dose-dependent pressor response, which was abolished by intravenous (iv) injection of prazosin. Microinjection of ATP into the NTS of awake rats also presented an intense bradycardic response, which was blocked by methyl-atropine. Therefore, the pattern of cardiovascular response to microinjection of ATP into the NTS of awake rats is quite similar to those produced by activation of the chemoreflex (de Paula et al., 2000). The difference of the responses to ATP in awake and anesthetized rats may be related to the effect of anesthetics on the processing of the cardiovascular reflex at the NTS level. Similar differences were observed in a previous study from our laboratory showing that microinjection of l-glutamate into the NTS of conscious rats produced a dose-related pressor and bradycardic response, in contrast to a dose-related depressor response observed when the same rats were studied under urethane or chloralose anesthesia (Machado and Bonagamba, 1992).

In order to explore the possible role of ATP and $P2_X$ and $P2_Y$ receptors in the processing of the chemoreflex we verified that microinjection of suramin, a non-selective $P2_X$ and $P2_Y$ receptors antagonist, into the caudal aspect of the commissural NTS produced a significant reduction in the pressor and bradycardic responses to chemoreflex activation. In this case, we were considering the possibility that ATP may act as a co-transmitter of l-glutamate because kynurenic acid, a non-selective ionotropic receptor antagonist, produced a significant reduction in the cardiovascular responses to microinjection of ATP into the NTS (de Paula et al., 2000). In a recent study by our group (Antunes and Machado, 2002) we verified that the microinjection of ATP into the NTS of awake rats produced apnea and no behavioral responses while chemoreflex activation in the same rats produced tachypnea and an important behavioral response of exploration of the environment. These data suggest that ATP may play a role in the processing of the cardiovascular responses to chemoreflex activation but is not involved in the processing of the respiratory and behavioral components of the complex physiological responses to chemoreflex activation.

In a recent study using the working heart brainstem preparation, another unanesthetized preparation, Paton et al. (2002) observed that microinjection of suramin into the NTS produced a selective blockade of the bradycardic component and no effect on reflex tachypnea following chemoreflex activation (Fig. 6).

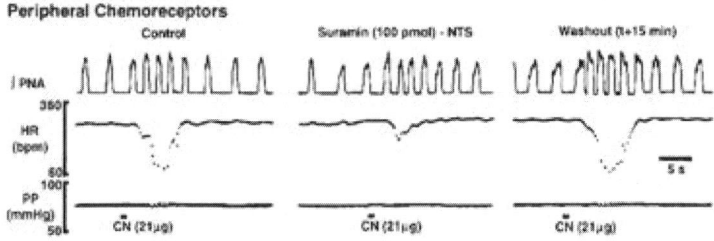

Figure 6. Suramin blockade of P2 receptors in the NTS selectively attenuates the reflex bradycardia but not the tachypneic response to chemoreflex activation. A representative WHBP in which cyanide (CN) stimulation of chemoreceptors evoked a decrease in phrenic nerve activity (PNA) cycle length and bradycardia. The bradycardiac, but not the respiratory component of this response, was attenuated by suramin (100 pmol). The attenuating effect of suramin on the chemoreceptor reflex bradycardia was abolished after 15 min. Data reprinted from Paton et al. (2002) with permission.

These findings support the hypothesis that NTS neurons mediating the respiratory component of the chemoreceptor reflex are distinct from those outputting to the cardiovascular system. In this case, P2 receptors would have to be distributed exclusively on those chemoreceptor afferents impinging on NTS neurons affecting cardiac vagal and sympathetic outflows, which is consistent with data obtained in awake rats showing that microinjection of ATP produced pressor and bradycardic but no ventilatory or behavioural responses (Antunes and Machado, 2002).

NEUROMODULATION OF THE CHEMOREFLEX IN THE NTS

With respect to the neuromodulation of the sympathoexcitatory component of the chemoreflex, Callera et al. (1999) reported that microinjection of muscimol, a $GABA_A$ receptor agonist, or baclofen, a $GABA_B$ receptor antagonist, into the rostral aspect of the commissural NTS produced a large increase in baseline MAP and in the case of baclofen, the magnitude of the pressor response to chemoreflex activation was reduced but not abolished, similarly to the results obtained with microinjections of kynurenic acid. On the other hand, microinjection of muscimol into the NTS produced a mild increase in baseline MAP and no significant changes in the pressor response to chemoreflex activation (Fig. 7). In this case it was suggested that the attenuation of the pressor response of the chemoreflex by a $GABA_B$ receptor agonist (baclofen) may be the consequence of inhibition of the sympathoexcitatory neurons in the NTS involved in the chemoreflex pathways or of the large increase in baseline MAP secondary to the blockade of sympathoinhibitory neurons of the NTS involved in the baroreflex pathways.

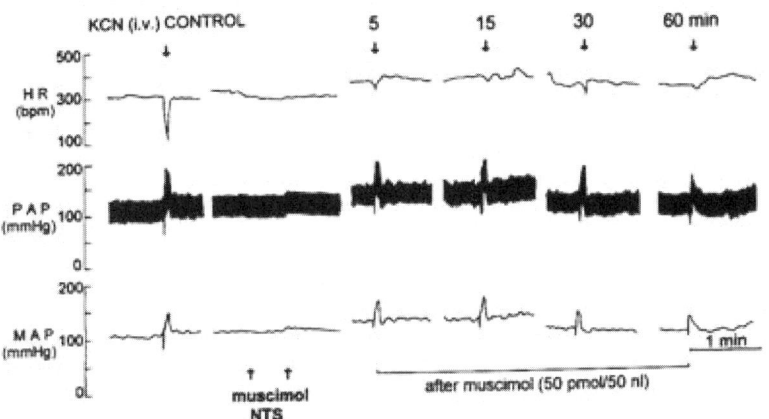

Figure 7. Tracings of a rat showing changes in heart rate (HR), pulsatile arterial pressure (PAP), and mean arterial pressure (MAP) in response to chemoreflex activation with KCN before and 5, 15, 30 and 60 min after bilateral microinjection of muscimol (50 pmol/50 nL) into NTS (arrows). Note that the bradycardic but not the pressor response was blocked by muscimol. Data reprinted from Callera et al. (1999) with permission.

In recent studies from our laboratory we verified that microinjection of glycine into the NTS produced a large increase in baseline MAP and a significant reduction in the magnitude of the pressor response to chemoreflex activation. However, when the baseline MAP was normalized by sodium nitroprusside infusion, the magnitude of the pressor response to chemoreflex activation was similar to control, indicating that glycine is also not an effective inhibitory neuromodulator of the sympathoexcitatory component of the chemoreflex at the NTS level.

The findings that the inhibitory amino acids GABA and glycine play no modulatory role in the sympathoexcitatory component of the chemoreflex at the NTS level suggest that the pathways involved in this autonomic response are not under inhibitory neuromodulation at this level. Considering that the arterial chemoreflex should be activated essentially under critical circumstances such as intense hypoxia, we may suggest that during a high risk or emergency situation the pathways involved in the processing of the sympathoexcitatory component of this reflex may not be under any inhibitory neuromodulation in order to allow an appropriate sympathoexcitatory response. We may consider that an abnormal chronic activation of the chemoreflex pathways, including the overexcitation of the sympathetic activity, may result in a pathophysiological process such as arterial hypertension.

PARAVENTRICULAR NUCLEUS OF THE HYPOTHALAMUS (PVN)

Another important aspect related to the sympathoexcitatory component of the chemoreflex that we must take into consideration is the possible involvement of other areas of the brain that play a role in its generation and/or modulation. Studies by Krukoff et al. (1994) showed that electrical and chemical stimulation of the PVN produced a significant increase in the number of neurons with Fos-like immunoreactivity in the lateral division of the PBN in the the pons and in the NTS and RVLM, indicating that these areas may receive direct projections from the PVN. On the other hand, there is anatomical and electrophysiological evidence of reciprocal connections between the NTS and the PVN (Ricardo and Koh, 1978; Ciriello and Caralesu, 1980; Silva-Carvalho et al., 1995a, 1995b) as well as projections from the PVN to the RVLM (Pyner and Coote, 1997, 1999; Yang and Coote, 1998) and to the spinal cord (Hosoya et al., 1991; Pyner and Coote, 1997). In addition, the PVN seems to play an important role in the modulation of sympathetic activity (Porter and Brody, 1986; Katafuchi et al., 1988; Martin and Haywood, 1992; Badoer and Merolli, 1998; Coote et al., 1998) and is

2. CHEMOREFLEX AND SYMPATHOEXCITATION

also involved in the cardiovascular responses to chemoreflex activation in anesthetized rats (Kubo et al., 1997).

In a recent study, Olivan et al. (2001) observed that bilateral electrolytic lesion of the PVN, including the parvo- and magnocellular neurons, produced a significant reduction in the magnitude and duration of the pressor response, indicating that PVN neurons are integral to the sympathoexcitatory component of the chemoreflex pathways (Fig. 8). It is plausible to suggest that the postsynaptic neurons in the NTS project to the PVN in order to excite sympathoexcitatory projections from the PVN to the RVLM (Yang and Coote, 1998; Pyner and Coote, 1999) or from the PVN directly to the intermediolateral column of the spinal cord (Hosoya et al., 1991; Pyner and Coote, 1997). The involvement of PVN in the generation of the sympathoexcitatory response to chemoreflex activation is consistent with previous studies showing that chemical or electrical stimulation of the PVN increases arterial pressure and the plasma level of norepinephrine and epinephrine as a consequence of the increase in sympathetic activity (Porter and Brody, 1986; Kannan et al., 1989; Martin and Haywood, 1992). The blockade of $GABA_A$ receptors in the PVN produced cardiovascular changes compatible with the concept that sympathoexcitatory projections of the PVN are under tonic inhibitory influence (Martin et al., 1991). The involvement of PVN in the sympathoexcitatory pathways of the chemoreflex is supported by the finding that rats submitted to chronic-intermittent hypoxia for 30 days exhibited intense c-Fos labeling in the PVN region (Sica et al., 2000a).

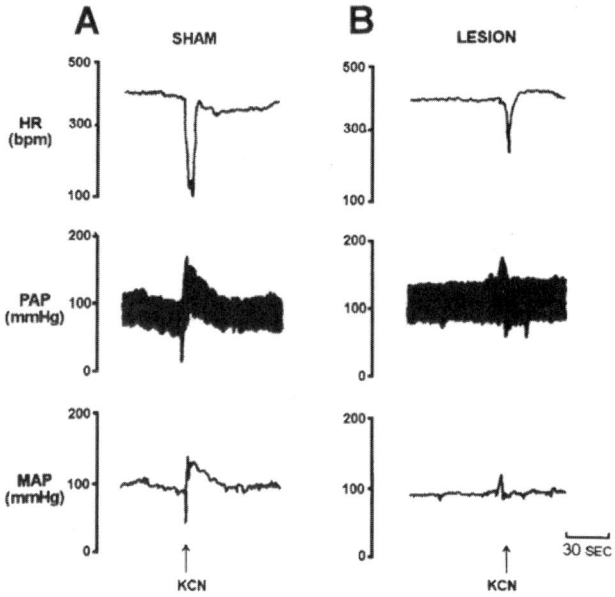

Figure 8. A: Typical tracings of one rat submitted to sham-lesion (SHAM) of the paraventricular nucleus of the hypothalamus (PVN) showing the changes in heart rate (HR), pulsatile arterial pressure (PAP) and mean arterial pressure (MAP) in response to chemoreflex activation with KCN (arrow). B: Typical tracings of one rat submitted to bilateral electrolytic lesion of the PVN (LESION) showing the changes in HR, PAP and MAP in response to chemoreflex activation with KCN (arrow). Note that the pressor response to chemoreflex activation was smaller in lesioned than in sham-lesioned rats. Data reprinted from Olivan et al. (2001) with permission.

With respect to the duration of the pressor response to chemoreflex activation, it is possible that part of the pressor response is dependent on the release of neurohormones such as vasopressin. In the study by Olivan et al. (2001) the duration of the pressor response to chemoreflex activation was also significantly reduced in comparison to the duration of the response in sham-lesioned rats or in rats with partial lesion of the PVN. In this case, it is possible that lesion of the PVN with a consequent absence of magnocellular neurons eliminates the secretion of vasopressin. In experiments performed on anesthetized rats with spinal cord transection, Kubo et al. (1997) documented that the pressor response to chemoreflex activation was due to the release of vasopressin. However, recent studies from our laboratory have documented that the magnitude and the duration of the pressor response to chemoreflex activation in awake rats was not altered by previous treatment with an AVP antagonist (iv) and subsequent injection of prazosin (iv) almost abolished these responses, indicating that the pressor responses to chemoreflex activation, as shown in Fig. 2, is essentially neural in origin.

It is also important to emphasize that partial electrolytic lesion of the PVN in a group of rats in the study by Olivan et al. (2001) produced no significant changes in the pressor response to chemoreflex activation, indicating that the remaining subpopulation of pre-autonomic neurons in the PVN seems to be sufficient to produce a sympathoexcitatory response when the peripheral chemoreceptors are activated. The neurotransmitters involved in the processing of the sympathoexcitatory projections from NTS to PVN and from PVN to RVLM is an important matter that requires further investigation. Therefore, the PVN neurons seems to play a key role in the generation of the sympathoexcitatory component (pressor response) of the chemoreflex and this area must be considered in the neural pathways involved in the complex cardiovascular, respiratory and behavioral responses to chemoreflex activation in awake rats.

PARABRACHIAL NUCLEUS (PBN)

Neuroanatomical studies have shown that the lateral portion of the PBN receives direct projections from several medullary, hypothalamic and forebrain structures implicated in cardiovascular control (Ricardo and Koh, 1978; Saper et al., 1979; Loewy et al., 1981), particularly afferents from specific regions of the NTS (Granata and Kitai, 1989; Herbert et al., 1990), which is the primary site of termination of chemoreceptor afferent fibers (Spyer, 1990; Ciriello et al., 1994). Therefore, the PBN is considered a relay for visceral afferent inputs coming from the NTS to the forebrain. Furthermore, the PBN provides the major source of descending pontine projections to regions of the RVLM, which are implicated in the control of sympathetic activity (Fulwiler and Saper, 1984). There is also evidence that electrical or chemical stimulation of the PBN produces cardiovascular responses characterized by increased arterial pressure resulting from the increase in sympathetic activity (Mraovitch et al., 1982; Ward, 1988; Lara et al., 1994).

The involvement of PBN in the chemoreflex pathways has been shown in previous studies. Studies using the expression of Fos protein (Erickson and Millhorn, 1994) have shown that stimulation of carotid chemoreceptors induced Fos-like immunoreactivity within neurons in the PBN. In addition, electrophysiological studies on anesthetized animals (Hayward and Felder, 1995) showed that PBN neurons were excited during chemoreceptor stimulation and that the increased activity of PBN neurons preceded the pressor response to chemoreflex activation, indicating an important role for this nucleus in the central integration of chemoreceptor inputs. In a series of recent experiments we verified that bilateral microinjection of lidocaine, a local anesthetic, into the dorsolateral PBN produced a significant reduction in the pressor response to chemoreflex activation (Fig. 9).

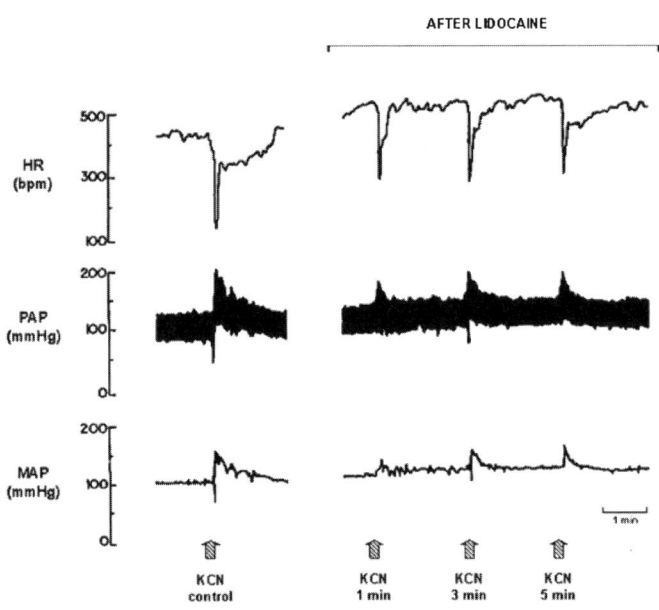

Figure 9. Tracings of one rat showing the changes in heart rate (HR), pulsatile arterial pressure (PAP) and mean arterial pressure (MAP) in response to chemoreflex activation before (KCN control) and 1, 3 and 5 min after bilateral microinjection of 2% lidocaine into the lateral parabrachial nucleus (PBN). Arrows indicate KCN injection. Data reprinted from Haibara et al. (2002) with permission.

In this case, we need to consider that the attenuation of the pressor response to chemoreflex activation after PBN blockade may be secondary to the blockade of fibers from the forebrain projecting to the PBN (Moga et al., 1990) and also that there is evidence of reciprocal projections between the hypothalamus/PVN and PBN (Thomas and Calaresu, 1972; Bester et al., 1997). It is possible that direct projections from the NTS to the PBN can activate PBN neurons projecting to the RVLM (Chamberlin and Saper, 1992) in order to increase the activity of sympathetic preganglionic neurons, with a consequent increase in MAP. Alternatively, we may suggest that projections from NTS to PVN and from NTS to PBN combined with reciprocal connections between PBN and PVN are integral to this complex network of the sympathoexcitatory component of the chemoreflex (Ciriello and Calaresu, 1980; Ciriello et al., 1984; Takeuchi and Hopkins, 1984; Granata, 1993; Bester et al., 1997). Studies by Granata (1993) have suggested that the PBN neurons integrate not only autonomic information, but also motor and some behavioral responses, such as defense reaction. Therefore, PBN neurons probably are not acting only as a relay in ascending and descending inputs, but are also able to play an integrative role in the control of sympathetic and other neurovegetative responses to chemoreflex activation.

An important aspect to be further investigated refers to the remaining pressor response after neuronal blockade of the dorsolateral aspect of the PBN. In this case we may suggest several possibilities such as direct projections from the NTS to RVLM (Urbanski and Sapru, 1988; Aicher et al., 1996) or from the PVN to the RVLM or even directly from the PVN to the intermediolateral column of the spinal cord (Coote et al., 1998). The remaining pressor response to chemoreflex even after PBN blockade is evidence of the complexity of the neural pathways that take part in the generation of sympathetic activity and indicates that several areas of the brain are integral to this network. The possibility that the volume of lidocaine microinjected into the PBN (100 nl) was not enough to block all PBN subnuclei also cannot be ruled out. Eventually, the blockade of the several subnuclei of the PBN may produce a greater reduction in the pressor response than that observed in the study by Haibara et al. (2002). Therefore, PBN neurons seem to be part of the pathways involved in the generation of sympathoexcitatory component of the chemoreflex. The specific neurotransmitter and/or neuromodulator in these synapses at the PBN level, as well as the projections to neurons in the ventral aspect of the brainstem involved in the generation of respiratory rhythm and sympathetic vasomotor tone are important features that require further investigation.

PATHOPHYSIOLOGICAL IMPLICATIONS OF OVERACTIVITY OF THE SYMPATHOEXCITATORY COMPONENT OF THE CHEMOREFLEX

The hypothesis that persistent activation of the chemoreflex may results in hypertension seems to be supported by the data obtained with the experimental model of chronic intermittent hypoxia, showing that rats submitted to hypoxia for several weeks presented a mild but significant increase in baseline MAP (Fletcher et al., 1992a, 2001). The understanding of the neural pathways involved in the generation of the complex cardiovascular, respiratory and behavioral responses to arterial chemoreceptor activation become even more important if we consider that some pathophysiological conditions such as hypoxia induced by the obstructive sleep apnea syndrome, may play a role in the genesis of hypertension in humans (Sommers et al., 1991, 1995; Pankow et al., 1997) as well as in experimental models (Fletcher et al., 1992a; Greenberg et al., 1999; Sica et al., 2000a). Therefore, the different experimental models of hypoxia became important tools for the understanding of the involvement of several areas and pathways in the brain with a complex pattern of responses to chemoreflex activation. Studies combining the hypoxic models such as moderate hypoxia or chronic intermittent hypoxia with the expression of c-Fos have shown a marked and significant increase in staining in several areas such as the NTS, lateral PBN, PVN, supra-optic nucleus, Kolliker-Fuse nucleus, locus coeruleus, periacqueductal gray-matter (Teppena et al., 1997; Berquin et al., 2000; Sica et al., 2000a). In addition, cortical circuits such as medial prefrontal, cingulated and insular cortices are also activated during chronic intermittent hypoxia, a fact that may also contribute to the regulation of sympathetic and cardiovascular function (Sica et al., 2000b).

In light of all these neural complexities related to the processing of the autonomic, ventilatory and behavioral responses to chemoreflex activation it is imperative to assume that the discovery of the role of arterial chemoreceptors by Corneille Heymans in the first half of the past century was just the beginning of a long history, which for sure will contribute to a complete understanding of the neural mechanisms of a critical reflex for our survival, particularly in emergency circumstances, and potentially involved in many autonomic dysfunctions.

ACKNOWLEDGEMENTS

I wish to thank all my colleagues that contributed to generate most of the data and ideas presented in this chapter. The studies performed in my laboratory were supported by grants from FAPESP, CNPq, CAPES and PRONEX.

REFERENCES

Aicher, S.A., Saravay, R.H., Cravo, S., Jeske, I., Morrison, S.F., Reis, D.J., Milner, T.A., 1996. Monosynaptic projections from the nucleus tractus solitarii to C1 adrenergic neurons in the rostral ventrolateral medulla: Comparison with input from the caudal ventrolateral medulla. J. Comp. Neurol. 373, 62-75.

Amano, M., Asari, T., Kubo, T., 1994. Excitatory amino acid receptors in the rostral ventrolateral medulla mediate hypertension induced by carotid body chemoreceptor stimulation. Naunyn-Schmiedeberg´s Arch. Pharmacol. 349, 549-554.

Antunes, V.R., Machado, B.H., 2002. Differential respiratory responses to microinjection of ATP into the nucleus tractus solitarii or to chemoreflex activation in awake rats (Abstract). FASEB J. 16, A828.

Badoer, E., Merolli, J., 1998. Neurons in the hypothalamic paraventricular nucleus that project to the rostral ventrolateral medulla are activated by haemorrhage. Brain Res. 791, 317-320.

Barraco, R.A., Walter G.A., Polasek, P.M., Phillis, J.W., 1991. Purine concentrations in the cerebrospinal fluid of unanesthetized rats during and after hypoxia. Neurochem. Int. 18: 243-248.

Barraco, R.A., O'Leary, D.S., Ergene, E., Scislo, T.J., 1996. Activation of purinergic receptor subtypes in the nucleus tractus solitarius elicits specific regional vascular response patterns. J. Auton. Nerv. Syst. 59, 113-124.

Barros, R.C.H., Bonagamba, L.G.H., Okamoto-Canesin, R., Oliveira, M., Branco, L.G.S., Machado, B.H., 2002. Cardiovascular responses to chemoreflex activation with potassium cyanide or hypoxic hypoxia in awake rats. Autonom. Neurosc.: Basic and Clinical 97, 110-115.

Bester, H., Besson, J.M., Bernard, J.F., 1997. Organization of efferent projections from the parabrachial area to the hypothalamus: a Phaseoulus vulgaris-leucoagglutinin study in the rat. J. Comp. Neurol. 383, 245-281.

Berquin, P., Bodineau, L., Gros, F., Larnicol, N., 2000. Brainstem and hypothalamic areas involved in respiratory chemoreflexes: a Fos study in adult rats. Brain Res. 857, 30-40.

Callera, J.C., Bonagamba, L.G.H., Nosjean, A., Laguzzi, R., Machado, B.H., 1999. Activation of GABAA but not GABAB receptors in the NTS blocked bradycardia of chemoreflex in awake rats. Am. J. Physiol. 276(45), H1902-H1910.

Chamberlin, N.L., Saper, C.B., 1992. Topographic organization of cardiovascular responses to electrical and glutamate microstimulation of the parabrachial nucleus in the rat. J. Comp. Neurol. 326, 215-262.

Ciriello, J., Calaresu, F.R., 1980. Monosynaptic pathway from cardiovascular neurons in the nucleus tractus solitarii to the paraventricular nucleus in the cat. Brain Res. 193, 529-533.

Ciriello, J., Lawrence, D., Pittman, Q.J., 1984. Electrophysiological identification of neurons in the parabrachial nucleus projecting directly to the hypothalamus in the rat. Brain Res. 322, 388-392.

Ciriello, J., Hochstenbach, S.L., Roder, S., 1994. Central projections of baroreceptor and chemoreceptor afferent fibers in the rat. In: Nucleus of the Solitary Tract, edited by Barraco, I.R.A., CRC, Orlando, p. 35-50.

Chitravanshi, V.C., Kachroo, A., Sapru, H.N., 1994. A midline area in the nucleus commissuralis of the NTS mediates the phrenic-nerve responses to carotid chemoreceptor stimulation. Brain Res. 662, 127-133.

Chitravanshi, V.C., Sapru, H.N., 1995. Chemoreceptor sensitive neurons in commissural subnucleus of nucleus tractus solitarius of the rat. Am. J. Physiol. 268 (Regulatory Integrative Comp. Physiol. 37), R851-R858.

Colombari, E., Bonagamba, L.G.H., Machado, B.H., 1994. Mechanisms of pressor and bradycardic responses to l-glutamate microinjected into the nucleus tractus solitarii of conscious rats. Am. J. Physiol. 266, R730-R738 (Regul. Integr. Comp. Physiol. 35).

Coote, J.H., Yang, Z., Pyner, S., Deering, J., 1998. Control of sympathetic outflows by the hypothalamic paraventricular nucleus. Clin. Exp. Pharmacol. Physiol. 25, 461-463.

Daly, M. de B., 1997. Discovery of the Respiratory Function. In: Peripheral Arterial Chemoreceptors and Respiratory-Cardiovascular Integration, edited by Daly, M. de B., Oxford University Press, Oxford, U.K., p. 134-160.

de Castro, F., 1928. Sur la structure et línnervation du sinus carotidien de l´homme et des mammifères. Nouveaux faits sur línnervation et la fonction du glomus caroticum. Études antomiques et physiologiques. Trabajos del Laboratora de Investigacion Biologicas de la Universidad de Madrid 25, 331-380.

de Paula, P.M., Bonagamba, L.G.H., Machado, B.H., 2000. Involvement of purinergic mechanisms in the neurotransmission of the chemoreflex in the nucleus tractus solitarius of awake rats (Abstract). Autonomic Neuroscience: Basic and Clinical 82, 55.

de Paula, P.M., Machado, B.H., 2001. Antagonism of adenosine A1 receptors in the NTS do not affect the chemoreflex in awake rats. Am. J. Physiol. 281, R2072-R2078 (Regul. Integr. Comp. Physiol.).

Ergene, E., Dunbar, J.C., O'Leary, D.S., Barraco, R.A., 1994. Activation of P2-purinoceptors in the nucleus tractus solitarius mediates depressor responses. Neurosci. Lett. 174, 188-192.

Erickson, J.T., Millhorn, D.E., 1994. Hypoxia and electrical stimulation of the carotid sinus nerve induce Fos-like immunoreactivity within catecholaminergic and serotoninergic neurons of the rat brainstem. J. Comp. Neuro. 348, 161-182.

Finley, J.C.W., Katz, D.M., 1992. The central organization of carotid body afferent projections to the brainstem of the rat. Brain Res. 572, 108-116.

Fletcher, E.C., Lesske, J., Behm, R., Miller, C., Staus, H., Unger, T., 1992a. Carotid chemoreceptors, systemic blood pressure and chronic episodic hypoxia mimicking sleep apnea. J. Appl. Physiol. 72, 1978-1984.

Fletcher, E.C., Lesske, J., Behm, R., Miller, C., Unger, T., 1992b. Repetitive episodic hypoxia causes diurnal elevations of systemic blood pressure in rats. Hypertension 19, 555-561.

Fletcher, E.C., Lesske, J., Culman, J., Miller, C., Unger, T., 1992c. Sympathetic denervation blocks blood pressure elevation in episodic hypoxia. Hypertension 20, 612-619.

Fletcher, E.C., 2001. Physiological consequences of intermittent hypoxia: systemic blood pressure. J. Appl. Physiol. 90, 1600-1605.

Franchini, K.G., Krieger. E.M., 1993. Cardiovascular responses of conscious rats to carotid body chemoreceptor stimulation by intravenous KCN. J. Auton. Nerv. Syst. 42, 63-70.

Fulwiler, C.E., Saper, C.B., 1984. Subnuclear organization of the efferent connections of the parabrachial nucleus in the rat. Brain Res. Rev. 7, 229-259.

Granata, A.R., Kitai, S.T., 1989. Intracellular study of nucleus parabrachialis and nucleus tractus solitarii interconnections. Brain Res. 492, 281-292.

Granata, A.R., 1993. Ascending and descending convergent inputs to neurons in the nucleus parabrachialis of the rat: an intracellular study. Brain Res. 600, 315-321.

Greenberg, H.E., Sica, A.L., Scharf, S.M., D.A. Ruggiero, D.A., 1999. Expression of c-fos in the rat brainstem after chronic intermitent hypoxia. Brain Res. 816, 638-645.

Haibara, A.S., Colombari, E., Chianca-Jr., D.A., Bonagamba, L.G.H., Machado, B.H., 1995. NMDA receptors in NTS are involved in bradycardic but not in pressor response to chemoreflex. Am. J. Physiol. 269 (Heart and Circ. Physiol., 38), H1421-H1427.

Haibara, A.S., Bonagamba, L.G.H., Machado, B.H., 1999. Sympathoexcitatory neurotransmission of the chemoreflex in the NTS of awake rats. Am. J. Physiol. 276 (Regulat. Integrative Comp. Physiol. 45), R69-R80.

Haibara, A.S., Tamashiro, E., Olivan, M.V., Bonagamba, L.G.H., Machado, B.H., 2002. Involvement of the parabrachial nucleus in the pressor response to chemoreflex activation in awake rats. Autonomic Neuroscience: Basic and Clinical 101, 60-67.

Hayward, L.F., Felder, R.B., 1995. Peripheral chemoreceptor inputs to the parabrachial nucleus of the rat. Am. J. Physiol. 268, R707-R714.

Herbert, H., Moga, M.M., Saper, C.B., 1990. Connections of the parabrachial nucleus with the nucleus of the solitary tract and the medullary reticular formation in the rat. J. Comp. Neurol. 293, 540-580.

Heymans, C., Bouckaert, J.J., Dautrebande, L., 1931a. Sinus carotidien et reflexes respiratoires. III. Sensibilité des sinus carotidiens aux substances chimiques. Action stimulante respiratoire réflexe du sulfure de sodium, do cyanure de potassium, de la nicotine et de la lobéline. Archives Internationales de Pharmacodynamie et de Thérapie 40, 54-91.

Heymans, C., Bouckaert, J.J., Dautrebande, L., 1931b. Au sujet du mécanisme de la bradycardie provoqueé par la nicotine, la lobéline, le cyanure, le sulfure de sodium, les nitrites et la morphine, et de la bradycardie asphyxique. Archives Internationales de Pharmacodynamie et de Thérapie 41, 261-289.

Heymans, C., Bouckaert, J.J., von Euler, U.S., Dautrebande, L., 1932. Sinus carotidiens et réflexes vasomoteurs. Archives Internationales de Pharmacodynamie et de Thérapie 41, 261-289.

Heymans, C., Bouckaert, J.J., Samaan, A., 1933. Action de lácide carbonique et de lóxygène sur le tonus et sur léxcitabilité reféxe et directe du système nerveux régulateur de la fréquence cardiaque. Comptes Rendus Hebdomadaires des Scéances et Memórias de la Société de Biologie et de sés Filiales 155, 423-425.

Heymans, C., Bouckaert, J.J., Samaan, A., 1934. Influences des variations de la teneur du sang oxygené et en CO2 sur la excitabilité réflexe et directe des elements centraux et periphériques des nerfs cardio-regulateurs. Archives Internationales de Pharmacodynamie et de Thérapie 48, 457-487.

Hosoya, Y., Sugiura, Y., Okado, N., Loewy, A.D., Kohno, K., 1991. Descending input from the hypothalamic paraventricular nucleus to sympathetic preganglionic neurons in the rat. Exp. Brain Res. 85, 10-20.

Kannan, H., Hayastuda, Y., Yamashita, H., 1989. Increase in sympathetic outflow by paraventricular nucleus stimulation in awake rats. Am. J. Physiol. 256, R1325-R1330.

Katafuchi, T., Oomura, Y., Kusowa, M., 1988. Effects of chemical stimulation of the paraventricular nucleus on adrenal and renal nerve activity in rats. Neurosc. Lett. 86, 195-200.

Koshiya, N., Huagfu, D., Guyenet, P.G., 1993. Ventrolateral medulla and sympathetic chemoreflex in the rat. Brain Res., 609, 174-184.

Koshiya, N., Guyenet, P.G., 1994. A5 noradrenergic neurons and the carotid sympathetic chemoreflex. Am. J. Physiol. (Regul. Integr. Comp. Physiol.) 267, R508-R518.

Koshiya, N., Guyenet, P.G., 1996. NTS neurons with carotid chemoreceptors inputs arborize in the rostral ventrolateral medulla. Am. J. Physiol. 270, R1273-R1278.

Kubo, T., Yanagihara, Y., Yamaguchi, H., Fukumori, R., 1997. Excitatory amino acid receptor in the paraventricular hypothalamic nucleus mediate pressor response induced by carotid body chemoreceptor stimulation in rats. Clin. Exp. Hypertension 19, 1117-1134.

Krukoff, T.L., Harris, K.H., Linetsky, E., Jhamandas, J.H., 1994. Expression of c-Fos protein in rat brain elicited by electrical and chemical stimulation of the hypothalamic paraventricular nucleus. Neuroendocrinol. 59, 590-602.

Lara, J.P., Parkes, M.J., Silva-Carvalho, L., Izzo, P., Dawid-Milner, M.S., Spyer, K.M., 1994. Cardiovascular and respiratory effects of stimulation of cell bodies of the parabrachial nuclei in the anaesthetized rat. J. Physiol. 477, 321-329.

Lawrence, A.J., Jarrott, B., 1996. Neurochemical modulation of cardiovascular control in the nucleus tractus solitarius. Prog. Neurobiol. 48, 21-53.

Leone, C., Gordon, F.J., 1989. Is l-glutamate a neurotransmitter of baroreceptor information in the nucleus tractus solitarius? J. Pharmacol. Exp. Ther. 250, 953-962.

Loewy, A.D., Wallach, J.H., McKellar, S., 1981. Efferent connections of the ventral medulla oblongata in the rat. Brain Res. Rev. 3, 63-80.

Loewy, A.D., 1990. Central autonomic pathways. In: Central Regulation of Autonomic Functions, edited by A.D. Loewy and K.M. Spyer. New York: Oxford University Press, p. 88-103.

Machado, B.H., Bonagamba, L.G.H., 1992. Microinjection of l-glutamate into the nucleus tractus solitarii increases arterial pressure in conscious rats. Brain Res. 576, 131-138.

Martin, D.S., Segura, T., Haywood, J.R., 1991. Cardiovascular responses to bicuculline in the paraventricular nucleus of the rat. Hypertension 18, 48-55.

Martin, D.S., Haywood, J.R., 1992. Sympathetic nervous system activation by glutamate injections into the paraventricular nucleus. Brain Res. 577, 262-267.

Mauad, H., Machado, B.H., 1998. Involvement of the ipsilateral rostral ventrolateral medulla in the pressor response to l-glutamate microinjection into the nucleus tractus solitarii of awake rats. J. Auton. Nerv. Syst. 74, 43-48.

Mauad, H., Machado, B.H., 2001. Pressor response to unilateral carotid chemoreceptor activation is not affected by ipsilateral antagonism of excitatory amino acid receptors in the rostral ventrolateral medulla of awake rats. Auton. Neurosc.: Basic and Clinical 91, 26-31.

Mifflin, S.W., 1992. Arterial chemoreceptor input to nucleus tractus solitarius, Am. J. Physiol. 263, R368-R375.

Moga, M.M., Herbert, H., Hurley, K.M., Yasui, Y., Gray, T.S., Saper, C.B., 1990. Organization of cortical, basal forebrain, and hypothalamic afferents to the parabrachial nucleus in the rat. J. Comp. Neuro. 295, 624-661.

Mosqueda-Garcia, R., Tseng, C.J., Appalsamy, M., Robertson, D., 1989. Modulatory effects of adenosine on baroreflex activation in the brainstem of normotensive rats. Eur. J. Pharmacol. 174, 119-122.

Mosqueda-Garcia, R., Tseng, C.J., Appalsamy, M., Beck, C., Robertson, D., 1991. Cardiovascular excitatory effects of adenosine in the nucleus of the solitary tract. Hypertension 18, 494-502.

Mraovitch, S., Kumada, M., Reis, D.J., 1982. Role of the nucleus parabrachialis in cardiovascular regulation in cat. Brain Res. 232, 57-75.

Olivan, M.V., Bonagamba, L.G.H., Machado, B.H., 2001. Involvement of the paraventricular nucleus of the hypothalamus in the pressor response to chemoreflex activation in awake rats. Brain Res. 895, 167-172.

Pankow, W., Nabe, B., Lies, A., Becker, H., Kohler, U., Kohl, F.V., Lohman, F.W., 1997. Influence of sleep apnea on 24-hour blood pressure. Chest 112, 1253-1258.

Paton, J.F.R., Kasparov, S., 2000. Sensory channel specific modulation in the nucleus of the solitary tract. J. Auton. Nerv. Syst. 80, 117-129.

Paton, J.F.R., Deuchars, J., Li, Y-W., Kasparov, S., 2001. Morphological and electrophysiological comparison of solitary tract neurones responding to peripheral chemoreceptor stimulation. Neurosci. 105, 231-248.

Paton, J.F.R., de Paula, P.M., Spyer, K.M., Machado, B.H., Boscan, P., 2002. Sensory afferent selective role of P2 receptors in the nucleus tractus solitarii for mediating the cardiac component of the peripheral chemoreceptor reflex. J. Physiol. 543, 995-1005.

Phillis, J. W., Scilo, T. J., O'Leary, D.S., 1997. Purines and the nucleus tractus solitarius: effects on cardiovascular and respiratory function. Clin. Exp. Pharmacol. Physiol. 24, 738–742.

Porter, J.P., Brody, M.J., 1986. A comparison of the hemodynamic effects produced by electrical stimulation of subnuclei of the paraventricular nucleus. Brain Res. 375, 20-29.

Pyner, S., Coote, J.H., 1997. The organization of the PVN projection to the RVLM and sympathetic preganglionic neurones in the spinal cord of rats. J. Physiol. 501P, P82-P83.

Pyner, S., Coote, J.H., 1999. Identification of an efferent projection from the paraventricular nucleus of the hypothalamus terminating close to spinally projecting rostral ventrolateral medulla neurons, Neurosci. 88, 949-957.

Ricardo, J., Koh, E.T., 1978. Anatomical evidence of direct projections from the nucleus of the solitary tract to the hypothalamus, amygdala, and other forebrain structures in the rat. Brain Res. 153, 1-26.

Ross, C.A., Ruggiero, D.A., Reis, D.J., 1985. Projections from the nucleus tractus solitarii to the rostral ventrolateral medulla. J. Comp. Neurol. 242, 511-534.

Saper, C.B., Swanson, L.W., Cowan, W.M., 1979. An autoradiographic study of the efferent connections of the lateral hypothalamic area in the rat. J. Comp. Neurol. 183, 689-706.

Scislo, T.J., Augustyniak, R.A., Barraco, R.A., Woodbury, D.J., O'Leary, D.S., 1997. Activation of P2X-purinoceptors in the nucleus tractus elicits differential inhibition of lumbar and renal sympathetic nerve activity. J. Auton. Nerv. Syst. 62, 103-110.

Scislo, T.J., Ergene, E., O'Leary, D.S., 1998. Imparied arterial baroreflex regulation of heart rate after blockade of P2-purinoceptors in nucleus tractus solitarius. Brain Res. Bull. 47(1), 67-67.

Scislo, T.J., O'Leary, D. S., 1998. Differential control of renal vs. adrenal sympathetic nerve activity by NTS A2a and P2x purinoceptors. Am. J. Physiol., Heart Circ. Physiol. 275, H2130-H2139.

Sica, A.L., Greenberg, H.E., Scharf, S.M., Ruggiero, D.A., 2000a. Immediate-early gene expression in cerebral cortex following exposure to chronic-intermittent hypoxia. Brain Res. 870, 204-210.

Sica, A.L., Greenberg, H.E., Scharf, S.M., Ruggiero, D.A., 2000b. Chronic-intermittent hypoxia induces immediate early gene expression in the midline thalamus and epithalamus. Brain Res. 883, 224-228.

Silva-Carvalho, L., Dawid-Milner, M., Spyer, K.M., 1995a. The pattern of excitatory inputs to the nucleus tractus solitarii evoked on stimulation in the hypothalamic defence area in the cat. J. Physiol. 487, 727-737.

Silva-Carvalho, L., Dawid-Milner, M.S., Spyer, K.M., 1995b. Hypothalamic modulation of the arterial chemoreceptor reflex in the anaesthetized cat: role of the nucleus tractus solitarii. J. Physiol. 487, 751-760.

Sommers, V.K., Mark, A., Abboud, F.M., 1991. Interaction of baroreceptor and chemoreceptor reflex control of sympathetic nerve activity in normal humans. J. Clin. Invest. 87, 1953-1957.

Sommers, V.K., Dyken, M.E., Clary, M., Abboud, F.M., 1995. Sympathetic neural mechanisms in obstructive sleep apnea. J. Clin. Invest. 96, 1897-1904.

Spyer, K.M., 1990. The central nervous organization of reflex circulatory control. In: Central Regulation of Autonomic Functions, edited by A.D. Loewy and K.M. Spyer. New York: Oxford University Press, p. 168-188.

Sun, M.K., Spyer K.M., 1991. Responses of rostroventrolateral medulla spinal vasomotor neurons to chemoreceptor stimulation in rats. J. Auton. Nerv. Syst. 33, 79-84.

Takeuchi, Y., Hopkins, D.A., 1984. Light and electron microscopic demonstration of hypothalamic projections to the parabrachial nuclei in the cat. Neurosci. Lett. 46, 53-58.

Thomas, M.R., Calaresu, F.R., 1972. Responses of single units in the medial hypothalamus to electrical stimulation of the carotid sinus nerve in the cat. Brain Res. 44, 49-62.

Talman, W.T., Perrone, M.H., Reis, D.J., 1980. Evidence for l-glutamate as the neurotransmitter of baroreceptor afferent nerve fibers. Sci. 209, 813-815.

Tao, S., Abdel-Rahman, A.A., 1993. Neuronal and cardiovascular responses to adenosine microinjection into the nucleus tractus solitarius. Brain Res. Bull. 32, 407-417.

Urbanski, R.W., Sapru, H.N., 1988. Evidence for a sympathoexcitatory pathway from the nucleus tractus solitarii to the ventrolateral medullary pressor area. J. Auton. Nerv. Syst. 23, 161-174.

Vardhan, A., Kachroo, A., Sapru, H.N., 1993. Excitatory amino acid receptors in the nucleus tractus solitarius mediate the responses to the stimulation of cardiopulmonary vagal afferent C fiber endings. Brain Res., 618, 23-31.

Ward, D., 1988. Stimulation of the parabrachial nuclei with monosodium glutamate increases arterial pressure. Brain Res. 462, 383-390.

Yang, Z., Coote, J.H., 1988. Influence of the hypothalamic paraventricular nucleus on cardiovascular neurones in the rostral ventrolateral meudulla of the rat. J. Physiol. 513, 521-530.

Zhang, W., Mifflin, S.W., 1993. Excitatory amino acid receptors within NTS mediate arterial chemoreceptor reflexes in rats. Am. J. Physiol. 265, H770-H773.

Zhang, C.H., Bonagamba, L.G.H., Machado, B.H., 2000. Blockade of NK-1 receptors in the nucleus tractus solitarii of awake rats produced no effect on cardiovascular responses to chemoreflex activation. Braz. J. Med. Biol. Res. 33, 1379-1385.

Chapter 3

CARDIOVASCULAR INTEGRATION IN THE NUCLEUS OF THE SOLITARY TRACT

Michael C. Andresen
Department of Physiology and Pharmacology, Oregon Health & Science University, Portland, OR 97239-3098, USA

Abstract: Afferents from the cardiovascular system including baroreceptors send information via cranial nerves to the nucleus tractus solitarius (NTS) where it is processed for homeostatic reflexes. The basic organization of autonomic pathways for cardiovascular control diverges beyond NTS with the cardiac parasympathetic arc as short as two central neurons. Significant modulation and integration occurs at the very first synapse between afferent endings and the second order NTS neuron. New evidence suggests important differentiation at both the pre- and postsynaptic parts of this first synapse and that such differences may contribute to differentiation in performance of the entire reflex pathways.

Key words: autonomic, baroreflex, baroreceptor, synaptic transmission, visceral afferent, integration

INTRODUCTION

Neural control systems are highly developed in mammals and confer unique regulatory functionality to overall control of the organism. For the cardiovascular system, autonomic reflex mechanisms provide both rapid responses and the capacity to integrate information from a wide range of sources including systems beyond the circulatory system. Such performance features importantly impact overall integration for homeostatic control. Since the central nervous system polls the status of multiple organ systems, these autonomic networks are uniquely situated to influence organism-wide homeostasis. Autonomic neurons orchestrate a broad range of efferent mechanisms to co-regulate these multiple organ systems in functionally complementary ways, e.g. cardiovascular adjustments simultaneous with respiratory regulation. Some of this complementary control appears to arise from common mechanisms and patterns of organization of these regulatory networks. This review will focus on the earliest phases of brain stem integration within the nucleus of the solitary tract (NTS) where sensory afferent information is transformed for use in cardiovascular regulation (Fig. 1).

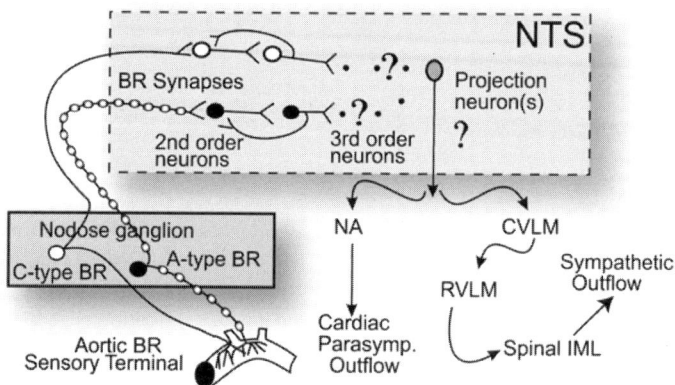

Figure 1. Basic organization of the core brain stem reflex networks for autonomic cardiovascular control. Minimal reflexes include peripheral sensory afferent neurons (e.g. nodose ganglion neurons, solid line box), contacting second order neurons in nucleus tractus solitarius (NTS) that then contact third order and/or projection neurons. Chief brain stem projection neurons are located in NTS, nucleus ambiguus (NA) and the rostral ventrolateral medulla (RVLM) for parasympathetic and sympathetic pathways respectively. This core circuit within the brain stem then projects either to peripheral parasympathetic postganglionic neurons (e.g. cardiac) or preganglionic sympathetic neurons in the spinal cord. Note that nodose neurons are of two types, those with myelinated axons (A-type, black) and those with unmyelinated axons (C-type, white). Reflex evidence suggests that A- and C-type afferents activate different pathways and new electrophysiological work suggests that these pathways are different even at the very first synapse on the second order NTS neuron. Little is known about the pathways beyond the second order neurons.

GENERAL ORGANIZATION OF CARDIOVASCULAR AUTONOMIC CONTROL

Cardiovascular neural networks are generally organized as negative feedback loops and begin with the afferent inputs (Andresen and Kunze, 1994) where they enter the brain stem to make synaptic contact at second order neurons within the NTS (Fig. 1). The most important regions for cardiovascular control are clearly concentrated in the caudal-most portions of the NTS and their distribution follows a generally viscerotopic pattern (Loewy, 1990). Aortic baroreceptors, for example, have cell bodies within the nodose ganglion (Fig. 1) and send central processes to make synaptic contacts that are concentrated in the medial portions of the dorsal caudal NTS (Dean and Seagard, 1995; Dean and Seagard, 1997; Mendelowitz et al., 1992; Chan et al., 2000a; Chan et al., 2000b; Doyle and Andresen, 2001). While this viscerotopic organization defined by afferent distribution and responses to local excitation of these subnuclei offers some general notion of destination within the caudal subnuclei of the NTS, anatomical and functional overlap is substantial. From an integrative standpoint, such proximity of visceral afferent inputs in a single relatively compact region of the nucleus likely facilitates contacts and interactions between relatively discrete afferent information sources that cross both modalities as well as organs and systems [e.g. carotid sinus baroreceptors and laryngeal mechanoreceptors (Mifflin, 1996)]. Thus broadly speaking, the first stage of these reflexes begins with input from sensory afferent neurons that transduce and then transmit information about local, peripheral conditions to central second order neurons within the NTS.

Generally in cardiovascular control, afferent information entering the NTS is broadcast widely to sites both within the brain stem as well as to supramedullary sites (Loewy, 1990; Dampney, 1994; Guyenet et al., 1996; Pilowsky and Goodchild, 2002). Interestingly, many of these central nuclei are reciprocally connected to the NTS (Loewy, 1990). The nature and function of these reciprocal links remain largely unexplored. The structural core of many such reflexes relies on pathways or relatively compact networks of neurons that appear to be restricted to the brain stem and spinal cord. For example, arterial baroreflex control of heart rate is well preserved following supramedullary decerebration (e.g. Lee et al., 2002). While the NTS location of second order neurons involved in cardiovascular regulation is well established, the location, nature and functional impact of third, fourth and higher order neurons are poorly understood. The final neural links for cardiovascular reflex pathways are motor neurons that project along fairly well described pathways effector targets such as the heart and blood vessels. Despite this solid foundation, much information remains described only in fairly general terms. Thus, clearly even in pathways whose core network is confined to brain stem and spinal cord, important details of how the neurons are interconnected and what are the nature or mechanisms of integration remain unclear.

CARDIAC PARASYMPATHETIC PATHWAY

The shortest cardiovascular reflex loop (Standish et al., 1994) and, therefore, perhaps one of the most tractable for study may be a fairly direct cardiac parasympathetic pathway (Standish et al., 1995; Irnaten et al., 2001) (Fig. 1). In this case, cardiovascular afferents such as aortic baroreceptors reach NTS second order neurons and may then project directly to vagal preganglionic neurons in nucleus ambiguus (Mendelowitz, 1999; Irnaten et al., 2001; Wang et al., 2001). This simple reflex path could contain as few as two central neurons since those vagal preganglionic neurons project directly to the cardiac postganglionic parasympathetic neurons within the heart. Thus, this short-loop reflex consists of four neurons and four synapses as a minimum pathway. Interestingly, unlike most spinal somatic reflexes, the baroreceptor reflexes are constantly conditioned by afferent baroreceptor information. Baroreceptor activity activates these pathways 24 hours a day with each cardiac cycle regardless of waking state or physical activity level of the individual. Given the constant cardiac cycling, cardiovascular afferents activate these reflex loops each second or more frequently. The arterial baroreceptor reflex exhibits/possesses dynamic sensitivity (Ead et al., 1952; Angell-James and Daly, 1970; Chapleau et al., 1989; Fan and Andresen, 1998; Fan et al., 1999) and use dependent resetting of performance (Kunze, 1986; Heesch and Carey, 1987; Fritsch et al., 1989; Tan and Zucker, 1989; Chapleau et al., 1991; Hayward et al., 1993). Thus, activity dependent processes are likely to contribute to the normal regulation and functional status of this reflex pathway and certainly experimental evidence suggests that the pathway is highly use-dependent. Processed sensory information must be combined with a wide range of signals originating from other parts of the nervous system and various hormonal signals to modulate cardiovascular control.

CARDIOVASCULAR SYMPATHETIC PATHWAY

In contrast to a relative simplicity of organization in the cardiac parasympathetic pathways, the sympathetic pathways of cardiovascular control are clearly and fundamentally more complicated (Fig. 1). Along the sympathetic arm of the arterial baroreflex, for example, a pool of inhibitory neurons in the caudal ventrolateral medulla (CVLM) are activated by increases in arterial baroreceptor activity via the NTS (Spyer, 1990; Pilowsky and Goodchild, 2002). These activated CVLM neurons then inhibit the premotor sympathetic neurons in the rostral ventrolateral medulla (RVLM) that project to the spinal cord (Fig. 1). These presympathetic neurons appear to be part of a tonically active network that gives rise to sustained activity so characteristic of the sympathetic branch of the autonomic nervous system, although the mechanism of tonic sympathetic drive remains controversial (Lipski et al., 1996; Sun and Guyenet, 1990; Sun et al., 1991; Kangrga and Loewy, 1995; Campos and McAllen, 1999). Recent attention to the caudal pressor area suggests an excitatory drive to RVLM (Gordon and McCann, 1988a; Gordon and McCann, 1988b; Natarajan and Morrison, 2000; Morrison, 2001) that is part of the regulation of cardiovascular sympathetic activity although the regulatory mechanisms are as yet not fully understood. Within cardiovascular regulation, reflex control mechanisms such as the baroreflex appear to offer a tonic level of sympathoinhibition that normally restrains sympathetic tone under resting or basal conditions.

From a regulatory viewpoint, neural reflexes provide the potential for highly differentiated performance and rapid actions (Pilowsky and Goodchild, 2002). Within cardiovascular autonomic regulation, this neural speed can reach fractions of a functional cycle time at the target. For example, in the cardiac parasympathetic baroreceptor reflex, afferent signals can evoke changes in cardiac performance within a single cardiac cycle – fractions of a second. Another powerful feature of neural regulation is the capacity to broadly integrate information. Afferent input is collected from a variety of sources important to the organism and the regulatory goal (e.g. blood pressure). This breadth of input integration includes integration across different qualities or modalities of inputs – sensory, hormonal, and even paracrine factors. Interestingly, even the simplest of these autonomic reflexes receive substantial modulation and display functionally important plasticity. The baroreceptor reflexes of humans and experimental animals are substantially modified by periods of hemodynamic alteration such as bed rest or zero gravity (Convertino et al., 1990; Eckberg and Fritsch, 1992; Moffitt et al., 1998; Cooke et al., 2000). The potential simplicity of at least some of these circuits limits the number of neurons that are required for these reflexes. Thus, within this rich context of function, the challenge is to identify the responsible mechanisms, even in some of the simpler systems that support these integrative responses. Clearly, the processing of cardiovascular sensory information is common to all of these cardiovascular reflexes and NTS may be a reasonable site to look.

INITIATION OF AFFERENT INTEGRATION

The processing of cranial visceral afferents for central autonomic control is analogous in fundamental neurobiological respects to that for general somatic spinal circuits. The NTS and its second order neurons are comparable to neurons within the superficial dorsal lamina of the spinal cord at which sensory neurons of the dorsal root ganglia (DRG) form synapses. In fact, some of these DRG neurons arise from viscera and this sensory information ascends via the cord and on to the brain stem including a portion to the NTS. Some of these connections are quite diffuse, but others appear more direct (Toney and Mifflin, 1995; Foreman, 2000; Boscan and Paton, 2002). The full contribution of such spinal afferent information to brain stem reflex mechanisms is only beginning to be revealed. The remainder of this review will focus on the processing of information arising from cranial visceral afferents at their first contact point within the NTS. Early evidence suggests that many of the most basic themes of neurotransmission reflect mechanisms that are common to both cranial visceral afferents and spinal somatic afferents, but that the afferent transmission at the NTS appears to present its own unique processing strategies.

STRONG AFFERENT SYNAPSES

As the afferent nerve enters the brainstem, it forms excitatory synapses on the NTS second order neuron. One approach to establishing whether a neuron is second or higher order has been to electrically activate the afferent nerve and examine the latency between stimulus shock and the response in the NTS, e.g. (Scheuer et al., 1996). Many such studies utilize activation of the aortic depressor nerve and extracellular recordings of action potentials in the NTS in the rat. Thus, in the narrowest interpretation, the findings apply to arterial baroreceptor transmission, although most conclusions seem more broadly relevant. Such afferent shocks activate postsynaptic action potentials in NTS neurons and responses evoked at the shortest delay appear to belong to the monosynaptically linked to second order neurons. Higher order NTS neurons are activated at longer latencies and often fail to be activated by closely paired stimuli. Such in vivo records indicate that afferent volleys of action potentials undergo considerable frequency dependent depression (Felder and Mifflin, 1994; Scheuer et al., 1996; Liu et al., 2000; Zhang and Mifflin, 2000). Thus, high frequencies of afferent action potentials are translated into substantially lower activities in the NTS neurons. For cardiovascular neurons, this depression may contribute to the observation that only small proportions of NTS neurons display obvious cardiac rhythmicity (e.g. Hayward and Felder, 1995) despite afferent inputs that are dominantly phasic with the cardiac cycle (Kunze and Andresen, 1991). The mechanisms responsible for depression are not fully understood but likely are common to most visceral afferent inputs. In vivo and neuroanatomical studies suggest that the second order NTS neurons are relatively small and this limits intracellular access presumably by the difficulty of impaling and maintaining the recording as the brain stem moves with breathing or arterial pulsations (Miura, 1975). As the details of afferent synaptic transmission in the NTS emerge, it is clear that very significant modulation and integration is initiated at the very first synapse of autonomic reflex pathways.

Certainly, second order NTS neurons receive prominent inhibitory inputs from several sources. Inhibition from gamma aminobutyric acid (GABA) neurons likely arises both locally from NTS interneurons (Andresen and Yang, 1990; Kunze and Andresen, 1991; Andresen and Mendelowitz, 1996) as well as projection neurons from within or beyond the brain stem (Felder and Mifflin, 1988; Jordan et al., 1988; Mifflin et al., 1988). In intracellular recordings from second order NTS neurons, afferent activation often triggers a GABAergic inhibitory postsynaptic potential (IPSP) following the initial excitatory postsynaptic potential (EPSP), e.g. (Mifflin and Felder, 1988). Presumably this might reflect actions of a local inhibitory interneuron. The timing of these connections suggests their potential to contribute to depression during sustained – even paired pulse – stimulation and represent a potential ambiguity for extracellular recordings since frequency dependent depression of the EPSP and coincidence of an IPSP and EPSP will result in the same failure of action potentials. The relative strength of inhibitory transmission is often enhanced by the presence of general anesthetics and this may affect in vivo results in the intact system. A number of mechanisms may be suggested to contribute to frequency dependent behaviors so prominent in NTS second order neurons. The advent of in vitro approaches offers new information on the cellular mechanisms of afferent integration.

PRESYNAPTIC DIFFERENTIATION

The two most fundamental components of visceral afferent synaptic transmission are the presynaptic endings and the postsynaptic NTS neurons (Fig. 2). Even this distilled view presented in simple terms requires a collage of evidence from multiple sources and experimental perspectives to reliably draw conclusions with sufficiently broad conviction. The identity of the neurotransmitter released from these afferents is the foundation of synaptic transmission and studies began in the whole animal. In vivo pharmacological strategies clearly suggest that glutamate mediates excitatory neurotransmission between sensory afferent and NTS neurons (Talman et al., 1984; Andresen and Kunze, 1994; Sved and Gordon, 1994; Lawrence and Jarrott, 1996). Broad-spectrum ionotropic glutamate receptor antagonists effectively block cardiovascular reflexes when placed in caudal portions of NTS (Leone and Gordon, 1989; Talman, 1989). Interpretation of changes in cardiovascular responses induced by local microinjection of drugs present important challenges that are not easily addressed experimentally. Clearly, the concentration and the spread of drugs are important factors in the interpretation of such results. Selectivity of antagonists is quite dependent on concentration. Other receptors outside the selectivity of specific antagonists often exist so that injection of antagonists may not prevent all actions of an agonist (Talman, 1989). The cardiovascular responses measured include blood pressure and/or heart rate and these require the recruitment of an unknown population of neurons locally together with neurons outside the area exposed to the drug. These pragmatic features produce potentially important confounders to interpretation and the potential for multiple and perhaps even offsetting effects. For example, a glutamate antagonist will block excitatory transmission to both second order NTS neurons as well as GABAergic interneurons and it is the net effect that is measured peripherally as a change in the relevant cardiovascular hemodynamics.

Focusing closer and more directly onto afferent transmission, in vivo, single-cell electrophysiology provides the focus on potential involvement of multiple cell types. Studies of short latency and, therefore, presumably second order neurons narrows the specific neuron type in question. Difficulties remain in the concentration aspects of drug specificity. Thus, despite the irreplaceable value of in vivo experiments in establishing how these NTS neurons generally operate in the intact state, technical problems in vivo importantly limit the nature of the questions that can be addressed.

As with other areas of neuroscience, important new in vitro initiatives have begun to offer avenues of experimental approach that can resolve many of the practical difficulties of experimental control of concentration and focus on single neurons. These approaches allow more direct assessments of the details of the cellular mechanisms of neurotransmission, connectivity, integration and excitability. The glutamate neurotransmission is an illustrative example of a starting point to discover how visceral afferents

3. CARDIOVASCULAR INTEGRATION IN THE NUCLEUS OF THE SOLITARY TRACT

affect NTS neurons. The cellular details are only beginning to reveal the greater complexity and the richness of mechanism that begins with this simple starting point.

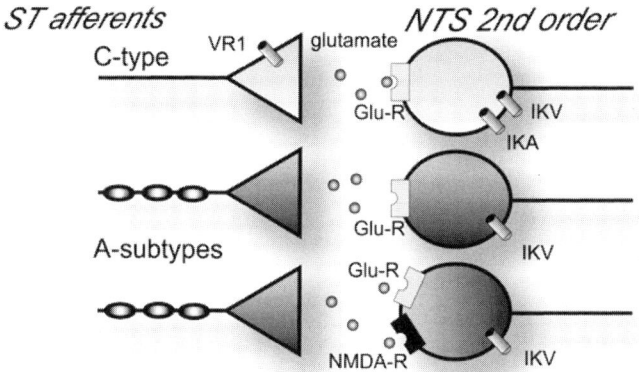

Figure 2. Central axons of visceral afferents including cardiovascular neurons form the solitary tract (ST) before forming synapses at NTS second order neurons. Activation of ST afferent axons releases glutamate that activates non-NMDA receptors (Glu-R) postsynaptically. These Glu-R responses are recorded under voltage clamp as excitatory postsynaptic currents (EPSCs). Presynaptic terminations of ST afferents are not uniform. Presynaptic release of glutamate can be evoked by capsaicin (CAP) in some neurons indicating the presence of vanilloid receptors (VR1) on a subset of ST terminals. Prolonged CAP application blocks ST transmission and these are considered coupled to C-type or unmyelinated afferents. CAP-resistant neurons are considered coupled to A-type axons. Interestingly, these two neuron types innervate different types of NTS neurons – i.e. these A- and C-type pathways are different beyond the afferent terminals. In a small subset of A-type neurons, ST released glutamate activates an additional small current via NMDA receptors. The C-type neurons express a prominent transient potassium current (IKA) in addition to the slow activating, steady outward currents (IKV) found in all second order NTS neurons.

To better understand the two neuron conversation between afferent and second order NTS neurons (Fig. 2), in vitro approaches isolate a portion of the brain stem in a stable and controlled environment (Miles, 1986; Andresen and Yang, 1990). Second order NTS neurons recorded in slices can be identified by electrophysiological criteria and/or anatomically by the visualization of sensory boutons on the neuron soma (Doyle and Andresen, 2001). In horizontal slices of caudal brain stem, patch electrodes are directed to contact NTS neuron cell bodies. In slices oriented in this manner, a concentric bipolar electrode is placed on the solitary tract (ST) that contains the axons of the cranial visceral afferents at substantial distance from the recording site as they traverse the brain stem. Electrical stimulation in this manner activates ST axons and this evokes short latency, highly consistent (low jitter) excitatory postsynaptic currents (EPSCs) under voltage clamp conditions (Doyle and Andresen, 2001). The low variability of EPSC latencies from shock to shock – typically with a standard deviation of about 100 µsec – likely reflects both the great strength of the synaptic coupling between visceral afferent and NTS second order neuron as well as the simplicity of the connection – axonal conduction and transmitter release. In such slices, increasing electrical stimulus intensity fails to recruit additional fibers – a finding consistent with a "unitary" synaptic response due to activation of a single afferent axon. Thus, despite an electrode tip with dimensions expected to contact many ST axons, the results are consistent with single afferent fibers contacting a single NTS neuron with little convergence of parallel afferent fibers. This may be a consistent artifact of the isolation procedure but is a striking finding. Despite evidence that single axons are responsible, EPSCs are characteristically quite large. This large amplitude of a unitary EPSC is consistent with the somatic contact of the afferent synapses of aortic baroreceptors (Mendelowitz et al., 1992; Balkowiec et al., 2000). Positioning of excitatory synapse close to the soma and initiation point of the axon favors a very high safety factor for transmission of afferent excitation to a conducted action potential in the postsynaptic neuron.

Application of antagonists selective for specific glutamate receptors to NTS containing slices fully blocked ST evoked EPSCs, a finding generally consistent with the in vivo results and the conclusion that visceral afferents release glutamate as their basic neurotransmitter (Fig. 2). This general conclusion is in keeping with the fact that the major excitatory neurotransmitter in the central nervous system is glutamate, a mechanism common to spinal sensory afferents. More precisely, however, only agents selective for the non-NMDA receptor subtype of glutamate receptor were effective at blocking excitatory synaptic transmission (Andresen and Yang, 1990; Doyle and Andresen, 2001). This is distinctly different than the pharmacological profile of antagonists injected into the NTS to evoke blood pressure changes or to modify baroreflex responses that requires a substantial role for NMDA receptors (Sved and Gordon, 1994). Together, the results emphasize the potential contribution in vivo of multiple sites of action that confound interpretation in the simplest terms. Clearly, NMDA receptors on higher order neurons might play a role in the intact situation but heterogeneity of responses is also possible.

An important potential source of heterogeneity may reside presynaptically at the synapses formed by the visceral afferents (Fig. 2). Presynaptic processing of incoming action potentials encompasses the mechanisms underlying axonal conduction, transmitter release, and any presynaptic modulation. This may importantly involve of multiple neurotransmitter receptors and additional nerve processes that release other transmitters - heterogeneity well supported anatomically at these afferent presynaptic endings (Andresen and Kunze, 1994). In the broadest cellular phenotypic classes of sensory neurons, these visceral afferent neurons belong to either myelinated or unmyelinated classes and generally resemble the divisions of DRG neurons (Lawson, 1992). Thus, studies of nodose neurons outline substantial and distinct differences in ion channels and discharge properties associated with these two myelination-based classes of visceral afferent neurons (Schild et al., 1994; Schild and Kunze, 1997; Glazebrook et al., 2002). Clearly, all of these afferents release glutamate since thus far no afferent synaptic responses have been recorded that are not blocked by broad-spectrum ionotropic glutamate antagonists.

Functional issues relevant to the afferent class are suggested from studies of arterial baroreceptors (BR) and baroreflexes. BR discharge responses to pressure fall into two patterns that are associated with two broad classes of afferent neurons that divide along classical lines by the axon fiber type (Seagard et al., 1990; Kunze and Andresen, 1991). Myelinated A-type BRs tend to have high rates of very regular discharge, are very sensitive to pressure and generally slowly adapting. C-type BRs conversely tend to have very irregular and modest rates of discharge, are less sensitive to pressure (high average thresholds and low sensitivities) and are often rapidly adapting (Kunze and Andresen, 1991). Thus, these two classes of neurons within a single modality (baroreceptive – mechanotransducer) give rise to functionally distinct discharge and performance patterns. Interestingly, selective activation of these BR subclasses by aortic nerve electrical stimulation suggest that the reflex effects of A-type and C-type BRs differ substantially in their frequency dependence (Fan and Andresen, 1998; Fan et al., 1999). The baroreflex responses to A-type BR activation are substantially larger when action potential trains follow a burst pattern than steady trains of shocks with the same number of stimuli, whereas phasic patterns of activation are no more effective than steady trains in C-type BRs (Fan et al., 1999). These dynamic response differences are present at the level of the second order NTS neurons and their synaptic transmission (Andresen and Yang, 1995; Seagard et al., 2001).

The cellular basis for the differences in discharge pattern is likely much more than simply the presence or absence of axonal myelination. The two groups of cranial visceral afferent neurons exhibit substantial differences in the expression of ion channels. The most obvious difference in nodose neurons is in the expression of sodium channels sensitive to TTX (Schild and Kunze, 1997; Li and Schild, 2002) – a myelinated/unmyelinated difference that is broadly reminiscent of spinal sensory DRG neurons (Lawson, 1992; Amaya et al., 2000). Interestingly, although the broad comparison between somatic afferent neurons and cranial visceral afferent neurons indicates general similarities, it is already clear that important functional differences exist. These critical details of ion channel expression may be responsible for interesting functional differences of these cranial afferents compared to their spinal cousins.

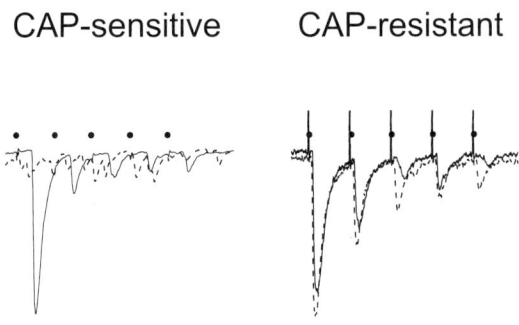

Figure 3. Electrical activation of solitary tract (ST) axons evokes excitatory postsynaptic currents (EPSCs) in voltage clamped (-60 mV) NTS neurons. Large dots mark each shock to ST in a train of five stimuli. EPSCs mediated by glutamate activation of non-NMDA receptors and these EPSCs are fully blocked by non-NMDA receptor antagonists (not shown) even in zero Mg++ conditions. Application of 100 nM capsaicin (CAP) eliminated CAP-sensitive EPSCs (left, broken line) but not CAP-resistant (right, broken line).

Recent investigations demonstrate one possible strategy for identifying NTS second order neurons innervated by unmyelinated afferents that is grounded in the cellular differences apparent in the afferent neuron phenotypes (Fig. 3). The vanilloid receptor VR1 is selectively expressed in a subset of spinal sensory neurons – typically those with unmyelinated or lightly myelinated axons (Stebbing et al., 1998; Szallasi and Blumberg, 1999; Caterina and Julius, 2001). VR1 couples to an ion channel with broad selectivity for cations and a substantial preference for calcium ions. Capsaicin (CAP), the active ingredient in a range of pungent peppers, selectively binds to VR1 and activates this ion channel. If CAP is painted onto the aortic depressor nerve, the conduction of C-type BR axons is completely and selectively blocked and reflex responses to C-fiber activation are eliminated (Fan and Andresen, 1998). CAP applied to brain stem slices initially increased spontaneous synaptic release of glutamate, facilitated ST-evoked EPSCs, but at high concentrations blocked ST synaptic transmission in a subgroup of second order neurons in medial NTS (Doyle et al., 2002). Such responses were found in NTS neurons bearing dye-labeled contacts from the aortic depressor nerve (Mendelowitz et al., 1992) as well as unlabeled neurons. CAP appears to trigger an initial burst in release of synaptic vesicles from the terminals of C-type ST axons and during prolonged exposure vesicles are either depleted or the presynaptic process becomes inactivated by CAP. Thus, the two classes of peripheral afferents appear to distribute ST axons and presynaptic processes that are pharmacologically distinct.

Based on such results, many other heterogeneities of transmitter, ion channel and receptor expression are to be expected. Presynaptic ion channels are targets for neuromodulation and such mechanisms control neurotransmitter release. The species of calcium channels that mediate ST synaptic transmission are closely correlated with the distributions of these channels found in nodose neuron cell bodies (Mendelowitz et al., 1995). In the case of peptides for example, angiotensin facilitates ST glutamatergic transmission to a minority of second order NTS neurons by a presynaptic mechanism (Barnes et al., 2003), a finding consistent with a selective expression of receptors to this peptide on a subset of the presynaptic ST afferent processes. Many other candidate receptors for peptides, glutamate and GABA are suggested to be located presynaptically on visceral afferent synaptic endings (Gao et al., 1992; Ding et al., 1998; Hoang and Hay, 2001). Such heterogeneity may underlie the specificity of actions of hormones and neurotransmitters to particular subpopulations of NTS neurons and, thus, to actions on specific autonomic pathways. At this very powerful synapse, these presynaptic modulators appear to reduce the probability of glutamate release so that synaptic depression appears to be the major mode of modulation of visceral afferent transmission (Hay and Kunze, 1994; Hay and Lindsley, 1995; Liu et al., 1998; Pamidimukkala and Hay, 2001; Chen et al., 2002).

POSTSYNAPTIC DIFFERENTIATION AND PROCESSING

Heterogeneity is not peculiar to the ST visceral afferent neurons. Different classes of NTS neurons have been described in both caudal and rostral NTS based on differences in discharge properties linked to potassium channels (Dekin and Getting, 1984; Bradley and Sweazey, 1992; Butcher and Paton, 1998). For the most part, the relation of such neurons to the ST or position in any reflex pathway has been unclear. As a result, despite the substantial impact of blockade of potassium channels on cardiopulmonary and arterial BR reflex responses (Butcher and Paton, 1998), variable, contrasting results (either increased or decreased excitability) were described at a cellular level from NTS neurons (Butcher et al., 1999).

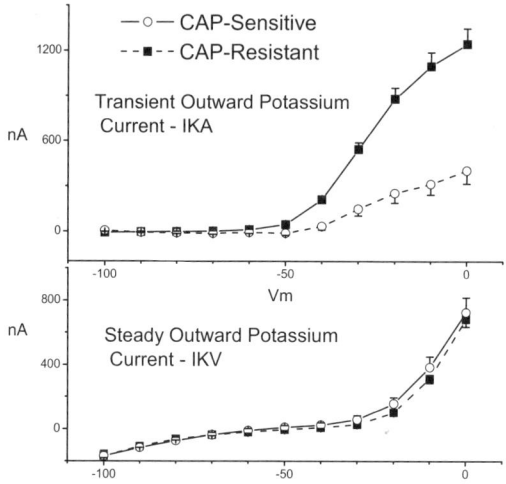

Figure 4. NTS neurons with capsaicin (CAP) sensitive ST synaptic responses have different potassium currents than CAP-resistant neurons. The C-type or CAP-sensitive neurons express a transient potassium current (IKA) in addition to the large early transient outward currents (IKV) found in all second order NTS neurons.

Differences in voltage gated potassium channels profoundly impact the discharge and adaptive characteristics of NTS neurons as well as the ability of excitatory synaptic potentials to elicit action potentials in second order neurons (Schild et al., 1993; Schild et al., 1995; Bailey et al., 2002). Surprising recent studies point to a possible systematic difference in the expression of potassium channels that is linked to the type of ST afferent innervation (Bailey et al., 2002). In second order NTS neurons with CAP-resistant ST evoked EPSCs (Fig. 4), depolarization evoked rather slowly activating voltage dependent potassium currents (IKV) that were sustained and sensitive to high concentrations of tetraethylammonium (TEA). In these CAP-resistant neurons, a variable but modest transient, rapidly inactivating potassium current (IKA) was blocked by 4-aminopyridine (4-AP). In contrast, second order NTS neurons with CAP-sensitive, ST-evoked EPSCs displayed large 4-AP sensitive IKAs and quite similar IKV currents. Such differences in potassium channel expression mediate a sensitivity of C-type

3. CARDIOVASCULAR INTEGRATION IN THE NUCLEUS OF THE SOLITARY TRACT

(CAP-sensitive) NTS neurons to membrane potential (Schild et al., 1993; Schild et al., 1995; Bailey et al., 2002). Hyperpolarizing conditions of C-type but not A-type NTS neurons reduced the discharge capacity of the cells. The presence of the transient IKA and its high voltage sensitivity near normal resting potentials means that the capacity of C-type second order NTS neurons to generate action potentials will be highly dependent on membrane potential. This process will in turn modulate the ability of these neurons to excite the axon and thereby transmit information out to higher order neurons along the C-type pathway. The mechanisms responsible for this phenotypic matching of pre- and postsynaptic neurons are not clear but could represent a trophic interaction between visceral afferent and the second order neuron via patterns of activity per se or by release of specific growth factors.

At many levels, the overall heterogeneity of the NTS is striking. Activation of different portions of caudal NTS can evoke different mixes of systemic cardiovascular, respiratory and other visceral responses. Immunohistochemically, the large number of candidate neurotransmitters identified within the NTS has yet to be fully appreciated (Van Giersbergen et al., 1992; Andresen and Kunze, 1994; Pilowsky and Goodchild, 2002). Patterns of differences across NTS neurons are only beginning to emerge but hold the promise of better understanding how these multiple mechanisms function together and where critical interactions give rise to new or better-controlled function. The recent surge of cellular information concerning the NTS is often discussed in terms of afferent transmission. However, truly little information exists concerning NTS neurons of higher than second order. One challenge will be to understand the ramifications for information processing within the NTS of the many inputs that descend upon the NTS from within the medulla or beyond. If the notion of a very strong afferent excitation driving the core pathways holds up, then one would anticipate that many of these modulatory inputs will involve mechanisms to directly depress this strong afferent drive either by depressing glutamate release at the afferent ending (presynaptic depression) or depressing second order neuron excitability (e.g. facilitating potassium channel activity). Recent reports continue to build an appreciation of the complexities but bridging cellular observations with systems information is quite daunting. New en bloc preparations may provide key connections (Paton et al., 1994; Paton, 1996; Chizh et al., 1997) between highly cellular observations and intact systems. The degree of differentiation and interaction requires careful attention to multiple mechanisms and the high potential for secondary and unanticipated contributions. Nonetheless, the NTS and in particular cardiovascular pathways offer unique opportunities to understand essential neurobiological mechanisms within the context of a fully functioning system.

In summary, cardiovascular homeostatic reflexes begin with afferents impinging on initial neurons in the NTS. New studies highlight functionally important differentiation at these afferent synapses and the second order neurons. Both the presynaptic regulation of glutamate release at afferent synapses as well as differences in postsynaptic receptors and ion channels appear likely to contribute to differentiation in performance of the entire reflex pathways.

REFERENCES

Amaya, F., Decosterd, I., Samad, T.A., Plumpton, C., Tate, S., Mannion, R.J., Costigan, M., Woolf, C.J., 2000. Diversity of expression of the sensory neuron-specific TTX-resistant voltage-gated sodium ion channels SNS and SNS2. Mol. Cell Neurosci. 15, 331-342.

Andresen, M.C., Kunze, D.L., 1994. Nucleus tractus solitarius: gateway to neural circulatory control. Annu. Rev. Physiol. 56, 93-116.

Andresen, M.C., Mendelowitz, D., 1996. Sensory afferent neurotransmission in caudal nucleus tractus solitarius - common denominators. Chemical Senses 21, 387-395.

Andresen, M.C., Yang, M., 1995. Dynamics of sensory afferent synaptic transmission in aortic baroreceptor regions of nucleus tractus solitarius. J. Neurophysiol. 74, 1518-1528.

Andresen, M.C., Yang, M., 1990. Non-NMDA receptors mediate sensory afferent synaptic transmission in medial nucleus tractus solitarius. Am. J. Physiol. 259, H1307-H1311.

Angell-James, J.E., Daly, M.d.B., 1970. Comparison of the reflex vasomotor responses to separate and combined stimulation of the carotid sinus and aortic arch baroreceptors by pulsatile and non-pulsatile pressure in the dog. J. Physiol. (Lond.) 209, 257-293.

Bailey, T.W., Jin, Y.-H., Doyle, M.W., Andresen, M.C., 2002. Vanilloid sensitive afferents activate neurons with prominent A-type potassium currents in nucleus tractus solitarius. J. Neurosci. 22, 8230-8237.

Balkowiec, A., Kunze, D.L., Katz, D.M., 2000. Brain-derived neurotrophic factor acutely inhibits AMPA-mediated currents in developing sensory relay neurons. J. Neurosci. 20, 1904-1911.

Barnes, K.L., DeWeese, D.M., Andresen, M.C., 2003. Angiotensin potentiates excitatory synaptic transmission to medial solitary tract nucleus neurons. Am. J. Physiol. Regul. Integr. Comp. Physiol. 284, R1340-R1353.

Boscan, P., Paton, J.F., 2002. Nociceptive afferents selectively modulate the cardiac component of the peripheral chemoreceptor reflex via actions within the solitary tract nucleus. Neurosci. 110, 319-328.

Bradley, R.M., Sweazey, R.D., 1992. Separation of neuron types in the gustatory zone of the nucleus tractus solitarii on the basis of intrinsic firing properties. J. Neurophysiol. 67, 1659-1668.

Butcher, J.W., Kasparov, S., Paton, J.F., 1999. Differential effects of apamin on neuronal excitability in the nucleus tractus solitarii of rats studied in vitro. J. Auton. Nerv. Syst. 77, 90-97.

Butcher, J.W., Paton, J.F.R., 1998. K^+ channel blockade in the NTS alters efficacy of two cardiorespiratory reflexes in vivo. Am. J. Physiol. 274, R677-R685.

Campos, R.R., McAllen, R.M., 1999. Tonic drive to sympathetic premotor neurons of rostral ventrolateral medulla from caudal pressor area neurons. Am. J. Physiol. 276, R1209-R1213.

Caterina, M.J., Julius, D., 2001. The Vanilloid Receptor: A Molecular Gateway to the Pain Pathway. Annu. Rev. Neurosci. 24, 487-517.

Chan, R.K., Jarvina, E.V., Sawchenko, P.E., 2000a. Effects of selective sinoaortic denervations on phenylephrine-induced activational responses in the nucleus of the solitary tract. Neurosci. 101, 165-178.

Chan, R.K., Peto, C.A., Sawchenko, P.E., 2000b. Fine structure and plasticity of barosensitive neurons in the nucleus of solitary tract. J. Comp. Neurol. 422, 338-351.

Chapleau, M.W., Hajduczok, G., Abboud, F.M., 1989. Pulsatile activation of baroreceptors causes central facilitation of baroreflex. Am. J. Physiol. 256, H1735-H1741.

Chapleau, M.W., Hajduczok, G., Abboud, F.M., 1991. Resetting of the arterial baroreflex: Peripheral and central mechanisms. In: Reflex Control of the Circulation (Zucker IH, Gilmore JP, eds), pp 167-194. Boca Raton: CRC Press.

Chen, C.Y., Ling Eh, E.H., Horowitz, J.M., Bonham, A.C., 2002. Synaptic transmission in nucleus tractus solitarius is depressed by Group II and III but not Group I presynaptic metabotropic glutamate receptors in rats. J. Physiol. 538, 773-786.

Chizh, B.A., Headley, P.M., Paton, J.F.R., 1997. An arterially-perfused trunk-hindquarters preparation of adult mouse in vitro. J. Neurosci. Methods 76, 177-182.

Convertino, V.A., Doerr, D.F., Eckberg, D.L., Fritsch, J.M., Vernikos-Danellis, J., 1990. Head-down bed rest impairs vagal baroreflex responses and provokes orthostatic hypotension. J. Appl. Physiol. 68, 1458-1464.

Cooke, W.H., Ames, J.E., IV, Crossman, A.A., Cox, J.F., Kuusela, T.A., Tahvanainen, K.U.O., Moon, L.B., Drescher, J., Baisch, F.J., Mano, T., Levine, B.D., Blomqvist, C.G., Eckberg, D.L., 2000. Nine months in space: effects on human autonomic cardiovascular regulation. J. Appl. Physiol. 89, 1039-1045.

Dampney, R.A.L., 1994. Functional organization of central pathways regulating the cardiovascular system. Physiol. Rev. 74, 323-364.

Dean, C., Seagard, J.L., 1995. Expression of *c-fos* protein in the nucleus tractus solitarius in response to physiological activation of carotid baroreceptors. Neurosci. 69, 249-257.

Dean, C., Seagard, J.L., 1997. Mapping of carotid baroreceptor subtype projections to the nucleus tractus solitarius using c-*fos* immunohistochemistry. Brain Res. 758, 201-208.

Dekin, M.S., Getting, P.A., 1984. Firing pattern of neurons in the nucleus tractus solitarius: modulation by membrane hyperpolarization. Brain Res. 324, 180-184.

Ding, Y.Q., Li, J.L., Lü, B.Z., Wang, D., Zhang, M.L., Li, J.S., 1998. Co-localization of m-opioid receptor-like immunoreactivity with substance P-LI, calcitonin gene-related peptide-LI and nitric oxide synthase-LI in vagal and glossopharyngeal afferent neurons of the rat. Brain Res. 792, 149-153.

Doyle, M.W., Andresen, M.C., 2001. Reliability of monosynaptic transmission in brain stem neurons in vitro. J. Neurophysiol. 85, 2213-2223.

Doyle, M.W., Bailey, T.W., Jin, Y.-H., Andresen, M.C., 2002. Vanilloid receptors presynaptically modulate visceral afferent synaptic transmission in nucleus tractus solitarius. J. Neurosci. 22, 8222-8229.

Ead, H.W., Green, J.N., Neil, E., 1952. A comparison of the effects of pulsatile and non-pulsatile blood flow through the carotid sinus on the reflexogenic activity of the sinus baroreceptors in the cat. J. Physiol. (Lond.) 118, 509-519.

Eckberg, D.L., Fritsch, J.M., 1992. Influence of ten-day head-down bedrest on human carotid baroreceptor-cardiac reflex function. Acta Physiol. Scand. 144 Suppl. 604, 69-76.

Fan, W., Andresen, M.C., 1998. Differential frequency-dependent reflex integration of myelinated and nonmyelinated rat aortic baroreceptors. Am. J. Physiol. 275, H632-H640.

Fan, W., Schild, J.H., Andresen, M.C., 1999. Graded and dynamic reflex summation of myelinated and unmyelinated rat aortic baroreceptors. Am. J. Physiol. 277, R748-R756.

Felder, R.B., Mifflin, S.W., 1988. Modulation of carotid sinus afferent input to nucleus tractus solitarius by parabrachial nucleus stimulation. Circ. Res. 63, 35-49.

Felder, R.B., Mifflin, S.W., 1994. Baroreceptor and chemoreceptor afferent processing in the solitary tract nucleus. In: Nucleus of the Solitary Tract. (Barraco RA, ed), pp 169-186. Boca Raton: CRC.

Foreman, R.D., 2000. Integration of viscerosomatic sensory input at the spinal level. Prog. Brain Res. 122, 209-221.

Fritsch, J.M., Rea, R.F., Eckberg, D.L., 1989. Carotid baroreflex resetting during drug-induced arterial pressure changes in humans. Am. J. Physiol. 256, R549-R553.

3. CARDIOVASCULAR INTEGRATION IN THE NUCLEUS OF THE SOLITARY TRACT

Gao, X., Phillips, P.A., Widdop, R.E., Trinder, D., Jarrott, B., Johnston, C.I., 1992. Presence of functional vasopressin V_1 receptors in rat vagal afferent neurones. Neurosci. Lett. 145, 79-82.

Glazebrook, P.A., Ramirez, A.N., Schild, J.H., Shieh, C.C., Doan, T., Wible, B.A., Kunze, D.L., 2002. Potassium channels Kv1.1, Kv1.2 and Kv1.6 influence excitability of rat visceral sensory neurons. J. Physiol. 541, 467-482.

Gordon, F.J., McCann, L.A., 1988b. Pressor responses evoked by microinjections of L-glutamate into the caudal ventrolateral medulla of the rat. Brain Res. 457, 251-258.

Gordon, F.J., McCann, L.A., 1988a. Pressor responses evoked by microinjections of L-glutamate into the caudal ventrolateral medulla of the rat. Brain Res. 457, 251-258.

Guyenet, P.G., Koshiya, N., Huangfu, D., Baraban, S.C., Stornetta, R.L., Li, Y.W., 1996. Role of medulla oblongata in generation of sympathetic and vagal outflows. Prog. Brain Res. 107, 127-144.

Hay, M., Kunze, D.L., 1994. Glutamate metabotropic receptor inhibition of voltage-gated calcium currents in visceral sensory neurons. J. Neurophysiol. 72, 421-430.

Hay, M., Lindsley, K.A., 1995. Metabotropic glutamate receptor inhibition of visceral afferent potassium currents. Brain Res. 698, 169-174.

Hayward, L., Hay, M., Felder, R.B., 1993. Acute resetting of the carotid sinus baroreflex by aortic depressor nerve stimulation. Am. J. Physiol. 264, H1215-H1222.

Hayward, L.F., Felder, R.B., 1995. Cardiac rhythmicity among NTS neurons and its relationship to sympathetic outflow in rabbits. Am. J. Physiol. 269, H923-H933.

Heesch, C.M., Carey, L.A., 1987. Acute resetting of arterial baroreflexes in hypertensive rats. Am. J. Physiol. 253, H974-H979.

Hoang, C.J., Hay, M., 2001. Expression of metabotropic glutamate receptors in nodose ganglia and the nucleus of the solitary tract. Am. J. Physiol. Heart Circ. Physiol. 281, H457-H462.

Irnaten, M., Neff, R.A., Wang, J., Loewy, A.D., Mettenleiter, T.C., Mendelowitz, D., 2001. Activity of cardiorespiratory networks revealed by transsynaptic virus expressing GFP. J. Neurophysiol. 85, 435-438.

Jordan, D., Mifflin, S.W., Spyer, K.M., 1988. Hypothalamic inhibition of neurons in the nucleus tractus solitarius of the cat is GABA mediated. J. Physiol. (Lond.) 399, 389-404.

Kangrga, I.M., Loewy, A.D., 1995. Whole-cell recordings from visualized C1 adrenergic bulbospinal neurons: Ionic mechanisms underlying vasomotor tone. Brain Res. 670, 215-232.

Kunze, D.L., 1986. Acute resetting of baroreceptor reflex in rabbits: a central component. Am. J. Physiol. 250, H866-H870.

Kunze, D.L., Andresen, M.C., 1991. Arterial baroreceptors: Excitation and modulation. In: Reflex Control of the Circulation (Zucker IH, Gilmore JP, eds), pp 141-166. Boca Raton: CRC Press.

Lawrence, A.J., Jarrott, B., 1996. Neurochemical modulation of cardiovascular control in the nucleus tractus solitarius. Prog. Neurobiol. 48, 21-53.

Lawson, S.N., 1992. Morphological and biochemical cell types of sensory neurons. In: Sensory neurons: diversity, development, and plasticity (Scott SA, ed), pp 27-59. New York: Oxford University Press.

Lee, J.S., Andresen, M.C., Morrow, D., Chang, K.S.K., 2002. Isoflurane depresses baroreflex control of heart rate in decerebrate rats. Anesthesiology 96, 1214-1222.

Leone, C., Gordon, F.J., 1989. Is L-glutamate a neurotransmitter of baroreceptor information in the nucleus of tractus solitarius? J. Pharmacol. Exp. Ther. 250, 953-962.

Li, B.Y., Schild, J.H., 2002. Patch clamp electrophysiology in the nodose ganglia of the adult rat. J. Neurosci. Meth. 115, 157-167.

Lipski, J., Kanjhan, R., Kruszewska, B., Rong, W.F., 1996. Properties of presympathetic neurones in the rostral ventrolateral medulla in the rat: An intracellular study '*in vivo*'. J. Physiol. 490, 729-744.

Liu, Z., Chen, C.Y., Bonham, A.C., 1998. Metabotropic glutamate receptors depress vagal and aortic baroreceptor signal transmission in the NTS. Am. J. Physiol. 275, H1682-H1694.

Liu, Z., Chen, C.Y., Bonham, A.C., 2000. Frequency limits on aortic baroreceptor input to nucleus tractus solitarii. Am. J. Physiol. Heart Circ. Physiol. 278, H577-H585.

Loewy, A.D., 1990. Central autonomic pathways. In: Central regulation of autonomic functions (Loewy AD, Spyer KM, eds), pp 88-103. New York: Oxford.

Mendelowitz, D., 1999. Advances in Parasympathetic Control of Heart Rate and Cardiac Function. News Physiol. Sci. 14, 155-161.

Mendelowitz, D., Yang, M., Andresen, M.C., Kunze, D.L., 1992. Localization and retention in vitro of fluorescently labeled aortic baroreceptor terminals on neurons from the nucleus tractus solitarius. Brain Res. 581, 339-343.

Mendelowitz, D., Yang, M., Reynolds, P.J., Andresen, M.C., 1995. Heterogeneous functional expression of calcium channels at sensory and synaptic regions in nodose neurons. J. Neurophysiol. 73, 872-875.

Mifflin, S.W., 1996. Convergent carotid sinus nerve and superior laryngeal nerve afferent inputs to neurons in the NTS. Am. J. Physiol. 271, R870-R880.

Mifflin, S.W., Felder, R.B., 1988. An intracellular study of time-dependent cardiovascular afferent interactions in nucleus tractus solitarius. J. Neurophysiol. 59, 1798-1813.

Mifflin, S.W., Spyer, K.M., Withington-Wray, D.J., 1988. Baroreceptor inputs to the nucleus tractus solitarius in the cat: modulation by the hypothalamus. J. Physiol. (Lond.) 399, 369-387.

Miles, R., 1986. Frequency dependence of synaptic transmission in nucleus of the solitary tract in vitro. J. Neurophysiol. 55, 1076-1090.

Miura, M., 1975. Postsynaptic potentials recorded from nucleus of the solitary tract and its subjacent reticular formation elicited by stimulation of the carotid sinus nerve. Brain Res. 100, 437-440.

Moffitt, J.A., Foley, C.M., Schadt, J.C., Laughlin, M.H., Hasser, E.M., 1998. Attenuated baroreflex control of sympathetic nerve activity after cardiovascular deconditioning in rats. Am. J. Physiol. 274, R1397-R1405.

Morrison, S.F., 2001. Differential control of sympathetic outflow. Am. J. Physiol. Regul. Integr. Comp. Physiol. 281, R683-R698.

Natarajan, M., Morrison, S.F., 2000. Sympathoexcitatory CVLM neurons mediate responses to caudal pressor area stimulation. Am. J. Physiol. Regul. Integr. Comp. Physiol. 279, R364-R374.

Pamidimukkala, J., Hay, M., 2001. Frequency dependence of endocytosis in aortic baroreceptor neurons and role of group III mGluRs. Am. J. Physiol. Heart Circ. Physiol. 281, H387-H395.

Paton, J.F.R., 1996. A working heart-brainstem preparation of the mouse. J. Neurosci. Methods 65, 63-68.

Paton, J.F.R., Ramirez, J.-M., Richter, D.W., 1994. Functionally intact in vitro preparation generating respiratory activity in neonatal and mature mammals. Pflügers Arch. 428, 250-260.

Pilowsky, P.M., Goodchild, A.K., 2002. Baroreceptor reflex pathways and neurotransmitters: 10 years on. J. Hypertension 20, 1675-1688.

Scheuer, D.A., Zhang, J., Toney, G.M., Mifflin, S.W., 1996. Temporal processing of aortic nerve evoked activity in the nucleus of the solitary tract. J. Neurophysiol. 76, 3750-3757.

Schild, J.H., Clark, J.W., Canavier, C.C., Kunze, D.L., Andresen, M.C., 1995. Afferent synaptic drive of rat medial nucleus tractus solitarius neurons: Dynamic simulation of graded vesicular mobilization, release, and non-NMDA receptor kinetics. J. Neurophysiol. 74, 1529-1548.

Schild, J.H., Clark, J.W., Hay, M., Mendelowitz, D., Andresen, M.C., Kunze, D.L., 1994. A- and C-type nodose sensory neurons: Model interpretations of dynamic discharge characteristics. J. Neurophysiol. 71, 2338-2358.

Schild, J.H., Khushalani, S., Clark, J.W., Andresen, M.C., Kunze, D.L., Yang, M., 1993. An ionic current model for neurons in the rat medial nucleus tractus solitarius receiving sensory afferent input. J. Physiol. (Lond.) 469, 341-363.

Schild, J.H., Kunze, D.L., 1997. Experimental and modeling study of Na^+ current heterogeneity in rat nodose neurons and its impact on neuronal discharge. J. Neurophysiol. 78, 3198-3209.

Seagard, J.L., Dean, C., Hopp, F.A., 2001. Properties of NTS neurons receiving input from barosensitive receptors. Ann. N.Y. Acad. Sci. 940, 142-156.

Seagard, J.L., Van Brederode, J.F.M., Dean, C., Hopp, F.A., Gallenberg, L.A., Kampine, J.P., 1990. Firing characteristics of single-fiber carotid sinus baroreceptors. Circ. Res. 66, 1499-1509.

Spyer, K.M., 1990. The central nervous organization of reflex circulatory control. In: Central Regulation of Autonomic Functions (Loewy AD, Spyer KM, eds), pp 168-188. New York: Oxford University Press.

Standish, A., Enquist, L.W., Escardo, J.A., Schwaber, J.S., 1995. Central neuronal circuit innervating the rat heart defined by transneuronal transport of pseudorabies virus. J. Neurosci. 15, 1998-2012.

Standish, A., Enquist, L.W., Schwaber, J.S., 1994. Innervation of the heart and its central medullary origin defined by viral tracing. Sci. 263, 232-234.

Stebbing, M.J., McLachlan, E.M., Sah, P., 1998. Are there functional P2X receptors on cell bodies in intact dorsal root ganglia of rats? Neurosci. 86, 1235-1244.

Sun, M.-K., Guyenet, P.G., 1990. Excitation of rostral medullary pacemaker neurons with putative sympathoexcitatory function by cyclic AMP and b-adrenoceptor agonists 'in vitro'. Brain Res. 511, 30-40.

Sun, M.-K., Stornetta, R.L., Guyenet, P.G., 1991. Morphology of rostral medullary neurons with intrinsic pacemaker activity in the rat. Brain Res. 556, 61-70.

Sved, A.F., Gordon, F.J., 1994. Amino acids as central neurotransmitters in the baroreceptor reflex pathway. News Physiol. Sci. 9, 243-246.

Szallasi, A., Blumberg, P.M., 1999. Vanilloid (Capsaicin) receptors and mechanisms. Pharmacol. Rev. 51, 159-212.

Talman, W.T., 1989. Kynurenic acid microinjected into the nucleus tractus solitarius of rat blocks the arterial baroreflex but not responses to glutamate. Neurosci. Lett. 102, 247-252.

Talman, W.T., Granata, A.R., Reis, D.J., 1984. Glutamatergic mechanisms in the nucleus tractus solitarius in blood pressure control. Fed. Proc. 43, 39-44.

Tan, W., Zucker, I.H., 1989. Acute carotid baroreflex resetting in conscious dogs. J. Physiol. (Lond.) 416, 557-569.

Toney, G.M., Mifflin, S.W., 1995. Time-dependent inhibition of hindlimb somatic afferent transmission within nucleus tractus solitarius: An *in vivo* intracellular recording study. Neurosci. 68, 445-453.

Van Giersbergen, P.L.M., Palkovits, M., De Jong, W., 1992. Involvement of neurotransmitters in the nucleus tractus solitarii in cardiovascular regulation. Physiol. Rev. 72, 789-824.

Wang, J., Irnaten, M., Neff, R.A., Venkatesan, P., Evans, C., Loewy, A.D., Mettenleiter, T.C., Mendelowitz, D., 2001. Synaptic and neurotransmitter activation of cardiac vagal neurons in the nucleus ambiguus. Ann. N.Y. Acad. Sci. 940, 237-246.

Zhang, J., Mifflin, S.W., 2000. Responses of aortic depressor nerve-evoked neurones in rat nucleus of the solitary tract to changes in blood pressure. J. Physiol. 529, 431-443.

Chapter 4

NEUROTRANSMITTERS IN THE NUCLEUS TRACTUS SOLITARIUS MEDIATING CARDIOVASCULAR FUNCTION

Hreday N. Sapru
UMDNJ - New Jersey Medical School, Newark, NJ 07103, USA

Abstract: The importance of the nucleus tractus solitarius (nTS) in the regulation of cardiovascular function has been known for a long time. However, the role of different neurotransmitters in this region in mediating cardiovascular functions is just beginning to be delineated. The carotid chemoreceptor afferents terminate predominantly in a midline area around the calamus scriptorius in the commissural subnucleus of the nTS while the baroreceptor and cardiopulmonary receptor afferents terminate in a region more rostral and lateral to the chemoreceptor projection site. There is a general consensus that glutamate is the neurotransmitter released at the terminals of baroreceptor, cardiopulmonary and chemoreceptor afferents in the nTS. However, cholinergic, GABAergic, and opioidergic mechanisms are also present in the nTS. Activation of glutamatergic and cholinergic mechanisms in the nTS elicits depressor responses while the activation of GABAergic and opioidergic mechanisms elicits pressor responses. Although, the precise physiological role of cholinergic, GABAergic, and opioidergic nTS mechanisms in regulating cardiovascular function remains to be elucidated, there is a general consensus that these mechanisms may play a neuromodulatory role in the nTS.

Key words: Blood pressure, bradycardia, caudal ventrolateral medullary depressor area, chemoreceptor projection site, depressor responses, heart rate, pressor responses, rostral ventrolateral medullary pressor area.

INTRODUCTION

The nucleus tractus solitarius (nTS) in the rat consists of a column of cells on each side of the fourth ventricle (Sapru, 1994). At the rostral edge of the area postrema, the columns of nTS-cells on the two sides merge in the midline to form the commissural subnucleus. The commissural subnucleus of nTS encompasses a region approximately 0.5 mm rostral to 1 mm caudal with reference to the calamus scriptorius (CS), 0-0.8 mm lateral to the midline on each side and 0-0.5 mm deep with reference to the dorsal medullary surface. The solitary tract lies between the medial and lateral subdivisions of nTS.

The essential role of the nTS in regulating cardiovascular function has been known for more than a century. However, interconnections between the nTS and other medullary neuronal groups involved in cardiovascular regulation have been identified with certainty only during the past two decades. Several neuroactive substances have been implicated as neurotransmitters or neuromodulators in the nTS (Van Giersbergen, 1992); it is beyond the scope of this chapter to review the role of all of them. Therefore, only a few neurotransmitter mechanisms (glutamatergic, cholinergic, GABAergic and opioidergic mechanisms) are discussed here. Although the main focus of the chapter is the nTS, other medullo-spinal areas are also mentioned where applicable for the sake of a coherent discussion. The literature cited includes mostly references to the studies on the anesthetized rat, which is the animal model commonly used in these investigations. Wherever applicable, references related to other experimental animals have also been included.

DEPRESSOR AND PRESSOR nTS SITES: IDENTIFICATION

L-glutamate (L-Glu) is known to stimulate neuronal cell bodies but not the fibers of passage. Therefore, microinjections of L-Glu have been used to identify neuronal pools involved in cardiovascular function. In anesthetized or mid-collicular decerebrate rats, microinjections of small volumes of L-Glu into a region of nTS 0.6 mm caudal to 1.4 mm rostral with reference to the calamus scriptorius, 0.4 – 1.2 mm lateral to the midline, and 0.4 – 1 mm deep from the dorsal medullary surface have been reported to elicit depressor and bradycardic responses. The caudal-most point of the floor of the fourth ventricle resembles a pen and is known as calamus scriptorius (Latin for nib of a pen). Maximum depressor and bradycardic responses to L-Glu were elicited from a region 0.6 mm rostral to the CS, 0.6 mm lateral to the midline and 0.6 mm deep from the dorsal medullary surface (Dhruva et al., 1998; Marchenko and Sapru, 2000). Pressor responses to microinjections of L-Glu were elicited only from the chemoreceptor projection site (CPS) located in a midline area caudal to the CS (Dhruva et al., 1998; Marchenko and Sapru, 2000; Vardhan et al., 1993a).

CPS is located 0.3-0.6 mm caudal to the calamus scriptorius, 0-0.5 mm lateral to the midline and 0.3-0.5 mm from the dorsal surface of the medulla. Although pressor responses were often accompanied with tachycardia, bradycardic responses are also observed in a region located 0.5 mm caudal to the CS and 0.5 mm lateral to the midline and 0.5 mm deep from the dorsal medullary surface.

GLUTAMATERGIC nTS MECHANISMS

Talman et al. (1980) were the first to report that microinjections of L-Glu into the nTS, in a region where baroreceptors terminate, elicit depressor and bradycardic responses. This observation has been repeatedly confirmed in anesthetized as well as decerebrate rats (Dhruva et al., 1998). L-Glu mediates fast excitatory neurotransmission in the nTS via ionotropic glutamate receptors (NMDA, AMPA, and Kainate receptors) (Bonham and Chen, 2002). Activation of different groups of mGLURs in the nTS also elicits bradycardia and hypotension (Foley et al., 1999; Viard and Sapru, 2002). The role played by glutamatergic and GABAergic mechanisms in the nTS in the mediation of cardiovascular reflexes is discussed later in the chapter.

GABAERGIC nTS MECHANISMS

GABA-containing neurons and nerve terminals are known to be present in the nTS (Blessing et al., 1984; Sved, 1994). GABA containing nTS neurons have been shown to project to the rostral ventrolateral medullary pressor (RVLM) and depressor (CVLM) areas (Suzuki et al., 1997). Radioactive ligand binding studies have revealed the presence of both GABA receptors in the nTS (Bowery et al., 1987). Bilateral microinjections of GABA receptor agonists or nipecotic acid (a GABA uptake blocker) into the nTS elicit variable heart rate and pressor responses (Kubo and Kihara, 1987; Sved, 1994). The pressor responses to microinjections of GABA receptor agonists have been reported to be the result of an inhibition of baroreceptor reflex (Florentino et al., 1990; Sved and Tsukamoto, 1992). Microinjections of GABA receptor antagonists into the nTS elicit depressor responses (Sved, 1994). Direct application of GABA receptor agonists and antagonists to barosensitive nTS neurons decreases and increases their firing, respectively (Bennett et al., 1987; Sved, 1994). Carotid occlusion has been reported to elicit a release of endogenous GABA in the nTS (Klausmair and

Philippu, 1989). GABA released in this manner is expected to inhibit nTS neurons and cause a pressor response characteristic of carotid occlusion.

Inhibition of nTS neurons by the stimulation of hypothalamus has been ascribed to the release of GABA in the nTS (Jordan et al., 1988). Collectively these reports suggest that endogenous GABA in the nTS may play a role in cardiovascular regulation. However, for the most part, the physiological relevance of GABAergic mechanisms in the nTS in the regulation of cardiovascular function remains to be delineated. For example, function of GABAergic projections from the nTS to the RVLM and CVLM (Suzuki et al., 1997) is not known. Similarly, the function of GABAergic projections to the nTS from other central nervous system (CNS) structures has not been delineated (Berlin et al., 1983; Sved, 1994).

CHOLINERGIC nTS MECHANISMS

In the rat brain, the presence of acetylcholine esterase, choline acetyltransferase, acetylcholine and muscarinic receptors has been demonstrated in the nTS (Helke et al., 1980; Watson et al., 1986; Van Giersbergen et al., 1992). Microinjections of acetylcholine, carbachol or selective M_2 muscarinic receptor agonists into the nTS have been reported to elicit depressor and bradycardic responses (Criscione et al., 1983; Sundaram et al., 1989; Sapru, 1994); these responses were blocked by AFDX-116 (M_2 receptor antagonist) but not pirenzepine (M_1 receptor antagonist). On the other hand, microinjections of McN-A343 (a M_1 muscarinic receptor agonist) or pirenzepine into the nTS failed to elicit a response (Sundaram et al., 1989; Sapru, 1994). Activation of nicotinic receptors in the nTS has also been reported to elicit depressor and bradycardic responses (Dhar et al., 2000); prior sequential microinjections of alpha-bungarotoxin (a specific toxin for nicotinic receptors containing alpha-7 subunits) and mecamylamine (a nicotinic receptor channel blocker as well as a competitive antagonist) into the nTS completely blocked the nicotine-induced responses. These results indicated that activation of muscarinic (predominantly of M_2 type) and nicotinic (predominantly containing $\alpha_3\beta_4$ subunits) receptors in the nTS elicits depressor and bradycardic responses. Microinjections of relatively large doses of atropine (a non-selective cholinergic receptor blocker) have been reported to increase resting blood pressure slightly and attenuate baroreflex (Criscione et al., 1983; Tsukamoto et al., 1994). Microinjections of AFDX-116, but not a nicotinic receptor blocker, into the nTS also elicited a slight rise in resting blood pressure (Sapru, 1994). Based on these observations, it is generally believed that the role of cholinergic receptors in mediating baroreflex is limited. However, these and other physiological

OPIOIDERGIC nTS MECHANISMS

The presence of opioid-peptide containing nerve terminals, neurons and opioid receptors in the nTS is well documented (Velley et al., 1991; Xia and Haddad, 1991; Bronstein et al., 1992). Opioid receptors are located both pre- and post-synaptically. Until recently, opioid receptors were divided into three subtypes, namely, mu, delta and kappa receptors (Uhl et al., 1994). The endogenous ligands for delta and kappa receptors are enkephalins and dynorphins, respectively. Recently, two opioid peptides (endomorphin-1 and endomorphin-2) have been identified as endogenous ligands for mu opioid receptors (Zadina et al., 1997; Sapru and Chitravanshi, 2002). A novel opioid receptor (ORLI or orphanin-Q receptor) and its endogenous ligand, nociceptin, have also been identified recently (Meunier et al., 1995; Neal et al., 1999; Sapru and Chitravanshi, 2002). The effects of endomorphins, but not nociceptin, are blocked by the traditional opioid receptor blocker, naloxone. Microinjections of mu and delta, but not kappa, receptor agonists into the nTS increase blood pressure in rats (Hassen et al., 1983; Gordon, 1990; Gordon, 1994). Unilateral microinjections of nociceptin into the nTS also elicit pressor and tachycardic responses in the rat (Sapru and Chitravanshi, 2002).

The physiological situations under which opioid peptides in the nTS come into play in the regulation of cardiovascular function remain to be identified. Since microinjection of naloxone alone had no effect on aortic baroreflex (Gordon, 1990), it appears that these peptides may have a limited role to play in the regulation of cardiovascular function under normal physiological conditions. However, a modulatory role of opioid peptides in the nTS in regulating cardiovascular function, especially under stress or in pathological conditions, cannot be excluded.

BAROREFLEX: AN OVERVIEW

Based on our current knowledge (Gordon and Sved, 2002; Pilowsky and Goodchild, 2002; Sapru 2002; Schreihofer and Guyenet, 2002), the mechanism of baroreflex responses can be briefly summarized as follows (Fig. 1). Activation of baroreceptor afferents results in activation of secondary nTS-neurons via excitatory amino acid (EAA) receptors.

Anatomical and functional studies have demonstrated the presence of a projection from the nTS to the CVLM (Urbanski and Sapru, 1988a,b; Aicher et al., 1995; Suzuki et al., 1997). Activation of the projection from the nTS to the CVLM mediates baroreflex-induced excitation of CVLM neurons via EAA receptors (Fig. 1, pathway 1). The identification CVLM in the rat and its importance in baroreflex is well documented (Willette et al., 1983a, 1984b). The subtypes of EAA receptors in the CVLM mediating baroreflex responses have not been firmly established. Either NMDA receptors alone (Gordon, 1987) or both NMDA and non-NMDA receptors have been implicated in the baroreflex-mediated activation of CVLM-neurons (Miyawaki et al., 1997). Activation of GABAergic neurons in the CVLM has been shown to inhibit sympathoexcitatory neurons located in the RVLM via GABA receptors (Willette et al., 1983b, 1984a; Sun and Guyenet, 1985; Agarwal et al., 1989) (Fig. 1, pathway 2). The identification of RVLM in the rat (Willette et al., 1983a; Ross et al., 1984) and its importance in baroreflex is well established. Brown and Guyenet (1985) were the first to carry out electrophysiological studies on the cardiovascular neurons in the RVLM and demonstrate monosynaptic projections to the intermediolateral cell column of the spinal cord (IML). Chemical stimulation of the RVLM activates the sympathetic preganglionic IML-neurons (SPGNs) via NMDA receptors (Bazil and Gordon, 1991; Sundaram and Sapru, 1991) and inhibition of sympathoexcitatory neurons in the RVLM results in withdrawal of normal excitatory input to the SPGNs in the IML (Fig. 1, pathway 3) and either pressor and tachycardic or depressor and bradycardic responses are elicited via increase or decrease in the activity of sympathetic nerves (Fig. 1, pathway 4). Baroreflex-induced bradycardia is primarily mediated via the activation of parasympathetic nervous system (Stornetta et al., 1987). The predominant pathway mediating bradycardia may involve projections from the nTS to the nucleus ambiguus (Stornetta et al., 1987; Neff et al., 1998) (Fig. 1, pathway 5). Chemical stimulation of preganglionic parasympathetic neurons in the nucleus ambiguus has been shown to elicit a vagally mediated bradycardia (Fig. 1, pathway 6) (Marchenko and Sapru, 2000).

4. NEUROTRANSMITTERS IN THE NUCLEUS TRACTUS SOLITARIUS MEDIATING CARDIOVASCULAR FUNCTION

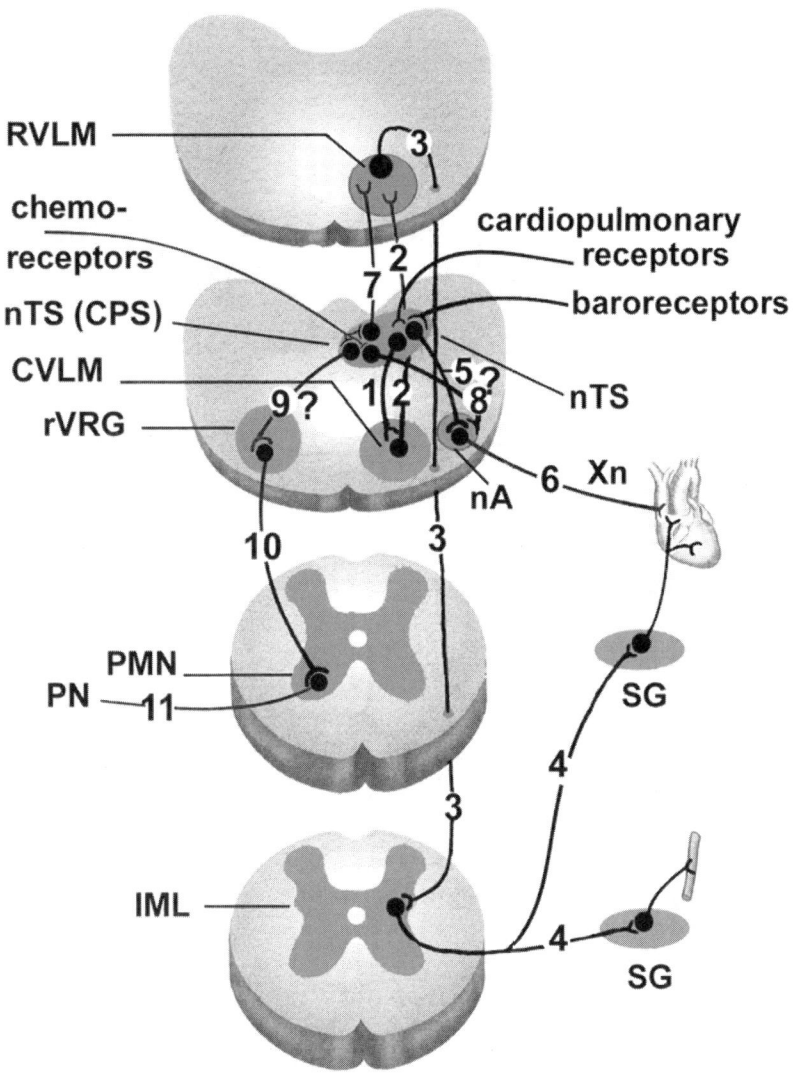

Figure 1. A schematic representation of selected medullo-apinal cardiovascular areas. Abbreviations: CPS, chemoreceptor projection site; CVLM, caudal ventrolateral medullary depressor area; IML, intermediolateral cell column of the thoraco-lumbar cord; nA, nucleus ambiguus; nTS, nucleus tractus solitarius; PMV, phrenic motor nucleus in the cervical spinal cord; PN, phrenic nerve; RVLM, rostral ventrolateral medullary pressor area; rVRG, rostral ventral respiratory group of neurons in the medulla; SG, sympathetic ganglion; Xn, vagus nerve. See text for a description of labeled pathways.

The contribution of nTS in mediating reflex responses to the stimulation of baroreceptors located at different vascular sites is described in the following sections.

CARDIOVASCULAR REFLEXES: ROLE OF nTS

Systemic Baroreflex Activation: Bradycardic Responses

Talman et al. (1980) were the first to report that L-glutamate may be the neurotransmitter released at the baroreceptor terminals in the nTS. One of the methods by which baroreceptors were stimulated involved elevation of systemic blood pressure by intravenous injections of vasoconstrictors, e.g., phenylephrine; this procedure is sometimes called "baroreceptor loading". Decreasing the activity of vascular baroreceptors using intravenous injections of vasodilators (e.g., sodium nitroprusside) is sometimes called "baroreceptor unloading". In-vivo microdialysis revealed that glutamate was released in the nTS when baroreflex was activated by intravenous infusion of phenylephrine (Ohta et al., 1996). Guyenet et al. (1987) demonstrated that reflex bradycardia elicited by intravenous injection of phenylephrine was blocked by bilateral microinjections of kynurenic acid (a broad-spectrum glutamate receptor blocker) into the nTS. Microinjections of a competitive NMDA receptor blocker (AP-5; 2-amino-5-phosphonopentanoic acid) or non-NMDA receptor blocker (CNQX; 6,7-dinitroquinoxaline-2, 3,-dione), or a combination of both antagonists into the nTS bilaterally attenuated the reflex bradycardia or tachycardia elicited by intravenous injections of phenylephrine or sodium nitroprusside, respectively. These studies suggested that both NMDA and non-NMDA receptors in the nTS may be involved in mediating the reflex bradycardic responses to baroreceptor stimulation (Ohta and Talman, 1994).

Aortic Baroreflex

Baroreceptor loading and unloading by intravenous injections of vasoconstrictors and vasodilators, respectively, is not specific for increasing or decreasing the activity of vascular baroreceptors because these procedures also involve atrial and ventricular mechanoreceptors. In order to avoid this complication, electrical stimulation of aortic nerve has been used to stimulate baroreceptors in several studies. This approach is reasonable

4. NEUROTRANSMITTERS IN THE NUCLEUS TRACTUS SOLITARIUS MEDIATING CARDIOVASCULAR FUNCTION

considering the fact that in the rat, the aortic nerve consists predominantly of baroreceptor afferents (Sapru and Krieger, 1977; Sapru et al., 1981b; Numao et al., 1985). Although, there are two reports suggesting the presence of a few chemoreceptor afferents in the aortic nerve of the rat (Cheng et al., 1997; Brophy et al., 1999), subsequent methodical studies have revealed that responses characteristic of chemoreceptor stimulation were completely lost by bilateral carotid sinus nerve denervation despite the fact that the aortic nerves remained intact in these rats (Kobayashi et al., 1999). Thus, there is a general consensus that aortic chemoreceptors in the rat may be of little functional significance.

Using electrical stimulation of the aortic nerve in the rat, Gordon and Leone (1991) reported that blockade of non-NMDA receptors in the nTS abolished cardiovascular responses to baroreceptor stimulation while the blockade of NMDA receptors affected these responses only when the aortic nerve was stimulated at high frequencies. Subsequent studies using electrical stimulation of the aortic nerve in the rat revealed that the synaptic inputs from the aortic nerve to monosynaptic neurons in the nTS, with faster rising time, were mediated by non-NMDA receptors while the synaptic integration of similar inputs to polysynaptic neurons in the nTS were mediated by both NMDA and non-NMDA receptors (Zhang and Mifflin, 1998). Andresen and Yang (1990) recorded nTS neurons intracellularly in rat brain slices and reported that stimulation of the solitary tract elicited excitatory postsynaptic potentials (EPSPs) with short latency in these neurons. Application of a specific non-NMDA receptor antagonist markedly suppressed these EPSPs. In another study, whole-cell patch-clamp recordings of nTS neurons were made in the rat coronal medullary slices. Stimulation of solitary tract in these preparations elicited EPSCs in the nTS-neurons. The fast component of these EPSCs was blocked by a non-NMDA receptor antagonist while the slow component was blocked by an NMDA receptor antagonist (Aylwin et al., 1997). These studies indicated that the sensory afferent synaptic transmission to nTS neurons is mediated by non-NMDA receptors while the NMDA receptors may play a modulatory role. Relatively few studies have been carried out to investigate the role of metabotropic glutamate receptors (mGLURs) in the nTS in mediating aortic baroreflex. In the rabbit, activity was recorded from nTS neurons extracellularly, and an mGLUR agonist was applied iontophoretically (Liu et al., 1998). Application of the mGLUR agonist attenuated the responses of the neuron that was activated monosynaptically by aortic nerve stimulation while the application of AMPA augmented these responses. Application of the mGLUR antagonist augmented the responses to the nTS neuron to high-frequency stimulation of the aortic nerve but had no effect on AMPA-evoked responses. It was concluded that mGLURs mediate frequency-dependent depression of aortic baroreceptor signal transmission in the nTS. This effect may be mediated by activation of presynaptic mGLURs in the nTS that results in inhibition of glutamate release and depression of glutamatergic transmission (Liu et al., 1998). Recently, whole-cell neuronal recordings were made from nTS neurons using coronal slices of the rat brainstem (Chen et al., 2002). EPSCs were evoked in these neurons by stimulating the solitary tract and changes in peak amplitude were studied after perfusion with selective mGLUR agonists. It was reported that Group II and III, but not Group I, agonists reduced the peak amplitude of synaptically evoked EPSCs but had no effect on the postsynaptic EPSCs evoked by perfusion of AMPA. These effects were blocked by Group II and III mGLUR antagonists. It was concluded that Group II and III mGLURs on the central terminals of afferent fibers mediate the depression of autonomic signal transmission by decreasing presynaptic release of glutamate (Chen et al., 2002).

Carotid Sinus Baroreflex

In view of the reports (Cheng et al., 1997; Brophy et al., 1999) indicating the presence of chemoreceptor afferents in the aortic nerve of the rat (albeit few), it is important to devise methods in which baroreceptors can be activated exclusively using a physiological stimulus. One such preparation is an in-situ carotid sinus with intact innervation, isolated from general circulation and perfused with physiological saline at different pressures. Using this preparation in the dog, both NMDA and non-NMDA receptor antagonists were reported to alter baroreceptor-induced changes in the nTS neuronal discharge. It was concluded that both types of ionotropic glutamate receptors mediated activation of nTS neurons in response to baroreceptor input but non-NMDA receptors played a greater role in mediating these responses (Seagard et al., 2001).

Identification of neurotransmitters mediating baroreflex in the nTS has not been studied in the rat using isolated carotid sinus technique perhaps due to technical difficulties.

Cardio-Pulmonary Reflex

Injections of chemicals like phenylbiguanide, capsaicin and opioid peptides intravenously or via the right atrium elicit a triad of responses characterized by apnea, bradycardia and hypotension. The receptors responding to these stimuli are located in the lungs and the heart and the responses are mediated by vagal C fiber afferents (Sapru et al., 1981a; Kwok and Dun, 1998). Others have referred to this reflex as the Bezold-Jarisch reflex (Verberne et al., 1992). A summary of the pathways involved in this reflex is as follows (Fig. 1, pathways 1-6). Cardiac and pulmonary vagal C-fiber afferents terminate in the nTS at sites similar those where the baroreceptor afferents make their primary synapse and EAAs in this region mediate the responses to cardiopulmonary receptor activation (Vardhan et al., 1993b). Activation of second order nTS neurons mediating cardiopulmonary reflex results in the activation of CVLM neurons. EAA receptors in the CVLM mediate the cardiovascular responses to cardiopulmonary receptor activation (Verberne et al., 1992). Furthermore, blockade of GABA receptors in the RVLM abolish the cardiovascular responses to cardiopulmonary receptor stimulation (Willette et al., 1983b). The pathways involved in the respiratory responses to the stimulation of cardiopulmonary receptors remain to be investigated.

Carotid Chemoreflex

Neural pathways involved in mediating carotid chemoreflex can be briefly summarized as follows (Fig. 1). Anatomical (Finley and Katz, 1990) and microinjection (Vardhan et al., 1993a; Chitravanshi et al., 1994) studies have shown that in the rat the afferents from the carotid body terminate predominantly in a midline area around the calamus scriptorius in the commissural subnucleus of the nTS. Neurons sensitive to chemoreceptor, but not to baroreceptor, input have been identified in this area (Chitravanshi and Sapru, 1995). Recent electrophysiological studies carried out in arterially perfused brainstem-spinal cord preparations of rats have confirmed that the carotid chemoreceptor afferents in the rat project to this midline area (Paton et al., 2002). This area has been designated as the CPS (Vardhan et al., 1993a; Chitravanshi et al., 1994; Chitravanshi and Sapru, 1995). Activation of chemoreceptor afferents results in the release of an EAA in the CPS, and both NMDA and non-NMDA receptors in this area have been implicated in mediating the pressor, bradycardic and respiratory responses to chemoreceptor stimulation (Vardhan et al., 1993a). Secondary nTS neurons are believed to mediate the sympathoexcitatory responses to the carotid chemoreflex stimulation via a direct projection from the nTS to the RVLM (Fig. 1, pathway 7) and EAA receptors in the RVLM mediate the sympathoexcitatory responses to chemoreceptor stimulation (Koshiya and Guyenet, 1996; Sun and Reis, 1995). The subtypes of EAA receptors in the RVLM mediating the carotid chemoreceptor reflex have not been established with certainty. Kubo et al. (1993) and Sun and Reis (1995) have implicated NMDA receptors while Miyawaki et al. (1996) have implicated non-NMDA receptors in these responses. Unlike in the baroreflex, the CVLM does not participate in mediating the cardiovascular responses to carotid chemoreceptor stimulation in the rat (Guyenet and Koshiya, 1995). Activation of RVLM neurons following the chemoreceptor stimulation leads to excitation of SPGNs in the IML (Fig. 1, pathway 3) and the activity of sympathetic nerves innervating the blood vessels increases (Fig. 1, pathway 4). The net result of activation of these pathways is an increase in blood pressure. The predominant cardiac effect of chemoreflex stimulation is bradycardia that may be mediated via a direct or indirect glutamatergic projection from the nTS to the nucleus ambiguus (Fig. 1, pathways 8 and 6) (Neff et al., 1998).

Neural circuits mediating the respiratory responses to chemoreceptor stimulation have not been firmly established. It is known that EAA receptors in the ventrolateral medulla containing the rostral ventral respiratory group of neurons (rVRG) are involved in these responses (Sun and Reis, 1996). It is, therefore, possible that activation of secondary CPS neurons results in the stimulation of rVRG neurons via EAA receptors (Fig. 1, pathway 9). The role of other medullary respiratory areas in mediating responses to peripheral chemoreceptor stimulation has not been established. In our studies, blockade of NMDA receptors in the phrenic motor nucleus (PMN) decreased the amplitude of phrenic nerve bursts indicating that, unlike in the newborn rat, NMDA receptors play a significant role in mediating the inspiratory drive to the phrenic motor neurons in the adult rat (Chitravanshi and Sapru, 1996). Blockade of both NMDA and non-NMDA receptors in the PMN abolished the phrenic nerve (PN) responses to chemoreceptor stimulation (Chitravanshi and Sapru, 1997). Thus, activation of rVRG stimulates the PMN (Fig. 1, pathway 10) and diaphragmatic movements are elicited via the activation of PNs (Fig. 1, pathway 11).

CONCLUSIONS

The nucleus tractus solitarius is the primary site where peripheral baroreceptors, chemoreceptors and cardiopulmonary receptors make their first synapse. In anesthetized rats, pressor responses are elicited by microinjections of L-glutamate into a midline area in the commissural subnucleus just caudal to the CS and depressor responses are elicited from sites located more rostral and lateral to this region. The presence of many putative neurotransmitters, ranging from small (e.g., acetylcholine, GABA, glutamate) to large molecules (e.g., opioid and other peptides) has been demonstrated in the nTS. Activation of glutamatergic and cholinergic nTS-mechanisms elicits depressor while activation of GABAergic and opioid mechanisms elicits pressor responses. There is a general consensus that the neurotransmitter in the nTS for mediating cardiovascular responses to the activation of baroreceptors, chemoreceptors, and cardiopulmonary receptors is glutamate. Glutamate is also the excitatory neurotransmitter in the CVLM and IML. However, the subtypes of glutamate receptors in the nTS involved in these reflexes have not been established firmly. GABA is the main inhibitory neurotransmitter in the projection from the CVLM to the RVLM. Despite the fact that characteristic cardiovascular responses are elicited by cholinergic, GABAergic and opioidergic mechanisms in the nTS, their exact role in mediating cardiovascular responses in physiological or pathological situations is not clear at present.

ACKNOWLEDGEMENTS

The work presented in this chapter was supported by a grant from NHLBI (HL 24347).

REFERENCES

Agarwal, S.K., Gelsema, A.J., Calaresu F.R., 1989. Neurons in rostral VLM are inhibited by chemical stimulation of caudal VLM in rats. Am. J. Physiol. 257, R265-R270.

Aicher, S.A., Kurucz, O.S., Reis, D.J., Milner, T.A., 1995. Nucleus tractus solitarius efferent terminals synapse on neurons in the caudal ventrolateral medulla that project to the rostral ventrolateral medulla. Brain Res. 693, 51-63.

Andresen, M.C., Yang M., 1990. Non-NMDA receptors mediate sensory afferent synaptic transmission in medial nucleus tractus solitarius. Am. J. Physiol. 259, H1307-H1311.

Aylwin, M.L., Horowitz, J.M., Bonham, A.C., 1997. NMDA receptors contribute to primary visceral afferent transmission in the nucleus of the solitary tract. J. Neurophysiol. 77, 2539-2548.

Bazil, M.K., Gordon F.J., 1991. Spinal NMDA receptors mediate pressor responses evoked from rostral ventrolateral medulla. Am. J. of Physiol. 260, H267-H275.

Bennet, J.A., McWilliam, P.N., Shepheard, S.L., 1987. A gamma-aminobutyric-acid-mediated inhibition of neurones in the nucleus tractus solitarius of the cat. J. Physiol. (Lond.) 392, 417-430.

Berlin, M.F., Nanopoulos, D., Didier, M., Aguera, M., Steinbusch, H., Verhofstad, A., Maitre, M., Pujol, J.F., 1983. Immunohistochemical evidence for the presence of gamma-aminobutyric acid and serotonin in one nerve cell. A study on the raphe nuclei of the rat using antibodies to glutamate decarboxylase and serotonin. Brain Res. 275, 329-339.

Blessing, W.W., Oertel, W.H., Willoughby, J.O., 1984. Glutamic acid decarboxylase immunoreactivity is present in perikaria of neurons in nucleus tractus solitarius of rat. Brain Res. 322, 346-350.

Bonham, A.C., Chen, C.Y., 2002. Glutamatergic neural transmission in the nucleus tractus solitarius: N-methyl-D-aspartate receptors. Clin. Exp. Pharmacol. Physiol. 29, 497-502.

Bowery, N.G., Hudson, A.L., Price, G.W., 1987. $GABA_A$ and $GABA_B$ receptor binding site distribution in the rat central nervous system. Neurosci. 20, 365-383.

Bronstein, D.M., Schafer, M.K.H., Watson, S.J., Akil, H., 1992. Evidence that beta-endorphin is synthesized in cells in the nucleus tractus solitarius: detection of POMC mRNA. Brain Res. 587, 269-275.

Brophy, S., Ford, T.W., Carey, M., Jones, J.F.X., 1999. Activity of aortic chemoreceptors in the anesthetized rat. J. Physiol. (Lond.) 514, 821-828.

Brown, D.L., Guyenet, P.G., 1985. Electrophysiological study of cardiovascular neurons in the rostral ventrolateral medulla in rats. Circ. Res. 56, 359-369.

Chen, C.Y., Ling, E.H., Horowitz, J.M., Bonham, A.C., 2002. Synaptic transmission in nucleus tractus solitarius is depressed by Group II and III but not Group I presynaptic metabotropic glutamate receptors in rats. J. Physiol. (Lond.) 538, 773-786.

Cheng, Z., Powley, T.L., Schwaber, J.S., Doyle, F.J., 1997. A laser confocal microscopic study of vagal afferent innervation of rat aortic arch: chemoreceptors as well as baroreceptors. J. Auton. Nerv. Syst. 67, 1-14.

Chitravanshi, V.C., Kachroo, A., Sapru, H.N., 1994. A midline area in the nucleus commissuralis of NTS mediates the phrenic nerve responses to carotid chemoreceptor stimulation. Brain Res. 662, 127-133.

Chitravanshi, V.C., Sapru, H.N., 1995. Chemoreceptor-sensitive neurons in commissural subnucleus of nucleus tractus solitarius of the rat. Am. J. Physiol. 268, R851-R858.

Chitravanshi, V.C., Sapru, H.N., 1996. NMDA as well as non-NMDA receptors mediate the neurotransmission of inspiratory drive to phrenic motoneurons in the adult rat. Brain Res. 715, 104-112.

Chitravanshi, V.C., Sapru, H.N., 1997. NMDA as well as non-NMDA receptors in phrenic nucleus mediate respiratory effects of carotid chemoreflex. Am. J. Physiol. 272, R302-R310.

Criscione, L., Reis, D.J., Talman, W.T., 1983. Cholinergic mechanisms in the nucleus tractus solitarii and cardiovascular regulation in the rat. Eur. J. Pharmacol. 88, 47-55.

Dhar, S., Nagy, F., McIntosh, J.M., Sapru, H.N., 2000. Receptor subtypes mediating depressor responses to microinjections of nicotine into the medial nTS of the rat. Am. J. Physiol. 279, R132-R140.

Dhruva, A., Bhatnagar, T., Sapru, H.N., 1998. Cardiovascular responses to microinjections of glutamate into the nucleus tractus solitarii of unanesthetized supracollicular decerebrate rats. Brain Res. 810, 88-100.

Finley, J.C.W., Katz, D.M., 1992. The central organization of carotid body afferent projections to the brainstem of the rat. Brain Res. 572, 108-116.

Florentino, A., Varga, K., Kunos, G., 1990. Mechanism of the cardiovascular effects of $GABA_B$ receptor activation in the nucleus tractus solitarii of the rat. Brain Res. 535, 264-270.

Foley, C.M., Vogl, H.W., Mueller, P.J., Hay, M., Hasser, E.M., 1999. Cardiovascular response to group I metabotropic glutamate receptor activation in NTS. Am. J. Physiol. 276, R1469-R1478.

Gordon, F.J., 1987. Aortic baroreceptor reflexes are mediated by NMDA receptors in caudal ventrolateral medulla. Am. J. Physiol. 252, R628-R633.

Gordon, F.J., 1990. Opioids and central baroreflex control. A site of action in the nucleus tractus solitarius. Peptides 11, 305-309.

Gordon, F.J., 1994. Opioids and the nucleus of the tractus solitarius: effects on cardiovascular and baroreflex function. In: Barraco, I.R.A. (Ed.), Nucleus of the Solitary Tract. CRC Press, Boca Raton, pp. 283-287.

Gordon, F.J., Leone, C., 1991. Non-NMDA receptors in the nucleus of the tractus solitarius play the predominant role in mediating aortic baroreceptor reflexes. Brain Res. 568, 319-322.

Gordon, F.J., Sved, A.F., 2002. Neurotransmitters in central cardiovascular regulation: Glutamate and GABA. Clin. Exp. Pharmacol. Physiol. 29, 522-524.

Guyenet, P.G., Filtz, T.M., Donaldson, S.R., 1987. Role of excitatory amino acids in rat vagal and sympathetic baroreflexes. Brain Res. 407, 272-284.

Guyenet, P.G., Koshiya, N., 1995. Working model of the sympathetic chemoreflex in rats. Clin. Exp. Hypertension 17, 167-179.

Hassen, A.H., Feuerstein, G., Faden, A.I., 1983. Differential cardiovascular effects mediated by mu and kappa opiate receptors in hindbrain nuclei. Peptides 4, 621-625.

Helke, C.J., Sohl, B.D., Jacobowitz, D.M., 1980. Choline acetyltransferase activity in discrete brain nuclei of DOCA-salt hypertensive rats. Brain Res. 193, 293-298.

Jordan, D., Mifflin, S.W., Spyer, K.M., 1988. Hypothalamic inhibition of neurones in the nucleus tractus solitarius of the cat is GABA mediated. J. Physiol. (Lond.) 399, 389-404.

Klausmair, A., Philippu, A., 1989. Carotid occlusion increases the release of endogenous GABA in the nucleus of the solitary tract. Naunyn-Schmiedeberg's Arch. Pharmacol. 340, 764-766.

Kobayashi, M., Cheng, Z.B., Tanaka, K., Nosaka, S., 1999. Is the aortic depressor nerve involved in arterial chemoreflexes in rats? J. Auton. Nerv. Syst. 78, 38-48.

Koshiya, N., Guyenet, P.G., 1996. NTS neurons with carotid chemoreceptor inputs arborize in the rostral ventrolateral medulla. Am. J. Physiol. 270, R1273-R1278.

Kubo, T., Kihara, M., 1987. Evidence for the presence of GABAergic and glycine-like systems responsible for cardiovascular control in the nucleus tractus solitarii of the rat. Neurosci. Lett. 74, 331-336.

Kubo, T., Amano, M., Asari, T., 1993. N-methyl-D-aspartate receptors but not non-N-methyl-D-aspartate receptors mediate hypertension induced by carotid body chemoreceptor stimulation in the rostral ventrolateral medulla of the rat. Neurosci. Lett. 164, 113-116.

Kwok, E.H., Dun, N.J., 1998. Endomorphins decrease heart rate and blood pressure possibly by activating vagal afferents in anesthetized rats. Brain Res. 803, 204-207.

Liu, Z., Chen, C.Y., Bonham, A.C., 1998. Metabotropic glutamate receptors depress vagal and aortic baroreceptor signal transmission in the NTS. Am. J. Physiol. 275, H1682-H1694.

Marchenko, V., Sapru, H.N., 2000. Different patterns of respiratory and cardiovascular responses elicited by chemical stimulation of dorsal medulla in the rat. Brain Res. 857, 99-109.

Meunier, J.C., Mollereau, C., Toll, L., Suaudeau, C., Moisand, C., Alvinerie, P., Butour, J.L., Guillemot, J.C., Ferrara, P., Monsarrat, B., Mazargull, H., Vassaart, G., Parmentier, M., Costentin, J., 1995. Isolation and structure of the endogenous agonist of opioid receptor-like ORL1 receptor. Nature 377, 532-535.

Miyawaki, T., Minson, J., Arnolda, L., Llewellyn-Smith, I., Chalmers, J., Pilowsky, P., 1996. AMPA/kainate receptors mediate sympathetic chemoreceptor reflex in the rostral ventrolateral medulla. Brain Res. 726, 64-68.

Miyawaki, T., Suzuki, S., Minson, J., Arnolda, L., Chalmers, J., Llewellyn-Smith, I., Pilowsky, P., 1997. Role of AMPA/kainate receptors in transmission of the sympathetic baroreflex in rat CVLM. Am. J. Physiol. 272, R800-R812.

Neal, C.R., Mansour, A., Reinscheid, R., Nothacker, H-P., Civelli, O., Watson, S.J., 1999. Localization of orphanin FQ (Nociceptin) peptide and messenger RNA in the central nervous system of the rat. J. Comp. Neurol. 406, 503-547.

Neff, R.A., Mihalevich, M., Mendelowitz, D., 1998. Stimulation of NTS activates NMDA and non-NMDA receptors in rat cardiac vagal neurons in the nucleus ambiguus. Brain Res. 792, 277-282.

Numao, Y., Siato, M., Terui, N., Kumada, M., 1985. The aortic nerve-sympathetic reflex in the rat. J. Auton. Nerv. Syst. 13, 65-79.

Ohta, H., Talman, W.T., 1994. Both NMDA and non-NMDA receptors in the NTS participate in the baroreceptor reflex in rats. Am. J. Physiol. 267, R1065-R1070.

Ohta, H., Li, X., Talman, W.T., 1996. Release of glutamate in the nucleus tractus solitarii in response to baroreflex activation in rats. Neurosci. 74, 29-37.

Paton, J.F.R., De Paula, P.M., Spyer, K.M., Machado, B.H., Boscan, P., 2002. Sensory afferent selective role of P2 receptors in the nucleus tractus solitarii for mediating the cardiac component of the peripheral chemoreceptor reflex in rats. J. Physiol. (Lond.) 543, 995-1005.

Pilowsky, P.M., Goodchild, A.K, 2002. Baroreceptor reflex pathways and neurotransmitters: 10 years on. J. Hypertension 20, 1675-1688.

Ross, C.A., Ruggiero, D.A., Park, D.H., Joh, T.H., Sved, A.F., Fernandez-Pardal, J., Saavedra, J.M., Reis, D.J., 1984. Tonic vasomotor control by the rostral ventrolateral medulla: effect of electrical or chemical stimulation of the area containing C_1 adrenaline neurons on arterial pressure, heart rate, and plasma catecholamines and vasopressin. J. Neurosci. 4, 474-494.

Sapru H.N., 1994. Transmitter/receptor mechanisms in cardiovascular control by the NTS: excitatory amino acids, acetylcholine and substance P. In: Barraco, I.R.A. (Ed.), Nucleus of the Solitary Tract. CRC Press, Boca Raton, pp. 267-281.

Sapru, H.N., 2002. Glutamate circuits in selected medullo-spinal areas regulating cardiovascular function. Clin. Exp. Pharmacol. Physiol. 29, 491-496.

Sapru, H.N., Chitravanshi, V.C., 2002. Responses to microinjections of endomorphin and nociceptin into the medullary cardiovascular areas. Clin. Exp. Pharmacol. Physiol. 29, 243-247.

Sapru, H.N., Krieger, A.J., 1977. Carotid and aortic chemoreceptor function in the rat. J. Appl. Physiol. 42, 344-348.

Sapru, H.N., Willette, R.N., Krieger, A.J., 1981a. Stimulation of pulmonary J receptors by an enkephalin-analog. J. Pharmacol. Exp. Ther. 217, 228-234.

Sapru, H.N., Gonzalez, E.R., Krieger, A.J., 1981b. Aortic nerve stimulation in the rat: cardiovascular and respiratory responses. Brain Res. Bull. 6, 393-398.

Seagard, J.L., Dean, C., Hopp, F.A., 2001. Properties of NTS neurons receiving input from barosensitive receptors. Ann. NY Acad. Sci. 940, 142-156.

Schreihofer, A.M., Guyenet, P.G., 2002. The baroreflex and beyond: Control of sympathetic vasomotor tone by GABAergic neurons in the ventrolateral medulla. Clin. Exp. Pharmacol. Physiol. 29, 514-521.

Stornetta, R.L., Guyenet, P.G., McCarty, R.C., 1987. Autonomic nervous system control of heart rate during baroreceptor activation in conscious and anesthetized rats. J. Auton. Nerv. Syst. 20, 121-127.

Sun, M.K., Guyenet, P.G., 1985. GABA-mediated baroreceptor inhibition of reticulospinal neurons. Am. J. Physiol. 249, R672-R680.

Sun, M.K., Reis, D.J., 1995. NMDA receptor-mediated sympathetic chemoreflex excitation of RVL-spinal vasomotor neurones in rats. J. Physiol. (Lond.) 482, 53-68.

Sun, M.K., Reis, D.J., 1996. Excitatory amino acid-mediated chemoreflex excitation of respiratory neurones in rostral ventrolateral medulla in rats. J. Physiol. (Lond.) 492, 559-571.

Sundaram, K., Watson, M., Sapru, H.N., 1989. M_2 muscarinic receptor agonists produce hypotension and bradycardia when injected into the nucleus tractus solitarii. Brain Res. 477, 358-362.

Sundaram, K., Sapru, H.N., 1991. NMDA receptors in the intermediolateral column of the spinal cord mediate sympathoexcitatory responses elicited from the ventrolateral medullary pressor area. Brain Res. 544, 33-41.

Suzuki, T., Takayama, K., Miura, M., 1997. Distribution and projection of the medullary cardiovascular control neurons containing glutamate, glutamic acid decarboxylase, tyrosine hydroxylase and phenylethanolamine N-methyltransferase in rats. Neurosci. Res. 27, 9-19.

Sved, A.F., 1994. GABA-mediated neural transmission in mechanisms of cardiovascular control by the NTS. In: Barraco, I.R.A. (Ed.), Nucleus of the Solitary Tract. CRC Press, Boca Raton, pp. 245-253.

Sved, A.F., Tsukamoto, K., 1992. Tonic stimulation of $GABA_B$ receptors in the nucleus tractus solitarius modulates the baroreceptor reflex. Brain Res. 592, 37-43.

Talman, W.T., Perrone, M.H., Reis, D.J., 1980. Evidence for L-glutamate as the neurotransmitter of baroreceptor afferent nerve fibers. Science 209, 813-815.

Tsukamoto, K., Yin, M., Sved, A.F., 1994. Effect of atropine injected into the nucleus tractus solitarius on the regulation of blood pressure. Brain Res. 648, 9-15.

Uhl, G.R., Childers, S., Pasternak, G., 1994. An opiate-receptor gene family reunion. Trends Neurosci. 17, 89-93.

Urbanski, R., Sapru, H.N., 1988a. Evidence for a sympathoexcitatory pathway from the nucleus tractus solitarius to the ventrolateral medullary pressor area, J. Auton. Nerv. Syst. 23, 161-174.

Urbanski, R., Sapru, H.N., 1988b. Putative neurotransmitters involved in medullary cardiovascular regulation. J. Auton. Nerv. Syst. 25, 181-193.

Van Giersbergen, P.L.M., Palkovits, M., De Jong, W., 1992. Involvement of neurotransmitters in the nucleus tractus solitarii in cardiovascular regulation. Physiol. Rev. 72, 789-824.

Vardhan, A., Kachroo, A., Sapru, H.N., 1993a. Excitatory amino acid receptors in the commissural nucleus of the NTS mediate carotid chemoreceptor responses. Am. J. Physiol., 264, R41-R50.

Vardhan, A., Kachroo, A., Sapru, H.N., 1993b. Excitatory amino acid receptors in the nTS mediate the responses to the stimulation of cardio-pulmonary vagal C fiber endings. Brain Res. 618, 23-31.

Velley, L., Milner, T.A., Chan, J., Morrison, S.F., Pickel, V.M., 1991. Relationship of met-enkephalin-like immunoreactivity to vagal afferents and motor dendrites in the nucleus of the solitary tract: A light and electron microscopic dual labeling study. Brain Res. 550, 298-312.

Verberne, A.J.M., Guyenet, P.G., 1992. Medullary pathway of the Bezold-Jarisch reflex in the rat. Am. J. Physiol. 263, R1195-R1202.

Viard, E., Sapru, H.N., 2002. Cardiovascular responses to activation of metabotropic glutamate receptors in the nTS of the rat. Brain Res. 952, 308-332.

Watson, M., Roeske, W.R., Vickroy, T.W., Smith, T.L., Akiyami, K., Gulya, K., Duckles, S.P., Serra, M., Adem, A., Nordberg, A., Gehlert, D.R., Wamsley, J.K., Yamamura, H.I., 1986. Biochemical and functional basis of putative muscarinic receptor subtypes and its implications. Trends Pharmacol. Sci. Suppl. 46-55.

Willette, R.N., Barcas, P.P., Krieger, A.J., Sapru, H.N., 1983a. Vasopressor and depressor areas in the rat medulla: identification by L-glutamate microinjections. Neuropharmacol. 22, 1071-1079.

Willette, R.N., Krieger, A.J., Barcas, P.P., Sapru, H.N., 1983b. Medullary GABA receptors and the regulation of blood pressure in the rat. J. Pharmacol. Exp. Ther. 226, 893-899.

Willette, R.N., Krieger, A.J., Barcas, P.P., Sapru, H.N., 1984a. Endogenous GABAergic mechanisms in the medulla and the regulation of blood pressure. J. Pharmacol. Exp. Ther. 230, 34-39.

Willette, R.N., Punnen, S., Krieger, A.J., Sapru, A.J., 1984b. Interdependence of rostral and caudal ventrolateral medullary areas in the control of blood pressure. Brain Res. 321, 169-174.

Xia, Y., Haddad, G.G., 1991. Ontogeny and distribution of opioid receptors in the rat brainstem. Brain Res. 549, 181-193.

Zadina, J.E., Hackler, L., Ge, L.J., Kastin, A.J., 1997. A potent and selective endogenous agonist for the mu-opiate receptor. Nature 386, 499-502.

Zhang, J., Mifflin, S.W., 1998. Differential roles for NMDA and non-NMDA receptor subtypes in baroreceptor afferent integration in the nucleus of the solitary tract of the rat. J. Physiol. (Lond.) 511.3, 733-745.

Chapter 5

CARDIOVASCULAR PATHWAYS REVEALED WITH FUNCTIONAL NEUROANATOMY

Teresa L. Krukoff
Department of Cell Biology and Center for Neuroscience, Faculty of Medicine and Dentistry, University of Alberta, Edmonton, AB, Canada T6G 2H7

Abstract: Expression of the immediate early gene, c-*fos*, has been used to contribute to the understanding of central autonomic pathways which are activated by changes in blood pressure (BP). This approach has been particularly useful in demonstrating multisynaptic pathways involved in cardiovascular regulation, for determining relative strengths of pathways involved and, in combination with other neuroanatomical techniques, for identifying the phenotypes of neurons involved in these functional pathways. Decreases in BP activate autonomic neurons throughout the neuraxis including in the brainstem (nucleus of the tractus solitarius, ventrolateral medulla, locus coeruleus, and parabrachial nucleus), in the forebrain (paraventricular nucleus, supraoptic nucleus, medial preoptic area, bed nucleus of the stria terminalis, and central nucleus of the amygdala), and sympathetic preganglionic neurons of the spinal cord. Increases in BP activate neurons in many of the same areas, with the notable exception of the paraventricular nucleus. The diversity of activated pathways, coupled with the phenotypic variety of neurons involved, reinforce the complexity of neuronal interactions which occur in the brain when BP homeostasis is disrupted.

Key words: c-*fos*, blood pressure, hypotension, hypertension, paraventricular nucleus, nucleus of the tractus solitarius, ventrolateral medulla

INTRODUCTION

Few subjects have been studied with c-*fos* functional neuroanatomy as extensively as have central regulatory pathways of the cardiovascular system during the 1990's. While traditional neuroanatomical tracing and electrophysiological studies had laid the groundwork for the basic understanding of these central pathways, application of c-*fos* technology has allowed a more complete elucidation of functional details of the pathways, the neurotransmitter phenotype of the neurons participating in these pathways, and the general complexity of these pathways. While other immediate early genes have been used occasionally for the same purposes, expression of c-*fos* has received the greatest attention.

This chapter will review the results obtained by localizing expression of c-*fos* in neurons of pathways activated by increases and decreases in blood pressure (BP), by electrically and/or chemically stimulating central sites which are important components of these pathways, and by stimuli which accompany establishment of hypertension in the spontaneously hypertensive rat. Except where needed for clarification, studies using more traditional mapping techniques (tracing, electrophysiology) will not be reviewed here.

DECREASED BLOOD PRESSURE (BP)

The most popular approach used to study central pathways activated by decreases in BP has been the administration of the peripheral vasodilator, sodium nitroprusside (NP). A second approach, but one which activates additional pathways (e.g. volume receptors, osmoreceptors) is hemorrhage. Comparisons of the results obtained from the two approaches have yielded useful information about central cardiovascular pathways.

Pharmacological Reduction of BP

Brainstem and Spinal Cord

Decreasing BP with infusions of NP has been shown to stimulate Fos expression in the nucleus of the tractus solitarius (NTS), area postrema, ventrolateral medulla (VLM), parabrachial nucleus (PBN), catecholamine cell group A5, and locus coeruleus of rat, rabbit, and pig (Chan and Sawchenko, 1994; Li and Dampney, 1994; Murphy et al., 1994; Sved et al., 1994; Krukoff et al., 1995; Ruggiero et al., 1996; Jhamandas et al., 1998). In one study, no Fos expression was observed in the area postrema of conscious rats receiving NP (Chan & Sawchenko, 1994), but this lack of response in the area postrema may have been due to the relatively short (10 min) stimulus time compared to the other studies where BP was decreased for approximately 1 h.

In the NTS of conscious rabbits, the largest number of activated neurons in a rostrocaudal direction was found near the obex (Li and Dampney, 1994).

In the rat, about one-quarter of the neurons projecting to the PVN were activated (Krukoff et al., 1995) and activated neurons projecting to the PVN or spinal cord were found to be primarily non-aminergic (Chan and Sawchenko, 1994). Only about 5% of activated neurons in the NTS also projected to the PVN (Krukoff et al., 1995); and, while this percentage is likely an underestimation because the PVN was not completely encompassed with injections of retrograde tracer, it does indicate that the majority of NTS neurons activated by NP project to targets other than the PVN. Finally, small numbers of neurons in the NTS activated by NP-induced decreases in BP were neuropeptide FF-containing neurons which projected to the pontine PBN (Jhamandas et al., 1998).

In the VLM, neurons activated by NP have been described throughout both the rostral VLM (RVLM) and caudal VLM (CVLM) (Chan and Sawchenko, 1994; Li and Dampney, 1994; Murphy et al., 1994; Sved et al., 1994; Krukoff et al., 1995; Ruggiero et al., 1996). In the CVLM, these neurons are found between the nucleus ambiguus and reticular nucleus; in the RVLM, they are found close to the ventral surface of the medulla (Li and Dampney, 1994). About one-third of PVN-projecting neurons in the VLM were activated, and the majority of PVN- or spinally-projecting neurons were aminergic (Chan and Sawchenko, 1994). About 17% of activated neurons in the VLM projected to the PVN (Krukoff et al., 1995), demonstrating that the projections of neurons responding to decreased BP are more concentrated towards the PVN from the VLM than from the NTS (see above).

Relatively large proportions of catecholaminergic neurons throughout the brainstem are activated by drug-induced decreases in BP, with about 50% activated by NP in the A1, A5, and A7 (Murphy et al., 1994), and about 80% of C1 neurons activated by hydralazine-induced hypotension (Sved et al., 1994). Neuropeptide Y (NPY) has also been identified as a peptidergic transmitter in VLM neurons which were activated by glyceryl trinitrate-induced hypotension (McLean et al., 1999).

Sympathetic preganglionic neurons throughout the thoracolumbar spinal cord of rat and rabbit were demonstrated to be activated by NP-induced hypotension (Li and Dampney, 1994; Minson et al., 2002). In addition, these neurons were generally apposed by monoamine- and neuropeptide-containing nerve fibers, including tyrosine hydroxylase, serotonin, substance P, and enkephalin (Minson et al., 2002). Neurons in the upper and middle thoracic spinal cord were apposed by NPY- and PNMT-containing fibers, whereas galanin was found in fibers apposing small numbers of preganglionic neurons in the middle to lower thoracic spinal cord (Minson et al., 2002). While the sources of these neuropeptidergic nerve fibers were not determined, these results suggest that inputs to neurons responsive to decreased BP receive a variety of phenotypically-identified inputs, including some which may be regionally specific to distinct levels of the thoracolumbar spinal cord.

Forebrain

Hypotension induced by NP infusion has been shown to stimulate Fos expression in all major subdivisions of the PVN (Krukoff et al., 1997), suggesting that decreases in BP activate all three major axes which the PVN regulates: the neural circuitry interconnecting the PVN with other autonomic centers, the hypothalamo-pituitary-adrenal axis, and the hypothalamo-neurohypophyseal axis. Thirty-seven percent and 92% of vasopressinergic neurons in the PVN and supraoptic nucleus (SON), respectively, were activated (Li and Dampney, 1994). Furthermore, relatively large proportions of activated PVN neurons were shown to be nitric oxide (NO)-producing neurons (Petrov et al., 1995a; Krukoff et al., 1997). We have studied the targets of activated PVN neurons to show that 3 to 4% project to the NTS or VLM (Krukoff et al., 1997). Of PVN neurons projecting to the NTS and VLM, 33% and 16%, respectively, were activated by decreased BP, and these included a subpopulation of NO-producing neurons. Finally, a small number of activated neurons in the PVN projected to both the NTS and VLM through collateral branches (Krukoff et al., 1997). Thus, hypotension activates large populations of neurons in the PVN, some of these neurons in turn project to autonomic centers in the brainstem, and NO may participate as a neurotransmitter in these pathways.

Other forebrain areas which have been shown to be stimulated by NP-induced hypotension are the SON, arcuate nucleus, medial preoptic area, bed nucleus of the stria terminalis, and central nucleus of the amygdala (Li and Dampney, 1994). As most of these centers can be considered to be autonomic/limbic in function, these results reinforce the concept that many areas in the brain participate in the integration of central responses to decreased BP.

Hemorrhage

Brainstem

Studies using hemorrhage have yielded results which are generally very similar to those described above using NP infusions or injections (McAllen et al., 1992; Dun et al., 1993). Additional information obtained with graded hemorrhage (between 15% and 25% blood loss) has shown that a subpopulation of activated neurons in the A1 and C1 cell groups produces both tyrosine hydroxylase and NPY and that expression of the genes encoding the two proteins in medullary cardiovascular cell groups increased in proportion to the strength of stimulus (Chan and Sawchenko, 1998a). In another study focused on the locus coeruleus (LC) and using graded hemorrhage, it was found that 20% (but not 10%, 30%, or 40%) blood loss activated neurons in the posterior LC and that the change in BP could be prevented with chronic lesions of the posterior LC (Anselmo-Franci et al., 1998). The authors interpreted these results to suggest that at 20% blood loss, the posterior LC is recruited to respond but that at greater degrees of stimulation, other mechanisms are recruited to compensate for the more severe drops in BP (Anselmo-Franci et al., 1998).

A few relatively subtle differences have been described for activation of neurons by hemorrhage compared to NP. We found Fos immunoreactive neurons in the area postrema (AP) of NP-treated, but not hemorrhaged rats (Krukoff et al., 1995) whereas opposite results for c-*fos* mRNA levels have been found in the AP (Chan and Sawchenko, 1994). As others have described Fos in the AP of NP-treated rats (Li and Dampney, 1994; Murphy et al., 1994), the lack of Fos expression in the AP of NP-treated rats in the study by Chan and Sawchenko (1994) may be due to the short stimulus time used (<15 min).

In the NTS and VLM, we found differences in the numbers of neurons activated by NP vs. hemorrhage only in the intermediate NTS, with numbers greater in NP-treated rats (Krukoff et al., 1995). This difference may be due to increased afferent information reaching the intermediate NTS via activated neurons of the AP in the same animals (Krukoff et al., 1995). On the other hand, hemorrhage was associated with activation of slightly more catecholaminergic neurons in the A1 cell group of the CVLM compared to NP (Chan and Sawchenko, 1994).

Forebrain

Decreasing BP with NP or hemorrhage made no difference to the numbers of PVN neurons activated, including the NO-producing neurons (Krukoff et al., 1997), suggesting that, at the PVN, the change in BP is the predominant stimulus associated with hemorrhage. It is interesting, however, that hypovolemia in the absence of hypotension is associated with neuronal activation in the PVN (Badoer et al., 1993). Therefore, the equal numbers of activated PVN neurons we have observed with NP and hemorrhage (Krukoff et al., 1997) may be due to the activation of (1) the same population of neurons, (2) similar but non-overlapping populations of neurons, or (3) a combination of the two possibilities. Finally, because baroreceptor denervation had no effect on the numbers of neurons in the hypothalamus (or medulla) activated by hemorrhage, it has been suggested that neuronal activation was due to stimulation of cardiac receptors rather than baroreceptors (Potts et al., 2000).

INCREASED BLOOD PRESSURE

Brainstem

Identification of neural pathways activated when baroreceptors are stimulated has been most commonly studied using intravenous injections of the peripheral vasoconstrictor, phenylephrine (PHE). The general pattern of neuronal activation is similar to that elicited by decreases in BP. Thus, increased BP stimulates Fos expression in neurons of the NTS, VLM, AP, A5 cell group, locus coeruleus, and PBN (Narvaez et al., 1993; Li and Dampney, 1994; Murphy et al., 1994; Polson et al., 1995; Kantor et al., 1996). Overall, only about 5% of activated neurons in the NTS, VLM, and PBN were catecholaminergic in conscious rabbits (Li and Dampney, 1994). In the A1, A5, and A7 cell groups of the rat, about 50% of catecholaminergic neurons were activated (Murphy et al., 1994).

In the NTS, activated neurons were numerous in the dorsal commissural nucleus and extending rostrally into the dorsal nucleus (Chan and Sawchenko, 1998b). The distribution of these neurons corresponds well with the terminal fields of the aortic depressor nerves (Ciriello, 1983) which carry baroreceptor information in the rat. Furthermore, sinoaortic denervation resulted in 90% reduction of numbers of Fos-positive neurons in the NTS (Chan et al., 2000). Of these activated NTS neurons, only a few were GABAergic and relatively few projected to the RVLM (Chan and

Sawchenko, 1998b). Of the neurons that did project to the RVLM, a high proportion was identified as NO-producing (Chan and Sawchenko, 1998b).

Not surprisingly, increased BP activates neurons in the VLM, predominantly the depressor CVLM. Substantial proportions (23% - 50%) of activated neurons in the CVLM projected to the RVLM (Polson et al., 1995; Minson et al., 1997) and about 40% of activated CVLM neurons were GABAergic (Minson et al., 1997). A triple-labeling study showed that activated GABAergic neurons in the CVLM project to the RVLM (Chan and Sawchenko, 1998b), illustrating that the inhibitory pathway from the CVLM to the RVLM is activated during stimulation of baroreceptors. The efficacy of this inhibitory pathway was demonstrated with injection of the $GABA_A$ receptor antagonist, muscimol, into the CVLM. These injections stimulated increases in BP and in the numbers of Fos-positive neurons in the RVLM, about half of which projected to the spinal cord (Minson et al., 1994).

Sinoartic denervation attenuated the numbers of neurons activated by PHE in NTS, VLM, and locus coeruleus (Potts et al., 1997; Grindstaff et al., 2000), illustrating that PHE injections activate brainstem neurons by stimulating baroreceptors. It was only in the LC, however, that denervation affected the activation of dopamine-β-hydroxylase-containing neurons (Grindstaff et al., 2000). Some of these neurons also projected to the diagonal band of Broca (DBB). Together with the earlier findings that the DBB sends projections to the vicinity of the supraoptic nucleus (Jhamandas et al., 1989) and that lesions of the LC attenuate responsiveness of vasopressinergic supraoptic neurons to baroreceptor activation (Grindstaff et al., 2000), these findings provide evidence that noradrenergic neurons in the LC participate in the baroreceptor activation of the DBB to regulate activity of supraoptic Arginine vasopressin (AVP) neurons (Grindstaff et al., 2000).

Forebrain

Whereas no induction of Fos expression was observed in the PVN or SON of PHE-treated, conscious rabbits (Li and Dampney, 1994), low numbers of Fos-positive neurons have been reported in the rat (McKinley et al., 1992). Other forebrain areas reported to contain activated neurons after PHE treatment are the central nucleus of the amygdala and the bed nucleus of the stria terminalis (McKinley et al., 1992; Li and Dampney, 1994).

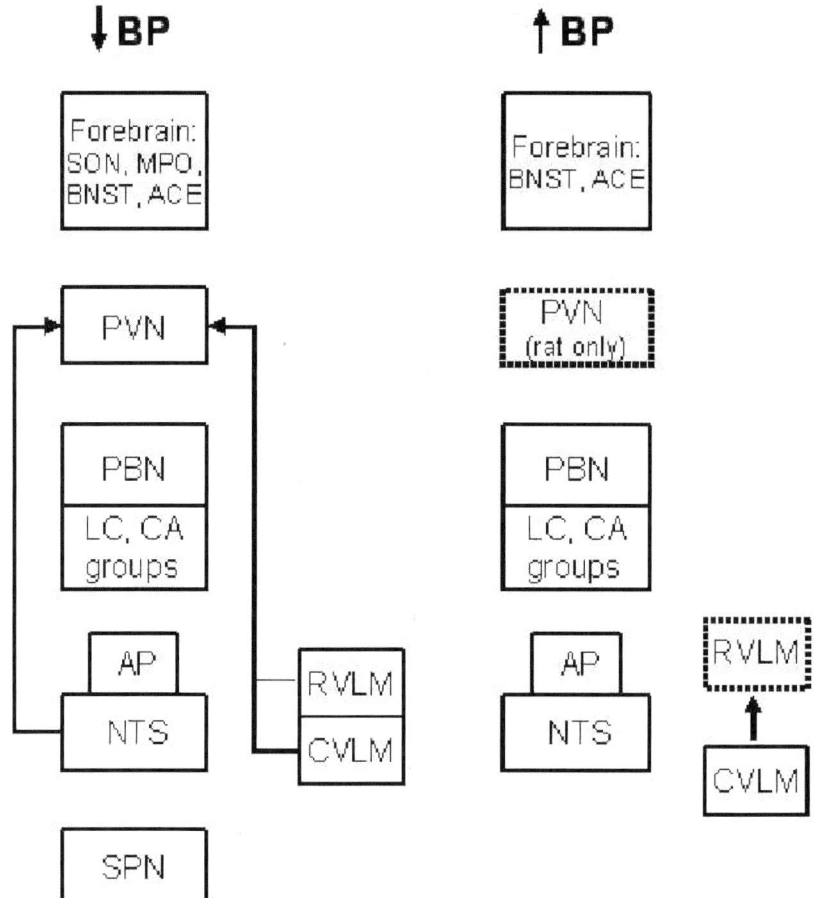

Figure 1. Summary of brain regions which show Fos expression due to decreased (left) or increased (right) blood pressure. Arrows illustrate demonstrated pathways of activated neurons. Dashed boxes respresent areas with no or species-limited Fos expression. Abbreviations: ACE, central nucleus of the amygdala; AP, area postrema; BNST, bed nucleus of the stria terminalis; CA, catecholaminergic; CVLM, caudal ventrolateral medulla; LC, locus coeruleus; MPO, medial preoptic area; NTS, nucleus of the tractus solitarius; PBN, parabrachial nucleus; PVN, paraventricular nucleus; RVLM, rostral ventrolateral medulla; SON, supraoptic nucleus; SPN, sympathetic preganglionic neurons.

DECREASED *VS.* INCREASED BLOOD PRESSURE

Fewer neurons expressing Fos have been found in the NTS and VLM of PHE-treated than NP-treated rats, and the neurons were primarily non-aminergic after PHE (Chan and Sawchenko, 1994; Li and Dampney, 1994). In the NTS, decreased BP activated more neurons in the A2 cell group than increased BP (Murphy et al., 1994). These authors suggested that, because A2 neurons project to the SON (Cunningham and Sawchenko, 1988; Day and Sibbald, 1988), these neurons may participate in stimulating vasopressin release during hypotension (Murphy et al., 1994). In contrast to A2 cell group, no differences were found in the numbers of activated C2 neurons after decreased and increased BP (Murphy et al., 1994).

Increased BP was associated with larger numbers of activated neurons in the rostral part of the CVLM whereas decreased BP activated more neurons in the caudal CVLM (Murphy et al., 1994), illustrating that, even within the CVLM, neurons may be functionally segregated. In the periaqueductal gray, separate columns of neurons were activated by decreased and increased BP in chloral hydrate-anesthetized rats (Murphy et al., 1995). The results are consistent with the role of the periaqueductal gray as an integrator of sensory, autonomic, limbic, and motor activities (Bandler and Shipley, 1994) and suggest that, like the VLM, separate components of the columnar nuclei participate in the responses to decreased and increased BP (Murphy et al., 1995).

Sinoaortic denervation reduced by about 90% the numbers of neurons in the NTS and VLM activated in hypo- and hypertension in conscious rabbits (Potts et al., 1997). In addition, neuronal activation was reduced by about 60% in the central nucleus of the amygdala and bed nucleus of the stria terminalis of PHE-treated rats, by about 90% in LC, periaqueductal gray, PVN, central nucleus of the amygdala, and bed nucleus of the stria terminalis of NP-treated rabbits, and by about 75% in the supraoptic nucleus of NP-treated rabbits (Potts et al., 1997). Thus the integrity of the baroreceptors is important to the activation of most central sites which participate in regulation of the blood pressure. Interestingly, denervation had no effects on neuronal activation in a few central sites of PHE- and NP-treated rabbits (Potts et al., 1997). Thus, continued activation of the PBN in denervated, PHE-treated rabbits may be due to inputs from the AP; the subfornical organ and organum vasculosum of the lamina terminalis (circumventricular organs) of NP-treated rabbits may continue to be stimulated by neuroendocrine pathways (e.g. circulating angiotensin II) (Potts et al., 1997).

ELECTRICAL OR CHEMICAL STIMULATION

Stimulation of afferent nerves and of central autonomic centers has been useful for more in-depth understanding of projections within the CNS which are important in autonomic and cardiovascular regulation. Stimulation experiments have also been successfully applied to the phenotypic identification of neurons which participate in central pathways. While the use of c-*fos* expression does not distinguish between direct and multisynaptic pathways, stimulation experiments have contributed useful data about central autonomic sites.

Afferent Nerves

In anesthetized rats, electrical stimulation of the aortic depressor nerve (ADN), which carries primarily baroreceptor information, resulted in decreased BP, increased Fos expression in the same brainstem and forebrain areas discussed above (McKitrick et al., 1992), and increased NGFI-A mRNA levels in the NTS and paratrigeminal nucleus (Rutherfurd et al., 1992). These results are complicated by the fact that anesthesia can itself be associated with increased Fos expression in these areas (Krukoff et al., 1992). Furthermore, the changes in BP elicited by nerve stimulation may have been the primary cause of Fos expression in some of these brain regions.

Similarly, electrical stimulation of the carotid sinus nerve (CSN) in anesthetized cats was found to result in Fos expression in the usual brainstem regions (Maqbool et al., 1997). Because the CSN carries both baro- and chemoreceptor information, Erickson & Millhorn (1991) compared results after CSN stimulation and hypoxia in awake rats. They found no differences in the distributions of Fos-positive neurons in the brainstem between the two groups of rats: in addition to label in the NTS, VLM, and AP, Fos neurons were also found in the nucleus raphe pallidus and along the ventral surface of the medulla (Erickson and Millhorn, 1991).

Brainstem Sites

Pressure injections of L-glutamate into the RVLM elicit Fos expression in sympathetic preganglionic neurons, some of which project to the adrenal glands (Leman et al., 2000). These manipulations elicit large increases in BP (about 70 mmHg), however, so that it is not possible to conclude that the preganglionic neurons were stimulated solely by the chemical stimulation of RVLM neurons. Indeed, the bilateral nature of the results in the spinal cord suggests that the changes in BP played a significant role in activation of preganglionic neurons.

Electrical stimulation of the PBN activated forebrain neurons in the ipsilateral dorsal SON, central nucleus of the amygdala, and the cerebral cortex (Krukoff et al., 1992). The ipsilateral nature of these responses suggests that the electrical stimulation was responsible. On the other hand, bilaterally activated neurons were found in the PVN, NTS, CVLM, and AP, and were probably due to the urethane anesthesia or to the increases in BP elicited by PBN stimulation (Krukoff et al., 1992).

The mesencephalic cuneiform nucleus of the pons is believed to play a role in modulation of stress-related cardiovascular responses (Korte et al., 1992). Electrical stimulation of this area resulted in increased ipsilateral *c-fos* mRNA levels in the PBN and Kölliker-Fuse nucleus (Lam et al., 1996).

Forebrain Sites

We have studied the effects of electrical and chemical (glutamate) stimulation of the PVN (Krukoff et al., 1994). First we found that electrical stimulation led to ipsilateral Fos expression in neurons of the cerebral cortex, medial amygdala, hypothalamus, medial PBN, NTS, and VLM. Second, comparison of the results for the two types of stimulation showed that, while numbers of neurons with Fos were higher after electrical stimulation, the qualitative results were similar in the ventromedial hypothalamus, arcuate nucleus, PBN, rostral NTS, and CVLM, and RVLM. Disparity of results occurred only in the NTS at the level of the AP, where electrical stimulation activated neurons but glutamate stimulation did not. This result indicates that activation of neurons in this part of the NTS was likely due to stimulation of fibers of passage in the PVN (Krukoff et al., 1994).

The central nucleus of the amygdala (CNA) has received the most attention with regard to stimulation experiments. Forebrain structures activated by electrical stimulation of the CNA were the ipsilateral PVN, SON, and arcuate nucleus (Petrov et al., 1994). In the magnocellular PVN, these neurons were both vasopressinergic and oxytocinergic (Petrov et al., 1994). In the brainstem, activated neurons, including those which produce catecholamines, were found in the NTS and VLM (Petrov et al., 1995b). Both activated catecholaminergic and non-catecholaminergic neurons in the NTS and VLM received projections from the PVN. Together, these results show that the CNA and PVN may act in synchrony to modulate output of brainstem autonomic neurons (Petrov et al., 1995b).

In a related study, neurons in the NTS and VLM which were activated by electrical stimulation of the CNA were found to be interconnected (Petrov et al., 1996). About 30% of activated neurons in the VLM were found to project to the NTS, and about 10% of activated neurons in the NTS were found to project to the VLM. Thus, neurons in these two centers may have the capacity to modulate each other's neuronal activity in response to stimulation of the CNA. Furthermore, these results demonstrate the complexity of central autonomic pathways because NTS and VLM neurons also project to the PVN and CNA (among other targets), including through branching collaterals (Petrov et al., 1993).

Chemical stimulation of the CNA using glutamate was associated with bilateral Fos expression in neurons of the CVLM and intermediate VLM (Salomé et al., 2001). Because Fos expression in these areas was not present when lower concentrations of glutamate (not causing changes in BP) were used, it is likely that neuronal activation in this study was primarily due to the changes in BP seen with the larger doses of glutamate.

Neurons in the NTS and VLM (Ba-M'Hamed et al., 1998), including catecholaminergic neurons (Viltart and Sequeira, 1999), were shown to express Fos after electrical stimulation of the motor cortex in rat. Some of these neurons may have been activated through direct projections, since direct projections have been demonstrated from the sensorimotor cortex to the dorsal vagal complex and RVLM (Ba-M'Hamed et al., 1993; Ba-M'Hamed et al., 1998). As discussed earlier, activation of these neurons may have also been elicited by concomitant decreases in BP or urethane anesthesia since similar numbers of Fos neurons were found bilaterally in these areas (Ba-M'Hamed et al., 1998).

PATHOLOGY

Hypertension in spontaneously hypertensive rats (SHRs) is believed to be the result of central changes which increase sympathetic drive to the periphery. No differences in numbers of Fos neurons have been found in fentanyl-anesthetized, resting SHRs compared to Wistar-Kyoto (WKY) rats (Xiong et al., 1997), but conscious SHRs demonstrated greater basal activation in bulbospinal neurons of the RVLM and in sympathetic preganglionic neurons which project to the adrenal glands (Minson et al., 1996). In response to decreased BP (nitroprusside), greater numbers of activated preganglionic neurons, but not RVLM neurons, were found in conscious SHRs suggesting that the inhibitory influence of the CVLM is attenuated in SHRs compared to WKY rats (Minson et al., 1996). On the other hand, it has been reported that increased BP (PHE) stimulated more neurons in the CVLM of fentanyl-anesthetized SHRs than in WKYs but more neurons in the RVLM (near the obex) in WKYs than in SHRs (Xiong et al., 1997). The activation of CVLM neurons in both experiments is difficult to reconcile, but may be related to the use of anesthetic in one of the studies (Xiong et al., 1997). In addition, the mechanisms underlying spontaneous hypertension are likely very complex and the identification of activated neurons with Fos expression may be too non-specific to elucidate cause and effect relationships in SHRs.

CONCLUSIONS

The use of c-*fos* immunohistochemistry has been particularly well-suited to the *in vivo* identification of neurons that belong to cardiovascular pathways for several reasons: (1) the technique illustrates multisynaptic pathways, providing a more complete understanding of the brain areas involved in regulating blood pressure; (2) as results can be easily quantitated, relative strengths of pathways subserving different functions have been determined; (3) it has been possible to compare results from different, but physiologically related functions (decreased BP vs. hemorrhage); and (4) methods have been successfully combined with other neuroanatomical techniques to visualize neurotransmitter phenotypes of cardiovascular neurons. On the other hand, care must be taken in the interpretation of results for the following reasons: (1) control experiments must be carefully planned to eliminate as many non-specific effects as possible which may contribute to stimulation of Fos expression; (2) pharmacological agents used to induce changes in BP may themselves have effects on neurons; and (3) activation of multisynaptic pathways may make it difficult to draw conclusions about direct or indirect activation of neurons. Notwithstanding the limitations of using c-*fos* expression for identification of functional pathways in the brain, the studies carried out during the last ten years have contributed substantially to the understanding of the complexity of central cardiovascular integrative pathways. The results reinforce the concept that brainstem and forebrain autonomic sites respond to changes in blood pressure via complex interconnections, including through collateral branching, and that a variety of neurotransmitters is involved in these integrative processes.

REFERENCES

Anselmo-Franci, J.A., Peres-Polon, V.L., da Rocha-Barros, V.M., Moreira, E.R., Franci, C.R., Rocha, M.J.A., 1998. C-fos expression and electrolytic lesions studies reveal activation of the posterior region of locus coeruleus during hemorrhage induced hypotension. Brain Res. 799, 278-284.

Badoer, E., McKinley, M.J., Oldfield, B.J., McAllen, R.M., 1993. A comparison of hypotensive and non-hypotensive hemorrhage on Fos expression in spinally projecting neurons of the paraventricular nucleus and rostral ventrolateral medulla. Brain Res. 610, 216-223.

Ba-M'Hamed, S., Sequeira, H., Poulain, P., Bennis, M., Roy, J.C., 1993. Sensorimotor cortex projections to the ventrolateral and the dorsomedial medulla oblongata in the rat. Neurosci. Lett. 164, 195-198.

Ba-M'Hamed, S., Viltart, O., Poulain, P., Sequeira, H., 1998. Distribution of cortical fibers and Fos immunoreactive neurons in ventrolateral medulla and in nucleus tractus solitarius following the motor cortex stimulation in the rat. Brain Res. 813, 411-415.

Bandler, R., Shipley, M.T., 1994. Columnar organization in the midbrain periaqueductal gray: modules for emotional expression. TINS 17, 379-389.

Chan, R.K.W., Jarvina, E.V., Sawchenko, P.E., 2000. Effects of selective sinoaortic denervations on phenylephrine-induced activational responses in the nucleus of the solitary tract. Neurosci. 101, 165-178.

Chan, R.K.W., Sawchenko, P.E., 1994. Spatially and temporally differentiated patterns of c-fos expression in brainstem catecholaminergic cell groups induced by cardiovascular challenges in the rat. J. Comp. Neurol. 348, 433-460.

Chan, R.K.W., Sawchenko, P.E., 1998a. Differential time- and dose-related effects of haemorrhage on tyrosine hydroxylase and neuropeptide Y mRNA expression in medullary catecholaminergic neurons. Eur. J. Neurosci. 10, 3747-3758.

Chan, R.K.W., Sawchenko, P.E., 1998b. Organization and transmitter specificity of medullary neurons activated by sustained hypertension: implications for understanding baroreceptor reflex circuitry. J. Neurosci. 18, 371-387.

Ciriello, J., 1983. Brainstem projections of aortic baroreceptor afferent fibers in the rat. Neurosci. Lett. 36, 37-42.

Cunningham, E.T., Sawchenko, P.E., 1988. Anatomical specificity of noradrenergic inputs to the paraventricular and supraoptic nuclei of the rat hypothalamus. J. Comp. Neurol. 274, 60-76.

Day, T.A., Sibbald, J.R., 1988. Direct catecholaminergic projections from nucleus tractus solitarius to supraoptic nucleus. Brain Res. 454, 387-392.

Dun, N.J., Dun, S.L., Chiaia, N.L., 1993. Hemorrhage induces Fos immunoreactivity in rat medullary catecholaminergic neurons. Brain Res. 608, 223-232.

Erickson, J.T., Millhorn, D.E., 1991. Fos-like protein is induced in neurons of the medulla oblongata after stimulation of the carotid sinus nerve in awake and anesthetized rats. Brain Res. 567, 11-24.

Grindstaff, R.J., Grindstaff, R.R., Sullivan, M.J., Cunningham, E.T., 2000. Role of the locus ceruleus in baroreceptor regulation of supraoptic vasopressin neurons in the rat. Am. J. Physiol. 279, R306-R319.

Jhamandas, J.H., Harris, K.H., Petrov, T., Yang, H.Y.T., Jhamandas, K.H., 1998. Activation of neuropeptide FF neurons in the brainstem nucleus tractus solitarius following cardiovascular challenge and opiate withdrawal. J. Comp. Neurol. 402, 210-221.

Jhamandas, J.H., Raby, W., Rogers, J., Buijs, R.M., Renaud, L.P., 1989. Diagonal band projection towards the hypothalamic supraoptic nucleus: light and electron microscopic observations in the rat. J. Comp. Neurol. 282, 15-23.

Kantor, R.K., Strauss, J.A., Sauro, M.D., 1996. Comparison of neurons in rat medulla oblongata with Fos immunoreactivity evoked by seizures, chemoreceptor, or baroreceptor stimulation. Neurosci. 73, 807-816.

Korte, S.M., Jaarsma, D., Luiten, P.G., Bohus, B., 1992. Mesencephalic cuneiform nucleus and its ascending and descending projections serve stress-related cardiovascular responses in the rat. J. Auton. Nerv. Syst. 41, 157-176.

Krukoff, T.L., Harris, K.H., Linetsky, E., Jhamandas, J.H., 1994. Expression of c-*fos* protein in rat brain elicited by electrical and chemical stimulation of the hypothalamic paraventricular nucleus. Neuroendocrinol. 59, 590-602.

Krukoff, T.L., MacTavish, D., Harris, K.H., Jhamandas, J.H., 1995. Changes in blood volume and pressure induced c-*fos* expression in brainstem neurons that project to the paraventricular nucleus of the hypothalamus. Molec. Brain Res. 34, 99-108.

Krukoff, T.L., MacTavish, D., Jhamandas, J.H., 1997. Activation by hypotension of neurons in the hypothalamic paraventricular nucleus that project to the brainstem. J. Comp. Neurol. 385, 285-296.

Krukoff, T.L., Morton, T.L., Harris, K.H., Jhamandas, J.H., 1992. Expression of c-fos protein in rat brain elicited by electrical stimulation of the pontine parabrachial nucleus. J. Neurosci. 12, 3582-3590.

Lam, W., Gundlach, A.L., Verberne, A.J.M., 1996. Increased nerve growth factor inducible-A gene and c-fos messenger RNA levels in the rat midbrain and hindbrain associated with the cardiovascular response to electrical stimulation of the mesencephalic cuneiform nucleus. Neurosci. 71, 193-211.

Leman, S., Viltart, O., Sequeira, H., 2000. Expression of Fos protein in adrenal preganglionic neurons following chemical stimulation of the rostral ventrolateral medulla of the rat. Brain Res. 854, 189-196.

Li, Y.-W., Dampney, R.A.L., 1994. Expression of Fos-like protein in brain following sustained hypertension and hypotension in conscious rabbits. Neurosci. 61, 613-634.

Maqbool, A., McWilliam, P.N., Batten, T.F.C., 1997. Co-localization of c-Fos and neurotransmitter immunoreactivities in the cat brain stem after carotid sinus nerve stimulation. J. Chem. Neuroanat. 13, 189-200.

McAllen, R.M., Badoer, E., Shafton, A.D., Oldfield, B.J., McKinley, M.J., 1992. Hemorrhage induces c-fos immunoreactivity in spinally projecting neurons of cat subretrofacial nucleus. Brain Res. 575, 329-332.

McKinley, M.J., Badoer, E., Oldfield, B.J., 1992. Intravenous angiotensin II induces Fos-immunoreactivity in circumventricular organs of the lamina terminalis. Brain Res. 594, 295-300.

McKitrick, D.J., Krukoff, T.L., Calaresu, F.R., 1992. Expression of c-fos protein in rat brain after electrical stimulation of the aortic depressor nerve. Brain Res. 599, 215-222.

McLean, K.J., Jarrott, B., Lawrence, A.J., 1999. Hypotension activates neuropeptide Y-containing neurons in the rat medulla oblongata. Neurosci. 92, 1377-1387.

Minson, J.B., Arnolda, L.F., Llewellyn-Smith, I.J., 2002. Neurochemistry of nerve fibers apposing sympathetic preganglionic neurons activated by sustained hypotension. J. Comp. Neurol. 449, 307-318.

Minson, J.B., Arnolda, L.F., Llewellyn-Smith, I.J., Pilowsky, P.M., Chalmers, J.P., 1996. Altered c-fos in rostral medulla and spinal cord of spontaneously hypertensive rats. Hypertension 27, 433-441.

Minson, J.B., Llewellyn-Smith, I.J., Arnolda, L.F., Pilowsky, P.M., Oliver, J.R., Chalmers, J.P., 1994. Disinhibition of the rostral ventral medulla increases blood pressure and fos expression in bulbospinal neurons. Brain Res. 646, 44-52.

Minson, J.B., Llewellyn-Smith, I.J., Chalmers, J.P., Pilowsky, P.M., Arnolda, L.F., 1997. c-fos identifies GABA-synthesizing barosensitive neurons in caudal ventrolateral medulla. NeuroReport 8, 3015-3021.

Murphy, A.Z., Ennis, M., Rizvi, T.A., Behbehani, M.M., Shipley, M.T., 1995. Fos expression induced by changes in arterial pressure is localized in distinct, longitudinally organized columns of neurons in the rat midbrain periaqueductal gray. J. Comp. Neurol. 360, 286-300.

Murphy, A.Z., Ennis, M., Shipley, M.T., Behbehani, M.M., 1994. Directionally specific changes in arterial pressure induce differential patterns of Fos expression in discrete areas of the rat brainstem: a double-labeling study for Fos and catecholamines. J. Comp. Neurol. 349, 36050.

Narvaez, J.A., Covenas, R., de Leon, M., Aguirre, J.A., Cintra, A., Goldstein, M., Fuxe, K., 1993. Induction of c-*fos* immunoreactivity in tyroseine hydroxylase and phenylethanolamine-N-methyl-transferase immunoreactive neurons of the medulla oblongata of the rat after phosphate-buffered saline load in the urethane-anesthetized rat. Brain Res. 602, 342-349.

Petrov, T., Harris, K.H., MacTavish, D., Krukoff, T.L., Jhamandas, J.H., 1995a. Hypotension induces Fos immunoreactivity in NADPH-diaphorase positive neurons in the paraventricular and supraoptic hypothalamic nuclei of the rat. Neuropharmacol. 34, 509-514.

Petrov, T., Jhamandas, J.H., Krukoff, T.L., 1994. Electrical stimulation of the central nucleus of the amygdala induces fos-like immunoreactivity in the hypothalamus of the rat: a quantitative study. Molec. Brain Res. 22, 333-340.

Petrov, T., Jhamandas, J.H., Krukoff, T.L., 1996. Connectivity between brainstem autonomic structures and expression of c-*fos* following electrical stimulation of the central nucleus of the amygdala in the rat. Cell Tiss. Res. 283, 367-374.

Petrov, T., Krukoff, T.L., Jhamandas, J.H., 1993. Branching projections of catecholaminergic brainstem neurons to the paraventricular hypothalamic nucleus and the central nucleus of the amygdala. Brain Res. 609, 81-92.

Petrov, T., Krukoff, T.L., Jhamandas, J.H., 1995b. Convergent influence of the central nucleus of the amygdala and the paraventricular hypothalamic nucleus upon brainstem autonomic neurons as revealed by c-*fos* expression and anatomical tracing. J. Neurosci. Res. 42, 835-845.

Polson, J.W., Potts, P.D., Li, Y.-W., Dampney, R.A.L., 1995. Fos expression in neurons projecting to the pressor region in the rostral ventrolateral medulla after sustained hypertension in conscious rabbits. Neurosci. 67, 107-123.

Potts, P.D., Ludbrook, J., Gillman-Gaspari, T.A., Horiuchi, J., Dampney, R.A.L., 2000. Activation of brain neurons following central hypervolaemia and hypovolaemia:contribution of baroreceptor and non-baroreceptor inputs. Neurosci. 95, 499-511.

Potts, P.D., Polson, J.W., Hirooka, Y., Dampney, R.A.L., 1997. Effects of sinoaortic denervation on Fos expression in the brain evoked by hypertension and hypotension in conscious rabbits. Neurosci. 77, 503-520.

Ruggiero, D.A., Tong, S., Anwar, M., Gootman, N., Gootman, P.M., 1996. Hypotension-induced expression of the c-*fos* gene in the medulla oblongata of piglets. Brain Res. 706, 199-209.

Rutherfurd, S.D., Widdop, R.E., Sannajust, F., Louis, W.J., Gundlach, A.L., 1992. Expression of c-fos and NGFI-A messenger RNA in the medulla oblongata of the anaesthetized rat following stimulation of vagal and cardiovascular afferents. Molec. Brain Res. 13, 301-312.

Salomé, N., Viltart, O., Leman, S., Sequeira, H., 2001. Activation of ventrolateral medullary neurons projecting to spinal autonomic areas after chemical stimulation of the central nucleus of amygdala: a neuroanatomical study in the rat. Brain Res. 890, 287-295.

Sved, A.F., Mancini, D.L., Graham, J.C., Schreihofer, A.M., Hoffman, G.E., 1994. PNMT-containing neurons of the C1 cell group express c-fos in response to changes in baroreceptor input. Am. J. Physiol. 266, R361-R367.

Viltart, O., Sequeira, H., 1999. Induction of c-Fos protein in bulbar catecholaminergic neurones by electrical stimulation of sensorimotor cortex in the rat. Neurosci. Lett. 260, 65-68.

Xiong, Y., Takayama, K., Miura, M., 1997. Differences in the density of barosensitive neurons in the medulla of spontaneously hypertensive and Wistar-Kyoto rats. Clin. Exper. Pharmacol. Physiol. 24, 398-402.

Chapter 6

THE HYPOTHALAMUS AND CARDIOVASCULAR REGULATION

John H. Coote
University of Birmingham, Department of Physiology,, Birmingham, United Kingdom, England

Abstract: The hypothalamus plays a pivotol role in homeostatic responses and in ensuring optimal physiological responses in different types of behaviour. The adequacy of this regulation is dependent on appropriate cardiovascular changes unique to each behaviour. This chapter surveys the literature describing the different neuronal phenotypes in the hypothalamus and their projections to cardiovascular neurones elsewhere in the brain and spinal cord. In this context projections from the paraventricular nucleus to the lateral parabrachial nucleus, the nucleus tractus solitarii, the cardiac vagal nuclei, the ventral medulla and to the sympathetic neurones in the spinal cord are described, together with significant projections from the lateral hypothalamus, the arcuate nucleus, the posterior hypothalamus, the dorsomedial hypothalamus and the preoptic area. An attempt is made to show how the different groups of neurones might give rise to non uniform patterns of vascular changes accompanying afferent physiological states. Knowledge of the paraventricular nucleus has expanded rapidly and its functional role in blood volume regulation is dealt with in depth. The significance of afferent input from specialised sensory neurones in orchestrating the pattern of sympathetic response is emphasised.

Key words: forebrain, brainstem, spinal cord, cardiovascular reflexes, hypothalamic nuclei, blood volume regulation, blood pressure, blood flow regulation, hypothalamic neuropeptides, sympathetic preganglionic neurones, homeostasis, autonomic nervous system, nitric oxide, GABA

INTRODUCTION

This chapter represents a selective overview of mainly recent literature dealing with the neuroanatomy and physiology of the hypothalamus. I shall attempt to use this as a basis to establish why its control of cardiovascular effectors is so important. It is well recognised that stimulation (electrical or chemical) at various sites in the hypothalamus can produce pressor or depressor responses involving virtually all the cardiovascular effectors in one way or another. Indeed, it was pointed out by Hilton (1980) that if nothing else was known about the brain control of heart and blood vessels one might conclude that this part of the forebrain was the principal "cardiovascular centre". However, the concept of localised regions with groups of neurones concerned with a particular autonomic function is generally not appealing nor in many ways acceptable. In fact the complexity in the detailed organisation emerging from modern elegant approaches shows that it is clearly misleading to talk of pressor areas or pathways for any single variable. The cardiovascular role of the hypothalamus can best be seen in the context of a set of basic biological responses which are essential for survival. From the work of pioneers like Cannon, Bard, Hess and Hilton it is already clear that this region of the brain is critically concerned with fluid and energy balance, temperature regulation, sleep, defence against threatening signals and reproduction. Specific patterns of cardiovascular adjustment accompany each behaviour and are uniquely directed to adequately respond to the metabolic demands of the behaviour. For example the circulatory adjustments to an environmentally-induced heat load results in a reduction in splanchnic blood flow and a vasodilatation in vessels supplying heat loss organs. In sleep, vasoconstriction occurs in skeletal muscle, which is relaxed and not needed, whereas sympathetic vasomotor activity is reduced to many other organs. The opposite happens during 'defence reactions' in the awake animal when skeletal muscle energy requirements increase dramatically for 'fight or flight'. In fact a detailed analysis of the repetoire of behaviour displayed by animals makes it obvious that an adequate delivery of fuel to specific tissues critically involved in each behaviour is the essential goal for optimum performance. The provision of oxygen and metabolic substrate inevitably requires a major role for the cardiovascular system in all types of activity. In fact under natural circumstances changes in the cardiovascular system scarcely ever occur in isolation but as part of a series of functional changes in response to each

6. THE HYPOTHALAMUS AND CARDIOVASCULAR REGULATION

stimulus. The integration of the numerous ongoing signals from both internal and external environment and the initiation of programmes which cater for a wide variety of needs each of which will require a unique pattern of cardiovascular changes is largely the role of the hypothalamus.

Despite the seeming enormity of this task, quantitatively the hypothalamus represents only a trifling portion of the whole brain lying below the thalamus on either side of the thin walled third ventricle from just anterior to the optic chiasma caudally to the mammillary bodies. Yet this small region is structurally complex, made up of several distinct groups of neurones (nuclei) as well as scattered neurones differing in morphology and neurochemistry. It has a network of internal and external connections including ones to endocrine organs via the pituitary gland. The complexity would at first sight appear to preclude making precise statements about the relation between structure and function but modern neuroanatomical, physiological and genetic approaches are providing a framework for understanding how the hypothalamus performs its functions. Since cardiovascular changes are pivotol to behaviour, the regions of the hypothalamus having connections to the brain and spinal cord neurones involved in control of the heart and circulation are a good starting point into unravelling the complexity.

'CARDIOVASCULAR' CONNECTING REGIONS OF THE HYPOTHALAMUS

It was the electrophysiological data provided by Magoun (1940) that gave the first clues to the presence of direct projections from the hypothalamus to autonomic areas in the lower brainstem and spinal cord. This was later confirmed in neuroanatomical studies by means of the retrograde chromatolytic cell degeneration technique or Nauta staining of degenerating axons (Smith, 1965; Szteyn et al., 1967). Using anterograde tracing of tritiated amino acid, Kuypers and Maisky (1975) demonstrated a projection to the spinal cord autonomic centres from the dorsal hypothalamus. Subsequently it was shown by retrograde transport of horseradish peroxidase that the paraventricular nucleus of the hypothalamus provided a major projection to sympathetic controlling regions of the spinal cord (Hancock, 1976; Saper et al., 1976; Hosoya and Matsushita, 1979). Subsequently the advent of better anterograde and retrograde (including retrograde transynaptic viral transport) tracing methods has led to more detail emerging on the extent of projections from different regions of the hypothalamus to brainstem and spinal cord areas involved in cardiovascular control (Swanson, 1987; Loewy, 1998). We are now aware that there are projections from the paraventricular nucleus, dorsomedial hypothalamus, lateral hypothalamus, posterior hypothalamus, arcuate nucleus, preoptic area and from neurones in the retrochiasmatic area to various groups of autonomic neurones in the brainstem and spinal cord (Holstege, 1987; Vertes and Crane, 1996). Furthermore, neuroimmunological methods have provided a description of multiple neuropeptides [arginine vasopressin, oxytocin, opioids, somatostatin, neurotensin, substance P, Orexin, cocaine and amphetamine regulated transcript (CART), agouti related protein (AGRP), and melatonin stimulating hormone (αMSH)] and dopamine associated with the extra hypothalamic projecting neurones (Buijs, 1978; Sawchenko and Swanson, 1982; De Vries et al., 1985; Strack et al., 1989a; Cechetto and Saper, 1988; Elias et al., 1998; van den Pol, 1999).

THE PARAVENTRICULAR NUCLEUS (PVN)

In the rat the PVN comprises approximately 21,500 neurons (Kiss et al., 1991) arranged into eight distinct groups of cells (subnuclei) (Swanson and Kuypers, 1980). The "large cell" division (magnocellular) consists of three subnuclei of 4500 cells of which 2700 are large cells that synthesize oxytocin (OT) and arginine vasopressin (AVP) (Armstrong et al., 1980; Sawchenko and Swanson, 1982; Kiss et al., 1991) and transport them to the posterior pituitary where they are released into the systemic circulation in response to appropriate stimuli. The synthesis and release of AVP is controlled by blood volume and tonicity and its major peripheral actions are to promote renal tubular reabsorption of water and to contract vascular smooth muscle. Oxytocin is best known in association with reproductive functions in the female animal, such as parturition and milk ejection, although it is co-secreted with AVP in response to osmotic and blood volume challenges (Hatton, 1990). However, it does contract blood vessels when present in relatively larger amounts (Richard et al., 1991) and hence can cause increases in blood pressure (Petty, 1987).

Smaller neurones form the parvocellular division of the PVN and can be divided into five subnuclei (anterior, medial, dorsal, lateral and periventricular) (Swanson and Kuypers, 1980). It is from amongst these neurones that projections arise to innervate different groups of autonomic neurones in the brainstem and spinal cord (Fig. 1A, B). The latter include the mid brain periaqueductal grey (PAG), parabrachial nucleus (PBN), rostral ventrolateral medulla (RVLM), caudal ventrolateral medulla (CVLM), nuclei tracti solitarii (NTS), dorsal vagal nucleus (DBN), nucleus ambiguus

6. THE HYPOTHALAMUS AND CARDIOVASCULAR REGULATION

and intermediolateral nucleus and associated sympathetic nuclei in the thoraco-lumbar spinal cord (Swanson and Kuypers, 1980; Luiten et al., 1985; Holstege, 1987).

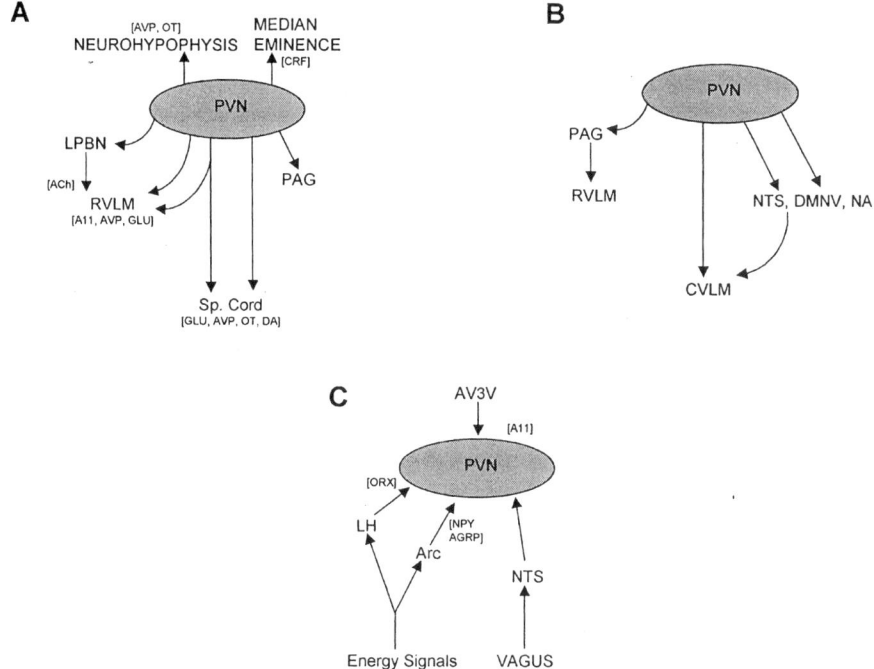

Figure 1. Diagram illustrating the principal autonomic efferent projections from the PVN A, B and the autonomic afferent inputs to PVN, C. PVN, paraventricular nucleus; LH, lateral hypothalamus; Arc, arcuate nucleus; AV3V, anteroventrolateral region of 3rd ventricle; LPBN, lateral parabrachial nucleus; RVLM, rostral ventrolateral medulla; CVLM caudal ventrolateral medulla; Sp Cord, thoraco-lumbar spinal cord; PAG, periaqueductal grey; NTS, nuclei tracti solitarii; DMNV, dorsal motor nucleus of vagus; NA, nucleus ambiguus; AVP, arginine vasopressin; CRF, corticotrophin releasing factor; AII, angiotensin II; GLU, glutamate; OT, oxytocin; DA, dopamine; ORX, Orexin, NPY, neuropeptide Y; AGRP, agouti related protein.

Some of the parvocellular neurones especially in the periventricular area contain corticotrophin releasing factor (CRF). Many of these project to the median eminence and release CRF into the hypophysial portal circulation to cause the release of adrenocortico tropic hormone (ACTH) from the adenohypophysis (Sawchenko, 1987a,b). Others act via projections within the brainstem to mediate or modulate transmission in pathways controlling cardiovascular function such as the RVLM (Milner et al., 1993).

PVN Projection to PAG

This projection is of interest because the mid brain PAG is clearly pivotol to the integration of a response to a threatening or noxious stimulus (Bandler et al., 1985; Hilton and Redfern, 1986; Yardley and Hilton, 1986; Lovick, 1991) and there is now clear evidence that the PVN participates in the response to stressful situations (Sawchenko et al., 1996). However, the evidence that the PVN plays a role in eliciting the classical defence reaction for 'fear, fight, flight' is sparse. The classical mapping of areas involved in the defence responses in the cat (Abrahams et al., 1960) using electrical stimulation revealed this was confined to a region surrounding the fornix particularly in the tuberal part of the hypothalamus. The distinctive pattern of cardiovascular changes associated with this response could not be obtained from the PVN. Since that time meticulous mapping of the hypothalamus with electrical or chemical stimulation has failed to evoke a cardiovascular response which is typical of that observed in defence-like behaviour (Bandler, 1982; Hilton and Redfern, 1986; Yardley and Hilton, 1986; Lovick, 1991). More recent data would suggest that the role of the PVN in emotional behaviour is to relay the signals of high stress conditions from the dorsomedial hypothalamic nucleus (DH) (Stotz-Potter et al., 1996) to the pituitary-adrenal cortical axis and the sympatho-adrenal axis (Heinrichs, 1999; Fontes et al., 2001).

PVN Projection to the Lateral Parabrachial Nucleus (LPBN)

The LPBN receives an innervation from the PVN, and ibotenic acid lesions of the LPBN or microinjections of the inhibitory amino acid GABA into LPBN reduce or abolish a blood pressure increase induced by PVN stimulation (Mortensen and Haywood, 1993; Kubo et al., 2000). This PVN-LPBN pressor effect is mediated via a cholinergic pathway from LPBN to RVLM (Kubo et al., 2000). Interestingly, microinjection of cholinergic antagonists into the RVLM prevents a defence type cardiovascular response elicited by electrical stimulation in the hypothalamus (Lin and Li, 1992; Li et al., 1995). Whether this response depended on a synapse in the LPBN or for that matter in the PAG, was not determined.

PVN Projection to the NTS

Parvocellular neurones of the PVN provide an extensive AVP and OT innervation of the NTS (Swanson, 1987). The majority of early evidence suggested this input was facilitatory to visceral afferent input from the abdominal vagus (Kannan and Yamashita, 1985; Rogers and Herman, 1985; Banks and Harris, 1987). However, there is also persuasive evidence that part of the PVN-NTS projection inhibits arterial baroreceptor reflex transmission since microinjection of the peptides AVP or OT causes an increase in blood pressure and heart rate (Matsuguchi et al., 1982). This result is a possible explanation for the very interesting observation that lesions of the PVN increase a baroreceptor-induced inhibition of lumbar sympathetic nerve activity (Darlington et al., 1988; Patel and Schmid, 1988). A further conclusion from these studies is that the PVN-NTS projection is tonically active. It was also shown to be selective since there was little effect on the cardiac components of the baroreceptor reflex. In this respect, recent studies appear to conflict. It was shown that vasopressin release in the NTS increases during exercise and this facilitates the tachycardia presumably by reducing the efficacy of the baroreceptor reflex (Dufloth et al., 1997). Additionally, a further study indicated that activation of oxytocin pathways to the NTS had the opposite effect during exercise (Braga et al., 2000).

PVN Projections to Cardiac Vagal Neurons

Microinjections of l-glutamate at a number of sites in the PVN parvocellular sub nuclei elicit a profound bradycardia which is markedly reduced by intravenous atropine and abolished by ganglion blockade, hence it is predominantly due to activation of cardiac vagal neurones (Darlington et al., 1989). This effect is likely due to activation of PVN-OT neurones which form a major projection to the cardiac vagal nuclei, dorsal motor nucleus of the vagus (DMNV) and nucleus ambiguus (NA) in the medulla (Buijs, 1978; Nilaver et al., 1980; Swanson and Kuypers, 1980; Sawchenko and Swanson, 1982; Lang et al., 1983; Luiten et al., 1985). It was shown by Rogers and Hermann (1985, 1986) that microinjections of OT but not AVP into the DMNV caused a marked decrease in heart rate, and this was mimicked by stimulation in the PVN; both effects being blocked by an OT antagonist.

PVN Projections to Ventral Medulla

The two regions in the ventral medulla that are important in cardiovascular control are the rostral and caudal groups of neurones. The RVLM is the site of the reticulo-spinal premotor neurones supplying vasomotor neurones in the spinal cord (Ross et al., 1984; Brown and Guyenet, 1985; Dampney, 1994a). These RVLM neurones are inhibited by neurones in the adjacent caudal medullary area, the CVLM, which form part of the baroreceptor reflex pathway (Argawal et al., 1991; Li et al., 1991; Masuda et al., 1991). The whole region receives a fairly heavy innervation from the PVN, however the first clear anatomical evidence that this was directed to RVLM vasomotor neurones came from a study showing a close association of labelled PVN terminals with identified RVLM spinal neurones retrogradely labelled from the thoracic cord (Pyner and Coote, 1999; 2000). This confirmed electrophysiological studies on RVLM-spinal vasomotor neurones which indicated a direct PVN synaptic projection (Yang and Coote, 1998) which was mediated via the release of vasopressin or glutamate (Yang et al., 2001). That the AVP innervation could be important in the RVLM was first shown by Gomez et al. (1993), who demonstrated AVP synapses on identified RVLM-spinal neurones at the ultrastructural level. As well as glutamate and vasopressin neurons, other PVN neuronal phenotypes appear to be important in the RVLM. Terminals of PVN neurones containing CRF have been demonstrated in the RVLM region, and bilateral microinjection of CRF into this region increases blood pressure (Milner et al., 1993) probably via CRF receptors on vasomotor neurones (De Souza et al., 1985). There is also evidence that the PVN can excite neurones in the RVLM by activation of angiotensin receptors (Tagawa and Dampney, 1999; Tagawa et al., 2000) but how direct this pathway is, is not clear. With regard to the CVLM, baroreceptor-excited neurones can be synaptically excited by stimulation of PVN neurones (Yang and Coote, 1999) indicating that the latter can influence the baroreceptor pathway at this site as well as at the NTS (see earlier).

PVN Projection to Spinal Sympathetic Neurons

A variety of studies have shown a substantial direct PVN projection to the spinal cord regions providing the sympathetic outflow to heart and blood vessels (Holstege, 1987; Swanson, 1987; Strack et al., 1989a, b; Hosoya et al., 1991; Shafton et al., 1998; Pyner and Coote, 2000) and this has been confirmed electrophysiologically by antidromic activation of PVN neurones from the spinal cord (Caverson et al., 1984; Yamashita et al., 1984; Lovick and Coote, 1988a, b). Definitive evidence that these PVN projections make monosynaptic connections has been obtained in an ultrastructural study of labelled superior cervical ganglion-projecting sympathetic preganglionic neurones (Hosoya et al., 1995) and close contacts have been shown at the light microscope level on soma and dendrites of sympathetic preganglionic neurones targeted to adrenal medulla, and stellate ganglia (Ranson et al., 1998; Motawei et al., 1999). The general picture has been confirmed in elegant studies of labelling in the PVN following retrograde transynaptic transport of pseudorabies virus developed by Loewy and his group (Strack and Loewy, 1990). Studies of this sort have revealed the target destination of PVN neurones and indicate there is some sort of topographic arrangement with PVN neurones projecting to the superior cervical ganglias, adrenal medulla, kidney and coeliac ganglia lying dorsally in the PVN, whereas those to stellate and heart located more ventrally and medially (Strack et al., 1989b; Schramm et al., 1993; Ter Horst et al., 1993; Jansen et al., 1995; Smith et al., 1998; Huang and Weiss 1999).

Immunohistochemical and in situ hybridisation studies have demonstrated the presence of several neurotransmitter candidates within PVN neurones that project to the spinal cord. Neurones expressing the peptides AVP (25-40%) and OT (20-30%) form a prominent part of this with enkephalin and dopamine contributing to a lesser extent (Sawchenko and Swanson, 1982; Cechetto and Saper, 1988; Hallbeck and Blomquist, 1999; Huang and Weiss, 1999).

There is now convincing evidence that some of these neurochemicals act as neurotransmitters in the PVN-spinal sympathetic pathways. In anaesthetised animals, AVP or OT applied iontophoretically alter the firing rate of cardiovascular-like sympathetic preganglionic neurones (Gilbey et al., 1982; Backman and Henry, 1984). Also, AVP or OT given intrathecally to the thoracic cord increase activity in renal sympathetic nerves (Porter and Brody, 1986; Tan and Tsou, 1986; Riphagen and Pittman, 1989; Malpas and Coote, 1994; Yang et al., 2002). Both peptides depolarise sympathetic preganglionic neurones recorded in vitro in slices of spinal cord (Ma and Dun, 1985; Sermasi and Coote, 1994; Desaulles et al., 1995; Kolaj and Renaud, 1998). These effects are produced by selective stimulation of the relevant V_{1A} and OT receptors expressed by sympathetic preganglionic neurones (Reiter et al., 1994; Sermasi et al., 1998). Release of an AVP-like peptide has been shown in perfusates of spinal cord following stimulation of the PVN (Pittman et al., 1984), and pressor responses or increases in renal sympathetic nerve activity elicited by PVN stimulation can be selectively blocked by intrathecally applied V_{1a} antagonist (Riphagen and Pittman,

1989; Malpas and Coote, 1994; Yang et al., 2002). Similar studies have so far failed to identify a PVN-oxytocin dependent effect on renal sympathetic outflow (Yang et al., 2002). It could be that the OT pathway terminates on different target specified sympathetic neurones compared to the AVP pathway. There is some evidence to support a selective innervation. Oxytocin fibres innervate sympathetic preganglionic neurones targeted to superior cervical ganglia (Hosoya et al., 1995) yet appear to avoid those targeted to the adrenal medulla (Appel and Elde, 1988). There also appears to be a difference in the sensitivity of sympathetic neurones to the peptides: some neurones in the upper thoracic segments display selective action to OT (Desaulles et al., 1995), whereas on sympathetic neurones in the lower thoracic segments which are especially sensitive to vasopressin, OT appears to only have a depolarising action via the V_{1a} receptor (Sermasi and Coote, 1994).

In regard to dopamine, a recent study indicates that sympatho-inhibitory effects, elicited by stimulation at a few sites in the caudal PVN, were blocked selectively at the spinal level by a dopamine D_1 antagonist (Yang et al., 2002).

THE LATERAL HYPOTHALAMUS

This region of the diencephalon extends from the lateral preoptic area to the ventral tegmental area lying in and around the medial forebrain bundle (Saper et al., 1979). It contains neurones projecting to the sympathetic lateral cell column in the spinal cord, vagal preganglionic nuclei (DMNV and NA) as well as other central autonomic sites such as the PAG, LPBN, NTS and RVLM (Hosoya and Matsushita, 1981; Ter Horst et al., 1984; Holstege, 1987; Allen and Cechetto, 1991; Horvath et al., 1999; van den Pol 1999; Date et al., 2000). A recently discovered neuropeptide, orexin, is expressed exclusively in neurones in the perifornical area and lateral hypothalamus, and orexin-containing neurones provide a substantial innervation of sympathetic preganglionic neurones in the spinal cord (De Lecea et al., 1998; van den Pol, 1999; Date et al., 2000) as well as projecting to the RVLM, NTS, DMNV and internally to several hypothalamic nuclei such as arcuate, PVN, ventromedial nucleus and dorsomedial area (Peyron et al., 1998; Horvath et al., 1999; van den Pol et al., 1999; Date et al., 2000) (Fig. 2B). There are no studies so far showing the effects of selective activation of the lateral hypothalamic neurones, however orexin applied to sympathetic preganglionic neurones recorded in vitro in slices of rat spinal cord depolarises them (van den Top et al., 2000; Antunes et al., 2001) and centrally administered orexins increase heart rate, arterial blood pressure and

6. THE HYPOTHALAMUS AND CARDIOVASCULAR REGULATION

sympathetic nerve activity (Swanson et al., 1999; Shirasaka et al., 1999; 2002; Chen et al., 2000; Antunes et al., 2001; Matsumura et al., 2001).

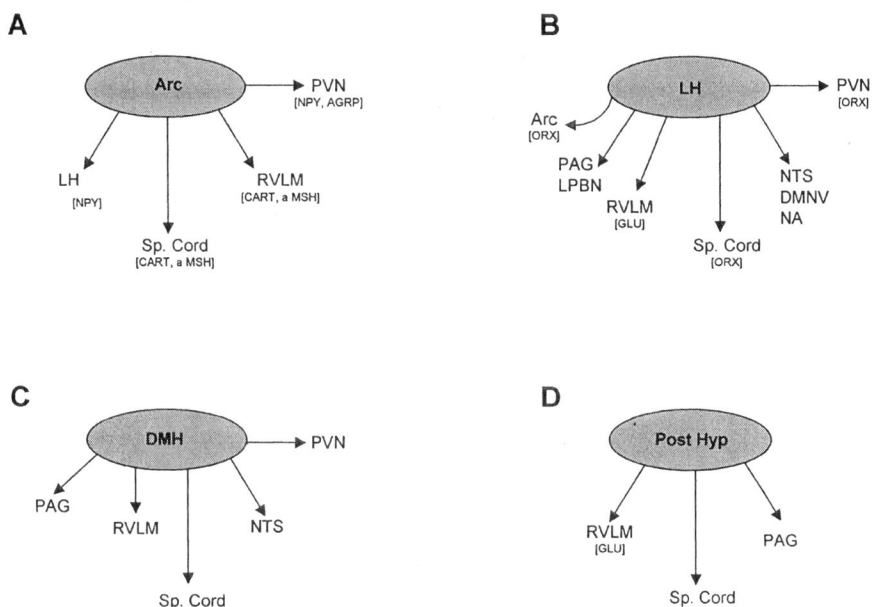

Figure 2. Diagram illustrating the principal cardiovascular-related efferent projections from the arcuate nucleus (Arc) A, the lateral hypothalamus (LH) B, the dorsomedial hypothalamic nucleus (DMH) C and the posterior hypothalamic region (post. Hyp) D. CART, cocaine and amphetamine regulated transcript; αMSH, αmelatonin stimulating hormone; other abbreviations as in Fig. 1.

Both pressor and depressor effects can be evoked by activating neurones in the lateral hypothalamus (Sun and Guyenet, 1986; Spencer et al., 1989; Allen and Cechetto, 1991, 1993). In the study by Sun and Guyenet (1986), an increase in blood pressure and lumbar sympathetic activity was accompanied by activation of RVLM-spinal vasomotor neurones and prevented by the glutamate receptor antagonist kynurenic acid applied at the RVLM synapse. In contrast, Spencer et al. (1989) studied depressor effects produced by stimulation of neurones in the tuberal region of the lateral hypothalamus and showed this effect was caused by a slowing of the heart and a decrease in cardiac output mainly due to activating cardiac vagal neurones most likely via the direct connections of this region with DMNV and NA in the medulla (Ter Horst et al., 1984).

The lateral hypothalamus is of particular interest because early studies in the cat showed that electrical stimulation in the perifornical area elicits the cardiovascular and behavioural components of the defence reaction (Hilton, 1980). However, these effects could not be repeated with chemical stimulation to excite neurones (Bandler, 1982; Hilton and Redfern, 1986) so the response to electrical stimulation was explained as due to activating fibres of passage on the way to the well described 'defence' area in the PAG. However, another possibility is that hypothalamic neurones involved with 'defence' response are tonically inhibited by GABA interneurones and these are further activated by microinjection of excitatory amino acids, so the response to direct excitation is depressed. Some support for this explanation is provided by a study of Di Scarla et al. (1984), who showed that microinjection of a GABA antagonist into the perifornical area of rats resulted in 'flight like' behaviour; unfortunately no cardiovascular data were obtained.

THE ARCUATE NUCLEUS

Although not so far implicated in circulatory control, the arcuate nucleus and the nearby retrochiasmatic area is of interest because it sends a direct projection to sympathetic neurones in the spinal cord (Cechetto and Saper, 1988) and RVLM (Fig. 2A). Recently this projection has been associated with a novel neuropeptide known as CART (Elias et al., 1998) which is expressed in terminals in the rat sympathetic nuclei of the spinal cord (Dun et al., 2000). Although there is still much to learn about the functional significance of this innervation, it was recently shown that the metabolic regulatory protein, leptin, activates these spinally-projecting CART neurones (Elias et al., 1998) and therefore this pathway is likely to be responsible for the pressor effects produced by intracerebroventricular administration of leptin (Casto et al., 1998) and increases in sympathetic activity associated with raised leptin levels.

The cardiovascular effects of activating arcuate neurones are probably secondary to the strategic role they possess in integrating a host of signals related to energy homeostasis (Oomura et al., 1969; Spanswick et al., 1997; 2001; Beck, 2001; Cowley et al., 2001).

THE POSTERIOR AREAS OF THE HYPOTHALAMUS

Early studies on the hypothalamus concluded that the posterior region was concerned with depressor responses but there is little to support that now (Dampney, 1994b).

The early view is surprising since it has long been known that stimulation at an ill-defined area in the posterior hypothalamus known as the Field of Forel, a sub thalamic area, elicits a pattern of cardiovascular changes accompanied by limb movements similar to that seen during voluntary exercise (Smith et al., 1960; Marshall and Timms, 1979; Eldridge et al., 1981, 1985; Ohta et al., 1985; Waldrop et al., 1988). There are also several studies showing that chemical activation of posterior hypothalamic neurones can increase blood pressure and heart rate (Di Micco et al., 1986; Di Micco and Abshire, 1987; Martin et al., 1988; Spencer et al., 1990). There are no data on how the cardiovascular changes are induced but it is quite likely that the direct projection of neurones in this area to the RVLM (Fig. 2D) are involved (Vertes and Crane, 1996).

DORSOMEDIAL HYPOTHALAMUS

Sympathetically-mediated increases in blood pressure and heart rate can be evoked by activation of neurones in the dorsomedial hypothalamus and these changes are accompanied by neuroendocrine, gastrointestinal and behavioural changes similar to those we associate with emotional stress (Di Micco et al., 1991, 1996; De Novellis et al., 1995; Greenwood and Di Micco, 1995; Stotz-Potter et al., 1996a, 1996b). Although there are neurones projecting to the PAG and NTS and even a few to the spinal cord (Hosoya et al., 1987; Ter Horst and Luiten, 1988; Thompson et al., 1996) as well as a heavy projection to the PVN (Fig. 2C), a recent study showed inhibition of these groups did not eliminate the cardiovascular changes produced by stimulating dorsomedial nuclei (Fontes et al., 2001). It was only when the RVLM was suppressed that the cardiovascular changes were attenuated. The authors concluded that the cardiovascular effects associated with the emotional stress in their experiments depended on a descending pathway from the dorsomedial hypothalamus to the RVLM, which they were able to demonstrate anatomically using retrograde tracing. However, the tachycardic component associated with DMH activation is more likely to depend on a direct projection from DMH to raphe pallidus neurones (Hosoya et al., 1987; Ter Horst and Luiten, 1988), connected to spinal cardioacceleratory neurones (Jansen et al., 1995), since it is strongly suppressed by microinjection of the inhibitory agonist muscimol into raphe pallidus (Samuels et al., 2002). The lack of direct involvement of the PVN in the responses confirmed a previous observation that the inhibitory agonist muscimol acts in the DH but not the PVN to suppress the cardiovascular effects of stress produced by an air jet (Stotz-Potter et al., 1996b). This conclusion is in contrast with that of several other studies (Callahan et al., 1989, 1992; Sawchenko et al., 1996) and has been explained as possibly due to lesions of the PVN in these studies causing collateral damage to the dorsomedial hypothalamus. Nonetheless oxytocin neurons in the PVN may well be activated during stress situations (Callahan et al., 1989; Piekut et al., 1996) as a consequence of inputs from dorsomedial hypothalamic nucleus.

Preoptic Area

It is well established that the preoptic area of the hypothalamus plays a critical role in thermoregulation (Kanosue et al., 1998). In terms of cardiovascular influence, neurones from this region make direct and indirect connections with the RVLM and raphe pallidus nucleus of the medulla, where there are cutaneous vasoconstrictor neurones (Rathner et al., 2001). It also has recently been shown that a vasodilator response in the tail circulation of the rat to thermal stimulation of the preoptic area, is mainly dependent on inhibition of raphe pallidus neurones (Tanaka et al., 2002).

Functional Role of the Paraventricular Nucleus

Of all the groups of neurones in the hypothalamus, those comprising the paraventricular nucleus are probably the most crucial for cardiovascular regulation. Their integrity is pivotal for blood volume regulation. Lesions which destroy a substantial portion of PVN-spinal parvocellular neurones reduce the renal response to plasma volume expansion (Lovick et al., 1993; Haselton et al., 1994). PVN-spinal neurones are activated or inhibited following stimulation of cardiac afferents or arterial baroreceptors (Lovick and Coote, 1988a, 1988b) or by circulating ANF (Lovick and Coote, 1989) and by changes in plasma osmolality or angiotensin II acting via the circumventricular organs (Schad and Seller, 1975; Ericson and Sjoquist, 1982; Ferrario et al., 1987; Weiss et al., 1996; May et al., 2000; Chen and Toney, 2001; Tanaka et al., 2000). Not all PVN-spinal neurones respond to the same afferent input nor in the same way to a single afferent input (Lovick and Coote, 1988a, b). Thus it is suggested that there are two groups, a vasomotor-like group and a non vasomotor-like group (Chen and Toney, 2003). Furthermore a recent study showed that the angiotensin AT1 receptor is present on neurones in the PVN but it appears not to be associated with the extra hypothalamic neurones but coexists strongly with neurones projecting to the median eminence (Ferguson 1988; Oldfield et al., 2001). Stimulation of volume receptors by distending the right atrial-caval junction with a balloon catheter or by plasma volume expansion, which appears to not activate NTS pathways directly projecting to the vasomotor neurones in the ventral medulla (Shafton et al., 1999), activates the early gene c-fos in PVN parvocellular neurones (Deng and Kaufman, 1995; Badoer et al., 1997; Pyner et al., 2002). Haemorrhage or hypotension and increases in plasma osmolality also activate a population of parvocellular neurones (Oldfield et al., 1991; Sharp et al., 1991; Badoer et al., 1992, 1993, 2002; McAllen et al., 1992; Shen et al., 1992; Chiu et al., 1994; Li and Dampney, 1994; Petrov et al., 1995; Smith et al., 1995; Krukoff et al., 1997; Badoer and Merolli, 1998; Ding et al., 1999; Xu et al., 2000), amongst which are vasopressin neurones that synapse in the RVLM (Gomez et al., 1993). Furthermore, activation of PVN neurones can either result in general excitation of a variety of sympathetic nerves supplying the heart and different vascular beds or lead to a 'volume load'-related pattern consisting of inhibition of activity in renal sympathetic nerves together with increases in activity in cardiac and other sympathetic nerves (Kannan et al., 1987; Katafuchi et al., 1988; Lu et al., 1990; Deering and Coote, 2000; Li et al., 2001). There is also now good evidence that PVN neurones projecting to the RVLM and spinal cord are excited by hyperosmotic solutions injected into the internal carotid artery (Chen and Toney, in press). This effect may be mediated via the subfornical input to PVN-spinal neurones (Bains and Ferguson, 1995).

These data support the contention that the PVN-parvocellular neurones (as well as the magnocellular neurones projecting to the posterior pituitary) are a target for afferent signals related to fluid balance which are then translated into an appropriate cardiovascular response (Haselton and Vari, 1998; Tanaka et al., 2000; Deering and Coote, 2000; Pyner et al., 2002). It is, therefore, hardly surprising that changes occur in the control of the PVN neurones in disturbances of fluid balance associated with pregnancy (Deng and Kaufman, 1995, 1998) during which there is a marked increase in blood volume (Lindheimer and Katz, 1985) accompanied by sodium retention (Atherton et al., 1982). Such a stimulus would normally, via atrial-caval distension, decrease renal sympathetic nerve activity (Kopp et al., 1987; Pyner et al., 2002) and increase urine output and sodium loss in virgin rats (Kaufman and Deng, 1993; Patel and Zhang, 1993). The pregnant animal thus fails to recognise an increased venous return to the heart suggesting there is a perceived underfilled state of the vascular compartment in pregnancy. Tam and Kaufman (2002) found that inhibition of NO synthesis with L-NAME given systemically via an osmotic minipump restored the renal response to atrial distension in pregnant rats, suggesting that upregulation of NO production was involved in decreasing the efficacy of the 'volume' reflex. Where in the reflex pathway this effect is occurring was not determined but it seems unlikely that it is at the level of the PVN since here NO via stimulation of GABA release reduces sympathetic activity to the kidney and this presumably would enhance the 'volume' reflex (Zhang et al., 1997; Zhang and Patel, 1998).

Further testament to the importance of the plasma volume regulatory role of the PVN for circulatory homeostasis is provided by studies of heart failure and hypertension. In a coronary occlusion rat model of heart failure there is a down regulation of NO synthesis and reduced GABA activity in the PVN resulting in a raised sympathetic activity (Zhang et al., 2001, 2002; Li et al., 2002). In genetic strains of hypertensive rats and in DOCA salt hypertensive rats, lesioning the PVN returns blood pressure and sympathetic activity to normal values (Ciriello et al., 1984; Nakata et al., 1989; Takeda et al., 1991). Again, reduction in the efficacy of GABA neurones in PVN are implicated (Unger et al., 1984; Allen, 2002) and the increased sympathetic drive appears to be due to PVN pathways activating angiotensin AT1 receptors in the RVLM (Tagawa et al., 2000; Di Bona and Jones, 2001; Allen, 2002).

These data emphasise the dominant influence the PVN has in plasma volume regulation. So why are PVN neurones also activated by systemic hypoxia (Sica et al., 2000; Olivan et al., 2001), by various stressors (Callahan et al., 1992; Senba et al., 1993; Piekut et al., 1996; Sawchenko et al., 1996), and by stimuli related to thermal and energy balance (Atrens and Menendez, 1993; Temple and Leibowitz, 1993; Dunn-Meynell et al., 1997; Dube et al., 1999; Shirasaka et al., 2002)?

A possible answer is that each of these challenges requires a redistribution of blood volume favouring those tissues or organs whose activity is strategically necessary for each situation. As a consequence there are a host of connections between groups of neurones within the hypothalamus and the PVN. Associated with the connections are a variety of neuropeptides such as, orexin and CART from the lateral hypothalamus; NPY, AGRP and αMSH from the arcuate nucleus; catecholaminergic inputs from the lower brainstem; angiotensin from the lamina terminalis and substance P from a variety of sources including the DH (McKellar and Loewy, 1981; Cechetto and Saper, 1988; Strack et al., 1989a; Bittencourt et al., 1991; Elias et al., 1998; Hallbeck and Blomqvist, 1999; van den Pol, 1999; Cole and Sawchenko, 2002). Such biochemically defined sub-populations of neurones projecting to PVN appear to correspond to defined functions, for example orexin to feeding, angiotensin to fluid balance and substance P to stress (Culman and Ungar, 1995; Maier et al., 1998; Swanson et al., 1999; May et al., 2000). However, the extent to which this apparent chemical coding explains the unique patterns of non-uniform sympathetic activity related to behaviour is lessened by the general excitatory action of the neuropeptides. There therefore must be an additional feature in the organisation that contributes to differential regulation of the sympathetic cardiovascular outflows. This could be activation of inhibitory interneurones to specific sympathetic pathways selected according to the source and function of an afferent nerve signal. Such an arrangement could explain the effect of stimulating atrial receptors by increases in plasma volume load. The response to this stimulus uniquely consists of an inhibition of renal sympathetic nerve activity accompanying an increase in other sympathetic activity (Deering and Coote, 2000; Pyner et al., 2002), which is mediated via the PVN (Lovick et al., 1993; Deng and Kaufman, 1995; Haselton et al., 1994; Pyner et al., 2002). Furthermore, preliminary data indicates that the inhibition of renal nerve activity is blocked by microinjecting the GABA antagonist bicuculline into the PVN (Yang and Coote, 2003).

More than 50% of the local synaptic input to PVN neurones is GABAergic (Decavel and van den Pol, 1990; Tasker and Dudek, 1993) and blocking this influence with bicuculline leads to increases in sympathetic activity in renal, splenic and lumbar nerves (Kenny et al., 2001). This indicates a non-discriminatory organisation of inhibitory interneurones in the PVN. Therefore non-uniform patterns of sympathetic activation/inhibition would require specific functional and anatomical connections of the afferent input. Hence the pattern of cardiovascular events associated with different behavioural responses which are integrated in the hypothalamus are likely to depend on different signals from modality-specific sources. Such an arrangement mirrors that established for somatic sensation where different stimulus modalities are transmitted by specific chains of neurones connecting to discrete regions of the thalamus and cerebral cortex.

Therefore it may be better to consider the groups of neurones in the hypothalamus as command centres, each responding to modality-specific afferent input. The output from such groups of neurons, which may be phenotypically distinguished by a chemical coding, could then ensure appropriate somatic and autonomic changes occur in target organs necessary for the fulfillment of the response. We should consider that this will also involve an appropriate pattern of cardiovascular changes which might indirectly or directly involve the PVN, since all types of behaviour require a redistribution of blood flow to the key target organs involved in the activity.

REFERENCES

Abrahams, C.V., Hilton, S.M., Zbrozyna, A., 1960. Active muscle vasodilatation produced by stimulation of the brainstem: its significance in the defence reaction. J. Physiol. 154, 491-513.

Allen, G.V., Cechetto, D.F., 1991. Functional and anatomical organisation of cardiovascular pressor and depressor sites in the lateral hypothalamic area 1. Descending projections. J. Comp. Neurol. 314, 1-20.

Allen, G.V., Cechetto, D.F., 1993. Functional and anatomical organisation of cardiovascular pressor and depressor sites in the lateral hypothalamic area. Ascending projections. J. Comp. Neurol. 330, 421-438.

Allen, R.M., 2002. Inhibition of the hypothalamic paraventricular nucleus in spontaneously hypertensive rats dramatically reduces sympathetic vasomotor tone. Hypertension 39, 275-280.

Antunes, V.R., Brailoiu, G.C., Kwok, E.H., Scruggs, P., Dun, N.J., 2001. Orexins/hypocretins excite rat sympathetic preganglionic neurones in vivo and in vitro. Am. J. Physiol. 281, R1801-R1807.

Appel, N.M., Elde, R.P., 1988. The intermediolateral cell column of the thoracic spinal cord is comprised of target specific subnuclei evidence from retrograde transport studies and immunohistochemistry. J. Neurosci. 8, 1767-1775.

Argawal, S., Gelsema, A.J., Calaresu, F.R., 1991. Inhibition of rostral VLM by baroreceptor activation is relayed through caudal VLM. Amer. J. Physiol. 258, R1271-R1278.

Armstrong, W.E., Warach, S., Hatton, G.I., McNeil, T.H., 1980. Subnuclei in the rat hypothalamic paraventricular nucleus: A cytoarchitectural horseradish peroxidase and immunocytochemical analysis. Neurosci. 5, 1931-1958.

Atherton, J.C., Dark, J.M., Garland, H.O., Morgan, M.R.A., Pigeon, J., Soni, S., 1982. Changes in water and electrolyte balance, plasma volume and composition during pregnancy in the rat. J. Physiol. 330, 81-93.

Atrens, D.M., Menendez, J.A., 1993. Metabolic modulation by amino acid stimulation of paraventricular nucleus of the hypothalamus. Pharmacol. Biochem. Behav. 46, 617-622.

Backman, S.B., Henry, J.L., 1984. Effect of oxytocin and vasopressin on thoracic sympathetic preganglionic neurones in the cat. Brain Res. Bull. 13, 679-684.

Badoer, E., McKinley, M.J., Oldfield, B., McAllen, R.M., 1992. Distribution of hypothalamic, medullary and lamina terminalis neurons expressing fos after haemorrhage in conscious rats. Brain Res. 582, 323-328.

Badoer, E., McKinley, M.J., Oldfield, B.J., McAllen, R.M., 1993. A comparison of hypotensive and non-hypotensive haemorrhage on fos expression in spinally projecting neurons of the paraventricular nucleus and rostral ventrolateral medulla. Brain Res. 610, 216-223.

Badoer, E., McKinlay, D., Trigg, L., McGrath, B.P., 1997. Distribution of activated neurons in the rabbit brain following volume load. Neurosci. 81, 1065-1077.

Badoer, E., Merolli, J., 1998. Neurons in the hypothalamic paraventricular nucleus that project to the rostral ventrolateral medulla are activated by haemorrhage. Brain Res. 792, 317-320.

Badoer, E., Ng, C.W., De Matteo, R., 2002. Tonic sympathoinhibition arising from the hypothalamic PVN in conscious rabbit. Brain Res. 947, 9-16.

Bains, J.S., Ferguson, A.V., 1995. Paraventricular nucleus neurons projecting to the spinal cord receive excitatory input from the subfornical organ. Am. J. Physiol. 268, R625-R633.

Bandler, R., 1982. Induction of 'rage' following microinjection of glutamate into midbrain but not hypothalamus of cats. Neurosci. Lett. 30, 183-188.

Bandler, R., Depaulis, A., Vergnes, M., 1985. Identification of midbrain neurones mediating defensive behaviour in the rat by microinjections of excitatory amino acids. Behav. Brain Res. 15, 107-119

Banks, D., Harris, M.C., 1987. Activation within dorsal medullary nuclei following stimulation in the hypothalamic paraventricular nucleus in rats. Pflugers Arch. 408, 619-627.

Beck, B., 2001. KO's and organisation of peptidergic feeding behaviour mechanisms. Neurosci. Behav. Rev. 25, 143-158.

Bittencourt, J.C., Benoit, R., Sawchenko, P.E., 1991. Distribution and origins of substance P-immunoreactive projections to the paraventricular and supraoptic nuclei – partial overlap with ascending catecholaminergic projections. J. Chem. Neuroanat. 4, 63-78.

Braga, D.C., Mori, E., Higa, K.T., Morris, M., Michelini, L.C., 2000. Central oxytocin modulates exercise-induced tachycardia. Am. J. Physiol. 278, R1474-R1482.

Brown, D.L., Guyenet, P.G., 1985. Electrophysiological study of cardioascular neurons in the rostral ventrolateral medulla in rats. Circ. Res. 56, 359-369.

Buijs, R.M., 1978. Intra- and extrahypothalamic vasopressin and oxytocin pathways in the rat. Cell. Tiss. Res. 192, 423-435.

Callahan, M.F., Kirby, R.F., Cunningham, J.T., Eskridge-Sloop, S.L., Johnson, A.K., McCarty, R., Gruber, K.A., 1989. Central oxytocin systems may mediate a cardiovascular response to acute stress in rats. Am. J. Physiol. 256, H1369-H1377.

Callahan, M.F., Thore, C.R., Sundberg, D.K., Gruber, K.A., O'Steen, K., Morris, M., 1992. Excitotoxin paraventricular nucleus lesions: stress and endocrine reactivity and oxytocin mRNA levels. Brain Res. 597, 8-15.

Casto, R.M., VanNess, J.M., Overton, J.M., 1998. Effects of central leptin administration on blood pressure in normotensive rats. Neurosci. Lett. 246, 29-32.

Caverson, M.M., Ciriello, J., Calaresu, F.R., 1984. Paraventricular nucleus of the hypothalamus: an electrophysiological investigation of neurones projecting directly to intermediolateral nucleus in the cat. Brain Res. 305, 380-383.

Cechetto, D.F., Saper, C.B., 1988. Neurochemical organisation of the hypothalamic projection to the spinal cord in the rat. J. Comp. Neurol. 272, 579-604.

Chen, Q.H., Toney, G.M., 2001. AT_1-receptor blockade in the hypothalamic PVN reduces central hyperosmolality-induced renal sympathoexcitation. Am. J. Physiol. 281, R844-R855.

Chen, Q.H., Toney, G.M., 2003. Identification and characterization of two functionally distinct groups of spinal cord-projecting PVN neurones with sympathetic related activity. Neurosci. 118, 797-807.

Chen, Q.H., Toney, G.M., 2003. Discharge properties and osmotic responsiveness of paraventricular nucleus neurons projecting to the rostral ventrolateral medulla. J. Neurophysiol. In Press.

Chen, C.T., Hwang, L.L., Chang, J.K., Dun, N.J., 2000. Pressure effects of orexins injected intracisternally and to rostral ventrolateral medulla of anaesthetised rats. Am. J. Physiol. 278, R692-R697.

Chiu, T.H., Dun, S.L., Tang, H., Dun, N.J., 1994. C-fos antisense attenuates Fos expression in rat central neurons induced by haemorrhage. Neuroreport 5, 2178-2180.

Ciriello, J., Kline, R.L., Zhang, T.X., Caverson, M.M., 1984. Lesions of the paraventricular nucleus alter the development of spontaneous hypertension in the rat. Brain Res. 310, 355-359.

Cole, R.L., Sawchenko, P.E., 2000. Neurotransmitter regulation of cellular activation and neuropeptide gene expression in the paraventricular nucleus of the hypothalamus. J. Neurosci. 22, 959-969.

Cowley, M.A., Smart, J.L., Rubinstein, M., Cerdan, M.G., Diano, S., Horvath, T.L., Cone, R.D., Low, M.J., 2001. Leptin activates anorexigenic POMC neurons through a neural network in the arcuate nucleus. Nature 411, 480-484.

Culman, J., Unger, T., 1995. Central tachykinins: mediators of defence reaction and stress reactions. Can. J. Physiol. Pharmacol. 73, 885-891.

Dampney, R.A.L., 1994a. The subretrofacial vasomotor nucleus: anatomical, chemical and pharmacological properties and role in cardiovascular regulation. Prog. Neurobiol. 42, 197-227.

Dampney, R.A.L., 1994b. Functional organisation of central pathways regulating the cardiovascular system. Physiol. Rev. 14, 323-374.

Darlington, D.N., Miyamoto, M., Keil, L.C., Dallman, M.F., 1989. Paraventricular stimulation with glutamate elicits bradycardia and pituitary responses. Am. J. Physiol. 256, R112-R119.

Darlington, D.N., Shinsako, J., Dallman, M.F., 1988. Paraventricular lesions: Hormonal and cardiovascular responses to haemorrhage. Brain Res. 439, 289-301.

Date, Y., Mondal, M.S., Matsukura, S., Nakazato, M., 2000. Distribution of orexin-A and orexin-B (hypocretins) in the rat spinal cord. Neurosci Lett. 288, 87-90.

De Lecea, L., Kilduff, T.S., Peyron, C., Gao, X.B., Foyer, P.E., Danielson, P.E., Fukuhara, C., Battenberg, E.L.F., Gautvik, V.T., Bartlett 11 F.S., Frankel, W.N., van den Pol, A.N., Bloom, F.E., Gautvik, K.M., Sutcliffe, J.G., 1998. The hypocretins: hypothalamics – specific peptides with neuroexcitatory activity. Proc. Natl. Acad. Sci. 95, 322-327.

De Novellis, V., Stotz-Potter, E.H., Morris, S.M., Rossi, F., Di Micco, J.A., 1995. Hypothalamic sites mediating cardiovascular effects of microinjected bicuculline and EAAs in rats. Am. J. Physiol. 269, R131-R140.

De Souza, E.B., Insel, T.R., Perrin, M.H., Rivier, J., Vale, W.N., Kuhar, M.J., 1985. Corticotropin-releasing factor receptors are widely distributed within the rat central nervous system: an autoradiographic study. J. Neurosci. 5, 3189-3203.

De Vries, G.J., Buijs, R.M., van Leeuwen, F.W., Caffe, A.R., Swaab, D.F., 1985. The vasopressinergic innervation of the brain in normal and castrated rats. J. Comp. Neurol. 233, 236-254.

Deering, J., Coote, J.H., 2000. Paraventricular neurones elicit a volume expansion-like change of activity in sympathetic nerves to heart and kidney in the rabbit. Exp. Physiol. 85, 177-186.

Deng, Y., Kaufman, S., 1995. Effect of pregnancy on activation of central pathways following atrial distension. Am. J. Physiol. 269, R552-R556.

Deng, Y., Kaufman, S., 1998. Pregnancy-induced changes in central response to atrial distension mimicked by progesterone metabolite. Am. J. Physiol. 275, R1875-R1877.

Desaulles, E., Reiter, M.K., Feltz, P., 1995. Electrophysiological evidence for oxytocin receptors on sympathetic preganglionic neurones – an in vitro study on the neonate rat. Brain Res. 699, 139-142.

Di Bona, G.F., Jones, S.Y., 2001. Effect of dietary sodium intake on the responses to bicuculline in the paraventricular nucleus of rats. Hypertension 38, 192-197.

6. THE HYPOTHALAMUS AND CARDIOVASCULAR REGULATION

Di Micco, J.A., Abshire, V.M., 1987. Evidence for GABAergic inhibition of hypothalamic sympathoexcitatory mechanism in anaesthetised rats. Brain Res. 402, 1-10.
Di Micco, J.A., Abshire, V.M., Hankins, K.D., Sample R.H.B., Wible, J.H., 1986. Microinjection of GABA antagonist into posterior hypothalamus elevates heart rate in anaesthetised rats. Neuropharmacol. 25, 1063-1066.
Di Micco, J.A., Soltis, R.P., Anderson, J.J., Wible, J.H., 1991. Hypothalamic mechanisms and the cardiovascular response to stress. In: Kurnos, G. and Ciriello, J. (Eds.), Central neural mechanisms in cardiovascular regulation. Birkhauser, Boston, Vol 2, pp. 52-79.
Di Micco, J.A., Stotz-Potter, E.H., Monroe, A.J., Morris, S.M., 1996. Role of the dorsomedial hypothalamus in the cardiovascular response to stress. Clin. Exp. Pharmacol. Physiol. 23, 171-176.
Ding, Y.Q., Lu, B.Z., Guan, Z.L., Wang, D.S., Xu, J.Q., Li, J.H., 1999. Neurokinin b receptor (NK3)-containing neurons in the paraventricular and supraoptic nuclei of the rat hypothalamus synthesize vasopressin and express Fos following intravenous injection of hypertonic saline. Neurosci. 91, 1077-1085.
Di Scarla, G., Schmitt, P., Karli, P., 1984. Flight induced by infusion of bicuculline methiodide into periventricular structures. Brain Res. 309, 199-208.
Dube, M., Kalra, S.P., Kalra, P.S., 1999. Food intake elicited by central administration of orexins/hypocretins: identification of hypothalamic sites of action. Brain Res. 842, 473-477.
Dufloth, D.L., Morris, M., Michelini, L.C., 1997. Modulation of exercise tachycardia by vasopressin in the nucleus tractus solitarii. Am. J. Physiol. 273, R1271-R1218.
Dunn-Meynell, A.A., Govek, E., Levin, B.E., 1997. Intracarotid glucose infusions selectively increase Fos-like immunoreactivity in paraventricular, ventromedial and dorsomedial nuclei neurons. Brain Res. 748, 100-106.
Dun, S.L., Chianca, Jr, Dun, N.J., Young, J., Chang, J.K., 2000. Differential expression of cocaine and amphetamine-regulated transcript immunoreactivity in the rat spinal preganglionic nuclei. Neurosci. Lett. 294, 143-146.
Eldridge, F.L., Millhorn, D.E., Kiley, J.P., Waldrop, T.G., 1985. Stimulation by central command of locomotion, respiration and circulation during exercise. Respir. Physiol. 59, 313-337.
Eldridge, F.L., Millhorn, D.E., Waldrop, T.G., 1981. Exercise hyperpnea and locomotion: parallel activation from the hypothalamus. Sci. 211, 844-846.
Elias, C.F., Lee, C., Kelly, J., Aschkenasi, C., Ahima, R.S., Coceyro, P.B., Kuhar, M.J., Saper, C.B., Elmquist, J.K., 1998. Leptin activates hypothalamic CART neurons projecting to the spinal cord. Neuron 21, 1375-1385.
Ericson, A.C., Sjoquist, M., 1982. Efferent renal nerve activity during intracarotid and intracerebroventricular infusions of hypertonic sodium chloride solutions and isotonic volume expansion in the rat. Acta Physiol. Scand. 114, 9-15.
Ferguson, A.V., 1988. Systemic angiotensin acts at the subfornical organ to control the activity of paraventricular nucleus neurons with identified projections to the median eminence. Neuroendocrinol. 47, 489-497.
Ferrario, C.M., Abe, I., Averill, D.B., 1987. Sodium and vasopressin modulation of renal sympathetic nerve activity. Clin. Exper. Theory Pract. A9, Suppl. 1, 59-74.
Fontes, M.A.P., Tagawa, T., Polson, J.W., Cavanagh, S.J., Dampney, R.A.L., 2001. Descending pathways mediating cardiovascular response from dorsomedial hypothalamic nucleus. Am. J. Physiol. 280, H2891-H2901.

Gilbey, M.P., Coote, J.H., Fleetwood-Walker, S.M., Peterson, D.F., 1982. The influence of the paraventriculo-spinal pathway, and oxytocin and vasopressin on sympathetic preganglionic neurons. Brain Res. 251, 283-290.

Gomez, R.E., Cannata, M.A., Milner, T.A., Anwar, M., Reis, D.J., Ruggiero, D.A., 1993. Vasopressinergic mechanisms in the nucleus reticularis lateralis in blood pressure control. Brain Res. 604, 90-105.

Greenwood, B., Di Micco, J.A., 1995. Activation of the hypothalamic dorsomedial nucleus stimulates intestinal motility in rats. Am. J. Physiol. 268, G514-G521.

Hallbeck, M., Blomqvist, A., 1999. Spinal cord-projecting vasopressinergic neurones in the rat paraventricular hypothalamus. J. Comp. Neurol. 411, 201-211.

Hancock, M.B., 1976. Cells of origin of hypothalamo-spinal projections in the rat. Neurosci. Lett. 3, 179-184.

Haselton, J.R., Goering, J., Patel, K.P., 1994. Parvocellular neurons of the paraventricular nucleus are involved in the reduction in renal nerve discharge during isotonic volume expansion. J. Auton. Nerv. Syst. 50, 1-11.

Haselton, J.R., Vari, R.C., 1998. Neuronal cell bodies in paraventricular nucleus affect renal haemodynamics and excretion via the renal nerves. Am. J. Physiol. 275, R1334-R1342.

Hatton, G.I., 1990. Emerging concepts of structure-function dynamics in adult brain. The hypothalamo-neurohypophysial system. Prog. Neurobiol. 34, 437-504.

Heinrichs, S.C., 1999. Stress-axis, coping and dementia: gene-manipulation studies. TIPS 20, 311-315.

Hilton, S.M., 1980. Central nervous origin of vasomotor tone. Adv. Physiol. Sci. 8, 1-12.

Hilton, S.M., Redfern, W.S., 1986. A search for brain stem cell groups integrating the defence reaction in the rat. J. Physiol. 378, 213-228.

Holstege, G., 1987. Some neuroanatomical observations on the projections from the hypothalamus to brainstem and spinal cord: An HRP and autoradiographic tracing study in the cat. J. Comp. Neurol. 260, 98-126.

Horvath, T.L., Peyron, C., Diano, S., Ivanova, A., Aston-Jones, G., Kilduff, T.S., van den Pol, A.N., 1999. Hypocretin (orexin) activation and synaptic innervation of the locus coeruleus noradrenergic system. J. Comp. Neurol. 415, 145-159.

Hosoya, Y., Ito, R., Kohno, K., 1987. The topographical organisation of neurones in the dorsal hypothalamic area that project to the spinal cord or to the nucleus raphe pallidus in the rat. Exp. Brain Res. 66, 500-506.

Hosoya, Y., Matsushita, M., 1981. Brainstem projections from the lateral hypothalamic area in the rat, as studied with autoradiography. Neurosci. Lett. 24, 111-116.

Hosoya, Y., Matsushita, M., 1979. Identification and distribution of the spinal and hypophyseal projection neurons in the paraventricular nucleus of the rat. A light and electron microscopic study with the horseradish peroxidase method. Exp. Brain Res. 35, 315-331.

Hosoya, Y., Matsukawa, M., Okado, N., Sugiura, Y., Kohnok, K., 1995. Oxytocinergic innervation to the upper thoracic sympathetic preganglionic neurons in the rat: a light and electron microscopical study using a combined retrograde transport and immunocytochemical technique. Exp. Brain Res. 107, 9-16.

Hosoya, Y., Sugiara, Y., Okado, N., Loewy, A.D., Kohnok, K., 1991. Descending input from the hypothalamic paraventricular nucleus to sympathetic preganglionic neurones in the rat. Exp. Brain Res. 85, 10-20.

Huang, J., Weiss, M.L., 1999. Characterisation of the central cell groups regulating the kidney in the rat. Brain Res. 845, 77-91.

6. THE HYPOTHALAMUS AND CARDIOVASCULAR REGULATION

Jansen, A.S.P., Wessendorf, M.W., Loewy, A.D., 1995. Transneuronal labelling of CNS neuropeptide and monoamine neurons after pseudorabies virus injections into the stellate ganglion. Brain Res. 683, 1-24.

Kannan, H., Yamashita, H., 1985. Connections of neurons in the region of the nucleus tractus solitarius with the hypothalamic paraventricular nucleus: their possible involvement in neural control of the cardiovascular system in rats. Brain Res. 329, 205-212.

Kannan, H., Nijima, A., Yamashita, H., 1987. Inhibition of renal sympathetic nerve activity by electrical stimulation of the hypothalamic paraventricular nucleus in anaesthetized rats. J. Auton. Nerv. Syst. 21, 83-86.

Kanosue, K., Hosono, T., Zhang, Y.H., Chen, X.M., 1998. Neuronal networks controlling thermoregulatory effectors. Prof. Brain Res. 115, 49-62.

Kaufman, S., Deng, Y., 1993. Renal response to atrial stretch during pregnancy in conscious rats. Am. J. Physiol. 265, R902-R906.

Kenney, M.J., Weiss, M.L., Patel, K.P., Wang, Y., Fels, R.J., 2001. Paraventricular nucleus bicuculline alters frequency components of sympathetic nerve discharge bursts. Am. J. Physiol. 281, H1233-H1241.

Kiss, J.Z., Martos, J., Palkovits, M., 1991. Hypothalamic paraventricular nucleus: A quantitative analysis of cytoarchitectonic subdivisions in the rat. J. Comp. Neurol. 313, 563-573.

Kolaj, M., Renaud, L.P., 1998. Vasopressin-induced currents in rat neonatal spinal lateral horn neurons are G-protein mediated and involve two conductances. J. Neurophysiol. 80, 1900-1910.

Kopp, U.C., Smith, L.A., Di Bona, F., 1987. Facilitatory role of efferent renal nerve activity on renal sensory receptors. Am. J. Physiol. 25, F767-F777.

Krukoff, T.L., MacTavish, D., Jhamandas, J.H., 1997. Activation by hypotension of neurons in the hypothalamic paraventricular nucleus that project to the brainstem. J. Comp. Neurol. 385, 285-296.

Kubo, T., Hagiwara, Y., Sekiya, D., Chiba, S., Fukumori, R., 2000. Cholinergic inputs to rostral ventrolateral medulla pressor neurons from the hypothalamus. Brain Res. Bull. 53, 275-282.

Kuypers, H.G.J.M., Maisky, V., 1975. Retrograde axonal transport of horseradish peroxidase from spinal cord to the brainstem cell groups in the cat. Neurosci. Lett. 1, 9-14.

Lang, R.E., Heil, J., Ganten, D., Hermann, K., Rascher, W., Unger, R., 1983. Effects of lesions in the paraventricular nucleus of the hypothalamus on vasopressin and oxytocin contents in brainstem and spinal cord of rat. Brain Res. 260, 326-329.

Li, P., Zhu, D.N., Kao, K.M., Lin, Q., Sun, S.Y., 1995. Role of acetylcholine, corticoids and opioids in the rostral ventrolateral medulla in stress-induced hypertensive rats. Biol. Signals 4, 124-132.

Li, Y.F., Roy, S.K., Channon, K.M., Zucker, I.H., Patel, K.P., 2001. Effect of in vivo gene transfer of nNOS in the PVN on renal nerve discharge in rats. Am. J. Physiol. 282, H594-H601.

Li, Y.W., Dampney, R.A.L., 1994. Expression of Fos-like protein in brain following sustained hypertension and hypotension in conscious rabbits. Neurosci. 61, 613-634.

Li, Y.W., Gievoba, Z.J., McAllen, R.M., Blessing, W.W., 1991. Neurons in rabbit caudal ventrolateral medulla inhibit bulbospinal barosensitive neurons in rostral medulla. Am. J. Physiol. 261, R44-R51.

Lin, Q., Li, P., 1992. A cholinergic mechanism in excitatory projections from hypothalamic and midbrain defence regions to rostral ventrolateral medulla. Chinese J. Physiol. Sci. 8, 198-207.

Lindheimer, M.D., Katz, A.I., 1985. Renal physiology in pregnancy. In: Seldin, D.W. and Gerbesch, G. (Eds.), The kidney: Physiology and Pathophysiology. Raven, New York, pp. 2017-2041.

Loewy, A.D., 1998. Viruses as transneuronal tracers for defining neural circuits. Neurosci. Behavioural Rev. 22, 679-684.

Lovick, T.A., 1991. Central nervous system integration of pain control and autonomic function. NIPS 6, 82-86.

Lovick, T.A., Coote, J.H., 1988a. Electrophysiological properties of paraventriculo-spinal neurones in the rat. Brain Res. 454, 123-130.

Lovick, T.A., Coote, J.H., 1988b. Effects of volume loading on paraventriculo-spinal neurones in the rat. J. Auton. Nerv. Syst. 25, 135-140.

Lovick, T.A., Coote, J.H., 1989. Circulating natriuretic factor acitvates vagal afferent inputs to paraventriculo-spinal neurones in the rat. J. Auton. Nerv. Syst. 26, 129-134.

Lovick, T.A., Malpas, S.C., Mahoney, M.T., 1993. Renal vasodilatation in response to acute volume load is attenuated following lesions of parvocellular neurones in the paraventricular nucleus in rats. J. Auton. Nerv. Syst. 43, 247-255.

Luiten, P.G.M., Ter Horst, G.J., Karst, H., Steffins, A.B., 1985. The course of paraventricular hypothalamic efferents to autonomic structures in medulla and spinal cord. Brain Res. 329, 374-378.

Ma, R.C., Dun, N.J., 1985. Vasopressin depolarises lateral horn cells of the neonatal rat spinal cord in vitro. Brain Res. 348, 36-43.

Magoun, H.W., 1940. Descending connections from the hypothalamus. Res. Publ. Ass. Nerv. Ment. Dis. 20, 270-285.

Maier, T., Dai, W.J., Csikos, T., Jirikowski, G.F., Unger, T., Culman, J., 1998. Oxytocin pathways mediate the cardiovascular and behavioural responses to substance P in the rat brain. Hypertension 31, 480-486.

Malpas, S.C., Coote, J.H., 1994. Role of vasopressin in sympathetic response to paraventricular nucleus stimulation in anaesthetised rats. Am. J. Physiol. 266, R228-R236.

Marshall, J.M., Timms, R.J., 1979. Experiments on the role of the subthalamus in the generation of the cardiovascular changes during locomotion in the cat. J. Physiol. 301, 92-93P.

Martin, J.R., Beinfeld, M.C., Westefall, T.C., 1988. Blood pressure increases after injection of neuropeptide Y into posterior hypothalamic nucleus. Am. J. Physiol. 254, H879-H888.

Masuda, N., Terui, N., Koshiya, N., Kumada, M., 1991. Neurons in the caudal ventrolateral medulla mediate the arterial baroreceptor reflex by inhibiting barosensitive reticulo-spinal neurones in the rostral ventrolateral medulla in rabbits. J. Auton. Nerv. Syst. 34, 103-117.

Matsuguchi, H., Sharabi, F.M., Gordon, F.J., Johnson, A.K., Schmid, P.G., 1982. Blood pressure and heart rate response to microinjection of vasopressin into the nucleus tractus solitarius regions of the rat. Neuropharmacol. 21, 687-693.

Matsumura, K., Tsuchihashi, T., Abe, I., 2001. Central orexin-A augments sympathoadrenal outflow in conscious rabbits. Hypertension 37, 1382-1387.

May, C.N., McAllen, R.M., McKinley, M.J., 2000. Renal nerve inhibition by central NaCl and AngII is abolished by lesions of the lamina terminalis. Am. J. Physiol. 279, R1827-R1833.

6. THE HYPOTHALAMUS AND CARDIOVASCULAR REGULATION

McAllen, R.M., Badoer, E., Shafton, A.D., Oldfield, B.J., McKinley, M.J., 1992. Haemorrhage induces c-fos immunoreactivity in spinally projecting neurons of cat subretrofacial nucleus. Brain Res. 575, 329-332.

McKellar, S., Loewy, A.D., 1981. Organisation of some brain stem afferents to the paraventricular nucleus of the hypothalamus in the rat. Brain Res. Bull. 217, 351-357.

Milner, T.A., Reis, D.J., Pickel, V.M., Aicher, S.A., Guiliano, R., 1993. Ultrastructural localisation and afferent sources of corticotropin-releasing factor in the rat rostral ventrolateral medulla: Implications for central cardiovascular regulation. J. Comp. Neurol. 333, 151-167.

Mortensen, L.H., Haywood, J.R., 1993. Lateral parabrachial nucleus ablation precludes paraventricular nucleus stimulated alterations in sympathoadrenal activity. FASEB J. 7 (4,II), A531.

Motawei, K., Pyner, S., Ranson, R.N., Kamel, M., Coote, J.H., 1999. Terminals of paraventricular spinal neurones are closely associated with adrenal medullary sympathetic preganglionic neurones: immunocytochemical evidence for vasopressin as a possible neurotransmitter in this pathway. Exp. Brain Res. 126, 68-76.

Nakata, T., Takeda, K., Itoh, H., Hirata, M., Kawasaki, S., Hayashi, J., Oguro, M., Sasaki, S., Nakayawa, M., 1989. Paraventricular nucleus lesions attenuate the development of hypertension in DOCA/salt-treated rats. Am. J. Hypertension 2, 625-630.

Nilaver, G., Zimmerman, E.A., Wilkins, J., Michaels, J., Hoffman, D., Silverman, A.J., 1980. Magnocellular hypothalamic projections to the lower brain stem and spinal cord of the rat. Immunocytochemical evidence for predominance of the oxytocin-neurophysin system compared to the vasopressin-neurophysin system. Neuroendocrinol. 30, 150-158.

Ohta, H., Nakamura, S., Watanabe, S., Ueki, S., 1985. Effect of L-Glutamate injected into the posterior hypothalamus on blood pressure and heart rate in unanaesthetised and unrestrained rats. Neuropharmacol. 24, 445-451.

Oldfield, B.J., Bicknell, R.J., McAllen, R.M., Weisinger, R.S., McKinley, M.J., 1991. Intravenous hypertonic saline induces Fos immunoreactivity in neurons throughout the lamina terminalis. Brain Res. 561, 151-156.

Oldfield, B.J., Daver, P.J., Giles, M.E., Allen, A.M., Badoer, E., McKinley, M.J., 2001. Efferent neural projections of angiotensin receptor (AT1) expressing neurones in the hypothalamic paraventricular nucleus of the rat. J. Neuroendocrinol. 13, 139-146.

Olivan, M.V., Bonagamba, L.G., Machado, B.H., 2001. Involvement of the paraventricular nucleus of the hypothalamus in the pressor response to chemoreflex activation in awake rats. Brain Res. 895, 167-172.

Oomura, Y., Ono, T., Wayner, M.J., 1969. Glucose and osmosensitive neurones of the rat hypothalamus. Nature 222, 282-284.

Patel, K.P., Schmid, P.G., 1988. Role of the paraventricular nucleus (PVH) in baroreflex mediated changes in lumbar sympathetic nerve activity and heart rate. J. Auton. Nerv. Syst. 22, 211-219.

Patel, K.P., Zhang, P.L., 1993. Role of renal nerves in renal responses to acute volume expansion during pregnancy in rats. Proc. Soc. Exp. Biol. Med. 03, 150-156.

Petrov, T., Harris, K.H., Mactavish, D., Krukoff, T.L., Jhamandas, J.H., 1995. Hypotension induces Fos immunoreactivity in NADPH-diaphorase positive neurons in the paraventricular and supraoptic hypothalamic nuclei in the rat. Neuropharmacol. 34, 509-514.

Petty, M.A., 1987. The cardiovascular effects of the neurohypophysial hormone oxytocin. J. Auton. Pharmac. 7, 97-104.

Peyron, C., Tighe, D.K., van den Pol, A.N., De Lecea, L., Heller, H.G., Sutcliffe, J.G., Kilduff, T.S., 1998. Neurones containing hypocretin (orexin) project to multiple neuronal systems. J. Neurosci. 18, 9996-10015.

Piekut, D.T., Pretel, S., Applegate, C.D., 1996. Activation of oxytocin-containing neurons of the paraventricular nucleus (PVN) following generalised seizures. Synapse 23, 312-320.

Pittman, Q.J., Riphagen, C.L., Lederis, K., 1984. Release of immunoassayable neurohypophyseal peptides from rat spinal cord in vivo. Brain Res. 300, 321-326.

Porter, J.P., Brody, M.J., 1986. Spinal vasopressin mechanisms of cardiovascular regulation. Am. J. Physiol. 251, R510-R517.

Pyner, S., Coote, J.H., 1999. Identification of an efferent projection from the paraventricular nucleus of the hypothalamus terminating closely to spinally projecting rostral ventrolateral medulla neurons. Neurosci. 8, 949-957.

Pyner, S., Coote, J.H., 2000. Identification of branching paraventricular neurons of the hypothalamus that project to the rostroventrolateral medulla and spinal cord. Neurosci. 100, 549-556.

Pyner, S., Deering, J., Coote, J.H., 2002. Right atrial stretch induces renal nerve inhibition and c-fos expression in parvocellular neurones of the paraventricular nucleus in rats. Exp. Physiol. 87, 25-32.

Ranson, R.N., Motawei, K., Pyner, S., Coote, J.H., 1998. The paraventricular nucleus of the hypothalamus sends efferents to the spinal cord of the rat that closely appose sympathetic preganglionic neurones projecting to the stellate ganglion. Exp. Brain Res. 120, 164-172.

Rathner, J.A., Owens, N.C., McAllen, R.M., 2001. Cold-activated raphe-spinal neurons in rats. J. Physiol. 535, 841-854.

Reiter, M.K., Kremarik, P., Freund-Mercier, M.J., Stoeckel, M.E., Desaulles, E., Feltz, P., 1994. Localisation of oxytocin binding sites in the thoracic and upper lumbar cord of the adult and postnatal rat: a histoautoradiographic study. Eur. J. Neurosci. 6, 98-104.

Richard, P., Moos, F., Freund-Mercier, M., 1991. Central effects of oxytocin. Physiological Soc. 71, 331-370.

Riphagen, C.L., Pittman, Q.J., 1989. Spinal arginine vasopressin elevates renal nerve activity in the rat. J. Neuroendocrinol. 1, 339-344.

Rogers, R.C., Hermann, G.E., 1985. Gastric-vagal solitary neurones excited by paraventricular nucleus microstimulation. J. Auton. Nerv. Syst. 14, 351-362.

Rogers, R.C., Hermann, G.E., 1986. Hypothalamic paraventricular nucleus stimulation-induced gastric acid secretion and bradycardia suppressed by oxytocin angatonist. Peptides 7, 695-700.

Ross, C.A., Ruggiero, D.A., Joh, T.H., Park, D.H., Reis, D.J., 1984. Rostral ventrolateral medulla: selective projections to the thoracic autonomic cell column from the region containing C1 adrenaline neurons. J. Comp. Neurol. 228, 168-185.

Samuels, B.C., Zaretsky, D.V., Di Micco, J.A., 2002. Tachycardia evoked by disinhibition of the dorsomedial hypothalamus in rats is mediated through medullary raphe. J. Physiol. 538, 941-946.

Saper, C.B., Loewy, A.D., Swanson, L.W., Cowan, W.M., 1976. Direct hypothalamic-autonomic connections. Brain Res. 117, 305-312.

Saper, C.B., Swanson, L.W., Cowan, W.M., 1979. An autoradiographic study of the efferent connections of the lateral hypothalamic area in the rat. J. Comp. Neurol. 183, 689-706.

Sawchenko, P.E., 1987a. Adrenalectomy-induced enhancement of CRF and vasopressin immunoreactivity in parvocellular neurosecretory neurons: Anatomic, peptide, and steroid specificity. J. Neurosci. 7, 1093-1106.

6. THE HYPOTHALAMUS AND CARDIOVASCULAR REGULATION

Sawchenko, P.E., 1987b. Evidence for differential regulation of corticotropin-releasing factor and vasopressin immunoreactivities in parvocellular neurosecretory and autonomic-related projections of the paraventricular nculeus. Brain Res. 437, 253-263.

Sawchenko, P.E., Brown, E.R., Chan, R.K.W., Ericsson, A., Li, H.Y., Roland, B.L., Kovacs, K.J., 1996. The paraventricular nucleus of the hypothalamus and the functional neuroanatomy of visceromotor response to stress. Prog. Brain Res. 107, 201-222.

Sawchenko, P.E., Swanson, L.W., 1982. Immunohistochemical identification of neurons in the paraventricular nucleus of the hypothalamus that project to the medulla or to the spinal cord in the rat. J. Comp. Neurol. 205, 260-272.

Schad, H., Seller, H., 1975. Influence of intracranial osmotic stimuli on renal nerve activity in anaesthetised cats. Pflugers Arch. 353, 107-121.

Schramm, L.P., Strack, A.M., Platt, K.B., Loewy, A.D., 1993. Peripheral and central pathways regulating the kidney: a study using psuedorabies virus. Brain Res. 616, 251-262.

Senba, E., Matsunaga, K., Tohyama, M., Noguchi, K., 1993. Stress induced c-fos expression in the rat brain: Activation mechanisms of sympathetic pathway. Brain Res. Bull. 31, 329-344.

Sermasi, E., Coote, J.H., 1994. Oxytocin acts at V_{1a} receptors to excite sympathetic preganglionic neurones in neonate rat spinal cord in vitro. Brain Res. 647, 323-332.

Sermasi, E. Howell, J., Wheatley, M., Coote, J.H., 1998. Localisation of arginine vasopressin V1a receptors on sympatho-adrenal preganglionic neurones. Exp. Brain Res. 119, 85-91.

Shafton, A.D., Ryan, A., Badoer, E., 1998. Neurons in the hypothalamic paraventricular nucleus send collaterals to the spinal cord and to the rostral ventrolateral medulla in the rat. Brain Res. 801, 239-243.

Shafton, A.D., Ryan, A., McGrath, B., Badoer, E., 1999. Volume expansion does not activate neuronal projections from the NTS or depressor VLM to the RVLM. Am. J. Physiol. 277, R39-R46.

Sharp, F.R., Sagar, S.M., Hicks, K., Lowenstein, D., Hisanaga, K., 1991. C-Fos mRNA, fos and fos-related antigen induction by hypertonic saline and stress. J. Neurosci. 11, 2321-2331.

Shen, E., Dun, S.L., Chen, R., Dun, N.J., 1992. Hypovolaemia induces Fos-like immunorectivity in neurons of the rat supraoptic and paraventricular nuclei. J. Auton. Nerv. Syst. 37, 227-230.

Shirasaka, T., Nakazato, M., Matsukura, S., Takasaki, M., Kannan, H., 1999. Sympathetic and cardiovascular actions of orexins in conscious rats. Am. J. Physiol. 277, R1780-R1785.

Shirasaka, T., Kunitake, T., Takasaki, M., Kannan, H., 2002. Neuronal effects of orexins: relevant to sympathetic and cardiovascular functions. Regulatory Peptides 104, 91-95.

Sica, A.L., Greenberg, S.M., Scharf, S.M., Ruggiero, D.A., 2000. Chronic intermittent hypoxia induces immediate early gene expression in the midline thalamus and epithalamus. Brain Res. 883, 224-228.

Smith, O.A., 1965. Anatomy of central neural pathways mediating cardiovascular functions. In: Randall, W.C. (Ed.), Nervous control of the heart. Williams and Wilkins, Baltimore. pp 334-353.

Smith, O.A., Rushmer, R.F., Lasher, E.P., 1960. Similarity of cardiovascular response to exercise and to diencephalic stimulation. Am. J. Physiol. 198, 1139-1142.

Smith, D.W., Sibbald, J.R., Khanna, S., Day, T.A., 1995. Rat vasopressin cell responses to simulated haemorrhage: stimulus dependent role for A1 noradrenergic neurons. Am. J. Physiol. 37, R1336-R1342.

Smith, J.E., Jansen, A.S.P., Gilbey, M.P., Loewy, A.D., 1998. CNS cell groups projecting to sympathetic outflow of tail artery: neural circuits involved in heat loss in the rat. Brain Res. 786, 153-164.

Spanswick, D., Smith, M.A., Groppi, V.E., Logan, S.D., Ashford, M.L., 1997. Leptin inhibits hypothalamic neurons by activation of ATP-sensitive potassium channels. Nature 390, 521-525.

Spanswick, D., Smith, M.A., Mirshamsi, S., Routh, V.H., Ashford, M.L., 2001. Insulin activates ATP-sensitive K+ channels in hypothalamic neurones of lean but not obese rats. Nature Neurosci. 3, 757-758.

Spencer, S.E., Sawyer, W.B., Loewy, A.D., 1989. Cardiovascular effects produced by L-glutamate stimulation of the lateral hypothalamic area. Am. J. Physiol. 257, H540-H552.

Spencer, S.E., Sawyer, W.B., Loewy, A.D., 1990. L-Glutamate mapping of cardioreactive areas in the rat posterior hypothalamus. Brain Res. 511, 149-157.

Stotz-Potter, E.H., Morris, S.M., Mi Micco, J.A., 1996a. Effect of microinjection of muscimol into the dorsomedial or paravantricular hypothalamic nucleus on air stress-induced neuroendocrine and cardiovascular changes in rats. Brain Res. 742, 219-224.

Stotz-Potter, E.H., Willis, L.R., Di Micco, J.A., 1996. Muscimol acts in dorsomedial but not paraventricular hypothalamic nucleus to suppress cardiovascular effects of stress. J. Neurosci. 16, 1173-1179.

Strack, A.M., Loewy, S.D., 1990. Pseudorabies virus: a highly specific transneuronal cell body marker in the sympathetic nervous system. J. Neurosci. 10, 2139-2147.

Strack, A.M., Sawyer, W.B., Hughes, J.H., Platt, K.B., Loewy, A.D., 1989b. A general pattern of CNS innervation of the sympathetic outflow demonstrated by transneuronal pseudorabies viral injections. Brain Res. 491, 156-162.

Strack, A.M., Sawyer, W.B., Platt, K.B., Loewy, A.D., 1989a. CNS cell groups regulating the sympathetic outflow to adrenal gland as revealed by transneuronal cell body labelling with pseudorabies virus. Brain Res. 491, 274-296.

Sun, M.K., Guyenet, P.G., 1986. Hypothalamic glutamatergic input to medullary sympathoexcitatory neurones in rats. Am. J. Physiol. 251, R798-R810.

Swanson, L.W., 1987. The Hypothalamus. In: Bjorkland, A., Hockfelt, T. and Swanson, L.W. (Eds.), Handbook of Chemical Neuroanatomy. Vol 5: Integrated systems of the CNS Part 1. Elsevier, pp. 1-123.

Swanson, L.W., Kuypers, H.G.J.M., 1980. The paraventricular nucleus of the hypothalamus: Cytoarchitectonic subdivisions and organisation of projections to the pituitary, dorsal vagal complex, and spinal cord as demonstrated by retrograde fluorescence double-labelling methods. J. Comp. Neurol. 194, 555-570.

Swanson, W.K., Gosnell, B., Chang, J.K., Resch, Z.T., Murphy, T.C., 1999. Cardiovascular regulatory actions of the hypocretins in brain. Brain Res. 831, 248-253.

Szteyn, S., Welento, J., Milart, Z., 1967. Centres of efferent tracts of the brainstem in the sheep. Anat. Anz. 121, 29-37.

Tagawa, T., Dampney, R.A.L., 1999. AT_1 receptors mediate excitatory inputs to rostral ventrolateral medulla pressor neurons from hypothalamus. Hypertension 34, 1301-1307.

Tagawa, T., Fontes, M.A.P., Potts, P.D., Allen, A.M., Dampney, R.A.L., 2000. The physiological role of AT1 receptors in the ventrolateral medulla. Brazilian J. Biol. Res. 33, 643-652.

Takeda, K., Nakata, T., Takesako, T., Itoh, H., Hirata, M., Kawasaki, S., Hayashi, J., Ogura, M., Sasaki, S., Nakagawa, M., 1991. Sympathetic inhibition and attenuation of spontaneous hypertension by PVN lesions in rats. Brain Res. 53, 286-300.

Tam, S.L., Kaufman, S., 2002. NOS inhibition restores renal responses to atrial distension during pregnancy. Am. J. Physiol. 282, R1364-R1367.

Tan, D.P., Tsou, K., 1986. New evidence for neuronal function of vasopressin: sympathetic mediation of intrathecal vasopressin-induced hypertension. Peptides 7, 569-572.

Tanaka, J., Hayashi, Y., Nomura, S., Miyakubo, H., Okumura, T., Sakamaki, K., 2000. Angiotensinergic and noradrenergic mechanisms in the hypothalamic paraventricular nucleus participate in the drinking response induced by activation of sub fornical organ in rats. Behav. Brain Res. 118, 117-122.

Tanaka, M., Nagashima, K., McAllen, R.M., Kanosue, K. 2002. Role of the medullary raphe in thermoregulatory vasomotor control in rats. J. Physiol. 540, 657-664.

Temple, D.L., Leibowitz, S.F., 1993. Glucocorticoid receptors in PVN: interactions with NE, NPY and GAL in relation to feeding. Am. J. Physiol. 265, E794-E800

Ter Horst, G.J., Luiten, P.G.M., 1988. The projections of the dorsomedial hypothalamic nucleus in the rat. Brain Res. Bull. 16, 231-248.

Ter Horst, G.J., Luiten, G.M., Kuypers, F., 1984. Descending pathways from hypothalamus to dorsal motor vagus and ambiguus nuclei in the rat. J. Auton. Nerv. Syst. 11, 59-75.

Ter Horst, G.J., van den Brink, A., Homminga, S.A., Hautvast, R.W.M., Rakhorst, G., Mettenleiter, T.C., De Jongste, M.J.L., Lie, K.I., Korf, J., 1993. Transneuronal viral labelling of rat heart left ventricle controlling pathways. Neuroreport 4, 1307-1310.

Thompson, R.H., Canteras, N.S., Swanson, L.W., 1996. Organisation of projections from the dorsomedial nucleus of the hypothalamus: a PHA-L study in the rat. J. Comp. Neurol. 376, 143-173.

Unger, T., Becker, H., Dietz, R., Ganten, D., Lang, R.E., Rettig, R., Schomig, A., Schwab, N.A., 1984. Antihypertensive effect of the GABA receptor agonist muscimol in spontaneously hypertensive rats. Role of the sympathoadrenal axis. Circulation Res. 54, 30-37.

van den Pol, A.N., 1999. Hypothalamic hypocretin (orexin): robust innervation of the spinal cord. J. Neurosci. 19, 3171-3182.

van den Top, M., Lee, K., Richardson, P.J., Spanswick, D., Nolan, M.F., 2000. Orexin depolarises rat sympathetic preganglionic neurones in vitro. J. Physiol. 528P, 107P-108P.

Vertes, R.P., Crane, A.M., 1996. Descending projections of the posterior nucleus of the hypothalamus: Phaseolus vulgaris Leucoaglutinin analysis in the rat. J. Comp. Neurol. 374, 607-631.

Waldrop, T.G., Bauer, R.M., Iwamoto, G.A., 1988. Microinjection of GABA antagonists into the posterior hypothalamus elicits locomotor movements and cardiorespiratory activation. Brain Res. 444, 84-94.

Weiss, M.L., Claassen, D.E., Hirai, T., Kenney, M.J., 1996. Non-uniform sympathetic nerve responses to intravenous hypertonic saline infusion. J. Auton. Nerv. Syst. 57, 109-115.

Xu, Z., Glenda, C., Day, L., Yao, J., Ross, M.G., 2000. Osmotic threshold and sensitivity for vasopressin release and Fos expression by hypertonic NACL in ovine fetus. Am. J. Physiol. 279, E1207-E1215.

Yamashita, H., Inenaga, K., Koizumi, K., 1984. Possible projections from regions of paraventricular and supraoptic nuclei to the spinal cord: electrophysiological studies. Brain Res. 96, 373-378.

Yang, Z., Bertram, D., Coote, J.H., 2001. The role of glutamate and vasopressin in the excitation of RVL neurones by paraventricular neurones. Brain Res. 908, 99-103.

Yang, Z., Coote, J.H., 1998. Influence of the hypothalamic paraventricular nucleus on cardiovascular neurones in the rostral ventrolateral medulla of the rat. J. Physiol. 513, 521-530.

Yang, Z., Coote, J.H., 1999. The influence of the paraventricular nucleus on baroreceptor dependent caudal ventrolateral medullary neurones of the rat. Pflugers Archiv. 438, 47-52.

Yang, Z., Coote, J.H., 2003. Role of GABA and NO in the PVN mediated reflex inhibition of renal sympathetic nerve activity following stimulation of right atrial receptors in the rat. Exp. Physiol. 88, 335-342.

Yang, Z., Wheatley, M., Coote, J.H., 2002. Neuropeptides, amines and amino acids as mediators of the sympathetic effects of paraventricular nucleus activation in the rat. Exp. Physiol. 87, 663-674.

Yardley, C.P., Hilton, S.M., 1986. The hypothalamic and brainstem areas from which the cardiovascular and behavioural components of the defence reaction are elicited in the rat. J. Auton. Nerv. Syst. 15, 227-244.

Zhang, K., Patel, K.P., 1998. Effect of nitric oxide within the paraventricular nucleus on renal sympathetic nerve discharge: role of GABA. Am. J. Physiol. 275, R728-R734.

Zhang, K., Mayhan, W.G., Patel, K.P., 1997. Nitric oxide within the paraventricular nucleus mediates changes in renal sympathetic nerve activity. Am. J. Physiol. 273, R864-R872.

Zhang, K., Li, Y.F., Patel, K.P., 2001. Blunted nitric oxide mediated inhibition of renal nerve discharge within the paraventricular nucleus of rats with heart failure. Am. J. Physiol. 281, H995-H1004.

Zhang, K., Li, Y.F., Patel, K.P., 2002. Reduced endogenous GABA-mediated inhibition in the PVN on renal nerve discharge in rats with heart failure. Am. J. Physiol. 282, R1006-R1015.

Chapter 7

CELLULAR PROPERTIES OF AUTONOMIC-RELATED NEURONS IN THE PARAVENTRICULAR NUCLEUS OF THE HYPOTHALAMUS

Javier Stern
Department of Pharmacology and Toxicology, Wright State University, Dayton, OH 45435, USA

Abstract: The paraventricular nucleus of the hypothalamus (PVN) is a complex area composed of functionally different subsets of neurons, including magnocellular neuroendocrine, parvocellular neuroendocrine and parvocellular pre-autonomic neurons. Accumulating evidence indicates that PVN pre-autonomic neurons play important roles in cardiovascular control, both under physiological and pathological conditions. Despite this growing evidence, fundamental information on the cellular mechanisms controlling neuronal excitability in PVN pre-autonomic neurons is still missing. Using a multilevel experimental approach that combines neuronal tract tracing, in vitro patch clamp recordings, three dimensional neuronal reconstruction and immunohistochemistry, we have recently started to characterize the cellular properties of pre-autonomic PVN neurons that innervate medullary autonomic centers. PVN neurons innervating the dorsal vagal complex are characterized by a prominent low threshold spike (LTS), mediated by the activation of Ni^{2+}-sensitive, low-threshold Ca^{2+} conductance (IT). The LTS can efficiently modulate the ability of these neurons to fire in different pattern modes. Morphologically, PVN pre-autonomic neurons have medium to large somata and a bipolar or multipolar dendritic configuration with a relatively high degree of branching. Interestingly, the axons of the majority of these neurons originated from a primary dendrite, instead of originating from the soma with a typical axon hillock. In summary, these studies provide for the first time a detailed characterization of the basic cellular properties underlying neuronal excitability in PVN pre-autonomic neurons, providing a basis for the study of their involvement in the patho-physiology of hypertension and congestive heart failure disorders.

Key words: autonomic control – paraventricular – hypothalamus - channels

INTRODUCTION

Despite the fact that cardiovascular diseases such as hypertension and congestive heart failure (CHF) are among the main public health problems in the United States, affecting almost 20% of the adult population, the basic pathophysiological mechanisms underlying these diseases remain largely unknown. An altered central autonomic function has been shown to play an important role in the pathophysiology of hypertension and congestive heart failure (Malliani and Pagani, 1983; Esler et al., 1995; Bousquet et al., 1998; Esler and Kaye, 1998). However, important information relative to the cellular mechanisms controlling physiological and pathological neuronal excitability of central pre-autonomic neurons is still missing.

Accumulating evidence indicates that pre-autonomic neurons in the hypothalamic paraventricular nucleus (PVN) play important roles in cardiovascular control under physiological and pathological conditions. Because of its well-defined organization and easy accessibility for in vitro studies, the PVN is an ideal model to address questions related to cellular mechanisms controlling excitability in central pre-autonomic neurons involved in cardiovascular diseases. This review will focus on our recent studies on the cellular mechanisms controlling neuronal excitability of autonomic-related neurons in the PVN.

CYTOARCHITECTURAL ORGANIZATION OF THE PVN

Cytoarchitecturally, the PVN is a complex nucleus composed of functionally different subsets of neurons, including magnocellular neuroendocrine, parvocellular neuroendocrine and parvocellular pre-autonomic neurons (van den Pol, 1982; Swanson and Sawchenko, 1983). Magnocellular neuroendocrine neurons are located in the posterior magnocellular subdivision and project to the posterior pituitary. Through the synthesis, and release of oxytocin and vasopressin hormones into the general circulation, they participate in the control of reproductive functions and fluid balance. Parvocellular neuroendocrine neurons are located in more periventricular aspects of the nucleus. These neurons project to the median eminence and participate in the regulation of hormone release from the anterior pituitary. Finally, parvocellular pre-autonomic neurons are located in the ventromedial, posterior and dorsal subnuclei. These neurons send long descending projections to brainstem and spinal cord autonomic centers (Armstrong et al., 1980; Sofroniew and Schrell, 1980; Swanson and Kuypers, 1980; Luiten et al., 1985; Hosoya et al., 1991) and are involved in

7. CELLULAR PROPERTIES OF AUTONOMIC-RELATED NEURONS IN THE PARAVENTRICULAR NUCLEUS OF THE HYPOTHALAMUS

autonomic control. See Fig. 1 for a diagrammatic description of the anatomical and functional organization of the PVN.

Neuroanatomical studies have shown that largely segregated populations of pre-autonomic neurons in the PVN project to parasympathetic- and sympathetic-related autonomic targets in the brainstem and spinal cord (Swanson et al., 1980; Portillo et al., 1996, 1998). Importantly, the activity of these segregated PVN neuronal populations can be selectively regulated in conditions of altered cardiovascular homeostasis, such as volume loading, hypotension and hemorraghe (Krukoff et al., 1997; Badoer and Merolli, 1998; Lovick and Coote, 1998; Horiuchi et al., 1999).

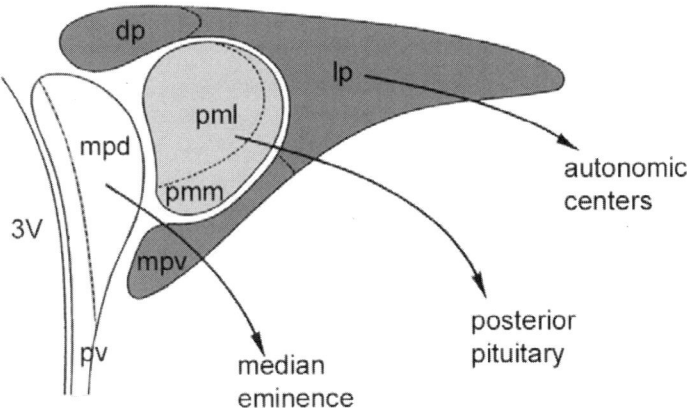

Figure 1. Functional and anatomical subdivisions of the PVN. Abbreviations: dp, dorsal parvocellular subnucleus; lp, lateral parvocellular subnucleus; mpd,v, dorsal and ventral subdivisions of the medial parvocellular subnucleus; pm,l,m, lateral (vasopressinergic) and medial (oxytocinergic) subdivisions of the posterior magnocellular subnucleus; pv, periventricular subnucleus; 3V, third ventricle. Modified from Swanson and Sawchenko, 1983 with permission.

THE PARAVENTRICULAR NUCLEUS OF THE HYPOTHALAMUS: ROLE IN CENTRAL AUTONOMIC CONTROL

Through reciprocal interconnections with autonomic centers in the brainstem and spinal cord, the PVN occupies a pivotal position within the circuitry involved in the integration of autonomic and endocrine responses that regulate the cardiovascular system (Palkovits, 1999). The status of body fluid and cardiovascular homeostasis is conveyed to the PVN through neural and humoral inputs. This information is processed and then translated into integrated visceral effector responses by means of neurohumoral pathways and neuronal inputs to lower autonomic centers located in the brainstem and spinal cord (Palkovits, 1999). Through projections to the rostral ventrolateral medulla (RVLM), spinal cord preganglionic neurons, or both (Shafton et al., 1998; Pyner and Coote, 2000), the PVN is able to influence sympathetic activity. Accordingly, activation of PVN neurons have been shown to induce sympathoexcitatory and pressor responses via excitatory connections with the RVLM (Coote et al., 1998; Yang and Coote, 1998; Tagawa and Dampney, 1999) and the spinal cord (Malpas and Coote, 1994).

Similarly, distinct descending projections to the dorsal motor nucleus of the vagus (dmnX), the nucleus ambiguus (NA) and the nucleus of the solitary tract (NTS) (Armstrong et al., 1980; Swanson and Kuypers, 1980; Portillo et al., 1998; Ranson et al., 1998) provide the anatomical substrate for PVN influences on parasympathetic activity and baroreflex control, respectively. For instance, PVN projections to the NTS have been shown to play a role in the control of the baroreceptor reflex (Zhang and Ciriello, 1985a, b), partly through direct modulation of NTS neuronal activity (Banks and Harris, 1987; Duan et al., 1999; Zhang et al., 1999). Thus, through modulatory actions on the baroreceptor reflex, sympathetic and parasympathetic nerve activity, the PVN is in a pivotal position to contribute to the short- and long-term control of the cardiovascular system.

IDENTIFICATION OF AUTONOMIC-RELATED PVN NEURONS FOR IN VITRO CELLULAR STUDIES

Despite the strong evidence for the PVN as an important site for cardiovascular control, little is known about the cytoarchitectural and physiological properties of identified pre-autonomic PVN neurons. Although previous in vitro studies have looked in detail into the general electrophysiological properties of PVN neurons, none of these studies were done in identified PVN neurons projecting to brainstem and/or spinal cord autonomic-related areas (Hoffman et al., 1991; Tasker and Dudek, 1991; Luther and Tasker, 2000).

Using a multilevel experimental approach that combines neuronal tract tracing, in vitro patch clamp recordings, three dimensional neuronal reconstruction and immunohistochemistry, we have recently started to characterize the cellular properties of pre-autonomic PVN neurons that innervate medullary autonomic centers (Stern, 2001). Using this approach, fluorescent retrograde tracers, such as DiI, Fluorogold or rhodamine beads

7. CELLULAR PROPERTIES OF AUTONOMIC-RELATED NEURONS IN THE PARAVENTRICULAR NUCLEUS OF THE HYPOTHALAMUS

are stereotaxically injected into the innervated target. Five to seven days after the tracer injections, acute hypothalamic slices containing the PVN are cut. Electrophysiological recordings are then obtained from fluorescently labeled PVN neurons, using infrared-DIC and conventional fluorescent videomicroscopy in combination. Furthermore, neurons are intracellularly filled with biocytin during recordings, allowing for subsequent identification of the recorded cell using a conventional avidin-biotin complex reaction. This approach in turn, enables us to obtain a three dimensional reconstruction of the neuronal dendritic arborization, and when possible, to immunohistochemically identify its neurochemical phenotype. Figure 2 shows an example of identified autonomic-related PVN neurons using these experimental approaches.

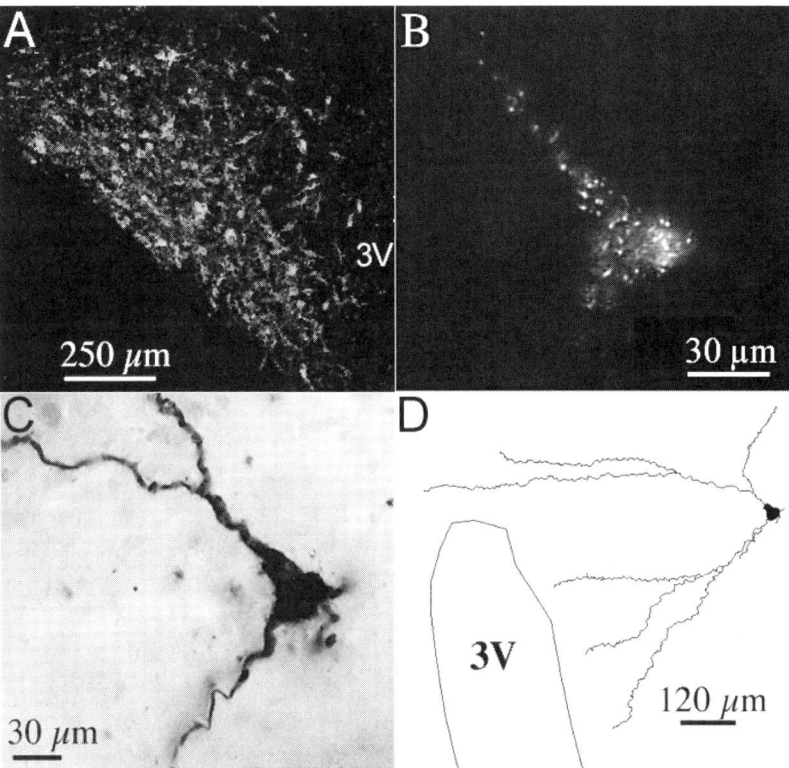

Figure 2. Identification of PVN preautonomic neurons in brain slices using retrograde labeling techniques. A, Confocal digital image (10X) showing DVC-projecting PVN neurons in a coronal hypothalamic slice (300 μm). Neurons were labeled following an injection of DiI

in the DVC. B, Example of a retrogradely labeled PVN neuron shown at higher magnification (60x). This neuron was intracellularly filled with biocytin during electrical recordings. C, Photomicrograph of the same neuron as in B after an ABC-DAB staining. D, The dendritic arborization of the same neuron was then traced and reconstructed in three-dimensions using a computer-assisted program. 3V: third ventricle. Modified from Stern, 2001.

ELECTROPHYSIOLOGICAL PROPERTIES OF IDENTIFIED PRE-AUTONOMIC PVN NEURONS

Using this multilevel experimental approach, we have recently obtained a detailed characterization of autonomic-related PVN neurons that innervate the dorsal vagal complex (DVC) (Stern, 2001). This area, considered to be a cardioinhibitory and vasodepressor center (Barnes et al., 1979; Nosaka et al., 1979), contains both parasympathetic preganglionic neurons (dmnX) (Nosaka et al., 1979; Ciriello and Calaresu, 1980; Standish et al., 1995) and baroreceptor sensitive neurons.

We observed that one of the most typical electrophysiological features of DVC-projecting PVN neurons is a prominent low threshold spike (LTS), which is able to evoke an overriding burst of action potentials (Fig. 3A1). This electrical property distinguishes pre-autonomic PVN neurons from their neighboring magnocellular vasopressin neurons, which upon depolarization display a strong transient outward rectification that delays the occurrence of the first spike (Fig. 3A2). These properties correspond very well with those previously described for type 1 (putative magnocellular) and type II/III (putative parvocellular) PVN neurons (Hoffman et al., 1991; Tasker and Dudek, 1991; Luther and Tasker, 2000). The LTS that we observed in PVN pre-autonomic neurons resembles that first described in inferior olivary neurons (Llinas and Yarom, 1981), which results from the activation of a Ni^{2+}-sensitive, low-threshold Ca^{2+} conductance (IT). In fact, voltage clamp recordings of isolated Ca^{2+} currents in acutely dissociated, retrogradely labeled pre-autonomic PVN neurons, revealed the presence of both low threshold and high threshold components (Fig. 3B1, 3B2), suggesting that IT mediates the LTS in this neuronal population as well.

One of the main properties of IT is that depending on its degree of activation/inactivation (determined for example by the resting membrane potential of the neuron), it contributes to the expression of different firing patterns. If neurons expressing IT are hyperpolarized enough to remove IT inactivation, a burst of action potentials will be triggered upon membrane depolarization (bursting mode). On the other hand, if neurons are initially depolarized so IT is inactivated, they will respond to the same depolarizing input with a tonic discharge of action potentials (tonic mode). These different firing modes can be in fact triggered in pre-autonomic PVN

7. CELLULAR PROPERTIES OF AUTONOMIC-RELATED NEURONS IN THE PARAVENTRICULAR NUCLEUS OF THE HYPOTHALAMUS

neurons, according to their initial membrane potentials (Fig. 4A). More importantly, both tonic and bursting modes can be observed in spontaneously active neurons (Fig. 4B). Thus, these data suggests that LTS appears to be a very important intrinsic membrane property controlling neuronal excitability in pre-autonomic PVN neurons.

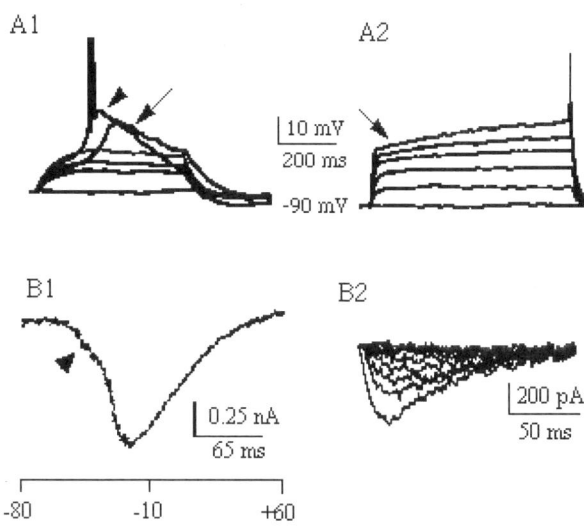

Figure 3. Preautonomic PVN neurons display a prominent low threshold spike (LTS) and express T-type Ca2+ channels. A1, When depolarized from relatively hyperpolarized membrane potentials, a slow depolarizing hump was evoked (LTS, arrow). Short bursts of action potentials riding on the slow depolarizing hump were evoked upon stronger depolarization (arrowhead). A2, On the other hand, neighboring magnocellular vasopressin neurons were characterized by the expression of a strong transient outward rectification (arrow) that delayed action potential firing. B, Example of Ca2+ channel currents recorded in an acutely dissociated, retrogradely labeled preautonomic PVN neuron. Currents were evoked by slow depolarizing voltage ramps (0.5 mV/ms, B1) or by increasing voltage steps to between -80 and -40 mV (B2). The presence of a low-threshold component (IT) is shown as a small current peak at relatively low membrane potentials during the voltage ramps (arrow, B1), as well as a small, rapidly inactivating current during the depolarizing rectangular pulses (B2).

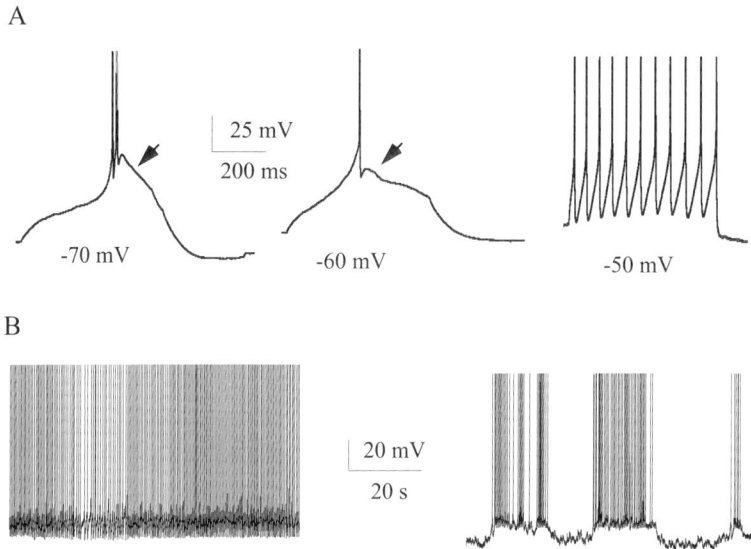

Figure 4. Preautonomic PVN neurons display different firing patterns. A, Voltage-dependency of LTS amplitude and evoked firing pattern. At hyperpolarized holding potentials (~ -70 mV) a depolarizing pulse evoked a fully activated LTS, which induced a short burst of action potentials (arrow, left panel). As the holding potential is depolarized, LTS and burst amplitude diminished (arrow, middle panel), and eventually disappeared, resulting in a tonic, rather than bursting firing activity (right panel). B, Spontaneously active preautonomic PVN neurons display either tonic (left panel) or bursting (right panel) firing modes. Modified from Stern, 2001.

MORPHOLOGICAL PROPERTIES OF IDENTIFIED PRE-AUTONOMIC PVN NEURONS

We have also characterized in details the morphological properties of identified pre-autonomic PVN neurons (Stern, 2001). In general, neurons had medium to large somata and a bipolar or multipolar dendritic configuration with a relatively high degree of branching. Dendrites were often varicose, and short spinous processes were occasionally observed. A common observation was that distal dendritic branches were not confined to their respective subnuclei. In some cases, dendrites were observed to cross to the contralateral side (see Fig. 2D). This finding is in general agreement with

7. CELLULAR PROPERTIES OF AUTONOMIC-RELATED NEURONS IN THE PARAVENTRICULAR NUCLEUS OF THE HYPOTHALAMUS

previous work, showing that dendrites from other neuronal populations within the PVN can also extend beyond their subnucleus boundaries (Armstrong et al., 1980; Rho and Swanson, 1989). Thus, though the PVN is in general anatomically compartmentalized, caution should be taken when considering the selectivity of a particular synaptic input that preferentially innervates a particular region of the PVN.

As previously noticed in magnocellular PVN neurons (Armstrong et al., 1980; van den Pol, 1982) dendrites from PVN pre-autonomic neurons tended to approach the walls of the third ventricle, indicating probably a secretory function in these dendrites. In this regard, it is now well established that dendrites from magnocellular neuroendocrine cells in the supraoptic and PVN nuclei are able to locally release peptides to pre- and postsynaptically modulate their own excitability (see Ludwig, 1998 for review). Whether dendrites from pre-autonomic PVN neurons have a similar secretory activity, and whether their neurotransmitter/neuropeptide content can be directly released into the cerebrospinal fluid are interesting questions that remain to be established.

Another interesting morphological feature of pre-autonomic PVN neurons is related to their axonal origins. In the majority of intracellularly filled neurons (~70%), axons were observed to originate from a primary dendrite, instead of originating from the soma with a typical axon hillock (see Fig. 5). The functional implication for this particular type of axonal origin is at present unknown. However, one could speculate that synaptic inputs impinging onto the dendrite from which the axon originates, will be more efficient in affecting neuronal firing output than those impinging in other segments of the dendritic arborization. Similarly, the efficacy of these synaptic inputs could be more readily modulated by dendritic invasion of back-propagating action potentials, as shown in other neuronal types (Magee and Johnston, 1997). These are important issues that deserve further investigation, and which will help understanding the overall input-output properties of this neuronal population. Interestingly, this morphological pattern has also been observed in other hypothalamic neurons, including a proportion of magnocellular neuroendocrine neurons (Armstrong et al., 1980; Armstrong et al., 1982; Hatton et al., 1985, Stern and Armstrong, 1998).

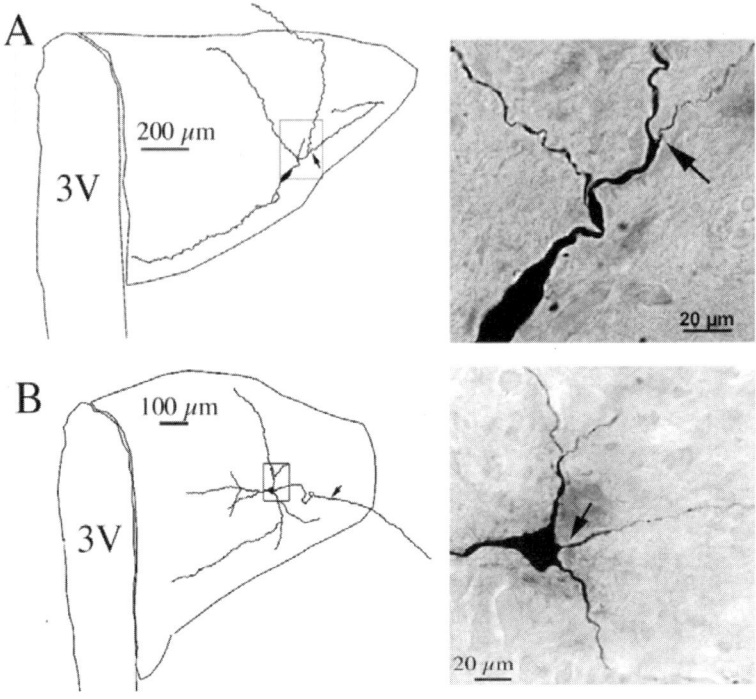

Figure 5. Examples of reconstructed DVC-projecting preautonomic PVN neurons displaying typical morphological features. Photomicrographs on right panels correspond to the boxed areas shown in the respective left panels. A and B show representative examples of cases in which axons (arrows) originated either from a proximal dendrite (A) or from the soma (B). Note also the beaded appearance of dendrites in A. Modified from Stern, 2001.

HETEROGENEITY IN PRE-AUTONOMIC PVN NEURONS

While pre-autonomic PVN neurons share some common general properties, such as the expression of low threshold spikes and a relatively complex dendritic arborization, a more detailed analysis revealed the presence of a high degree of morphological and electrophysiological heterogeneity within this population. In the case of DVC-projecting PVN neurons, cellular heterogeneity is in part related to their topographical distribution within the PVN (Stern, 2001). Moreover, we have recently found that the properties of RVLM-projecting PVN neurons differ from those of DVC-projecting neurons (unpublished observations). Thus, it appears that the cellular properties of pre-autonomic PVN neurons are also dependent upon their innervated targets.

7. CELLULAR PROPERTIES OF AUTONOMIC-RELATED NEURONS IN THE PARAVENTRICULAR NUCLEUS OF THE HYPOTHALAMUS

Finally, long descending PVN neurons are known to express a variety of neuroactive substances, including oxytocin, vasopressin, angiotensin, somatostatin and enkephalin, among others (Sawchenko and Swanson, 1982), all of which are known to differentially affect the cardiovascular system (Sun and Guyenet, 1989; Ishizuka et al., 1993; Okada et al., 1994; Michelini and Morris, 1999; Tagawa and Dampney, 1999; Braga et al., 2000). Thus, whether heterogeneity in pre-autonomic PVN neurons is also dependent upon their neurochemical phenotype remains to be determined. Thus, uncovering the sources, and more importantly the functional implications of cellular heterogeneity among pre-autonomic PVN neurons will be an important piece in the puzzle of understanding their specific role in the control of the autonomic and cardiovascular functions.

CONCLUSIONS

Neuroendocrine and autonomic homeostasis requires the coordinated and integrative activity of distinct neuronal populations within the PVN. Among them, pre-autonomic parvocellular neurons have been shown to play an important role in the control of the autonomic and cardiovascular systems under physiological as well as pathological conditions. In this sense, there is growing consensus for the involvement of pre-autonomic PVN neurons in cardiovascular disorders, such as hypertension (Goto, et al., 1981; Ciriello, et al., 1984) and congestive heart failure (Patel and Zhang, 1996; Patel, et al., 2000). In order to understand the central neuronal mechanisms contributing to these pathological processes, it will be necessary first to shed light into the basic cellular mechanisms controlling neuronal excitability of functionally relevant neuronal populations under physiological conditions. In vitro approaches such as those described in these studies will contribute to obtain this essential information.

REFERENCES

Armstrong, W.E., Warach, S., Hatton, G.I., McNeill, T.H., 1980. Subnuclei in the rat hypothalamic paraventricular nucleus: a cytoarchitectural, horseradish peroxidase and immunocytochemical analysis. Neurosci. 5, 1931-1958.

Armstrong, W.E., Scholer, J., McNeill, T.H., 1982. Immunocytochemical, Golgi and electron microscopic characterization of putative dendrites in the ventral glial lamina of the rat supraoptic nucleus. Neurosci. 7, 679-694.

Badoer, E., Merolli, J., 1998. Neurons in the hypothalamic paraventricular nucleus that project to the rostral ventrolateral medulla are activated by haemorrhage. Brain Res. 791, 317-320.

Banks, D., Harris, M.C., 1987. Activation within dorsal medullary nuclei following stimulation in the hypothalamic paraventricular nucleus in rats. Pflugers Arch. 408, 619-627.

Barnes, K.L., Ferrario, C.M., Conomy, J.P., 1979. Comparison of the hemodynamic changes produced by electrical stimulation of the area postrema and nucleus tractus solitarii in the dog. Circ. Res. 45, 136-143.

Bousquet, P., Monassier, L., Feldman, J., 1998. Autonomic nervous system as a target for cardiovascular drugs. Clin. Exp. Pharmacol. Physiol. 25, 446-448.

Braga, D.C., Mori, E., Higa, K.T., Morris, M., Michelini, L.C., 2000. Central oxytocin modulates exercise-induced tachycardia. Am. J. Physiol. Regul. Integr. Comp. Physiol. 278, R1474-1482.

Ciriello, J., Calaresu, F.R., 1980. Distribution of vagal cardioinhibitory neurons in the medulla of the cat. Am. J. Physiol. 238, R57-64.

Ciriello J., Kline, R.L., Zhang, T.X., Caverson, M.M., 1984. Lesions of the paraventricular nucleus alter the development of spontaneous hypertension in the rat. Brain Res. 310, 355-359.

Coote, J.H., Yang, Z., Pyner, S., Deering, J., 1998. Control of sympathetic outflows by the hypothalamic paraventricular nucleus. Clin. Exp. Pharmacol. Physiol. 25, 461-463.

Duan, Y.F., Kopin, I.J., Goldstein, D.S., 1999. Stimulation of the paraventricular nucleus modulates firing of neurons in the nucleus of the solitary tract. Am. J. Physiol. 277, R403-411.

Esler, M.D., Lambert, G.W., Ferrier, C., Kaye, D.M., Wallin, B.G., Kalff, V., Kelly, M.J., Jennings, G.L., 1995. Central nervous system noradrenergic control of sympathetic outflow in normotensive and hypertensive humans. Clin. Exp. Hypertension 17, 409-423.

Esler, M., Kaye, D., 1998. Increased sympathetic nervous system activity and its therapeutic reduction in arterial hypertension, portal hypertension and heart failure. J. Auton. Nerv. Syst. 72, 210-219.

Goto, A., Ikeda, T., Tobian, L., Iwai, J., Johnson, M.A., 1981. Brain lesions in the paraventricular nuclei and catecholaminergic neurons minimize salt hypertension in Dahl salt-sensitive rats. Clin. Sci. 61 Suppl. 7, 53s-55s.

Hatton, G.I., Cobbett, P., Salm, A.K., 1985. Extranuclear axon collaterals of paraventricular neurons in the rat hypothalamus: intracellular staining, immunocytochemistry and electrophysiology. Brain Res. Bull. 14, 123-132.

Hoffman, N., Tasker, J., Dudek, E., 1991. Immunohistochemical differentiation of electrophysiologically defined neuronal populations in the region of the rat hypothalamic paraventricular nucleus. J, Comp. Neurol. 307, 405-416.

Horiuchi, J., Potts, P.D., Polson, J.W., Dampney, R.A., 1999. Distribution of neurons projecting to the rostral ventrolateral medullary pressor region that are activated by sustained hypotension. Neurosci. 89, 1319-1329.

7. CELLULAR PROPERTIES OF AUTONOMIC-RELATED NEURONS IN THE PARAVENTRICULAR NUCLEUS OF THE HYPOTHALAMUS

Hosoya, Y., Sugiura, Y., Okado, N., Loewy, A.D., Kohno, K., 1991. Descending input from the hypothalamic paraventricular nucleus to sympathetic preganglionic neurons in the rat. Exp. Brain Res. 85, 10-20.

Ishizuka, T., Wei, X., Kubo, T., 1993. Cardiovascular effects of microinjections of thyrotropin-releasing hormone, oxytocin and other neuropeptides into the rostral ventrolateral medulla of the rat. Arch. Int. Pharmacodyn. Ther. 322, 35-44.

Krukoff, T.L., Mactavish, D., Jhamandas, J.H., 1997. Activation by hypotension of neurons in the hypothalamic paraventricular nucleus that project to the brainstem. J. Comp. Neurol. 385, 285-296.

Llinas, R., Yarom, Y., 1981. Properties and distribution of ionic conductances generating electroresponsiveness of mammalian inferior olivary neurons in vitro. J. Physiol. 315, 569-584.

Lovick, T.A., Coote, J.H., 1988. Effects of volume loading on paraventriculo-spinal neurones in the rat. J. Auton. Nerv. Syst. 25, 135-140.

Ludwig, M., 1998. Dendritic release of vasopressin and oxytocin. J. Neuroendocrinol. 10, 881-895.

Luiten, P.G., ter Horst, G.J., Karst, H., Steffens, A.B., 1985. The course of paraventricular hypothalamic efferents to autonomic structures in medulla and spinal cord. Brain Res. 329, 374-378.

Luther, J.A., Tasker, J.G. 2000. Voltage-gated currents distinguish parvocellular from magnocellular neurones in the rat hypothalamic paraventricular nucleus. J. Physiol. (Lond.) 523, 193-209.

Magee, J., Johnston, D., 1997. A synaptically controlled, associative signal for hebbian plasticity in hippocampal neurons. Sci. 275, 209-212.

Malliani, A., Pagani, M., 1983. The role of the sympathetic nervous system in congestive heart failure. Eur. Heart J. 4 Suppl. A 49-54.

Malpas, S.C., Coote, J.H., 1994. Role of vasopressin in sympathetic response to paraventricular nucleus stimulation in anesthetized rats. Am. J. Physiol. 266, R228-236.

Michelini, L.C., Morris, M., 1999. Endogenous vasopressin modulates the cardiovascular responses to exercise. Ann. N.Y. Acad. Sci. 897, 198-211.

Nosaka, S., Yamamoto, T., Yasunaga, K., 1979. Localization of vagal cardioinhibitory preganglionic neurons with rat brain stem. J. Comp. Neurol. 186, 79-92.

Okada, J., Takayama, K., Xiong, Y., Miura, M., 1994. Influence of humoral control peptides on medullary vasomotor control neurons: microstimulation and double-labeling studies using SHR and WKY rats. J. Auton. Nerv. Syst. 49, 171-182.

Palkovits, M., 1999. Interconnections between the neuroendocrine hypothalamus and the central autonomic system. Geoffrey Harris Memorial Lecture, Kitakyushu, Japan, October 1998. Front. Neuroendocrinol. 20, 270-295.

Patel, K.P., Zhang, K., 1996. Neurohumoral activation in heart failure: role of paraventricular nucleus. Clin. Exp. Pharmacol. Physiol. 23, 722-726.

Patel, K.P., Zhang, K., Kenney, M.J., Weiss, M., Mayhan, W.G., 2000. Neuronal expression of Fos protein in the hypothalamus of rats with heart failure. Brain Res. 865, 27-34.

Portillo, F., Carrasco, M., Vallo, J.J., 1996. Hypothalamic neuron projection to autonomic preganglionic levels related with glucose metabolism: a fluorescent labelling study in the rat. Neurosci. Lett. 210, 197-200.

Portillo, F., Carrasco, M., Vallo, J.J., 1998. Separate populations of neurons within the paraventricular hypothalamic nucleus of the rat project to vagal and thoracic autonomic preganglionic levels and express c-Fos protein induced by lithium chloride. J. Chem. Neuroanat. 14, 95-102.

Pyner, S., Coote, J.H., 2000. Identification of branching paraventricular neurons of the hypothalamus that project to the rostroventrolateral medulla and spinal cord. Neurosci. 100, 549-556.

Ranson, R.N., Motawei, K., Pyner, S., Coote, J.H., 1998. The paraventricular nucleus of the hypothalamus sends efferents to the spinal cord of the rat that closely appose sympathetic preganglionic neurones projecting to the stellate ganglion. Exp. Brain Res. 120, 164-172.

Rho, J.H., Swanson, L.W., 1989. A morphometric analysis of functionally defined subpopulations of neurons in the paraventricular nucleus of the rat with observations on the effects of colchicine. J. Neurosci. 9, 1375-1388.

Sawchenko, P.E., Swanson, L.W., 1982. Immunohistochemical identification of neurons in the paraventricular nucleus of the hypothalamus that project to the medulla or to the spinal cord in the rat. J. Comp. Neurol. 205, 260-272.

Shafton, A.D., Ryan, A., Badoer, E., 1998. Neurons in the hypothalamic paraventricular nucleus send collaterals to the spinal cord and to the rostral ventrolateral medulla in the rat. Brain Res. 801, 239-243.

Sofroniew, M.V., Schrell, U., 1980. Hypothalamic neurons projecting to the rat caudal medulla oblongata, examined by immunoperoxidase staining of retrogradely transported horseradish peroxidase. Neurosci. Lett. 19, 257-263.

Standish, A., Enquist, L.W., Escardo, J.A., Schwaber, J.S., 1995. Central neuronal circuit innervating the rat heart defined by transneuronal transport of pseudorabies virus. J. Neurosci. 15, 1998-2012.

Stern, J.E., Armstrong, W.E., 1998. Reorganization of the dendritic trees of oxytocin and vasopressin neurons of the rat supraoptic nucleus during lactation. J. Neurosci. 18, 841-853.

Stern, J.E., 2001. Electrophysiological and morphological properties of pre-autonomic neurones in the rat hypothalamic paraventricular nucleus. J. Physiol. 537, 161-177.

Sun, M.K., Guyenet, P.G., 1989. Effects of vasopressin and other neuropeptides on rostral medullary sympathoexcitatory neurons 'in vitro'. Brain Res. 492, 261-270.

Swanson, L.W., Kuypers, H.G., 1980. The paraventricular nucleus of the hypothalamus: cytoarchitectonic subdivisions and organization of projections to the pituitary, dorsal vagal complex, and spinal cord as demonstrated by retrograde fluorescence double-labeling methods. J. Comp. Neurol. 194, 555-570.

Swanson, L.W., Sawchenko, P.E., Wiegand, S.J., Price, J.L., 1980. Separate neurons in the paraventricular nucleus project to the median eminence and to the medulla or spinal cord. Brain Res. 198, 190-195.

Swanson, L.W., Sawchenko, P.E., 1983. Hypothalamic integration: organization of the paraventricular and supraoptic nuclei. Annu. Rev. Neurosci. 6, 269-324.

Tagawa, T., Dampney, R.A., 1999. AT(1) receptors mediate excitatory inputs to rostral ventrolateral medulla pressor neurons from hypothalamus. Hypertension 34, 1301-1307.

Tasker, J., Dudek, E., 1991. Electrophysiological properties of neurones in the region of the paraventricular nucleus in slices of rat hypothalamus. J. Physiol. 434, 271-93.

van den Pol, A.N., 1982. The magnocellular and parvocellular paraventricular nucleus of rat: intrinsic organization. J. Comp. Neurol. 206, 317-345.

Yang, Z., Coote, J.H., 1998. Influence of the hypothalamic paraventricular nucleus on cardiovascular neurones in the rostral ventrolateral medulla of the rat. J. Physiol. (Lond.) 513, 521-530.

Zhang, T.X., Ciriello, J., 1985a. Effect of paraventricular nucleus lesions on arterial pressure and heart rate after aortic baroreceptor denervation in the rat. Brain Res. 341, 101-109.

7. CELLULAR PROPERTIES OF AUTONOMIC-RELATED NEURONS IN THE PARAVENTRICULAR NUCLEUS OF THE HYPOTHALAMUS

Zhang, T.X., Ciriello, J., 1985b. Kainic acid lesions of paraventricular nucleus neurons reverse the elevated arterial pressure after aortic baroreceptor denervation in the rat. Brain Res. 358, 334-338.

Zhang, X., Fogel, R., Renehan, W.E., 1999. Stimulation of the paraventricular nucleus modulates the activity of gut-sensitive neurons in the vagal complex. Am. J. Physiol. 277, G79-90.

Chapter 8

THE ANTERIOR HYPOTHALAMUS AND SALT-SENSITIVE HYPERTENSION

[1]Suzanne Oparil and [2]J. Michael Wyss
[1]*Department of Medicine, Physiology and Vascular Biology, University of Alabama at Birmingham, Birmingham, AL 35294 and* [2]*Department of Cell Biology Biophysics, University of Alabama at Birmingham, Birmingham, AL 35294, USA*

Abstract: Early studies in our laboratory suggested that in young (3-12 weeks of age) male spontaneously hypertensive rats (SHR), ingestion of a high NaCl diet exacerbates hypertension (Fig. 1), increases peripheral sympathetic nervous system activity and alters the norepinephrine (NE) content of a hypothalamic nucleus known to be involved blood pressure (BP) regulation (Winternitz et al., 1982a, b; Wyss et al., 1987). Exposure to a high (8% compared to 0.6%) NaCl diet for 2 weeks causes increases in both arterial pressure and peripheral sympathetic nervous system activity and a decrease in NE stores in anterior hypothalamic area but not in other brain regions tested in these animals. None of these NaCl-induced alterations occur in NaCl resistant normotensive Wistar-Kyoto (WKY) rats or NaCl resistant SHR (Wyss et al., 1997). Subsequent studies demonstrated that exposure to a high NaCl diet selectively reduces the turnover of NE in the anterior hypothalamic nucleus (AHN), but not in other hypothalamic nuclei of young, NaCl sensitive male SHR (Chen et al., 1988). These findings are consistent with the hypothesis that dietary NaCl supplementation exacerbates hypertension in SHR by decreasing the release of NE from nerve terminals in the AHN, thus reducing the activity of the AHN neurons and decreasing the inhibition of peripheral sympathetic nervous system activity (Fig. 2). NaCl-resistant SHR (SHR-R) and WKY have a genetically based resistance to these effects of dietary NaCl and as a result do not display these responses to a high NaCl diet (Wyss et al., 1997). The push-pull microperfusion technique was then used to provide direct evidence of selective reduction in NE release from nerve terminals in the AHN of NaCl supplemented SHR (Chen et al., 1991) (Fig. 3). Dietary NaCl had no effect on NE release in other brain regions tested in SHR and no effect on NE release in the AHN of NaCl-resistant WKY. Subsequent studies focused on defining the functional significance of the selective NaCl-induced reduction in NE release in the AHN of SHR and identifying the local and systemic mechanisms responsible.

A combination of ligand binding and functional studies with administration of α_2 adrenergic agonists systematically or locally into the AHN was used to test the hypothesis that reduction in NE release in the AHN of NaCl supplemented SHR results in functionally significant local up-regulation of postsynaptic α_2 adrenoreceptors. Binding of p-[^{3}H]aminoclonidine in the AHN, assessed by both membrane binding techniques and autoradiography, was increased selectively in SHR but not in WKY rats in response to dietary NaCl supplementation (Klangkalya et al., 1988), and depressor and bradycardic responses to locally injected α_2 adrenoreceptor agonists were greater in NaCl supplemented SHR than in SHR on a basal diet or in WKY rats on either diet (Wyss et al., 1988). Chronic administration of α_2 adrenergic agonists either systematically (Jin et al., 1989) or locally via microcannula into the AHN (Jin et al., 1991b) prevented the NaCl-induced increases in BP, left ventricular weight and plasma NE levels in SHR (Fig. 4). These findings support the hypothesis that NaCl-induced hypertension in this model is associated with diminished sympathoinhibitory function of central α_2 adrenergic receptors due to reduced release of endogenous NE from nerve terminals in the AHN. Studies in the mouse with homozygous deletion of the α_{2A} adrenergic receptor have confirmed that activation of this receptor subtype in the AHN mediates sympatoinhibitory and depressor/bradycardic responses (Peng et al., 2003a).

> The regional selectivity of the reduction in NE release in the AHN of SHR during dietary NaCl supplementation can be explained by either selective diminution in noradrenergic input into the region, as by blunted bororeflex function, or by local overexpression of inhibitory neuromodulators. We have observed blunting of both arterial (Calhoun et al., 1991) and cardiopulmonary reflex gain (Thornton et al., 1989; Nakamura et al., 1993) in SHR that is exaggerated by dietary NaCl supplementation and have demonstrated that a high NaCl diet attenuates baroreflex-induced NE release in the AHN of SHR (Peng et al., 1995). Further studies have defined a role for atrial natriuretic peptide (ANP) as a neuromodulator that tonically inhibits NE release from nerve terminals in the AHN, thus reducing activation of sympathoinhibitory neurons in this brain region, resulting in increases in sympathetic outflow and BP (Oparil et al., 1996). Peripheral and central neural mechanisms linking dietary NaCl with tonic regulation of BP by the AHN in SHR are described in detail in this review.

Key words: angiotensin II, anterior hypothalamus, atrial natriuretic hormone, NaCl-sensitive hypertension, norepinephrine, organum vasculosum of the lamina terminalis

8. THE ANTERIOR HYPOTHALAMUS AND SALT-SENSITIVE HYPERTENSION

FIGURE 1

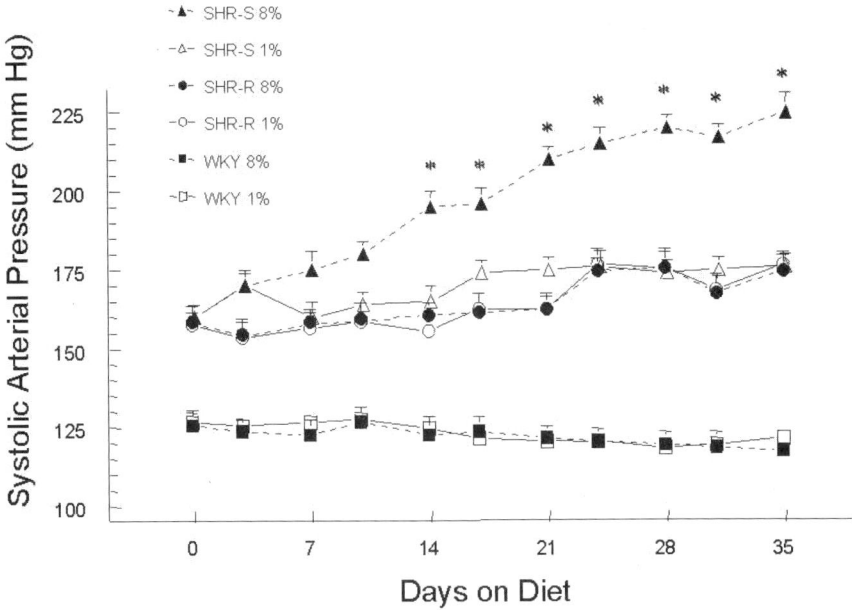

Figure 1. Blood pressure (tail cuff) profiles of young male salt-sensitive or salt-resistant spontaneously hypertensive rats (SHR-S, SHR-R) and normotensive salt resistant Wistar Kyoto (WKY) rats fed basal (1% NaCl) or high salt (8% NaCl) diets. Values are means ±SEM. N=8-10/group.

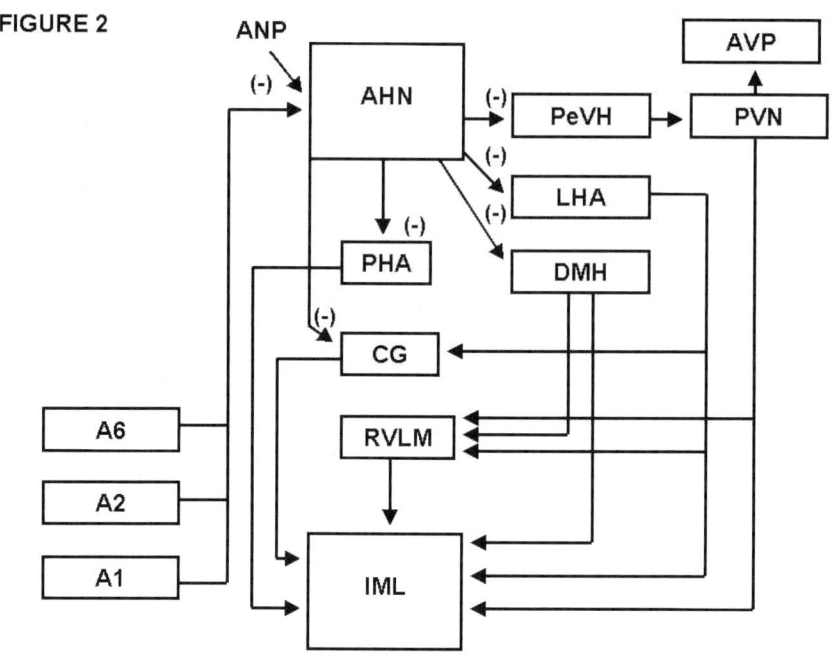

Figure 2. Schematic diagram of neural pathways mediating salt sensitive hypertension in SHR. Nerve terminals that arise from cells in the brainstem (A1-A6) release NE onto the AHN neurons, thereby increasing the AHN mediated inhibition of surrounding sympathoexcitatory neurons in the dorsomedial (DMH), lateral (LHA), paraventricular (PVN), periventricular (PeVH), and posterior (PHA) hypothalamic nuclei and the central gray (CG) of the brainstem. Inhibition of these neurons leads to a decrease in drive to neurons in the reticular and rostroventrolateral nuclei of the medulla (RVLM) and the intermediolateral nuclei (IML) of the spinal cord (both direct and indirect, e.g., via the RVLM), thus decreasing sympathetic drive to the periphery. The AHN also inhibits the release of vasopressin (AVP) from the PVN, thus augmenting the sympathoinhibition.

8. THE ANTERIOR HYPOTHALAMUS AND SALT-SENSITIVE HYPERTENSION

FIGURE 3

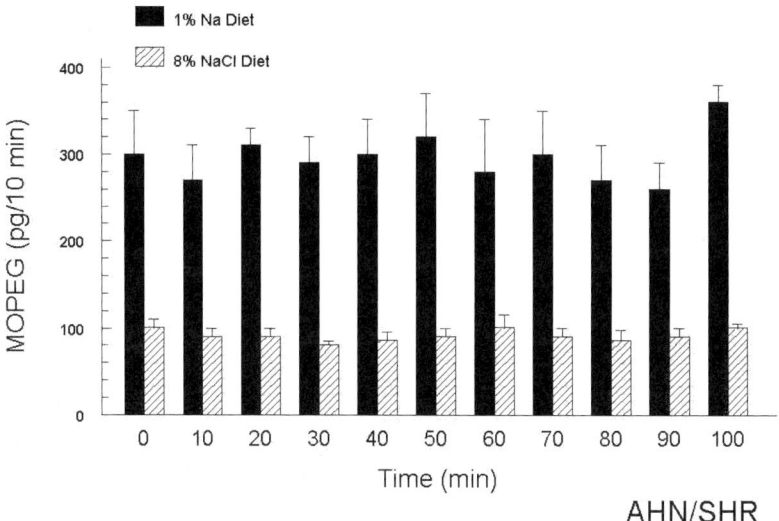

Figure 3. The 3-hydroxy-4-methoxyphenylglycol (MOPEG) content of each 10 minute aliquot of AHA perfusate in 1% or 8% NaCl-fed SHR-S. Each vertical bar represents the mean ± SEM (Chen et al., 1991).

THE AHN HAS A SYMPATHOINHIBITORY FUNCTION

Several lines of evidence suggest that the AHN has a sympathoinhibitory and and depressor function. Electrical stimulation of the AHN reduces BP and heart rate in both normotensive and hypertensive rats, while electrical stimulation of surrounding nuclei, such as the paraventricular, ventromedial and posterior hypothalamic nuclei, increases BP and heart rate (Folkow et al., 1959; Spyer, 1972; Mitchell et al., 1988). Large lesions of the anterior hypothalamic region produce fulminating hypertension in normotensive rats (Nathan and Reis, 1975), and in SHR, bilateral, neurotoxin-induced lesions that are restricted to the AHN result in BP increases equal to those induced by high NaCl diets (Wyss et al., 1990). This is in marked contrast to lesions in the posterior hypothalamic area or anteroventral third ventricle area (AV3V), both of which tend to decrease BP (Hartle and Brody, 1982). Miyajima and Bunag suggest that both vagal and sympathetic nerve activity are controlled by the AHN, since destruction of this area results in increased pressor and decreased bradycardic responses to i.v. injection of phenylephrine (Miyajima and Bunag, 1985). ICV NaCl loading reduces the sympathoinhibitory responses elicited by anterior the AHN stimulation (Miyajima and Bunag, 1984, 1985). Further, the discharge rate of anterior hypothalamic neurons is increased by carotid sinus nerve stimulation (Calaresu and Ciriello, 1980) or baroreceptor activation in isolated carotid sinus preparations (Hilton and Spyer, 1971). Neurons in this region respond to atrial stretch (Grizzle et al., 1975), and NE release is increased in the AHN during baroreceptor activation (Peng et al., 1995).

Large (>200nl) injections of NE or α_2-adrenoceptor agonists into the anterior hypothalamus lower BP in normotensive rats (Folkow et al., 1959), and small, discrete (<50 nl) injections of clonidine into the AHN reduce BP and heart rate in both normotensive and hypertensive rats (Wyss et al., 1988). In contrast, microinjections of clonidine into the lateral or posterior hypothalamic areas cause increases in BP and heart rate. Studies in the rabbit indicate that NE turnover in the AHN is increased during baroreflex activation and decreased during baroreceptor unloading (Robinson et al., 1983). Together, these findings suggest that the neurons of the AHN subserve a sympathoinhibitory function and are activated by local release of NE.

CHRONIC CLONIDINE MICROINFUSION INTO THE AHN PREVENTS NaCl-SENSITIVE HYPERTENSION IN SHR

In order to assess the functional role of the AHN in the pathogenesis of NaCl-sensitive hypertension in SHR, we tested two related hypotheses: (1) that chronic lesions of the AHN would prevent the NaCl-induced BP increase and (2) that chronic microinfusion of an α_2-adrenergic receptor agonist into the AHN would prevent NaCl-sensitive hypertension. The results supported both hypotheses. Lesions of the AHN cause a ~25 mm Hg rise in BP in SHR consuming a basal NaCl diet, but following the lesion dietary NaCl supplementation no longer elicits a hypertensive response (Wyss et al., 1990). In normotensive, NaCl-resistant WKY, the lesions have little effect. Further, in SHR fed an 8% NaCl diet, chronic microinfusion of clonidine into the AHN prevents the hypertensive effect of the dietary NaCl supplementation and the related increase in left ventricular weight and plasma NE concentration (Jin et al., 1991b) (Fig. 4). In contrast, chronic microinfusion of clonidine into the AHN does not significantly affect any of these parameters in SHR on a 1% NaCl diet. Further, intravenous infusion of clonidine at the rate used for the AHN infusion does not alter any of these measures in 8% NaCl-fed SHR-S (Fig. 4). Together, these data support the hypothesis that NaCl-sensitive hypertension in SHR is associated with diminished sympathoinhibitory function of the AHN and more specifically, of central α_2-adrenergic receptors.

8. THE ANTERIOR HYPOTHALAMUS AND SALT-SENSITIVE HYPERTENSION

FIGURE 4

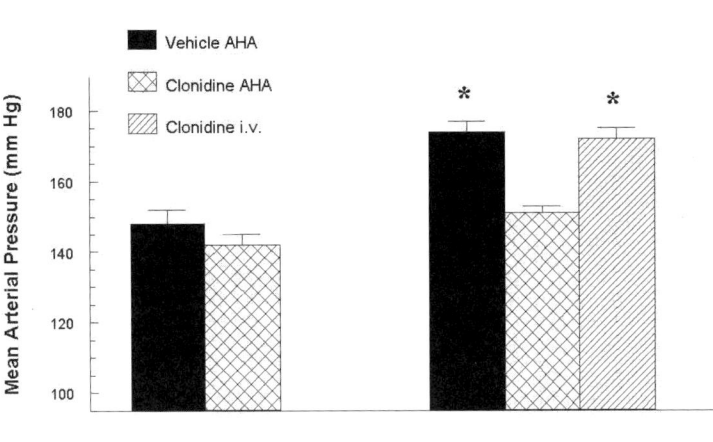

Figure 4. Effects of chronic microinfusion of clonidine into the AHN or femoral vein on mean arterial pressure. * p<0.01 compared with both 1% NaCl groups and the 8% NaCl group receiving clonidine infusion in the AHN (Jin et al., 1991b).

DIETARY NaCl EXCESS ALTERS PLASMA SODIUM REGULATION

One possible mechanism by which alterations in dietary NaCl can modulate sympathetic nervous system activity and BP is by altering plasma Na^+ concentration. We have demonstrated that the response of plasma Na^+ to dietary NaCl supplementation is altered in NaCl-sensitive SHR, and that this alteration is reflected in BP regulation (Fang et al., 2000). Arterial pressure displays a circadian rhythm that is entrained to the light/dark cycle, resulting in the highest BPs during the night when rats are most active and the lowest BPs during the day when rats typically sleep (Calhoun et al., 1994; Carlson et al., 1998). Ingestive behavior follows a similar pattern, i.e., rats consume food primarily during the night (Madrid et al., 1998). Therefore, we tested the hypothesis that plasma Na^+ displays a 24-hour rhythm parallel to the BP rhythm and that a dysregulation of this rhythm may contribute to changes in BP observed throughout the 24-hour cycle. In contrast to our predictions, the plasma Na^+ and arterial pressure rhythms are inversely correlated (Fig. 5). Further, plasma Na^+ is elevated in SHR (compared to WKY) on a basal NaCl diet without any differential effect on diurnal rhythms (Fig. 6). Dietary NaCl excess elevates average plasma Na^+ similarly in both strains, but blunts the normal diurnal rhythm of plasma Na$^+$ in SHR only (Fig. 6). This change leaves plasma Na$^+$ very high at the time that natriuretic hormones (e.g., aldosterone) (Hilfenhaus, 1976; Jensen and Pedersen, 1997; Charloux et al., 1999), vasopressin and oxytocin (Yasin et al., 1993; Morawska-Barszczewska et al., 1996; Pasqualetti et al., 1998), and the renin-angiotensin system (i.e., angiotensin II) (Hilfenhaus, 1976; Jensen and Pedersen, 1997) and sympathetic nervous system activity are high, thus potentially contributing to the NaCl-sensitive rise in arterial pressure in SHR.

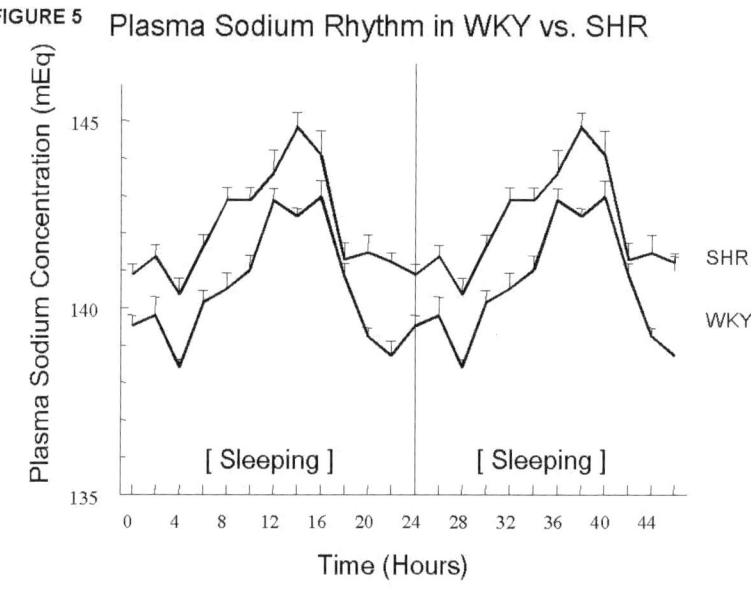

Figure 5. A comparison of plasma Na+ rhythms in young male SHR and WKY on basal NaCl diets. The rhythms were analyzed by Pharmfit and are shown for a 48-hour period. Plasma Na+ concentration in SHR compared to WKY is from 0.5-2 mEq/liter higher throughout the day (sleeping period) (Fang et al., 2000).

Our previous studies have addressed the question of whether the 4-5 mEq increase in plasma Na$^+$ that we have observed in awake SHR on a high NaCl diet is capable of elevating BP by 20 mm Hg, the usual NaCl-induced increment in BP in this strain. This increment in plasma Na$^+$ can be sensed by the brain, since arginine vasopressin release from the hypothalamus has been shown to increase in response to a Na$^+$ stimulus of this magnitude (Thrasher, 1985). Further, an infusion of hypertonic saline that causes an acute 4 mEq/L increase in plasma Na$^+$ elevates BP by ~20 mmHg and reduces

8. THE ANTERIOR HYPOTHALAMUS AND SALT-SENSITIVE HYPERTENSION

the AHN NE release in SHR but not in control WKY rats (Peng et al., 1996a). Li and associates demonstrated that in normotensive control rats, drinking hypertonic saline increases plasma Na^+ by ~4 mEq/L but does not alter BP. In contrast, in mRen rats, which overexpress renin, drinking hypertonic saline is associated with increases in both plasma Na^+ concentration and BP (Li et al., 1998). Together, these findings suggest that small increases in plasma Na^+ can increase BP, especially in at risk animals.

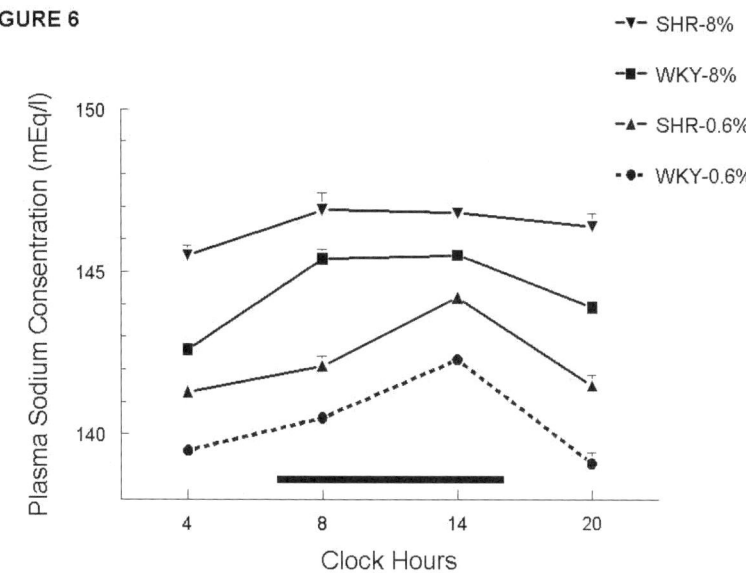

Figure 6. Plasma Na+ concentrations in WKY and SHR fed either a basal or a high NaCl diet. Both strains showed a significant increase in plasma Na+ when placed on the high NaCl diet, but only the SHR showed a significantly blunted diurnal rhythm. The black bar above the timeline indicates the daytime hours during which rats are typically asleep (Fang et al., 2000).

DETECTION OF PLASMA SODIUM CONCENTRATION BY THE BRAIN

Plasma Na^+ can be detected by the nervous system via a number of mechanisms, including Na^+ receptive neurons in the kidney and liver, changes in intercerebroventricular Na^+ and monitoring of plasma Na^+ by areas of the brain that lie outside of the blood-brain barrier. Several lines of evidence suggest that peripheral osmoreceptors responding to alterations in dietary NaCl can adjust renal sympathetic nerve activity (Morita et al., 1993; Hosomi and Morita, 1996; Morita et al., 1997). Further, hepatic and renal nerve feedback to the brain has been shown to be blunted in SHR, leading to a decreased ability of the brain to orchestrate an adequate response to high dietary NaCl. However, the impact of these mechanisms on chronic regulation of mean arterial pressure was not elucidated by these studies. Results of our own studies indicate that in the rat, hepatic osmoreceptors participate in chronic arterial pressure regulation overall, but do not contribute appreciably to dietary NaCl-induced increases in arterial pressure in SHR (Carlson et al., 1998). We demonstrated that removal of the sensory nerves to the liver elevates BP in normotensive, NaCl-resistant WKY, but does not induce NaCl sensitive hypertension in these rats (Carlson et al., 1998), nor does it alter arginine vasopressin release in WKY or SHR (Carlson and Wyss, 1999). Further, renal sensory denervation does not alter the BP response to dietary NaCl excess in either SHR or WKY, and changes in intracerebroventricular Na^+ do not appear to contribute significantly to NaCl-sensitive hypertension in SHR (Sripairojthikoon et al., 1989; Mozaffari et al., 1990; Hosomi and Morita, 1996).

Another important mechanism for detecting plasma Na^+ concentration is the organum vasculosum of the lamina terminalis (OVLT) (Thrasher et al., 1982; Oldfield et al., 1991; Renauld et al., 1993; Han and Rowland, 1995; Bourke and Richard, 2001). In the rat, the OVLT appears to contain the majority of CNS osmoreceptive neurons that lie outside of the blood-brain barrier. Bourque and associates demonstrated that osmotic challenge to the OVLT changes the firing rate of neurons in the hypothalamic supraoptic nucleus (Bourque and Oliet, 1997; Richard and Bourque, 1995). Further, lesions of the OVLT eliminate drinking responses to increased plasma osmolarity (McKinley et al., 1982; Thrasher et al., 1982). We hypothesized that the OVLT modifies the activity of neurons in the AHN in response to alterations in plasma Na^+ concentration, thereby decreasing the AHN NE release in situations where plasma Na^+ concentration is chronically elevated. To test the importance of this mechanism in the SHR, we lesioned the polysynaptic pathway from the OVLT to the AHN and then challenged the rats with a hypertonic saline infusion (Fig. 7) (Peng et al., 2000). As predicted, hypertonic saline infusion resulted in a decrease in the AHN NE release on the side of the brain in which the pathway between the OVLT and the AHN was intact; however, on the side that was lesioned, the hypertonic saline infusion resulted in an increase in release of the AHN NE. In contrast, the lesion of the pathway between the OVLT and the AHN had no effect on responses of the AHN NE release to increases in arterial pressure induced by activation of peripheral α_1 adrenergic receptors, as the latter response is mediated by the nucleus of the solitary tract. Together with data from other studies, these results suggest that the OVLT plays a critical role in mediating

8. THE ANTERIOR HYPOTHALAMUS AND SALT-SENSITIVE HYPERTENSION

responses in the AHN NE release to alterations in plasma Na^+ concentration. Interestingly, in NaCl-resistant WKY there is only a small transient decrease in the AHN NE release in response to hypertonic saline infusion. Together, these findings suggest that the sensitivity of NE release mechanisms in the AHN to alterations in plasma Na^+ concentration is increased in SHR compared to WKY.

FIGURE 7

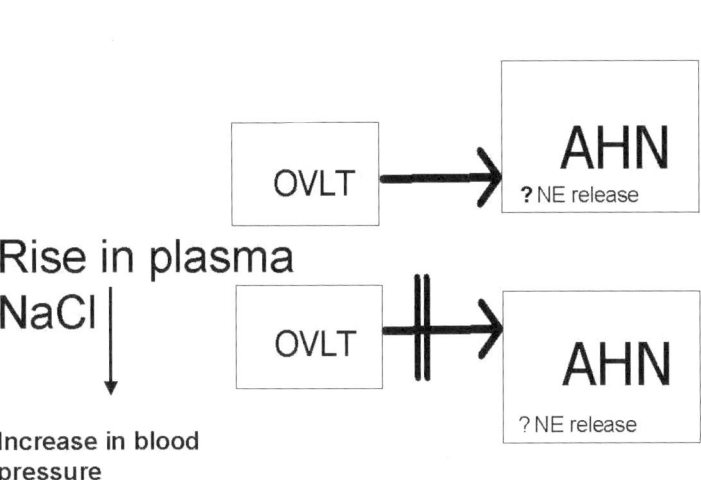

Figure 7. Diagrammatic summary of the pathways by which plasma Na+ regulates the release of NE in the AHN. In this experiment, the polysynaptic connection between OVLT and the AHN is severed on one side of the brain, but intact on the other side. In this condition, the infusion of hypertonic saline elicits a rise in BP, and in the intact the AHN it causes a decrease in the AHN NE release. In contrast, in the contralateral AHN (which is cut off from the OVLT), the saline infusion and associated rise in BP increase local NE release, suggesting that OVLT is part of the circuit that decreases the AHN NE in the SHR in response to a high NaCl diet (Peng et al., 2000).

A rise in plasma Na^+ of ~4 mEq/L is sufficient to alter the AHN NE release. Our studies cited above suggest that such a rise in plasma Na^+ is physiologically relevant (Peng et al., 1996a). In SHR and WKY an 8% NaCl diet elevates plasma Na^+ by ~4 mEq/L within 48 hours, and the increase in plasma Na^+ persists for several weeks (Fang et al., 1999). In SHR chronic exposure to a high NaCl diet (Peng et al., 1995, 1996a) decreases the AHN NE release by ~60%, similar to the effect of the acute hypertonic saline infusion.

BRAIN ANGIOTENSIN II ALTERS THE AHN NE RESPONSES

Angiotensin II plays a role in the cascade of events that leads to NaCl-sensitive hypertension in SHR. Our previous studies demonstrate that microinjection of the selective angiotensin II type 1 (AT_1) receptor blocker losartan into the AHN causes a significant, acute, dose-related fall in BP in SHR but not in NaCl-resistant WKY (Yang et al., 1992a). The hypotensive effects of AT_1 blockade in the AHN of SHR are exaggerated by dietary NaCl supplementation (Oparil et al., 1994). Together with findings that the AT_1 receptor numbers in the AHN are upregulated in SHR compared to WKY, these data suggest that activation of AT_1 receptors in the AHN participates in the pathogenesis of NaCl-sensitive hypertension in SHR. In contrast, microinjections of losartan into the posterior hypothalamic nucleus and of the selective angiotensin II type 2 receptor blocker (PD 123319) into the AHN has no effect on BP in either strain. Axons from the circumventricular organs course through the AHN en route to other hypothalamic nuclei and transection of this pathway eliminates BP responses to angiotensin II infusion in rats, supporting the functional significance of this pathway (Hartle and Brody, 1984). Further, as detailed above, the decrease in NE release observed in the AHN following hypertonic saline administration is mediated by a pathway from the OVLT to the AHN (Peng et al., 2000). Thus, these data suggest that angiotensin II via AT_1 receptors may modulate the transfer of information from the OVLT to the AHN, thereby participating in the cascade of events leading to NaCl-sensitive hypertension in SHR.

Further, we have found that brainstem angiotensin II is also involved in dietary NaCl-sensitive hypertension in SHR. We injected the AT_1 receptor blocker candesartan or vehicle bilaterally into the rostral ventrolateral medulla of α-chloralose anesthetized SHR, immediately preceding an iv infusion of hypertonic or normal saline. BP and heart rate were significantly elevated (21 ± 2 mm Hg) by the hypertonic (but not isotonic) saline infusion in vehicle-treated rats, but this hypertensive response was prevented in the candesartan-treated group. Together, these studies indicate that in SHR the NaCl-induced decrease in the AHN NE release is transmitted to the sympathetic nervous system by way of an angiotensin II mechanism in the brainstem. These data are consistent with the work of several groups that have demonstrated the importance of angiotensin II in modulating basal sympathetic nervous system activity, thereby affecting arterial pressure (Dampney et al., 2002; Ito et al., 2003; Mayorov and Head, 2003).

ANP IN THE HYPOTHALAMUS CONTRIBUTES TO THE INHIBITION OF NOREPINEPHRINE RELEASE IN THE AHN

While circulating atrial ANP plays an important role in the regulation of BP and fluid and electrolyte homeostasis (De Bold et al., 1981), neuronally produced ANP can directly modulate neuronal activity in the brain and thereby modify BP and other cardiovascular responses (Chang et al., 1989; Phillips et al., 1989; Ermirio et al., 1990; Vatta et al., 1994). Altered synthesis and/or release of ANP in the brain appear to contribute to hypertension in several rodent models, including the SHR (Imada et al., 1985; Jin et al., 1989, 1991a). Our laboratory demonstrated for the first time a functional role for endogenous ANP in the AHN and nucleus tractus solitarius in BP control in SHR (Oparil and Wyss, 1993; Oparil et al., 1996). We showed that microinjection of a blocking monoclonal antibody to ANP into the caudal nucleus tractus solitarius, the site of termination of baroreceptor and chemoreceptor afferents, elicits significant pressor responses in SHR (but not in WKY controls) independent of dietary NaCl (Yang et al., 1992a, b). Further, the antibody injections enhance the sensitivity of baroreflex control of heart rate and lumbar sympathetic nerve activity in SHR; microinjection of exogenous ANP blunts these baroreflex control mechanisms in the hypertensive model but not in normotensive controls (Zhu et al., 1996). Together, these studies provide the first demonstration that endogenous ANP in the nucleus tractus solitarius modulates any aspect of baroreflex function in hypertension.

Subsequent studies demonstrated that in the AHN, endogenous ANP participates in cardiovascular homeostasis by negatively modulating NE release. In SHR on a basal NaCl diet, local microperfusion of ANP into the AHN causes a significant, concentration-related decrease in local extracellular NE concentration, reflecting reduced release (Peng et al., 1996b). These data indicate that receptors for ANP are present in the AHN and that their activation can negatively modulate local release of NE from nerve terminals. Microperfusion of the ANP-C receptor agonist (ANP 4-23) into the AHN results in an enhanced reduction in extracellular NE metabolites in the AHN, suggesting that endogenous ANP in the AHN can be elevated by blockade of the ANP-C receptors, with resultant inhibition of the AHN NE release. In contrast, neither microperfusion of exogenous ANP nor blockade of the ANP clearance receptors effects a significant reduction in the AHN NE concentration in SHR on a high NaCl diet or in WKY on a basal NaCl diet, suggesting that in these rats, tonic inhibition of the AHN NE release by ANP is already maximal.

These data provide strong evidence that ANP in the AHN of SHR is an inhibitory neuromodulator of NE release. ANP likely regulates NE release in the hypothalamus by acting as an inhibitory neuromodulator of the presynaptic release of NE (Nakamaru and Inagami, 1986; Holtz et al., 1987; Rankin et al., 1987; Drewett et al., 1988). In PC-12 cells in vitro and peripheral nerve terminals in vivo, ANP has a significant, direct inhibitory effect on the release of NE. In man, ANP inhibits sympathetic nervous system activity (Floras, 1995), and the depressor effect of ANP is related to the ability of circulating ANP to modify noradrenergic neurotransmission (Lang et al., 1992). In the rat hypothalamus, ANP regulates the pressor action of angiotensin II (Shibata et al., 1993) and inhibits neuronal firing (Yamashita and Kannan, 1992). The effect of ANP on extracellular NE concentration in the brain is likely mediated by two direct actions of ANP on noradrenergic nerve terminals: (1) inhibition of NE release from nerve terminals, as documented above, and (2) enhancement of the reuptake of NE by nerve terminals (Fernandez et al., 1993; Vatta et al., 1994). Further, in SHR compared to WKY, ANP concentration is increased selectively in the AHN but not in other hypothalamic nuclei (Jin et al., 1991a).

ANP in the AHN plays a role in the tonic regulation of arterial pressure in SHR but not in WKY (Yang et al., 1990). Microinjection of a blocking monoclonal antibody to ANP directly into the AHN causes a significant decrease in arterial pressure and heart rate in SHR but not in WKY. Neither microinjection of IgG into the AHN nor microinjection of ANP antibody into the posterior hypothalamic area alters BP in either strain. Together these observations support the hypothesis that the increased stores of ANP in the AHN of SHR are functionally significant in the tonic.

To further test the specificity of this mechanism, we microinjected ANP into the AHN of conscious C57BL/6 mice in which the α_2-adrenergic receptor was functionally deleted by a single point mutation (Peng et al., 2003a). In control mice, microinjection of either clonidine or guanabenz (10^{-3} to 10^{-7} M) caused a rapid fall in mean arterial pressure that lasted for several minutes, suggesting that the noradrenergic mechanism in the AHN are comparable in rat and mouse. In the knockout mice the responses to the adrenergic agonists were blocked. Microinjection of ANP (10^{-6} to 10^{-7}M) caused a rapid increase in BP in the control mice that was similar to the responses previously observed in WKY rats (Fig. 8). In contrast, the ANP microinjections did not significantly alter BP in the knockout mice. These experiments demonstrate that in the AHN of the mouse (and also likely in the rat), α_{2A}-adrenergic receptors mediate both sympathoinhibitory responses to α_2-adrenergic receptor agonists and the action of ANP.

8. THE ANTERIOR HYPOTHALAMUS AND SALT-SENSITIVE HYPERTENSION

ESTROGEN IN FEMALE SHR GREATLY BLUNTS NaCL-SENSITIVE HYPERTENSION VIA NEURAL MECHANISMS

Compared to men, premenopausal women have a lower prevalence of hypertension, but after menopause this relative protection appears to be lost (Ross et al., 1981; Kotchen et al., 1982). Our studies in the SHR model provide evidence that this may be related to loss of endogenous ovarian estrogen. Our studies demonstrate that young female SHR (compared to age-matched males) have a lower BP and a reduced hypertensive response to excess dietary NaCl (Fang et al., 2001). Elimination of most endogenous estrogen by ovariectomy causes no significant change in BP and only a modest alteration in NaCl-sensitive hypertension, but simultaneous removal of both endogenous and dietary estrogens (phytoestrogens in soy-based rodent chow) leads to a small rise in baseline BP and a very large increase in the hypertensive response to a high NaCl diet (Fig. 9) (Fang et al., 2001).

FIGURE 8

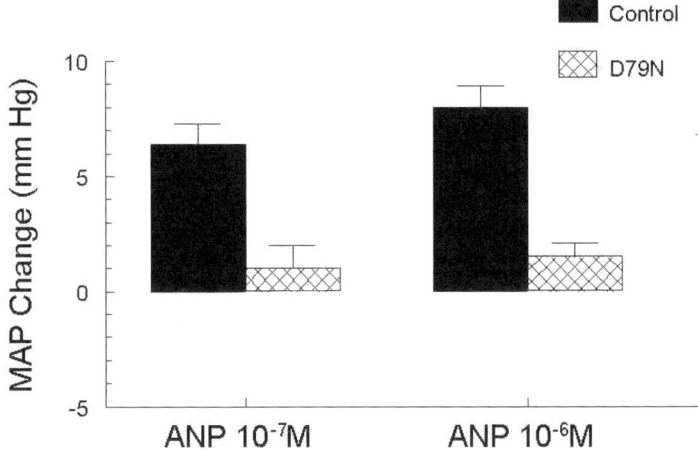

Figure 8. Changes in mean arterial pressure (MAP) in wild-type control and D79N mice after microinjection of ANP into the AHN (Peng et al., 2003a).

Further, in male SHR, a high NaCl diet activates the sympathetic nervous system and, thereby, increases BP (Chen et al., 1988, 1991). A key gender difference in the response to the dietary NaCl excess is that in male (but not female) SHR, the high NaCl diet decreases NE release in the AHN (Chen et al., 1991). In the young female SHR, the synergy between NaCl excess and estrogen depletion is clearly the result of overactivity of the sympathetic nervous system (Fang et al., 2001). We found that in middle-aged female SHR the loss of estrogen allows dietary NaCl excess to decrease NE release in the AHN, and thereby leads to NaCl-sensitive hypertension (Peng et al., 2003b).

The results of our most recent study suggest that the mechanisms that underlie NaCl-sensitive hypertension in intact middle-aged female SHR are, at least in some respects, similar to the mechanisms that underlie NaCl-sensitive hypertension in young male and young estrogen-depleted female SHR (Peng et al., 2003b). As discussed previously, in young male SHR, the ability of a high NaCl diet to increase BP is dependent on the sympathetic nervous system and a decrease in the AHN NE release. NaCl-sensitive hypertension in middle-aged female SHR appears to be related to a similar NaCl-induced reduction in NE release in the AHN, and estrogen depletion may increase NaCl-sensitivity through the same neural mechanism (Peng et al., 2003b). On a basal NaCl diet, middle-aged, estrogen depleted female rats display slightly increased BP and decreased the AHN NE (~20%) release. Both effects are reversed by chronic estrogen replacement with 17β-estradiol. Importantly, there is a significant inverse correlation between BP and the AHN NE. These data suggest that both dietary NaCl excess and estrogen depletion raise BP in middle-aged female SHR by a mechanism that involves reduced the AHN-mediated sympathoinhibition.

8. THE ANTERIOR HYPOTHALAMUS AND SALT-SENSITIVE HYPERTENSION

FIGURE 9

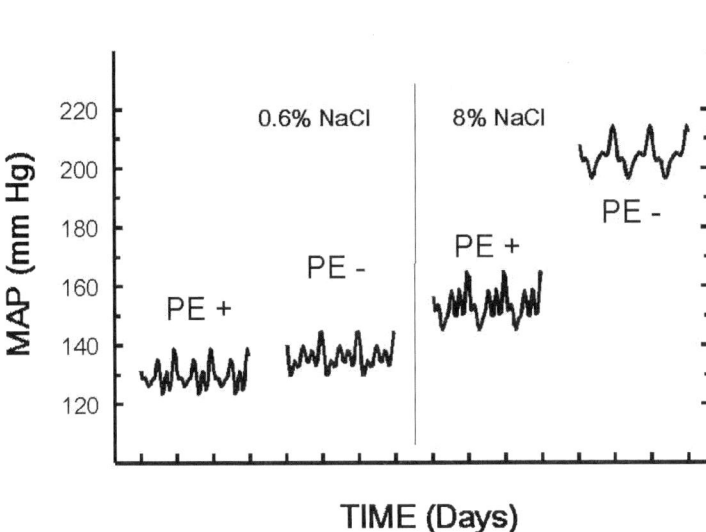

Figure 9. Twenty-four-hour circadian rhythm (PHARMFIT analysis) of BP in ovariectomized SHR fed either a basal or high NaCl diet with (PE+) or without (PE-) dietary phytoestrogens. The high NaCl diet elevated BP in the PE- group, and removal of dietary phytoestrogens exacerbated the BP response to NaCl. Elimination of phytoestrogens did not significantly affect BP in rats fed the basal NaCl diet. (Fang et al., 2001).

CONCLUSIONS

Our studies indicate that in SHR dietary NaCl-sensitive hypertension is mediated by a cascade that begins with a plasma Na^+ imbalance (Fig. 10). The increased plasma Na^+ appears to be detected by the OVLT, which in turn indirectly transmits this information to NE terminals in the AHN. This cascade releases ANP into the AHN, thereby inhibiting local NE release and decreasing the firing rate of AHN neurons. This leads to a withdrawal of the AHN-mediated inhibition of hypothalamic and/or brainstem sympathoexcitatory neurons and a resulting increase in drive to the RVLM, mediated, at least in part, by angiotensin II. The increased firing rate of the RVLM neurons increases sympathetic nerve activity and vasoconstriction, leading to chronically elevated BP. Finally, estrogen blunts this cascade at a point proximal to the AHN in young SHR, but is less effective in decreasing NaCl-sensitive hypertension in aged SHR. The extent to which any part of this cascade underlies dietary NaCl-sensitive hypertension in humans is unclear, but several lines of evidence suggest that withdrawal of centrally mediated sympathoinhibition occurs in clinical hypertension.

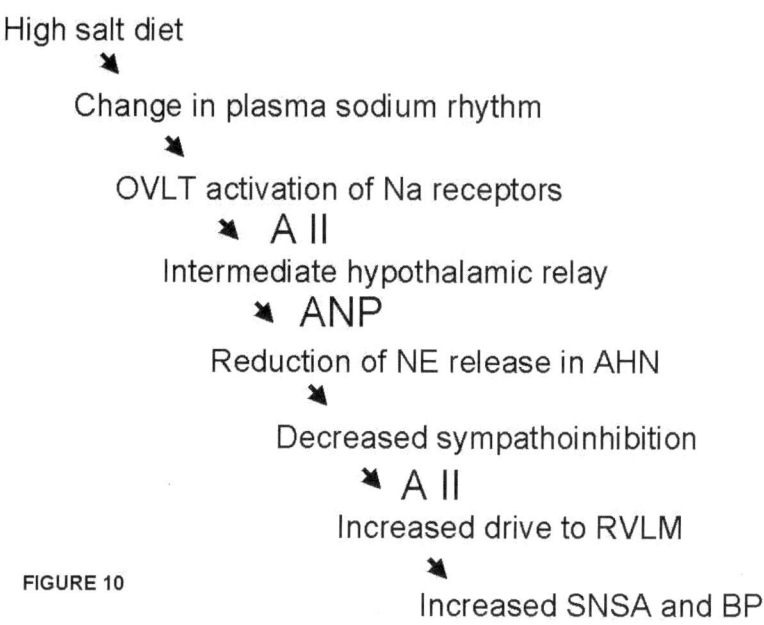

FIGURE 10

Figure 10. Schematic diagram of the mechanism by which dietary NaCl supplementation increases sympathetic nervous system activity and BP. A high NaCl diet elevates and dysregulates plasma Na+ concentration, leading to an activation of sodium sensitive neurons in the OVLT which via a polysynaptic pathway decrease the release of NE AHN. Two of the neurotransmitters in this pathway are angiotensin II (AII) and ANP. The reduction in the AHN NE leads to disinhibition of the rostral ventrolateral medulla (RVLM), thereby increasing sympathetic nervous system activity and BP. AII is important for the transmission of this information to the RVLM.

ACKNOWLEDGMENTS

This study was supported, in part, by the National Institutes of Health grants HL 37722, HL 47081, HL 07457.

REFERENCES

Bourque, C.W., Oliet, S.H., 1997. Osmoreceptors in the central nervous system. Annu. Rev. Physiol. 59, 601-619.

Calaresu, F.R., Ciriello, J., 1980. Projections to the hypothalamus from buffer nerves and nucleus tractus solitarius in the cat. Am. J. Physiol. 239, R130-R136.

Calhoun, D.A., Wyss, J.M., Oparil, S., 1991. High NaCl diet enhances arterial baroreceptor reflex in NaCl-sensitive spontaneously hypertensive rats. Hypertension 17, 363-368.

Calhoun, D.A., Zhu, S., Wyss, J.M., Oparil, S., 1994. Diurnal blood pressure variation and dietary NaCl in spontaneously hypertensive rats. Hypertension 24, 1-7.

Carlson, S.H., Osborn, J.W., Wyss, J.M., 1998. Hepatic denervation chronically elevates arterial pressure in Wistar-Kyoto rats. Hypertension 32, 46-51.

Carlson, S.H., Wyss, J.M., 1999. Hepatic denervation does not affect plasma vasopressin response to intragastric hypertonic saline in conscious rats. Am. J. Physiol. 277, E161-E167.

Chang, M.S., Lowe, D.G., Lewis, M., Hellmiss, R., Chen, E., Goeddel, D.V., 1989. Differential activation by atrial and brain natriuretic peptides of two different receptor guanylate cyclases. Nature 341(6237), 68-72.

Charloux, A., Gronfier, C., Lonsdorfer-Wolf, E., Piquard F., Brandenberger, G., 1999. Aldosterone release during the sleep-wake cycle in humans. Am. J. Physiol. (Endocrinol. Metab.) 276(1 Pt. 1), E43-E49.

Chen, C.W., Chen, Y.F., Meng, Q.C., Wyss, J.M., Oparil, S., 1991. Decreased norepinephrine release in anterior hypothalamus of NaCl-sensitive spontaneously hypertensive rats during high NaCl intake. Brain Res. 565(1), 135-141.

Chen, Y.F., Meng, Q.C., Wyss, J.M., Jin, H., Oparil, S., 1988. High NaCl diet reduces hypothalamic norepinephrine turnover in hypertensive rats. Hypertension 11(1), 55-62.

Dampney, R.A., Fontes, M.A., Hirooka, Y., Horiuchi, J., Potts, P.D., Tagawa, T., 2002. Role of angiotensin II receptors in the regulation of vasomotor neurons in the ventrolateral medulla. Clin. Exp. Pharmacol. Physiol. 29, 467-472.

De Bold, A.J., Borenstein, H.B., Veress, A.T., Sonnenberg, H., 1981. A rapid and potent natriuretic response to intravenous injection of atrial myocardial extract in rats. Life Sci. 28, 89-94.

Drewett, J., Marchand, G., Ziegler, R., Trachte, G., 1988. Atrial natriuretic factor inhibits norepinephrine release in an adrenergic clonal cell line (PC12). Eur. J. Pharmacol. 150, 175-179.

Ermirio, R., Ruggeri, P., Cogo, C.E., Molinari, C., Bergaglio, M., Calaresu, F.R., 1990. The role of ANF in functional properties of the central pathways of cardiovascular reflexes. J. Auton. Nerv. Syst. 30(Suppl.), S51-S53.

Fang, Z., Carlson, S.H., Chen, Y.F., Oparil, S., Wyss, J.M., 2001. Estrogen depletion induces NaCl-sensitive hypertension in female spontaneously hypertensive rats. Am. J. Regul. Integr. Comp. 281, R1934-R1939.

Fang, Z., Carlson, S.H., Peng, N., Wyss, J.M., 2000. Circadian rhythm of plasma sodium is disrupted in spontaneously hypertensive rats fed a high-NaCl diet. Am. J. Physiol. Regul. Integr. Comp. Physiol. 278, R1490-R1495.

Fang, Z., Sripairojthikoon, W., Calhoun, D.A., Zhu, S., Berecek, K.H., Wyss, J.M., 1999. Interaction between lifetime captopril treatment and NaCl-sensitive hypertension in spontaneously hypertensive rats and Wistar-Kyoto rats. J. Hypertension 17, 983-991.

Fernandez, B.E., Vatta, M.S., Bianciotti, L.G., 1993. Comparative effects of bradykinin and atrial natriuretic factor on neuronal and non-neuronal NE uptake in the central nervous system of the rat. Arch. Int. Physiol. Biochim. Biophys. 101, 337-340.

Floras, J.S., 1995. Inhibitory effect of atrial natriuretic factor on sympathetic ganglionic neurotransmission in humans. Am. J. Physiol. 269, R406-R412.

Folkow, B., Johansson, B., Oberg, B., 1959. A hypothalamic structure with a marked inhibitory effect on tonic sympathetic activity. Acta Physiol. Scand. 47, 262-270.

Grizzle, W.E., Johnson, R.N., Schramm, L.P., Gann, D.S., 1975. Hypothalamic cells in an area mediating ACTH release respond to right atrial stretch. Am. J. Physiol. 228, 1039-1045.

Han, L., Rowland, N.E., 1995. Sodium depletion and Fos-immunoreactivity in lamina terminalis. Neurosci. Lett. 193, 173-176.

Hartle, D.K., Brody, M.J., 1982. Hypothalamic vasomotor pathways mediating the development of hypertension in the rat. Hypertension 4, III68-III71.

Hartle, D.K., Brody, M.J., 1984. The angiotensin II pressor system of the rat forebrain. Circ. Res. 54, 355-366.

Hilfenhaus, M., 1976. Circadian rhythm of the renin-angiotensin-aldosterone system in the rat. Arch. Toxicol. 36, 305-316.

Hilton, S.M., Spyer, K.M., 1971. Participation of the anterior hypothalamus in the baroreceptor reflex. J. Physiol. (Lond.) 218, 271-293.

Holtz, J., Sommer, O., Bassenge, E., 1987. Inhibition of sympathoadrenal activity by atrial natriuretic factor in dogs. Hypertension 9, 350-354.

Hosomi, H., Morita, H., 1996. Hepatorenal and hepatointestinal reflexes in sodium homeostasis. NIPS 11, 103-107.

Imada, T., Takayanagi, R., Inagami, T., 1985. Changes in the content of atrial natiuretic factor with the progression of hypertension in spontaneously hypertensive rats. Biochem. Biophys. Res. Commun. 133, 759-765.

Ito, S., Hiratsuka, M., Komatsu, K., Tsukamoto, K., Kanmatsuse, K., Sved, A.F, 2003. Ventrolateral medulla AT1 receptors support arterial pressure in Dahl salt-sensitive rats. Hypertension 41, 744-750.

Jensen, L.W., Pedersen, E.B., 1997. Nocturnal blood pressure and relation to vasoactive hormones and renal function in hypertension and chronic renal failure. Blood Pressure 6, 332-342.

Jin, H., Yang, R.H., Chen, Y.F., Wyss, J.M., Oparil, S., 1989. Dietary NaCl loading enhances antihypertensive effect of guanabenz in spontaneously hypertensive rats. Am. J. Hypertension 2, 435-439.

Jin, H., Yang, R.H., Chen, Y.-F., Wyss, J.M., 1991a. Altered stores of atrial natriuretic peptide in specific brain nuclei of NaCl-sensitive spontaneously hypertensive rats. Am. J. Hypertension 4, 449-455.

Jin, H.K., Yang, R.H., Wyss, J.M., Chen, Y.F., Oparil, S., 1991b. Intrahypothalamic clonidine infusion prevents NaCl-sensitive hypertension. Hypertension 18, 224-229.

Klangkalya, B., Sripairojthikoon, W., Oparil, S., Wyss, J.M., 1988. High NaCl diets increases anterior hypothalamic α_2 adrenoceptors in SHR. Brain Res. 451, 77-84.

Kotchen, J.M., McKean, H.E., Kotchen, T.A., 1982. Blood pressure trends with aging. Hypertension 4, III128-III134.

Lang, C.C., Choy, A.M., Balfour, D.J., Struthers, A.D., 1992. Prazosin attenuates the natriuretic response to atrial natriuretic factor in man. Kidney Int. 42, 433-441.

Li, P., Morris, M., Ferrario, C.M., Barrett, C., Ganten, D., Callahan, M.F., 1998. Cardiovascular, endocrine, and body fluid-electrolyte responses to salt loading in mRen-2 transgenic rats. Am. J. Physiol. (Heart Circ.Physiol.) 275, H1130-H1137.

8. THE ANTERIOR HYPOTHALAMUS AND SALT-SENSITIVE HYPERTENSION

Madrid, J.A., Sanchez-Vazquez, F.J., Lax, P., Matas, P., Cuenca, E.M., Zamora, S., 1998. Feeding behavior and entrainment limits in the circadian system of the rat. Am. J. Physiol. 275, R372-R383.

Mayorov, D.N., Head, G.A., 2003. AT1 receptors in the RVLM mediate pressor responses to emotional stress in rabbits. Hypertension 41, 1168-1173.

McKinley, M.J., Denton, D.A., Leksell, L.G., Mouw, D.R., Scoggins, B.A., Smith, M.H., Weisinger, R.S., Wright, R.D., 1982. Osmoregulatory thirst in sheep is disrupted by ablation of the anterior wall of the optic recess. Brain Res. 236, 210-215.

Mitchell, V., Oparil, S., Wyss, J.M., 1988. Sympathoinhibitory response of hypothalamic neurons to electrical stimulation is not altered by NaCl loading in salt sensitive spontaneously hypertensive rats. Clin. Res. 36, 38A.

Miyajima, E., Bunag, R.D., 1984. Chronic cerebroventricular infusion of hypertonic sodium chloride in rats reduces hypothalamic sympatho-inhibition and elevates blood pressure. Circ. Res. 566-575.

Miyajima, E., Bunag, R.D., 1985. Anterior hypothalamic lesions impair reflex bradycardia selectively in rats. Am. J. Physiol. 248, H937-H944.

Morawska-Barszczewska, J., Guzek, J.W., Kaczorowska-Skora, J., 1996. Cholecystokinin octapeptide and the daily rhythm of vasopressin and oxytocin release. Exper. Clin. Endocrin. Diabetes 104, 164-171.

Morita, H., Matsuda, T., Furuya F., Khanchowdhury, M., Hosomi, H., 1993. Hepatorenal reflex plays an important role in natriuresis after high NaCl food intake in conscious dogs. Circ. Res. 72, 552-559.

Morita, H., Yamashita, Y., Nishida, Y., Tokuda, M., Hatase, O., Hosomi H., 1997. Fos induction in rat brain neurons after stimulation of the hepatoportal Na-sensitive mechanism. Am. J. Physiol. 272, R913-R923.

Mozaffari, M.S., Jirakulsomchok, S., Oparil, S., Wyss, J.M., 1990. Changes in cerebrospinal fluid Na+ concentration do not underlie hypertensive responses to dietary NaCl in spontaneously hypertensive rats. Brain Res. 506, 149-152.

Nakamura, Y., Calhoun, D.A., Chen, Y.F., Wyss, J.M., Oparil, S., 1993. Excitatory sympathetic reflex in NaCl-sensitive spontaneously hypertensive rats. Hypertension 22, 285-291.

Nakamaru, M., Inagami, T., 1986. Atrial natriuretic factor inhibits norepinephrine release evoked by sympathetic nerve stimulation in isolated perfused rat mesenteric arteries. Eur. J. Pharmacol. 123, 459-461.

Nathan, M.A., Reis, D.J., 1975. Fulminating arterial hypertension with pulmonary edema from release of adrenomedullary catecholamines after lesions of the anterior hypothalamus in the rat. Circ. Res. 37, 226-235.

Oldfield, B.J., Bicknell, R.J., McAllen, R.M., Weisingerm, R.S., McKinley, M.J., 1991. Intravenous hypertonic saline induces Fos immunoreactivity in neurons throughout the lamina terminalis. Brain Res. 561, 151-156.

Oparil, S., Chen, Y.F., Peng, N., Wyss, J.M., 1996. Anterior hypothalamic norepinephrine, atrial natriuretic peptide, and hypertension. Front. Neuroendocrinol. 17, 212-246.

Oparil, S., Wyss, J.M., 1993. Atrial natriuretic factor in central cardiovascular control. News Physiol. Sci. 8, 223-228.

Oparil, S., Yang, R.H., Jin, H.K., Chen, S.J., Meng, Q.C., Berecek, K.H., Wyss, J.M., 1994. Role of anterior hypothalamic angiotensin II in the pathogenesis of salt sensitive hypertension in the spontaneously hypertensive rat. Am. J. Med. Sci. 307, S26-S37.

Pasqualetti, P., Festuccia, V., Collacciani, A., Acitelli, P., Casale, R., 1998. Circadian rhythm of arginine vasopressin in hepatorenal syndrome. Nephron. 78, 33-37.

Peng, N., Chambless, B.D., Oparil, S., Wyss, J.M., 2003a. Alpha 2A-adrenergic receptors mediate sympathoinhibitory responses to atrial natriuretic peptide in the anterior hypothalamic nucleus of the mouse. Hypertension 41, 571-575.

Peng, N., Clark, J.T., Wei, C.C., Wyss, J.M., 2003b. Estrogen depletion increases blood pressure and hypothalamic norepinephrine in middle-aged spontaneously hypertensive rats. Hypertension 41, 1164-1167.

Peng, N., Meng, Q.C., King, K., Oparil, S., Wyss, J.M., 1995. Acute hypertension increases norepinephrine release in the anterior hypothalamic area. Hypertension 25, 828-833.

Peng, N., Meng, Q.C., Oparil, S., Wyss, J.M., 1996a. Acute saline infusion decreases NE release in the anterior hypothalamic area. Hypertension 27, 578-583.

Peng, N., Oparil, S., Meng, Q.C., Wyss, J.M., 1996b. Atrial natriuretic peptide regulation of NE release in the anterior hypothalamic area of spontaneously hypertensive rats. J. Clin. Invest. 98, 2060-2065.

Peng, N., Wei, C.C., Oparil, S., Wyss, J.M., 2000. The organum vasculosum of the lamina terminalis regulates NE release in the anterior hypothalamic nucleus. Neurosci. 99, 149-156.

Phillips, M.I., Kimura, B., Wang, H., Hoffman, W.E., 1989. Effect of vagotomy on brain and plasma atrial natriuretic peptide during hemorrhage. Am. J. Physiol. 257, R1393-R1399.

Rankin, A.J., Wilson, N., Ledsome, J.R., 1987. Effects of autonomic stimulation on plasma immunoreactive atrial natriuretic peptide in the anesthetized rabbit. Can. J. Physiol. Pharmacol. 65, 532-537.

Renauld, L.P., Cunningham, J.T., Nissen, R., Yang, C.R., 1993. Electrophysiology of central pathways controlling release of nerohypophysial hormones: Focus on the lamina terminalis and diagonal band inputs to the supraoptic nucleus. Ann. NY Acad. Sci. 689, 122-132.

Richard, D., Bourque, C.W., 1995. Synaptic control of rat supraoptic neurones during osmotic stimulation of the organum vasculosum lamina terminalis in vitro. J. Physiol. (Lond.) 489, (Pt 2), 567-577.

Robinson, R., Dietl, H., Bald, M., Kraus, A., Phillipu, A., 1983. Effects of short-lasting and long-lasting blood pressure changes on the release of endogenous catecholamines in the hypothalamus of the conscious, freely moving rabbit. Naunyn-Schmiedeberg's Arch. Pharmacol. 322, 203-209.

Ross, R.K., Paganini-Hill, A., Mack, T.M., Arthur, M., Henderson, B.E., 1981. Menopausal estrogen therapy and protection from death from ischaemic heart disease. Lancet 18, 858-860.

Shibata, K., Sakimura, M., Furukawa, T., 1993. Antagonism of central pressor response to angiotensin II by alpha-human atrial natriuretic polypeptide at the preoptic area and posterior hypothalamus in rats. Neuropharm. 32, 175-184.

Spyer, K.M., 1972. Baroreceptor sensitive neurons in the anterior hypothalamus of the cat. J. Physiol. (Lond.) 224, 245-257.

Sripairojthikoon, W., Oparil, S., Wyss, J.M., 1989. Renal nerve contribution to NaCl-exacerbated hypertension in spontaneously hypertensive rats. Hypertension 14, 184-190.

Thornton, R.M., Wyss, J.M., Oparil, S., 1989. Impaired reflex response to volume expansion in NaCl- sensitive spontaneously hypertensive rats. Hypertension 14, 518-523.

Thrasher, T.N., 1985. Circumventricular organs, thirst, and vasopressin secretion, In: Schrier R.W., editor. Vasopressin, ed. New York: Raven Press. p. 311-318.

8. THE ANTERIOR HYPOTHALAMUS AND SALT-SENSITIVE HYPERTENSION

Thrasher, T.N., Keil, L.C., Ramsay, D.J., 1982. Lesions of the organum vasculosum of the lamina terminalis (OVLT) attenuate osmotically-induced drinking and vasopressin secretion in the dog. Endocrinology 110, 1837-1839.

Vatta, M., Travaglianti, M., Bianciotti, L., Coll, C., Perazzo, J., Fernandez, B., 1994. Atrial natriuretic factor effects on norepinephrine uptake in discrete telencephalic and diencephalic nuclei of the rat. Brain Res. 646, 324-326.

Winternitz, S.R., Oparil, S., 1982a. Sodium-neural interactions in the development of spontaneous hypertension. Clin. Exp. Hypertension A4, 751-760.

Winternitz, S.R., Wyss, J.M., Meadows, J.R., Oparil, S., 1982b. Increased norepinephrine content of hypothalamic nuclei in association with worsening of hypertension following high sodium intake in the young spontaneous hypertensive rat. Clin. Sci. Mol. Med. 63, 339s-342s.

Wyss, J.M., Chen, Y.F., Jin, H., Gist, R., Oparil, S., 1987. Spontaneously hypertensive rats exhibit reduced hypothalamic noradrenergic input after NaCl loading. Hypertension 10, 313-320.

Wyss, J.M., Peng, N., Meng, Q.C., Chen, Y.-F., Oparil, S., 1997. The role of anterior hypothalamic area NE release in salt-sensitive hypertension in spontaneously hypertensive rats. Fundam. Clin. Pharmacol. 11, 31s-35s.

Wyss, J.M., Yang, R.H., Jin, H.K., Oparil, S., 1988. Hypothalamic microinjection of alpha 2-adrenoceptor agonists causes greater sympathoinhibition in spontaneously hypertensive rats on high NaCl diets. J. Hypertension 6, 805-813.

Wyss, J.M., Yang, R.H., Oparil, S., 1990. Lesions of the anterior hypothalamic area increase arterial pressure in NaCl-sensitive spontaneously hypertensive rats. J. Auton. Nerv. Syst. 31, 21-30.

Yamashita, H., Kannan, H., 1992. Inhibition of hypothalamic neurons by the atrial natriuretic peptide family. News Physiol. Sci. 7, 75-79.

Yang, R.H., Jin, H.K., Chen, Y.F., Wyss, J.M., Oparil, S., 1990. Blockade of endogenous anterior hypothalamic atrial natriuretic peptide with monoclonal antibody lowers blood pressure in spontaneously hypertensive rats. J. Clin. Invest. 86, 1985-1990.

Yang, R.H., Jin, H., Wyss, J.M., Chen, Y.F., Oparil, S., 1992a. Pressor effect of blocking atrial natriuretic peptide in nucleus tractus solitarii. Hypertension 19, 198-205.

Yang, R.H., Jin, H., Wyss, J.M., Oparil, S., 1992b. Depressor effect of blocking angiotensin subtype 1 receptors in anterior hypothalamus. Hypertension 19, 475-481.

Yasin, S.A., Costa, A., Besser, G.M., Hucks, D., Grossman, A., Forsling, M.L., 1993. Melatonin and its analogs inhibit the basal and stimulated release of hypothalamic vasopressin and oxytocin in vitro. Endocrinology 132, 1329-1336.

Zhu, S.T., Chen, Y.F., Wyss, J.M., Nakao, K., Imura, H., Oparil, S., Calhoun, D., 1996. Atrial natriuretic peptide blunts arterial baroreflex in spontaneously hypertensive rats. Hypertension 27, 297-302.

Chapter 9

THE PRESYMPATHETIC CELLS OF THE ROSTRAL VENTROLATERAL MEDULLA (RVLM): ANATOMY, PHYSIOLOGY AND ROLE IN THE CONTROL OF CIRCULATION

Patrice G. Guyenet and Ruth L. Stornetta
University of Virginia, Department of Pharmacology, Charlottesville, VA 22908, USA

Abstract: The ventrolateral aspect of the medulla oblongata (RVLM) contains presympathetic (PS) neurons that are highly active and receive strong inhibitory inputs from baroreceptors. These bulbospinal and barosensitive (BSBS) neurons are glutamatergic and express several other phenotypes including adrenergic (C1 cells). BSBS neurons selectively target sympathetic preganglionic neurons (SPGNs) that control the heart, kidney, adrenal medulla and the resistance vessels and are therefore essential for blood pressure maintenance and stabilization. In tissue slices, RVLM PS neurons display intrinsic beating properties that can be upregulated by slow transmitters such as angiotensin II. Glutamate, GABA and glycine are the best known mediators of fast synaptic transmission in these cells. Both glutamatergic and GABAergic inputs to BSBS neurons are regulated presynaptically by catecholamines, opioids and, probably, serotonin. The on-going activity of RVLM BSBS neurons *in vivo* is probably due to inotropic synaptic inputs and slow transmitters that upregulate the intrinsic beating properties of the cells. *In vivo,* RVLM BSBS neurons are subject to an intense GABAergic inhibitory tone indicating that dishinhibition could be crucial for the regulation of sympathetic tone. Mono or oligosynaptic inputs to BSBS neurons originate from numerous regions of the spinal cord, medulla oblongata, pons, midbrain and hypothalamus. The relative weight of these various inputs depends on the behavior or physiological status of the animal. Under anesthesia some of these brain regions are active causing excitation or disinhibition of RVLM BSBS neurons. Increased activity of RVLM BSBS neurons is suspected to contribute to the increased sympathetic tone associated with hypertension and, possibly, heart failure.

Key words: RVLM, C1 adrenergic cells, epinephrine, neural control of circulation, blood pressure, medulla oblongata.

INTRODUCTION

The rate and pattern of discharge of sympathetic preganglionic neurons (SPGNs) are the major determinants of the sympathetic tone to blood vessels and heart. The discharge of SPGNs ultimately results from the integration by these cells of all inotropic and metabotropic signals released by neurons antecedent to them, henceforth called presympathetic (PS) neurons. The retrograde transsynaptic migration of the pseudorabies virus (PRV) has provided a detailed and probably exhaustive map of the location of PS neurons (Strack and Loewy, 1990; Jansen et al., 1992; Jansen et al., 1995a; Loewy, 1998).

This chapter focuses on a specific PS pathway located in the rostral ventrolateral medulla (RVLM). This glutamatergic and adrenergic pathway plays a dominant role in regulating the activity of the SPGNs that control the heart, the kidneys and major resistance vessels. The topic has been reviewed previously (Dampney, 1994; Guyenet et al., 1996; Sun, 1996a; Blessing, 1997; Guyenet et al., 1998; Guyenet, 2000; Dampney et al., 2000; Dampney et al., 2002).

BULBOSPINAL C1 CELLS AND BLOOD PRESSURE CONTROL: A BRIEF RETROSPECTIVE

The earliest evidence that the adrenergic cells of the medulla oblongata regulate sympathetic tone can be traced to the 1973-1974 immunohistochemical work of Hökfelt and colleagues describing the location of phenylethanolamine-N-methyl transferase (PNMT), the enzyme that converts noradrenaline into adrenaline (Hökfelt et al., 1974). This work showed that, in the spinal cord, PNMT is confined to nerve terminals that are located almost exclusively within the intermediolateral cell column; whereas, PNMT-ir cell bodies, the source of these terminals, are found exclusively in the medulla oblongata. In 1981 the adrenergic innervation of the spinal cord was shown to originate mostly from the rostral end of the C1 group of adrenergic neurons located in the RVLM (Ross et al., 1981). Based upon work done in the cat, this region of the medulla oblongata was already suspected to play a key role in blood pressure control (Guertzenstein et al., 1974; Amendt et al., 1978). Using precise exploratory methods Reis and his colleagues demonstrated that the region of the medulla oblongata that is the most essential for generation of the sympathetic vasomotor tone and blood pressure homeostasis coincides with the rostral cluster of C1 cells that innervate the intermediolateral cell column (for review see Blessing, 1997). In a landmark 1984 paper (Ross et al., 1984), these authors proposed that

9. THE PRESYMPATHETIC CELLS OF THE ROSTRAL VENTROLATERAL MEDULLA (RVLM): ANATOMY, PHYSIOLOGY AND ROLE IN THE CONTROL OF CIRCULATION

the bulbospinal C1 cells may be excitatory neurons whose ongoing discharges are essential to maintain resting sympathetic tone and blood pressure. They also proposed that these cells may serve as a major relay for supraspinal vasomotor sympathetic reflexes (Reis et al., 1984; Ross et al., 1984). This theory has been extensively tested since and proven to be largely correct. Beginning in 1984, electrophysiologists reported that the C1 region of the rostral ventrolateral medulla contains bulbospinal neurons with properties congruent with Reis's concept (Brown et al., 1984; Barman et al., 1985; Brown et al., 1985; for review see Sun, 1996a; Guyenet et al., 1998). These neurons are indeed highly active at rest and powerfully inhibited by activation of arterial baroreceptors as expected of excitatory PS cells that generate the resting sympathetic vasomotor tone and mediate the baroreflex. Furthermore, in most instances, these cells display a discharge pattern that is highly correlated with that of the sympathetic efferents that innervate the heart, kidney or blood vessels of the skeletal muscles and splanchnic area (Sun, 1996a). Although clearly located in the C1 region (Morrison et al., 1988), definitive evidence that some of these barosensitive bulbospinal (BSBS) RVLM cells are C1 neurons had to await the late 90s when investigators finally succeeded in labelling BSBS cells following their neurophysiological characterization *in vivo* (Lipski et al., 1995, 1996; Schreihofer and Guyenet, 1997). The specific contribution of bulbospinal C1 cells to the generation of sympathetic vasomotor tone was recently investigated by examining the deficits caused by destroying these cells with the toxin anti-DBH-saporin (Madden et al., 1999; Schreihofer and Guyenet, 2000; Schreihofer et al., 2000). The results of these studies are consistent with the notion that the bulbospinal C1 cells have a sympathoexcitatory and vasomotor role, a hypothesis also supported by the fact that these cells express Fos under conditions known to elevate sympathetic vasomotor tone (Ceccatelli et al., 1989; McAllen et al., 1992; Chan et al., 1994 among others). Animals with up to 85% loss of bulbospinal C1 cells have weak baroreflexes and deficient sympathoexcitatory responses in response to hypotension, activation of peripheral chemoreceptors or direct stimulation of the RVLM (Madden et al., 1999; Schreihofer and Guyenet, 2000; Schreihofer et al., 2000). However, these animals still maintain a normal blood pressure and have a substantial level of splanchnic sympathetic tone that is still generated by the activity of RVLM neurons (Schreihofer and Guyenet, 2000). The residual sympathetic tone present in the lesioned rats is probably generated by the non-adrenergic BSBS neurons (see below) and by the small percentage of bulbospinal C1 cells that escape destruction by the saporin conjugate.

The glutamatergic nature of the C1 cells, suspected since about 1989 on the basis of indirect anatomical and pharmacological evidence (Morrison et al., 1989a,b), was more firmly established in 2002 when these cells were shown to contain high levels of the mRNA that encodes the vesicular glutamate transporter 2, a diagnostic marker of glutamatergic cells (Stornetta et al., 2002a,b). This observation calls for a reinterpretation of the significance of the C1 adrenergic phenotype since it now appears that this phenotype is just one of many exhibited by the glutamatergic PS neurons of RVLM that control the circulation (Stornetta et al., 1999, 2001, 2002a).

Electrophysiological Characterization of RVLM BSBS Neurons *In Vivo* and Phenotypes

The barosensitive and bulbospinal (BSBS) neurons of the RVLM are a class of PS neurons identified *in vivo* by their high resting discharge rate, pronounced inhibition by arterial baroreceptor activation and pulse rhythmicity (Brown et al., 1985). These properties underlie the current assumption that these cells control selectively the sympathetic efferents that exhibit the same characteristics i.e. renal, splanchnic and muscle vasoconstrictor and cardiac efferents. In contrast, it is thought that RVLM BSBS neurons exert little influence on the activity of sympathetic cutaneous vasoconstrictor (CVC), brown fat and other sympathetic efferents that are not regulated by baroreceptors (Morrison, 2001; Ootsuka et al., 2002).

The percentage of RVLM PS neurons that exhibit barosensitivity is uncertain, may be species dependent and, of course, depends on how restrictively the RVLM is defined. Based on the fraction of bulbospinal or rostral C1 cells that express Fos when animals are subjected to drug-induced hypotension or hemorrhage, up to 80% of RVLM PS cells may receive baroreceptor inputs (Li and Dampney, 1994; Sved et al., 1994; Dun et al., 1995; Chan et al., 1998). However this percentage could be overestimated because prolonged hypotension or hemorrhage may induce Fos expression by mechanisms other than baroreceptor unloading (e.g. stress). The figure could also be underestimated because all activated PS cells may not express Fos. In addition the calculation does not take into consideration the PS neurons that are not C1 cells. Electrophysiological evidence for the presence of baroinsensitive PS neurons in the RVLM of the rat remains anecdotal or circumstantial. The PS neurons that control the secretion of adrenaline by the adrenal medulla (Morrison et al., 2000, 2001) are among the baroinsensitive PS neurons thought to reside in the region of the RVLM (Ritter et al., 1998, 2001). The PS neurons with presumed CVC function are another potential population of baroinsensitive RVLM PS cells. In the rabbit, these cells are

9. THE PRESYMPATHETIC CELLS OF THE ROSTRAL VENTROLATERAL MEDULLA (RVLM): ANATOMY, PHYSIOLOGY AND ROLE IN THE CONTROL OF CIRCULATION

generally found medial to the BSBS neurons, in a region that would clearly qualify as rostral ventromedial medulla (RVMM), not RVLM (Ootsuka et al., 2002) but their location in the rat is not known. In brief, up to 80% of bulbospinal C1 cells may receive strong baroreceptor inputs but the RVLM may also contain PS excitatory neurons that are not barosensitive.

BSBS neurons display a wide range of axonal conduction velocities. One group has unmyelinated axons with conduction velocities in the range of 0.4-0.8 m/s. These cells express very high levels of tyrosine hydroxylase (TH) (Schreihofer and Guyenet, 1997). A second group of neurons has higher axonal conduction velocities (2-7 m/s in the rat) and consists of cells whose TH immunoreactivity ranges from strong to light to undetectable (Schreihofer and Guyenet, 1997). On average, one third of the electrophysiologically characterized BSBS neurons lack detectable TH immunoreactivity (Lipski et al., 1995; Schreihofer and Guyenet, 1997). Many of these cells express preproenkephalin (PPE) mRNA, a property that they share with some of the fast conducting C1 cells (Stornetta et al., 2001). The peptide neuropeptide Y (NPY) is made by most C1 cells with hypothalamic projections, but it is expressed by a relatively small proportion of the BSBS neurons, all of which are of the C1 variety (Stornetta et al., 1999). Based on immunohistochemical evidence, CART (cocaine and amphetamine-related transcript peptide) is expressed by most C1 cells strongly suggesting that it must be present in many BSBS neurons of the C1 variety (Dun et al., 2002). Substance P and somatostatin are expressed by some PS neurons of the RVMM / RVLM region (Jansen et al., 1995b) but there is no evidence that these peptides are present in either C1 or BSBS cells. The non-adrenergic BSBS neurons are resistant to anti- dopamine beta-hydroxylase–saporin, a toxin that selectively destroys neurons that exteriorize the enzyme dopamine beta-hydroxylase (Schreihofer et al., 2000a). This evidence reinforces the conclusion that the BSBS neurons that do not express detectable levels of TH are indeed non-catecholaminergic, not merely a false negative histological result.

Vesicular glutamate transporter 2 (VGLUT2) mRNA has been detected in at least 70% of the bulbospinal C1 cells and in virtually all electrophysiologically identified BSBS neurons regardless of whether they also contain TH immunoreactivity (Stornetta et al., 2002a,b). Since VGLUT2 mRNA encodes a protein that is diagnostic of glutamatergic neurons (for references see: Stornetta et al., 2002a), the presence of this mRNA suggests that virtually all BSBS neurons are glutamatergic. The glutamatergic nature of the BSBS cells extends prior physiological evidence that a powerful glutamatergic input to SPGNs may originate from the region of the RVLM (Morrison et al., 1989a,b; Huangfu et al., 1994; Deuchars et al., 1995).

PS Neurons of the RVLM: Functional Organization and Collateralization

It seems clear that RVLM BSBS neurons innervate selectively the sympathetic vasomotor efferents that are under strong baroreceptor control; whereas, other PS neurons provide the dominant excitatory input to other sympathetic efferents (Morrison, 2001; Ootsuka et al., 2002). This well documented dichotomy provides some of the strongest support to the organotopic theory which states that medullary PS neurons are organized into subgroups that control distinct anatomical targets (Dampney et al., 1987; McAllen et al., 1995). The evidence reviewed below suggests that pools of BSBS neurons influence sympathetic cardiovascular efferents to varying degrees. However, the data falls short of demonstrating that visceral vasoconstrictor, muscle vasoconstrictor, renal and cardiac sympathetic efferents are regulated by entirely different pools of BSBS neurons.

The notion that the heart, kidney, adrenal medulla etc. are driven by distinct pools of BSBS neurons is primarily based on the fact that small amounts of glutamate or GABA produce somewhat different effects on regional blood flows or on the discharge of simultaneously recorded sympathetic efferents depending on which region of the VLM is targeted (Lovick, 1987; Carrive et al., 1989; Campos et al., 1997; Ootsuka et al., 1997). These experiments are not without interpretative difficulties because the neurons that contribute to the responses elicited by the injected chemicals are essentially unknown and likely to be very heterogeneous e.g. barosensitive RVLM neurons, other types of PS neurons, interneurons, all of the above. For instance, baroinsensitive PS neurons of the rostromedial RVLM control skin blood flow and brown fat in the context of thermoregulation (Morrison, 2001). These neurons probably also regulate cardiac sympathetic efferents since both thermogenesis and increased cutaneous flow require adjustments to the cardiac output. Glutamate injections into the rostral RVLM of cat could therefore exert a larger effect on cardiac sympathetic efferents than on muscle vasoconstrictor fibers (Campos et al., 1997) by recruiting baroinsensitive PS neurons specialized in thermoregulation along with unspecialized RVLM BSBS neurons that drive cardiac efferents and muscle vasoconstrictors equally.

Electrophysiological evidence for specialization of BSBS neurons is modest and also difficult to interpret. Under anesthesia, many vasomotor reflexes such as the Von-Bezold Jarisch reflex, baroreflex,

9. THE PRESYMPATHETIC CELLS OF THE ROSTRAL VENTROLATERAL MEDULLA (RVLM): ANATOMY, PHYSIOLOGY AND ROLE IN THE CONTROL OF CIRCULATION

somatosympathetic reflex, reflex due to nasal mucosa stimulation, peripheral chemoreflex and reflex caused by stimulation of cardiac afferents, produce qualitatively similar changes in the activity of peripheral sympathetic nerves with barosensitive discharges (Morrison et al., 1989c; Verberne et al., 1992; Koshiya et al., 1993; McCulloch et al., 1999; Cao et al., 2001). Not surprisingly, these stimuli affect all RVLM BSBS neurons in much the same manner. Few stimuli, including increased central respiratory drive, stimulation of the periaqueductal gray matter and peripheral injection of cholecystokinin-8 (CCK8), produce dramatically different effects on the various sympathetic vasomotor nerves that are under baroreceptor control; but, these stimuli differentially affect the discharge of RVLM BSBS neurons (Darnall et al., 1990; Lovick, 1992; Kishi et al., 2000; Ootsuka et al., 2002; Sartor et al., 2002). For instance, like most barosensitive sympathetic ganglionic neurons, the BSBS cells of the RVLM have a respiratory modulation caused by inputs from the central respiratory network (McAllen, 1987; Haselton et al., 1989; for reviews see Guyenet et al., 1992; Habler et al., 1994; Miyawaki et al., 1995; Guyenet, 2000). The same respiratory patterns can be detected in separate populations of barosensitive sympathetic ganglionic neurons suggesting that subgroups of BSBS neurons exert their control over only a fraction of the vasomotor SPGNs efferents (Numao et al., 1987; Haselton et al., 1989; Darnall et al., 1990). However these respiratory patterns are widespread in sympathetic nerves and do not neatly segregate by organ. Injection of CCK-8 inhibits only about half the BSBS neurons and tends to excite the rest (Sartor et al., 2002). Since CCK-8 inhibits splanchnic sympathetic nerve activity (SNA) without inhibiting lumbar SNA, it is tempting to conclude that the BSBS cells that are inhibited by the peptide only control the mesenteric circulation (Sartor et al., 2002). However, here again, selectivity is being judged by measuring only two sympathetic targets (lumbar and splanchnic nerve) out of a multitude of other possible sympathetic efferents.

Other evidence argues against the notion that BSBS neurons are organized in a strict organ-specific manner. Vestibulo-sympathetic reflexes in the cat affect hindlimb and forelimb resistance differentially suggesting that a somatotopic organization of muscle blood vessel control may also exist, in addition to the postulated organotopic one (Kerman et al., 2000a,b). More importantly perhaps, experiments based on the retrograde transsynaptic migration of PRV variants suggest that the adrenal medulla and the stellate ganglion share appreciable numbers of RVLM PS neurons, including many C1 neurons (Jansen et al., 1995a). Assuming that the retrograde migration of PRV is 100% transsynaptic, this data (Jansen et al.,

1995a) suggests the existence of RVLM PS neurons with divergent projections to unrelated organs.

Finally, RVLM BSBS neurons have extensive collaterals that target most of the major brainstem autonomic centers (solitary tract nucleus, dorsolateral pons, central gray or, more rarely, the hypothalamus (Haselton et al., 1990; Lipski et al., 1995). Therefore the BSBS neurons relay a "copy" of the information sent to the SPGNs to these various autonomic centers. This information which, if the organotopic theory is correct, encodes the state of activation of the various branches of the vasomotor sympathetic outflow (renal, cardiac etc.) may be essential for the hypothalamus to orchestrate the global neuroendocrine regulation of blood pressure. A rarely considered consequence of their collaterization is that RVLM PS cells could influence the respiratory centers or may control sympathetic tone by multiple indirect routes in addition to their monosynaptic inputs to SPGNs (Granata et al., 1998). Finally, the collaterals of the RVLM PS cells may also be a building block of brainstem loops implicated in the slow rhythms of sympathetic nerve discharge that are present in larger species (Zhong et al., 1993).

In short, the RVLM contains groups of BSBS neurons with distinctive patterns of respiratory modulation and differentiated responses to selected stimuli. The bulk of the evidence suggests that these various groups of PS neurons contribute unequally to the activation of sympathetic vasomotor efferents, especially splanchnic and lumbar vasoconstrictor neurons, however the precise pattern of connectivity between BSBS cells and SPGNs remains to be determined.

Inputs to BSBS Neurons: *In Vivo* Studies

Inhibitory Drives to BSBS Neurons In Vivo: Control of Vasomotor Tone by Disinhibition

BSBS cells are tonically inhibited by GABAergic inputs (Sun et al., 1985; for review: Blessing, 1997). One of these GABAergic inputs is largely dependent on the discharges of arterial baroreceptors (Sun et al., 1985; Lipski et al., 1996) and is presumed to originate from interneurons located somewhat caudal to the BSBS neurons within the caudal VLM (CVLM) region (Terui et al., 1990; Kumada et al., 1990; for recent review: Schreihofer and Guyenet, 2002). BSBS neurons also receive a tonic baroreceptor-independent GABAergic inhibitory drive most clearly revealed in debuffered animals or when resting BP is below the threshold for activation of arterial baroreceptors (Sun et al., 1985; reviewed in Sun, 1995;

Sun, 1996b; Blessing, 1997). The origin of this baroreceptor "independent" input is unclear. It could originate, at least in part, from some of the same CVLM GABAergic neurons that convey baroreceptor inputs to the BSBS cells since many of these CVLM cells retain significant activity below baroreceptor threshold (Jeske et al., 1993; Jeske et al., 1995; for reviews see Blessing, 1997; Schreihofer and Guyenet, 2002). Neurons located caudal to the baroactivated CVLM neurons may also contribute some of the baroreceptor independent GABAergic inhibition of the BSBS neurons (Cravo et al., 1991). These unidentified cells could be part of the ventral medullary respiratory network given the prominent respiratory modulation of the BSBS cells (McAllen, 1987; Haselton et al., 1989). In functional terms, the existence of a tonic inhibitory drive to BSBS neurons suggests that disinhibition is probably an important mechanism for their activation, hence for BP elevation.

Excitatory Drives to RVLM BSBS Neurons: The Origin of the Sympathetic Vasomotor Tone Under Anesthesia

The mechanisms responsible for the discharges of RVLM PS cells *in vivo* are the cornerstone of sympathetic vasomotor generation and the regulation of blood pressure (for prior reviews see: Sun, 1996a; Guyenet, 2000; Dampney et al., 2000). This problem is still unsolved as more layers of complexity are uncovered.

In vitro experiments described below suggest that many RVLM PS neurons have intrinsic beating properties that can be enhanced by a variety of metabotropic (slow) transmitters such as angiotensin II, substance P and many others. Mechanisms independent of inotropic glutamatergic transmission clearly contribute to the discharges of RVLM BSBS neurons *in vivo* in anesthetized rats. The key evidence is that the broad spectrum inotropic glutamate receptor antagonist kynurenate or mixtures of AMPA/kainate and NMDA receptor antagonists reduce sympathetic tone very modestly when injected into RVLM; whereas, the same treatments virtually block sympathoexcitatory reflexes (e.g. Sun et al., 1986a, 1987; Koshiya et al., 1993; Miyawaki et al., 1996a,b; Ito et al., 1997; McCulloch et al., 1999).

Intracellular recordings *in vivo* indicate that the action potentials from BSBS cells are often preceded and triggered by depolarizing events that resemble postsynaptic potentials (PSPs) (Lipski et al., 1996). These PSP-like events have been taken as evidence that the BSBS cells could not derive any of their activity from intrinsic beating properties or slow transmitters (Lipski et al., 1996). However, the persistence of these PSP-like events was not demonstrated in cells that were silenced by bias current. Therefore, it is possible that these relatively large events were not PSPs but prepotentials mediated by voltage-activated conductance such as low-voltage activated (LVA) calcium currents or persistent sodium current. Events of this type are clearly present in C1 cells with low discharge rate recorded in slices *in vitro* (Li et al., 1995). Also Lipski et al. (1996) did not show that the frequency of the PSP-like events was high enough to account for the typical high discharge rate of undisturbed BSBS neurons. This point is crucial. In some preparations, glutamate receptor blockade in RVLM does reduce sympathetic tone (e.g. Ito et al., 2000) and one would certainly expect glutamatergic PSPs to make a finite contribution to the discharge of BSBS neurons even "at rest". For instance, even at rest, BSBS cells receive substantial kynurenic acid-sensitive inputs from the central respiratory generator if the latter is active (Guyenet et al., 1990; Miyawaki et al., 1996a). Low level nociceptive inputs from tissue damage due to the surgery also could contribute additional glutamatergic inputs depending on the type of anesthetic used and depth of anesthesia. In brief, the currently available *in vivo* intracellular recording evidence leaves very adequate room for the possibility that BSBS cells could derive a large part of their activity from depolarizations that are intrinsic to the cells or amplified by slow transmitters.

Additional complexity stems from the observation that the contribution of glutamate to the discharge of RVLM PS neurons, though minor at rest, may become predominant under specific conditions. For instance, microinjection of kynurenic acid into the RVLM lowers blood pressure massively after the CVLM has been previously injected with the inhibitory drug muscimol (Ito et al., 1997; Ito et al., 2000). This challenging result suggests that inhibition of the CVLM may cause the discharges of BSBS neurons to shift from a glutamate independent mechanism to a state where discharges are completely dependent on a glutamatergic drive. One interpretation of these observations is that the discharges of RVLM BS neurons and thus basal sympathetic tone can be sustained by several interlocking and redundant circuits. This notion is also supported by other data. For instance, though mid-collicular transection does not appreciably change blood pressure and sympathetic tone (e.g. Koshiya et al., 1994), lesions caudal to the paraventricular nucleus of the hypothalamus (PVH) or microinjections of GABA directly into PVH decrease sympathetic tone substantially (Allen, 2002; de Almeida Colombari et al., 2002). Thus, under anesthesia, a comparable sympathetic output can be generated with or without supra-collicular drives that are individually strong but cancel each other. The observations of Allen, de Almeida Colombari or Ito and Sved highlight the difficulty in identifying the precise sources of excitatory inputs to RVLM BSBS neurons using intraparenchymal microinjections of excitatory or inhibitory agents. Nevertheless, based on this type of study and the unit recording work of Barman et al. (2000), it appears that, under anesthesia, tonically active "excitatory" drives to the BSBS cells may originate from the lateral tegmental field (in cats), the hypothalamus, the caudal NTS and from the so-called caudal pressor area, an ill-defined region located at the ventromedial edge of the spinal trigeminal nucleus (Possas et al., 1994; Campos et al., 1999; Natarajan et al., 2000). Most of these areas, with possible exception of the one originating in the LTF, probably control RVLM neurons by polysynaptic routes. In most instances it is not clear whether the tonic "excitatory" influence exerted by these regions over RVLM is due to activation of excitatory synaptic inputs to the BSBS neurons or to tonic dishinhibition of these neurons. Both mechanisms have been invoked to account for the tonic sympathoexcitatory influence of the caudal pressor area (Campos Junior et al., 1994; Natarajan et al., 2000).

RVLM BSBS Neurons and Sympathetic Reflexes

Under anesthesia, inhibition or excitation of RVLM BSBS neurons makes a major contribution to the sympathetic reflexes investigated so far including somato-, baro-, chemo- cardiosympathetic reflexes and diving response (Sun et al., 1985, 1987; Morrison et al., 1989c; Verberne et al., 1992; Koshiya et al., 1993; Sun, 1996a; McCulloch et al., 1999; Li and Pan, 2000). This conclusion always derives from the same types of experiments. First, these reflexes are greatly if not completely attenuated by blocking glutamate (sympathoexcitatory reflexes) or GABA receptors (sympathoinhibitory reflexes) in the region of the RVLM with appropriate receptor antagonists. Secondly, the discharge of a substantial portion of the BSBS neurons displays variations that are consistent with the direction of the changes in sympathetic efferent activity. This evidence is often interpreted as meaning that RVLM BSBS cells mediate these reflexes entirely but some caution is in order. First, drug microinjection into the RVLM may affect other PS cells (e.g. cells located in the nearby RVMM) and blockade of the reflexes by drugs is often incomplete. Secondly, the correlation between unit activity in RVLM and the discharge of sympathetic nerves does not prove that the BSBS cells are the sole contributors to the observed changes in sympathetic tone. For instance, a descending inhibitory drive to SPGNs may also contribute to the baroreflex (Barman et al., 1988; Lewis et al., 1995; Lewis et al., 1996).

Under anesthesia, inhibition or excitation of RVLM BSBS neurons also makes a major contribution to the sympathetic vasomotor responses evoked by stimulating a variety of brain areas such as the hypothalamus (Sun et al., 1986a; Allen, 2002), the CVLM (Li et al., 1991), and the caudal pressor area (Possas et al., 1994; Campos et al., 1999; Natarajan et al., 2000).

Intrinsic and Synaptic Properties of RVLM PS Neurons

Intrinsic Properties

The intrinsic properties of RVLM PS neurons have been examined in brain slices from immature rats (P2-P14) and in dissociated neurons harvested from 13-19 day old rats. The data is subject to the interpretative limitations inherent to the fact that immature neurons may exhibit properties that are different from that of mature cells (e.g. Berger et al., 1996).

In slices recorded at room temperature, a majority of the C1 and surrounding bulbospinal cells of the RVLM have a low level of discharge (2.5 spikes / s; range <1 to 5) that is irregular in the case of slower-firing neurons (0- 2Hz) and regular in the more active ones (Kangrga et al., 1995; Li et al., 1995). Their resting discharge persists after simultaneous blockade of AMPA-kainate and $GABA_A$ receptors with a drug mixture that also eliminates all spontaneous postsynaptic currents (PSCs) observable at a holding potential of –60 mV or during perfusion with a low Ca^{+2}/high Mg^+ medium (Li et al., 1995; Hayar et al., 1998). Finally, ongoing activity is also observed in juxtacellularly recorded cells i.e. in cells whose intracellular content is unperturbed by whole cell recording (Li et al., 1995). According to this evidence, in slices, the activity of many PS cells including the C1 cells is triggered neither by conventional fast PSCs nor by calcium-dependent exocytotic release of transmitters. Thus, either the discharge of the cells is due purely to their intrinsic properties or it is induced by the presence of one or more chemicals whose release is not due to a calcium-dependent process.

The input resistance of neonate C1 cells in slices is high (around 900 megaohms) and their interspike or resting potential (if silent) is somewhat depolarized (around -58mV). Inward current in the low picoamp range is therefore sufficient to bring these cells to their action potential threshold. Some of the inward current required for their on-going discharges may be contributed by LVA-Ca^{+2} current and/or by a persistent voltage-activated sodium current ($I_{Na}p$) that are both present in these cells (Kangrga et al.,

1995; Li et al., 1995a; Li et al., 1998). A combination of $I_{Na}p$ and LVA calcium current may also underlie the slow oscillations of membrane potential (pre-potentials) that have occasionally been observed in slowly-discharging C1 neurons in slices (Li et al., 1995). Immature bulbospinal RVLM neurons, including C1 cells, also have a hyperpolarization-activated current (I_h) that may contribute to their intrinsic beating properties given that cesium slows the rate of membrane depolarization between spikes (Kangrga et al., 1995). Various potassium currents have also been identified in these cells including an A-type current sensitive to 4-aminopyridine and a delayed rectifyer (I_K). TASK-(TWIK-related acid-sensitive) channels have been identified by anatomical methods (see below).

Mechanically and enzymatically dissociated C1 bulbospinal neurons obtained from the same RVLM region in 13-19 day-old rats are not active and do not exhibit membrane oscillations even though $I_{Na}p$ and, possibly, LVA calcium current are detectable (Lipski et al., 1998). Their input resistance is lower than their counterpart in slices (430 vs 900 mohms) and their resting membrane potential is slightly more hyperpolarized (MP: -62.5 mV vs –58 mV). The absence of spontaneous activity in dissociated bulbospinal C1 cells has been interpreted by Lipski and collaborators as evidence that the activity of C1 cells in slice cannot be due solely to their intrinsic properties. However, according to biophysical models all that is needed for a neuron to be autoactive is a proper balance between membrane resistance, leak currents and voltage–activated conductances (Del Negro et al., 2002 and many others). This balance is likely to be altered by mechanical/enzymatic dissociation which could account for the lack of spontaneous discharges in isolated C1 cells. Yet it remains possible that the "spontaneous" discharge of C1 and other putative PS cells in slices could be conditional on the presence of signaling molecules that are lost after enzymatic dissociation. Given the lack of effect of a low Ca^{+2}/high Mg^+ medium on the discharge of the cells in slices, these hypothetical factors would have to be unconventional.

Synaptic Properties of RVLM PS Neurons

Spontaneous PSPs and PSCs evoked by focal stimulation in RVLM PS neurons of the neonate rat brain have been examined using whole cell patch recordings (Hayar et al., 1998, 1999). This method has also been used to record evoked IPSPs from identified C1 bulbospinal neurons in an *en bloc* preparation of the neonate rat medulla plus cord (Lin et al., 1998). Finally two studies in the rat *in vivo* describe synaptic events recorded in adult RVLM PS neurons using sharp electrodes (Dembowsky and McAllen, 1990; Lipski et al., 1996).

In coronal slices, most of the spontaneous PSCs present in C1 and other RVLM bulbospinal neurons were miniature, TTX-resistant PSCs (mPSCs) (Hayar et al., 1998). About two-thirds of the mPSCs were glutamatergic and the rest GABAergic. As elsewhere in the brain, glutamatergic mPSCs had a fast decay ($t_{1/2}$: 4.7 ms); whereas, GABAergic mPSCs were typically longer-lasting ($t_{1/2}$: ~20 ms) (Hayar et al., 1998). All mPSCs were eliminated by a combination of DNQX (an AMPA/kainate receptor antagonist) and gabazine (a selective $GABA_A$ receptor blocker). Also in slices, PSCs produced by focal stimulation at sites dorsal to the putative PS cells were also found to be either glutamatergic or GABAergic (Hayar et al., 1998, 1999). In the *en bloc* preparation, IPSPs recorded from RVLM bulbospinal neurons were also predominantly GABAergic, but others were glycinergic or had a mixed pharmacology (Lin et al., 1998). Glycinergic miniature or evoked PSCs were not observed in retrogradely labelled putative PS neurons in slices (Hayar et al., 1998). The discrepancy between this work and that of Lin et al., (1998) is probably due to the fact that focal stimulation recruited different sets of inputs to the PS neurons in the two preparations.

In summary, according to the work performed in neonate brains, most of the fast synaptic inputs received by the PS neurons are glutamatergic, GABAergic or glycinergic. Anatomical evidence at the EM level confirms that RVLM BSBS neurons in the adult receive glutamatergic, GABAergic and glycinergic synapses (Llewellyn-Smith et al., 2001). Pharmacological data *in vivo* indicate that most vasomotor sympathetic reflexes are attenuated by administration of antagonists of ionotropic glutamatergic or GABAergic receptors to the vicinity of the RVLM PS neurons (Sun, 1996a). However, the glycine receptor antagonist strychnine does not change the sympathetic outflow or the activity of RVLM PS neurons *in vivo* (Ross et al., 1984b; Sun et al., 1985; Sun, 1996a; Guyenet et al., 1990) suggesting that the glycinergic input to these cells is inactive under anesthesia.

Though nicotinic transmission does not appear to mediate the spontaneous or evoked PSCs that can be recorded in RVLM PS neurons these cells express some form of nicotinic receptors (Huangfu et al., 1997; Hayar et al., 1998). Carbachol also increases the PSP frequency in these cells suggesting the existence of presynaptic effects of still undisclosed pharmacology. RVLM BSBS neurons are excited by iontophoretic applications of ATP *in vivo* suggesting the presence of purinergic receptors, conceivably of the P2X variety given their rapid desensitization, their blockade by suramin and the efficacy of $\alpha\beta$ methylene ATP (Sun, 1996a). There is no clear evidence that these receptors are postsynaptic.

9. THE PRESYMPATHETIC CELLS OF THE ROSTRAL VENTROLATERAL MEDULLA (RVLM): ANATOMY, PHYSIOLOGY AND ROLE IN THE CONTROL OF CIRCULATION

Effect of Catecholamines and Serotonin on RVLM PS Neurons

Systemic administration of alpha-2 adrenergic receptor (alpha-2AR) agonists reduces sympathetic tone and similar effects are produced when these drugs are injected into the RVLM (for reviews see Guyenet, 1997; Guyenet et al., 1998). These agents, e.g. clonidine, inhibit RVLM BSBS neurons when applied by iontophoresis *in vivo* (for reviews see Guyenet, 1997, Guyenet et al., 1998). In slices, alpha-2AR agonists produce in C1 cells the two classic postsynaptic effects associated with this type of receptor, namely, activation of an inwardly rectifying potassium conductance sensitive to low concentrations of barium and a reduction in high-voltage activated (HVA) calcium current (Li et al., 1995, 1998). Alpha-2AR agonists also reduce the frequency of mEPSCs by a barium-independent mechanism indicative of presynaptic inhibition (Hayar et al., 1999). Electron microscopy (EM) data confirmed the existence of pre-and postsynaptic alpha-2ARs in RVLM C1 cells (Milner et al., 1999). Presynaptic alpha-2ARs are also present on some of the GABAergic inputs to the PS cells (Hayar et al., 1999), a conclusion also supported by EM evidence (Milner et al., 1999). The presence of alpha-2ARs in the RVLM probably underlies some of the hypotensive effects of centrally active sympatholytics of the clonidine class (Guyenet, 1997). Most likely, a mix of postsynaptic and presynaptic inhibition accounts for the powerful inhibition of the PS neurons of the RVLM that follows systemic administration of alpha-2AR agonists or their introduction directly into the RVLM. Alpha-2AR agonists that have affinity for imidazoline-1 binding sites (moxonidine) exert the same effects as noradrenaline on C1 cells in slices (Hayar et al., 2000). Although imidazoline binding sites may be present in the RVLM (Reis et al., 1997), electrophysiological effects attributable specifically to an imidazoline receptor have not been uncovered yet.

C1 cells receive a dense serotonergic innervation (Nicholas and Hancock, 1988). RVLM BSBS neurons are typically inhibited by 5-HT1A receptor agonists and excited by 5-HT2 receptor agonists (Wang et al., 1992a,b). The presence of 5-HT1A receptors on some C1 cells has been confirmed by imunohistochemistry (Helke et al., 1997). In brain slices from neonate rats, the postsynaptic effects of serotonin include hyperpolarization and depolarization. The former is due to a barium sensitive inwardly-rectifying gK with 5-HT1A-like pharmacology (Hwang and Dun, 1998), the latter to a decrease in gK and an increase in an unspecified cationic conductance with 5-HT2 pharmacology (Hwang and Dun, 1998). Finally, in slices, serotonin causes powerful presynaptic inhibition via receptors with 5-HT1B or 1D pharmacology. This study did not focus on RVLM cells with PS function but its conclusions are congruent with the iontophoretic data of Wang et al., (1992a, b). *In vivo,* 5-HT1A receptors may also be presynaptic to BSBS neurons (Miyawaki et al., 2001). Serotonin appears to be released in the RVLM during severe hemorrhage and may depress RVLM PS neurons via 5-HT1A receptors (Dean et al., 2002). Hemorrhage may activate serotonergic cells via inputs from the periaqueductal gray matter (Henderson et al., 1998a; Henderson et al., 1998b; Bago et al., 2001).

Effect of Peptides and Nitric Oxide on RVLM PS Neurons

Many peptides change blood pressure when injected into the RVLM. These substances are therefore suspected to directly or indirectly modulate the activity of RVLM PS cells. Many of these peptides originate from the hypothalamus, especially the PVH.

Angiotensin II

Angiotensin-1 receptors (AT1R) are present in the RVLM of all mammals (Allen et al., 1998). Comparative anatomy provided the first clue that some of these receptors might be expressed by C1 adrenergic cells (Allen et al., 1988). Direct evidence that functional AT1Rs are expressed by bulbospinal C1 neurons was obtained in the neonate rat by electrophysiological methods (Li and Guyenet, 1995; Li and Guyenet, 1996). In these cells angiotensin II (Ang II) closes a non-rectifying potassium conductance by activating postsynaptic AT1Rs (Li and Guyenet, 1996). This effect produces an increase in input resistance, a slight depolarization and an increase in the discharge rate of the responsive cells. TASK channels are a known effector of AT1Rs in adrenal glomerulosa cells (Czirjak et al., 2000). TASK channel closure may also account for the effect of Ang II on C1 cells since both TASK-1 and TASK-3 mRNAs are present in these cells (unpublished data of Washburn and Guyenet). A mechanism similar to that found in the RVLM accounts for the postsynaptic effects of Ang II in neonatal spinal motoneurons (Oz et al., 2002).

Consistent with an AT1R-mediated excitatory effect on PS neurons, microinjection of Ang II into the RVLM produces mild pressor effects that are attenuated by antagonists of the losartan family (Hirooka et al., 1997).

Moreover, fourth ventricular injections of Ang II cause Fos expression in C1 cells in awake rabbits (Hirooka et al., 1996a). Because injection of Ang II into RVLM raises BP without changing respiratory outflow, AT1Rs are probably expressed selectively by RVLM cells involved in blood pressure control (Dampney et al., 1996). Losartan and related AT1R blockers produce little or no effect when administered by themselves into the RVLM of anesthetized animals suggesting that Ang II is not being released under those conditions. In contrast, large drops in blood pressure result from injections of sarcosine-containing AT1R antagonists such as saralasin or sarthran in similar preparations (Hirooka et al., 1997; Ito and Sved, 2000). Though the severe hypotension evoked by sarthran disappears when this peptide is coadministered with either Ang II, Ang (1-7) or Ang (3-8), it seems unrelated to the activation of an angiotensin receptor (Ito and Sved, 2000).

Though Ang II is not released in the RVLM at rest under anesthesia, the peptide contributes to the pressor effects caused by activating the PVH in rats (Tagawa et al., 1999). This pressor effect is unaffected by even bilateral injection of the wide-spectrum glutamate receptor antagonist kynurenate into RVLM, illustrating the powerful role that metabotropic transmission can play in regulating the discharge of BSBS neurons (Tagawa et al., 1999).

Opioid Peptides

Based on anatomical evidence, μ-opioid receptors (MORs) are expressed by a large proportion of adult RVLM BSBS neurons (reviewed in Guyenet et al., 2002); and, accordingly, most of these cells are inhibited by iontophoretic application of morphine and enkephalin *in vivo* (Baraban et al., 1995). Postsynaptic MORs are responsible for some of this inhibitory effect (Hayar et al., 1998). MOR agonists also reduce the amplitude of EPSCs elicited by focal stimulation in RVLM PS cells and they reduce the frequency of TTX-resistant glutamatergic mEPSCs without changing their amplitude (Hayar et al., 1998). Therefore MORs located on glutamatergic terminals that contact the BSBS neurons cause presynaptic inhibition i.e. disfacilitation by reducing the release of glutamate. Endomorphin-1 also reduces the frequency of TTX-resistant GABAergic mIPSCs indicating that endogenous opioids can also cause disinhibition of the PS cells (Hayar et al., 1998). The ultimate effect of MOR agonists on the discharge of the PS neurons *in vivo* is therefore quite complex and includes postsynaptic inhibition in about two-thirds of the cells along with presynaptic disfacilitation and presynaptic disinhibition (Guyenet et al., 2002). Adding to this complexity, presynaptic δ-opioid receptors (DOR) are also present in the RVLM, including on inputs to the C1 cells (Milner et al., 2002). Microinjection of a specific DOR agonists into RVLM selectively attenuates the somatosympathetic reflex without changing resting sympathetic tone, suggesting that DORs are located selectively on some of the glutamatergic inputs to the PS cells (Stornetta et al., 1989; Morrison et al., 1989c; Miyawaki et al., 2002). The sources of opioid peptide input to the RVLM are equally complex. Judging from the presence of preproenkephalin mRNA, enkephalins are made by a fraction of the BSBS neurons (Stornetta et al., 2001); and, in addition, enkephalin immunoreactivity is frequently present in boutons that are in synaptic contact with C1 and identified PS neurons (Milner et al., 1990; Llewellyn-Smith et al., 2001). All these synapses also contain an amino acid (Llewellyn-Smith et al., 2001), which unexpectedly, is as likely to be glutamate as GABA or glycine (Llewellyn-Smith et al., 2001). Thus, in the RVLM, enkephalin is potentially released by excitatory as well as inhibitory neurons. The presence of enkephalins in excitatory neurons is not uncommon in the brainstem. The BSBS neurons of the RVLM and the phrenic premotor neurons of the rVRG are two specific examples of glutamatergic neurons that also express the preproenkephalin gene (Stornetta et al., 2001, 2003). The origin and function of the enkephalinergic inputs to the PS cells are not known.

Substance P (SP)

In the adult rat, SP immunoreactivity is present in many terminals located on the C1 neurons (Milner et al., 1989). Microinjection of this peptide into the RVLM increases blood pressure suggesting a direct or indirect excitatory effect of the peptide on the PS cells (Urbanski et al., 1989). Consistent with this interpretation, SP depolarizes the putative PS neurons of neonate rat slices by a postsynaptic mechanism (Li and Guyenet, 1997). The inward current induced by SP is due to the reduction of a potassium conductance and to an unspecified cationic current (Li and Guyenet, 1997). The effect of SP on RVLM neurons has NK1 receptor (NK1R)-like pharmacology (Li and Guyenet, 1997); but, in the adult, NK1R immunoreactivity cannot be detected in C1 cells using available antibodies and the C1 cells are resistant to saporin conjugated with a selective NK1R agonist, a neurotoxin that destroys the surrounding neurons that do express NK1R immunoreactivity (Wang et al., 2002). This discrepancy suggests that the C1 cells may express NK1Rs only during the neonatal period or that the receptor responsible for the effect of SP in the neonate has been incorrectly identified.

Other Peptides

The orexins (hypocretins), peptides synthesized by a group of hypothalamic neurons involved in sleep and feeding, are present in the RVLM as a light plexus of nerve terminals (Peyron et al., 1998). When injected into this region, the peptide produces large increases in blood pressure and sympathetic nerve activity presumably mediated at some point by activation of the PS cells (Chen et al., 2000; Dun et al., 2000; Machado et al., 2002).

Terminals immunoreactive for corticotrophin-releasing factor (CRF) are present in the RVLM and most likely originate from neurons located in the parvocellular subdivision of the PVH, although CRF-ir neurons have also been detected in the rostral medulla (Sawchenko et al., 1993; Milner et al., 1993). Microinjection of CRF into the RVLM of anesthetized rats elevates blood pressure by still unknown mechanisms (Milner et al., 1993). Autonomic adjustments, including a rise in blood pressure, are a prominent feature of stress, a behavioral state in which the CRF neurons of the hypothalamus play a key role. The RVLM is among the many autonomic centers targeted by these neurons (Sawchenko et al., 1993).

Oxytocin-immunoreactive terminals originating from the paraventricular nucleus of the hypothalamus are also present in the RVLM (Sofroniew et al., 1981) where they make close appositions (possible synapses) with the C1 cells (Hancock et al., 1987). Microinjection of vasopressin into RVLM increases blood pressure (Andreatta-Van Leyen et al., 1990). Oxytocin and vasopressin increase the firing rate of a group of RVLM neurons that discharge regularly at about 9 Hz active in adult rat brain slices maintained at 31°C (Sun et al., 1989). These active neurons with putatively intrinsic beating properties ("pacemaker" neurons) are intermingled with the C1 cells but not demonstrably catecholaminergic (Sun et al., 1988a,b). Some of them were shown to be bulbospinal and therefore are likely to have been PS neurons of the non-C1 variety (Sun et al., 1988a). The RVLM "pacemaker" neurons were also activated by TRH, CGRP, NPY and SP but neither by CRF nor by Ang II (Sun et al., 1989). The lack of effect of CRF and Ang II suggest that these peptides may selectively target PS neurons of the C1 variety. Both TRH and SP terminals (Batten, 1995) are likely to originate from subsets of raphe serotonergic neurons (Helke et al., 1986; Sun et al., 1996), though other potential sources exist (Tsuruo et al., 1987).

Glucagon-like peptide-1 (GLP-1), a satiety peptide released by the gut, is also found in the CNS where it controls feeding and drinking behavior among other things. GLP-1 administered centrally increases blood pressure and heart rate and induced c-fos expression in over 40% of C1 neurons with documented spinal projections. It is speculated that the activation of the RVLM neurons may occur directly via GLP-1 inputs from the caudal NTS and indirectly by activation of hypothalamic inputs to RVLM (Yamamoto et al., 2002). Apelin-13, a peptide of unknown physiological function in brain also produces some increase in BP when injected into the VLM (Seyebadi et al., 2002).

Role of Nitric Oxide (NO) in the RVLM

Intense NADPH-diaphorase, a general marker of NO-synthesizing neurons, is present in a small proportion of RVLM neurons of the rabbit and other species (Hirooka et al., 1996b). In the rat, NADPH-diaphorase or neuronal nitric oxide synthase (nNOS) immunoreactivity is found in non-catecholaminergic neurons located somewhat medial to the bulk of the C1 PS cells (Iadecola et al., 1993; Ohta et al., 1993). Microinjection of NO donors, NO scavengers or inhibitors of NOS into the RVLM has produced a puzzling variety of effects on blood pressure and sympathetic nerve discharge (SND). In pentobarbitone-anesthetized rabbits with denervated baroreceptors, NO donors increase BP and SND; whereas, the NOS inhibitor L-NAME lowers BP and SND (Hirooka et al., 1996b). According to these results, endogenously produced NO has a pressor and sympathoactivating action at rest that is further enhanced by exogenously applied NO (Hirooka et al., 1996b). In anesthetized rats, NO appears to have a tonic sympathoinhibitory effect that is enhanced by exogenously applied NO. For instance, overexpression of e-NOS in the RVLM via adenovirus transfection lowers the BP of conscious rats (Kishi et al., 2001). The inhibitory effect of NO in the RVLM may be mediated by an increase in GABA release (Ishide et al., 2000; Kishi et al., 2001; Nauli et al., 2001), a mechanism that has also been invoked to explain the inhibitory effect of NO in the NTS (Paton et al., 2001) or the PVH (Li et al., 2002b). In freely behaving rats, the pressor effect produced by microinjection of glutamate into the RVLM is attenuated by coadministration of an NO scavenger or by an inhibitor of guanylate cyclase (GC) (Martins-Pinge et al., 1999). According to these results, the pressor effect of exogenously applied glutamate may be mediated in part via the activation of soluble GC by NO (Martins-Pinge et al., 1999). It is unclear why NO produces either pressor or depressor effects depending on the experiment. The most interesting possibility is that NO may have multiple targets within the RVLM and that its overall effect on BP depends on what type of synaptic input contributes most to the activity of the PS cells. The work of Pilowsky and colleagues illustrates that some RVLM transmitters such as μ or δ opiate agonists and serotonin can inhibit very specific inputs to the RVLM while sparing others (Miyawaki et al., 2001; Miyawaki et al., 2002). Finally, both nNOS and iNOS may be expressed in the RVLM; and, according to one study, nNOS-dependent NO production may raise BP; whereas, NO originating from iNOS may have the opposite effect (Chan et al., 2001). Further understanding of the effects of NO in the RVLM will require a more cellular level of analysis based on *in vitro* neurophysiological methods.

RVLM PS Neurons, Hypertension and Heart Failure

Upregulation of the activity of PNMT in the spontaneously hypertensive rat (SHR) was identified in the 70s (Saavedra et al., 1976) and upregulation of the mRNAs encoding TH and PNMT in the C1 region is now well established for this strain (Reja et al., 2002). Under chloralose anesthesia, kynurenic acid injections into RVLM produce more hypotension in the SHR than in a close genetic match, the WKY rat (Ito et al., 2000; Sved et al., 2002). Also, inhibition of the PVH or the commissural nucleus of the NTS causes more hypotension in SHRs than in WKY rats (Sato et al., 2001; Allen, 2002). Finally, inhibition of the RVLM also causes far more hypotension in the SHR than in WKY; whereas, stimulation, electrical or with glutamate, causes greater rises in blood pressure in the SHR (reviewed by Arnolda et al., 1997). These results are usually interpreted as evidence that RVLM BSBS neurons are more active in SHR than in WKY rats, either because they receive more excitatory inputs in the case of the SHRs (Ito et al., 2000) or less GABA inhibition in the stroke-prone SHR (SHR-SP) (Kishi et al., 2002). Similar types of evidence suggest that heightened glutamatergic inputs may upregulate RVLM PS neurons in salt-fed salt-sensitive Dahl rats (Ito et al., 2001). While these interpretations may be correct, their tentative nature cannot be stressed enough. Suffice it to say that direct neurophysiological evidence of increased activity of RVLM BSBS neurons in these hypertension models is usually unavailable, and when available, the evidence is contradictory (Sun et al., 1986b; Chan et al., 1991). Fos studies in awake animals are especially ambiguous. At rest, more Fos positive cells were found in the RVLM of SHRs than WKY rats but the numbers were smaller in SHRs than in WKY rats after sustained hypotension (Minson et al., 1996). Chronic overexpression of NO synthase in the RVLM decreases blood pressure more in the SHR-SP than the WKY rat (Kishi et al., 2002). This result illustrates brilliantly that the RVLM plays a role in the long-term regulation of blood pressure but it provides little additional evidence that this structure is abnormally active in the SHR-SP given the vast difference in vascular reactivity between the two strains.

Heart failure causes a major rise in sympathetic tone, especially to the heart (reviewed in Dampney et al., 2002; Weiss et al., 2003). Increased neuronal activity within the PVH appears to contribute to the sympathoactivation (Zhang et al., 2002). The mechanism may involve dishinhibition of PVH output neurons via increased NO production (Li et al., 2002a; Zhang et al., 2002). Since PVH neurons presumably excite RVLM neurons (Allen, 2002), the sympathoexcitation of heart failure may be mediated in part by increases in the activity of RVLM BSBS neurons.

CONCLUSIONS

The BSBS neurons of the RVLM provide a crucial excitatory drive to the sympathetic efferents that control the heart, the kidneys and major resistance vessels. These cells release glutamate and several other transmitters including catecholamines. *In vitro*, the on-going activity of BSBS neurons derives primarily from intrinsic beating properties; whereas, *in vivo,* many metabotropic and inotropic transmitters contribute to their activity. These transmitters originate from a complex and only partially understood network of neurons located in the lower brainstem, midbrain and hypothalamus.

9. THE PRESYMPATHETIC CELLS OF THE ROSTRAL VENTROLATERAL MEDULLA (RVLM): ANATOMY, PHYSIOLOGY AND ROLE IN THE CONTROL OF CIRCULATION

REFERENCES

Allen, A.M., 2002. Inhibition of the hypothalamic paraventricular nucleus in spontaneously hypertensive rats dramatically reduces sympathetic vasomotor tone. Hypertension 39, 275-280.

Allen, A.M., McKinley, M.J., Oldfield, B.J., Dampney, R.A., Mendelsohn, F.A., 1988. Angiotensin II receptor binding and the baroreflex pathway. Clinical & Experimental Hypertension - Part A, Theory & Practice 10 Suppl 1, 63-78.

Allen, A.M., Moeller, I., Jenkins, T.A., Zhuo, J., Aldred, G.P., Chai, S.Y., Mendelsohn, F.A.O., 1998. Angiotensin receptors in the nervous system. Brain Res. Bull 47, 17-28.

Amendt, K., Czachurski, J., Dembowsky, K., Seller, H., 1978. Neurones within the "chemosensitive area" on the ventral surface of the brainstem which project to the intermediolateral column. Pflugers Archiv. - Eur. J. Physiol. 375, 289-292.

Andreatta-Van Leyen, S., Averill, D.B., Ferrario, C.M., 1990. Cardiovascular actions of vasopressin at the ventrolateral medulla. Hypertension 15 (2 Suppl), I102-I106.

Arnolda, L., Wang, H.H., Minson, J., Llewellyn-Smith, I., Suzuki, S., Pilowsky, P., Chalmers, J., 1997. Central control mechanisms in hypertension. Austr. & New Zealand J. Medicine 27(4), 474-478.

Bago, M., Dean, C., 2001. Sympathoinhibition from ventrolateral periaqueductal gray mediated by 5-HT_{1A} receptors in the RVLM. Am. J. Physiol.- Regul. Integr. Comp.Physiol. 280, R976-R984.

Baraban, S.C., Stornetta, R.L., Guyenet, P.G., 1995. Effects of morphine and morphine withdrawal on adrenergic neurons of the rat rostral ventrolateral medulla. Brain Res. 676, 245-257.

Barman, S.M., Gebber, G.L., 1985. Axonal projection patterns of ventrolateral medullospinal sympathoexcitatory neurons. J. Neurophys. 53, 1551-1566.

Barman, S.M., Gebber, G.L., 1988. The axons of raphespinal sympathoinhibitory neurons branch in the cervical spinal cord. Brain Res. 441, 371-376.

Barman, S.M., Gebber, G.L., Orer, H.S., 2000. Medullary lateral tegmental field: an important source of basal sympathetic nerve discharge in the cat. Am. J. Physiol.-Reg. Integr. Comp. Physiol. 278, R995-R1004.

Batten, T.F.C., 1995. Immunolocalization of putative neurotransmitters innervating autonomic regulating neurones of cat ventral medulla. Brain Res. Bull. 37, 487-506.

Berger, A.J., Bayliss, D,A,, Viana, F., 1996. Development of hypoglossal motoneurons. J. Appl. Physiol. 81, 1039-1048.

Blessing, W.W., 1997. Arterial pressure and blood flow to the tissues. The lower brainstem and bodily homeostasis. New York: Oxford University Press, p 165-268.

Brown, D.L., Guyenet, P.G., 1984. Cardiovascular neurons of brain stem with projections to spinal cord. Am. J. Physiol. 247, R1009-R1016.

Brown, D.L., Guyenet, P.G., 1985. Electrophysiological study of cardiovascular neurons in the rostral ventrolateral medulla in rats. Circ. Res. 56, 359-369.

Campos Junior, R.R., Possasd O.S., Cravo, S.L., Lopes, O.U., Guertzenstein, P.G., 1994. Putative pathways involved in cardiovascular responses evoked from the caudal pressor area. Brazilian J. Med. & Biol. Res. 27, 2467-2479.

Campos, R.R., McAllen, R.M., 1997. Cardiac sympathetic premotor neurons. Am. J. Physiol.-Regul. Integr. Comp. Physiol. 272, R615-R620.

Campos, R.R., McAllen, R.M., 1999. Tonic drive to sympathetic premotor neurons of rostral ventrolateral medulla from caudal pressor area neurons. Am. J. Physiol.-Reg. Integr. Comp. Physiol. 276, R1209-R1213.

Cao, W.H., Morrison, S.F., 2001. Differential chemoreceptor reflex responses of adrenal preganglionic neurons. Am. J. Physiol.- Regul. Integr. Comp. Physiol. 281, R1825-R1832.

Carrive, P., Bandler, R., Dampney, R.A., 1989. Viscerotopic control of regional vascular beds by discrete groups of neurons within the midbrain periaqueductal gray. Brain Res. 493, 385-390.

Ceccatelli, S., Villar, M.J., Goldstein, M., Hökfelt, T., 1989. Expression of c-Fos immunoreactivity in transmitter- characterized neurons after stress. Proc. Nat. Acad. Sci. (USA) 86, 9569-9573.

Chan, R.K.W., Chan, Y.S., Wong, T.M., 1991. Electrophysiological properties of neurons in the rostral ventrolateral medulla of normotensive and spontaneously hypertensive rats. Brain Res. 549, 118-126.

Chan, R.K.W., Sawchenko, P.E., 1994. Spatially and temporally differentiated patterns of c-fos expression in the brainstem catecholaminergic cell groups induced by cardiovascular challenges in the rat. J. Comp. Neurol. 348, 433-460.

Chan, R.K.W., Sawchenko, P.E., 1998. Organization and transmitter specificity of medullary neurons activated by sustained hypertension: implications for understanding baroreceptor reflex circuitry. J. Neurosci. 18, 371-387.

Chan, S.H.H., Wang, L.L., Wang, S.H., Chan, J.Y.H., 2001. Differential cardiovascular responses to blockade of nNOS or iNOS in rostral ventrolateral medulla of the rat. Br. J. Pharmacol. 133, 606-614.

Chen, C.T., Hwang, L.L., Chang, J.K., Dun, N.J., 2000. Pressor effects of orexins injected intracisternally and to rostral ventrolateral medulla of anesthetized rats. Am. J. Physiol.-Reg. Integr. Comp. Physiol. 278, R692-R697.

Cravo, S.L., Morrison, S.F., Reis, D.J., 1991. Differentiation of two cardiovascular regions within caudal ventrolateral medulla. Am. J. Physiol. 261, R985-R994.

Czirjak, G., Fischer, T., Spat, A., Lesage, F., Enyedi, P., 2000. TASK (TWIK-related acid-sensitive K+ channel) is expressed in glomerulosa cells of rat adrenal cortex and inhibited by angiotensin II. Molecular Endocrinology 14, 863-874.

Dampney, R.A., Goodchild, A.K., McAllen, R.M., 1987. Vasomotor control by subretrofacial neurones in the rostral ventrolateral medulla. Can. J. Physiol. Pharmacol. 65, 1572-1579.

Dampney, R.A.L., 1994. Functional organization of central pathways regulating the cardiovascular system. Physiol. Rev. 74, 323-364.

Dampney, R.A.L., Coleman, M.J., Fontes, M.A.P., Hirooka, Y., Horiuchi, J., Li, Y.W., Polson, J.W., Potts, P.D., Tagawa, T., 2002. Central mechanisms underlying short- and long-term regulation of the cardiovascular system. Clin. Exp. Pharmacol. Physiol. 29, 261-268.

Dampney, R.A.L., Hirooka, Y., Potts, P.D., Head, G.A., 1996. Functions of angiotensin peptides in the rostral ventrolateral medulla. Clin. Exp. Pharmacol. Physiol. 23, S105-S111.

Dampney, R.A.L., Tagawa, T., Horiuchi, J., Potts, P.D., Fontes, M., Polson, J.W., 2000. What drives the tonic activity of presympathetic neurons in the rostral ventrolateral medulla? Clin. Exp. Pharmacol. Physiol. 27, 1049-1053.

Darnall, R.A., Guyenet, P.G., 1990. Respiratory modulation of pre- and postganglionic lumbar vasomotor sympathetic neurons in the rat. Neurosci. Lett. 119, 148-152.

de Almeida Colombari, D.S., Portelinha, L.S., Campos, R.R. Jr., Lopes, O.U., 2002. Haemodynamic effects of hypothalamic disconnection in anaesthetized rats. Auton. Neurosci. 28, 51-54.

Dean, C., Bago, M., 2002. Renal sympathoinhibition mediated by 5-HT1A receptors in the RVLM during severe hemorrhage in rats. Am. J. Physiol.- Regul. Integr. Comp. Physiol. 282, R122-R130.

Del Negro, C.A., Koshiya, N., Butera, R.J., Jr., Smith, J.C., 2002. Persistent sodium current, membrane properties and bursting behavior of pre-Botzinger complex inspiratory neurons in vitro. J. Neurophys. 88, 2242-2250.

Dembowsky, K., McAllen, R.M., 1990. Baroreceptor inhibition of subretrofacial neurons: evidence from intracellular recordings in the cat. Neurosci. Lett. 111, 139-143.

Deuchars, S.A., Morrison, S.F., Gilbey, M.P., 1995. Medullary-evoked EPSPs in neonatal rat sympathetic preganglionic neurones *in vitro*. J. Physiol. (Lond.) 487, 453-463.

Dun, N.J., Dun, S.L., Shen, E., Tang, H., Huang, R., Chiu, T.H., 1995. C-fos expression as a marker of central cardiovascular neurons. Biol. Signals 4, 117-123.

9. THE PRESYMPATHETIC CELLS OF THE ROSTRAL VENTROLATERAL MEDULLA (RVLM): ANATOMY, PHYSIOLOGY AND ROLE IN THE CONTROL OF CIRCULATION

Dun, N.J., Dun, S.L., Chen, C.T., Hwang, L.L., Kwok, E.H., Chang, J.K., 2000. Orexins: a role in medullary sympathetic outflow. Regul. Pept. 96, 65-70.

Dun, S.L., Ng, Y.K., Brailoiu, G.C., Ling, E.A., Dun, N.J., 2002. Cocaine- and amphetamine-regulated transcript peptide-immunoreactivity in adrenergic C1 neurons projecting to the intermediolateral cell column of the rat. J. Chem. Neuroanat. 23, 123-132.

Granata, A.R., Ruggiero, D.A., 1998. Evidence of disynaptic projections from the rostral ventrolateral medulla to the thoracic spinal cord. Brain Res. 781, 329-334.

Guertzenstein, P.G., Silver, A., 1974. Fall in blood pressure produced from discrete regions of the ventral surface of the medulla by glycine and lesions. J. Physiol. (Lond.) 242, 489-503.

Guyenet, P.G., 1997. Is the hypotensive effect of clonidine and related drugs due to imidazoline binding sites? Am. J. Physiol.-Reg. Integr. Comp. Physiol. 273, R1580-R1584.

Guyenet, P.G., 2000. Neural structures that mediate sympathoexcitation during hypoxia. Respir. Physiol. 121, 147-162.

Guyenet, P.G., Darnall, R.A., Riley, T.A., 1990. Rostral ventrolateral medulla and sympathorespiratory integration in rats. Am. J. Physiol. 259, 1063-R1074.

Guyenet, P.G., Koshiya, N., 1992. Respiratory-sympathetic integration in the medulla oblongata. In: Kunos, G., Ciriello, J., editors. Central neural mechanisms in cardiovascular regulation, vol. II. Boston: Birkhauser, p 226-247.

Guyenet, P.G., Koshiya, N., Huangfu, D., Baraban, S.C., Stornetta, R.L., Li, Y.-W., 1996. Role of medulla oblongata in generation of sympathetic and vagal outflows. Prog. Brain Res. 107, 127-144.

Guyenet, P.G., Stornetta, R.L., 1998. Central nervous system regulation of the sympathetic and cardiovagal vasomotor outflows. In: Yaksh, T.L., Lynch, C., Zapol, W.M., Maze, M., Biebuyck, J.F., Saidman, L.J., editors. Anesthesia: biologic foundations. Philadelphia: Lippincott-Raven, p 1205-1232.

Guyenet, P.G., Stornetta, R.L., Schreihofer, A.M., Pelaez, N.M., Hayar, A., Aicher, S., Llewellyn-Smith, I.J., 2002. Opioid signalling in the rat rostral ventrolateral medulla. Clin. Exp. Pharmacol. Physiol. 29, 238-242.

Habler, H.J., Janig, W., Michaelis, M., 1994. Respiratory modulation in the activity of sympathetic neurones. Prog. Neurobiol. 43, 567-606.

Hancock, M.B., Nicholas, A.P., 1987. Oxytocin-immunoreactive projections onto medullary adrenaline neurons. Brain Res. Bull. 18, 13-220.

Haselton, J.R., Guyenet, P.G., 1989. Central respiratory modulation of medullary sympathoexcitatory neurons in rat. Am. J. Physiol.-Reg. Integr. Comp. Physiol. 256, R739-R750.

Haselton, J.R., Guyenet, P.G., 1990. Ascending collaterals of medullary barosensitive neurons and C1 cells in rats. Am. J. Physiol.-Reg. Integr. Comp. Physiol. 258, R1051-R1063.

Hayar, A., Guyenet, P.G., 1998. Pre- and postsynaptic inhibitory actions of methionine- enkephalin on identified bulbospinal neurons of the rat RVL. J. Neurophysiol. 80, 2003-2014.

Hayar, A., Guyenet, P.G., 1999. α_2-adrenoceptor-mediated presynaptic inhibition in bulbospinal neurons of rostral ventrolateral medulla. Am. J. Physiol. Heart Circ. Physiol. 277, H1069-H1080.

Hayar, A., Guyenet, P.G., 2000. Prototypical imidazoline-1 receptor ligand moxonidine activates alpha2-adrenoceptors in bulbospinal neurons of the RVL. J. Neurophys. 83, 766-776.

Helke, C.J., Capuano, S., Tran, N., Zhuo, H., 1997. Immunocytochemical studies of the 5-HT$_{1A}$ receptor in ventral medullary neurons that project to the intermediolateral cell column and contain serotonin or tyrosine hydroxylase immunoreactivity. J. Comp. Neurol. 379, 261-270.

Helke, C.J., Sayson, S.C., Keeler, J.R., Charlton, C.G., 1986. Thyrotropin-releasing hormone-immunoreactive neurons project from the ventral medulla to the intermediolateral cell column: partial coexistence with serotonin. Brain Res. 381, 1-7.

Henderson, L.A., Keay, K.A., Bandler, R., 1998b. Hypotension following acute hypovolemia depends on the caudal midline medulla. NeuroReport 9, 1839-1844.

Henderson, L.A., Keay, K.A., Bandler, R., 1998a. The ventrolateral periaqueductal gray projects to caudal brainstem depressor regions, a functional-anatomical and physiological study. Neurosci. 82, 201-221.

Hirooka, Y., Head, G.A., Potts, P.D., Godwin, S.J., Bendle, R.D., Dampney, R.A.L., 1996a. Medullary neurons activated by angiotensin II in the conscious rabbit. Hypertension 27, 287-296.

Hirooka, Y., Polson, J.W., Dampney, R.A.L., 1996b. Pressor and sympathoexcitatory effects of nitric oxide in the rostral ventrolateral medulla. J. Hypertension 14, 1317-1324.

Hirooka, Y., Potts, P.D., Dampney, R.A.L., 1997. Role of angiotensin II receptor subtypes in mediating the sympathoexcitatory effects of exogenous and endogenous angiotensin peptides in the rostral ventrolateral medulla of the rabbit. Brain Res. 772, 107-114.

Hökfelt, T., Fuxe, K., Goldstein, M., Johansson, O., 1974. Immunohistochemical evidence for the existence of adrenaline neurons in the rat brain. Brain Res. 66, 235-251.

Huangfu, D., Hwang, L.J., Riley, T.A., Guyenet, P.G., 1994. Role of serotonin and catecholamines in sympathetic responses evoked by stimulation of rostral medulla. Am. J. Physiol.- Regul. Integr. Comp. Physiol. 266, R338-R352.

Huangfu, D., Schreihofer, A.M., Guyenet, P.G., 1997. Effect of cholinergic agonists on bulbospinal C1 neurons in rats. Am. J. Physiol.-Reg. Integr. Comp. Physiol. 272, R249-R258.

Hwang, L.L., Dun, N.J., 1998. 5-hydroxytryptamine responses in immature rat rostral ventrolateral medulla neurons in vitro. J. Neurophys. 80, 1033-1041.

Iadecola, C., Faris, P.L., Hartman, B.K., Xu, X.H., 1993. Localization of NADPH diaphorase in neurons of the rostral ventral medulla--possible role of nitric oxide in central autonomic regulation and oxygen chemoreception. Brain Res. 603, 173-179.

Ishide, T., Hara, Y., Maher, T.J., Ally, A., 2000. Glutamate neurotransmission and nitric oxide interaction within the ventrolateral medulla during cardiovascular responses to muscle contraction. Brain Res. 874, 107-115.

Ito, S., Komatsu, K., Tsukamoto, K., Sved, A.F., 2000. Excitatory amino acids in the rostral ventrolateral medulla support blood pressure in spontaneously hypertensive rats. Hypertension 35, 413-417.

Ito, S., Komatsu, K., Tsukamoto, K., Sved, A.F., 2001. Tonic excitatory input to the rostral ventrolateral medulla in Dahl salt-sensitive rats. Hypertension 37(2 PT 2), 687-691.

Ito, S., Sved, A.F., 1997. Tonic glutamate-mediated control of rostral ventrolateral medulla and sympathetic vasomotor tone. Am. J. Physiol.-Reg. Integr. Comp. Physiol. 273, R487-R494.

Ito, S., Sved, A.F., 2000. Pharmacological profile of depressor response elicited by sarthran in rat ventrolateral medulla. Am. J. Physiol. Heart Circ. Physiol. 279, H2961-H2966.

Jansen, A.S., Ter Horst, G.J., Mettenleiter, T.C., Loewy, A.D., 1992. CNS cell groups projecting to the submandibular parasympathetic preganglionic neurons in the rat: a retrograde transneuronal viral cell body labeling study. Brain Res. 572, 253-260.

Jansen, A.S.P., Nguyen, X.V., Karpitskiy, V., Mettenleiter, T.C., Loewy, A.D., 1995a. Central command neurons of the sympathetic nervous system, basis of the fight-or flight response. Science 270, 644-646.

Jansen, A.S.P., Wessendorf, M.W., Loewy, A.D., 1995b. Transneuronal labeling of CNS neuropeptide and monoamine neurons after pseudorabies virus injections into the stellate ganglion. Brain Res. 683, 1-24.

Jeske, I., Morrison, S.F., Cravo, S.L., Reis, D.J., 1993. Identification of baroreceptor reflex interneurons in the caudal ventrolateral medulla. Am. J. Physiol.-Reg. Integr. Comp. Physiol. 264, R169-R178.

Jeske, I., Reis, D.J., Milner, T.A., 1995. Neurons in the barosensory area of the caudal ventrolateral medulla project monosynaptically on to sympathoexcitatory bulbospinal neurons in the rostral ventrolateral medulla. Neurosci. 65, 343-353.

9. THE PRESYMPATHETIC CELLS OF THE ROSTRAL VENTROLATERAL MEDULLA (RVLM): ANATOMY, PHYSIOLOGY AND ROLE IN THE CONTROL OF CIRCULATION

Kangrga, I.M., Loewy, A.D., 1995. Whole-cell recordings from visualized C1 adrenergic bulbospinal neurons, Ionic mechanisms underlying vasomotor tone. Brain Res. 670, 215-232.

Kerman, I.A., McAllen, R.M., Yates, B.J., 2000a. Patterning of sympathetic nerve activity in response to vestibular stimulation. Brain Res. Bull. 53, 11-16.

Kerman, I.A., Yates, B.J., McAllen, R.M., 2000b. Anatomic patterning in the expression of vestibulosympathetic reflexes. Am. J. Physiol. Regul. Integr. Comp. Physiol. 279, R109-R117.

Kishi, E., Ootsuka, Y.O., Terui, N., 2000. Different cardiovascular neuron groups in the ventral reticular formation of the rostral medulla in rabbits, single neurone studies. J. Auton. Nerv. Syst. 79, 74-83.

Kishi, T., Hirooka, Y., Ito, K., Sakai, K., Shimokawa, H., Takeshita, A., 2002. Cardiovascular effects of overexpression of endothelial nitric oxide synthase in the rostral ventrolateral medulla in stroke-prone spontaneously hypertensive rats. Hypertension 39, 264-268.

Kishi, T., Hirooka, Y., Sakai, K., Shigematsu, H., Shimokawa, H., Takeshita, A., 2001. Overexpression of eNOS in the RVLM causes hypotension and bradycardia via GABA release. Hypertension 38, 896-901.

Koshiya, N., Guyenet, P.G., 1994. Role of the pons in the carotid sympathetic chemoreflex. Am. J. Physiol. 267, R508-R518.

Koshiya, N., Huangfu, D., Guyenet, P.G., 1993. Ventrolateral medulla and sympathetic chemoreflex in the rat. Brain Res. 609, 174-184.

Kumada, M., Terui, N., Kuwaki, T., 1990. Arterial baroreceptor reflex, its central and peripheral neural mechanisms. Prog. in Neurobiol. 35, 331-361.

Lewis, D.I., Coote, J.H., 1995. Chemical mediators of spinal inhibition of rat sympathetic neurones on stimulation in the nucleus tractus solitarii. J. Physiol. (Lond.) 486, 483-494.

Lewis, D.I., Coote, J.H., 1996. Baroreceptor-induced inhibition of sympathetic neurons by GABA acting at a spinal site. Am. J. Physiol. Heart Circ. Physiol. 270, H1885-H1892.

Li, D.P., Chen, S.R., Pan, H.L., 2002b. Nitric oxide inhibits spinally projecting paraventricular neurons through potentiation of presynaptic GABA release. J. Neurophys. 88, 2664-2674.

Li, D.P., Pan, H.L., 2000. Responses of neurons in rostral ventrolateral medulla to activation of cardiac receptors in rats. Am. J. Physiol. Heart Circ. Physiol. 279, H2549-H2557.

Li, Y.F., Roy, S.K., Channon, K.M., Zucker, I.H., Patel, K.P., 2002a. Effect of in vivo gene transfer of nNOS in the PVN on renal nerve discharge in rats. Am. J. Physiol. Heart Circ. Physiol. 282, H594-H601.

Li, Y.W., Bayliss, D.A., Guyenet, P.G., 1995. C1 neurons of neonatal rats, Intrinsic beating properties and alpha$_2$-adrenergic receptors. Am. J. Physiol.-Reg. Integr. Comp. Physiol. 269, R1356-R1369.

Li, Y.W., Dampney, R.A.L., 1994. Expression of Fos-like protein in the brain following sustained hypertension and hypotension in conscious rabbits. Neurosci. 61, 613-634.

Li, Y.W., Gieroba, Z.J., McAllen, R.M., Blessing, W.W., 1991. Neurons in rabbit caudal ventrolateral medulla inhibit bulbospinal barosensitive neurons in rostral medulla. Am. J. Physiol.-Reg. Integr. Comp. Physiol. 261, R44-R51.

Li, Y.W., Guyenet, P.G., 1995. Neuronal excitation by angiotensin II in the rostral ventrolateral medulla of the rat in vitro. Am. J. Physiol. 268, R272-R277.

Li, Y.W., Guyenet, P.G., 1996. Angiotensin II decreases a resting K$^+$ conductance in rat bulbospinal neurons of the C1 area. Circ. Res. 78, 274-282.

Li, Y.W., Guyenet, P.G., 1997. Effect of substance P on C1 and other bulbospinal cells of the RVLM in neonatal rats. Am. J. Physiol.-Reg. Integr. Comp. Physiol. 273, R805-R813.

Li, Y.W., Guyenet, P.G., Bayliss, D.A., 1998. Voltage-dependent calcium currents in bulbospinal neurons of neonatal rat rostral ventrolateral medulla, Modulation by alpha$_2$- adrenergic receptors. J. Neurophys. 79, 583-594.

Lin, H.H., Wu, S.Y., Lai, C.C., Dun, N.J., 1998. GABA- and glycine-mediated inhibitory postsynaptic potentials in neonatal rat rostral ventrolateral medulla neurons *in vitro*. Neurosci. 82, 429-442.

Lipski, J., Kanjhan, R., Kruszewska, B., Rong, W.F., 1996. Properties of presympathetic neurones in the rostral ventrolateral medulla in the rat, An intracellular study '*in vivo*'. J. Physiol. (Lond.) 490, 729-744.

Lipski, J., Kanjhan, R., Kruszewska, B., Smith, M., 1995. Barosensitive neurons in the rostral ventrolateral medulla of the rat *in vivo*, Morphological properties and relationship to C1 adrenergic neurons. Neurosci. 69, 601-618.

Lipski, J., Kawai, Y., Qi, J., Comer, A., Win, J., 1998. Whole cell patch-clamp study of putative vasomotor neurons isolated from the rostral ventrolateral medulla. Am. J. Physiol. Regul. Integr. Comp. Physiol. 274, R1099-R1110.

Llewellyn-Smith, I.J., Schreihofer, A.M., Guyenet, P.G., 2001. Distribution and amino acid content of enkephalin-immunoreactive inputs onto juxtacellularly-labeled bulbospinal barosensitive neurons in rat rostral ventrolateral medulla. Neurosci. 108, 307-322.

Loewy, A.D., 1998. Viruses as transneuronal tracers for defining neural circuits. Neurosci. Biobehav. Rev. 22, 679-684.

Lovick, T.A., 1987. Differential control of cardiac and vasomotor activity by neurones in nucleus paragigantocellularis lateralis in the cat. J. Physiol. (London) 389, 23-36.

Lovick, T.A., 1992. Midbrain influences on ventrolateral medullo-spinal neurones in the rat. Exp. Brain Res. 90, 147-152.

Machado, B.H., Bonagamba, L.G., Dun, S.L., Kwok, E.H., Dun, N.J., 2002. Pressor response to microinjection of orexin/hypocretin into rostral ventrolateral medulla of awake rats. Regul. Pept. 104, 75-81.

Madden, C.J., Ito, S., Rinaman, L., Wiley, R.G., Sved, A.F., 1999. Lesions of the C1 catecholaminergic neurons of the ventrolateral medulla in rats using anti-dopamine beta-hydroxylase-saporin. Am. J. Physiol. 277, R1063-R1075.

Martins-Pinge, M.C., Araújo, G.C., Lopes, O.U., 1999. Nitric oxide-dependent guanylyl cyclase participates in the glutamatergic neurotransmission within the rostral ventrolateral medulla of awake rats. Hypertension 34, 748-751.

McAllen, R.M., 1987. Central respiratory modulation of subretrofacial bulbospinal neurons in the cat. J. Physiol. (London) 388, 533-545.

McAllen, R.M., Badoer, E., Shafton, A.D., Oldfield, B.J., McKinley, M.J., 1992. Hemorrhage induces c-fos immunoreactivity in spinally projecting neurons of cat subretrofacial nucleus. Brain Res. 575, 329-332.

McAllen, R.M., May, C.N., Shafton, A.D., 1995. Functional anatomy of sympathetic premotor cell groups in the medulla. Clin. Exp. Hypertension 17, 209-221.

McCulloch, P.F., Panneton, W.M., Guyenet, P.G., 1999. The rostral ventrolateral medulla mediates the sympathoactivation produced by chemical stimulation of the rat nasal mucosa. J. Physiol. (Lond.) 516, 471-484.

Milner, T.A., Drake, C.T., Aicher, S.A., 2002. C1 adrenergic neurons are contacted by presynaptic profiles containing delta-opioid receptor immunoreactivity. Neurosci. 110, 691-701.

Milner, T.A., Pickel, V.M., Morrison, S.F., Reis, D.J., 1989. Adrenergic neurons in the rostral ventrolateral medulla: ultrastructure and synaptic relations with other transmitter- identified neurons. Prog. Brain Res. 81, 29-47.

Milner, T.A., Pickel, V.M., Reis, D.J., 1990. Tyrosine hydroxylase and enkephalin in the rostral ventrolateral medulla: major synaptic contacts from opioid terminals on catecholaminergic neurons. Prog .Clin. Biol. Res. 328, 195-198.

Milner, T.A., Reis, D.J., Pickel, V.M., Aicher, S.A., Giuliano, R., 1993. Ultrastructural localization and afferent sources of corticotropin-releasing factor in the rat rostral ventrolateral medulla, implications for central cardiovascular regulation. J. Comp. Neurol. 333, 151-167.

Milner, T.A., Rosin, D.L., Lee, A., Aicher, S.A., 1999. Alpha2A-adrenergic receptors are primarily presynaptic heteroreceptors in the C1 area of the rat rostral ventrolateral medulla. Brain Res. 821, 200-211.

Minson, J., Arnolda, L., Llewellyn-Smith, I., Pilowsky, P., Chalmers, J., 1996. Altered c-*fos* in rostral medulla and spinal cord of spontaneously hypertensive rats. Hypertension 27, 433-441.

Miyawaki, T., Goodchild, A.K., Pilowsky, P.M., 2001. Rostral ventral medulla 5-HT_{1A} receptors selectively inhibit the somatosympathetic reflex. Am. J. Physiol. Regul. Integr. Comp. Physiol. 280, R1261-R1268.

Miyawaki, T., Goodchild, A.K., Pilowsky, P.M., 2002. Activation of mu-opioid receptors in rat ventrolateral medulla selectively blocks baroreceptor reflexes while activation of delta opioid receptors blocks somato-sympathetic reflexes. Neurosci. 109, 133-144.

Miyawaki, T., Minson, J., Arnolda, L., Chalmers, J., Llewellyn-Smith, I., Pilowsky, P., 1996a. Role of excitatory amino acid receptors in cardiorespiratory coupling in ventrolateral medulla. Am. J. Physiol.-Reg. Integr. Comp. Physiol. 271, R1221-R1230.

Miyawaki, T., Minson, J., Arnolda, L., Chalmers, J., Llewellyn-Smith, I., Pilowsky, P., 1996b. AMPA/kainate receptors mediate sympathetic chemoreceptor reflex in the rostral ventrolateral medulla. Brain Res. 726, 64-68.

Miyawaki, T., Pilowsky, P., Sun, Q.J., Minson, J., Suzuki, S., Arnolda, L., Llewellyn-Smith, I., Chalmers, J., 1995. Central inspiration increases barosensitivity of neurons in rat rostral ventrolateral medulla. Am. J. Physiol. 268, R909-R918.

Morrison, S.F., 2001. Differential control of sympathetic outflow. Am. J. Physiol. Regul. Integr. Comp. Physiol. 281, R683-R698.

Morrison, S.F., Callaway, J., Milner, T.A., Reis, D.J., 1989b. Glutamate in the spinal sympathetic intermediolateral nucleus: localization by light and electron microscopy. Brain Res. 503, 5-15.

Morrison, S.F., Cao, W.H., 2000. Different adrenal sympathetic preganglionic neurons regulate epinephrine and norepinephrine secretion. Am. J. Physiol. Regul. Integr. Comp. Physiol. 279, R1763-R1775.

Morrison, S.F., Ernsberger, P., Milner, T.A., Callaway, J., Gong, A., Reis, D.J., 1989a. A glutamate mechanism in the intermediolateral nucleus mediates sympathoexcitatory responses to stimulation of the rostral ventrolateral medulla. Prog. Brain Res. 81, 159-169.

Morrison, S.F., Milner, T.A., Reis, D.J., 1988. Reticulospinal vasomotor neurons of the rat rostral ventrolateral medulla: relationship to sympathetic nerve activity and the C1 adrenergic cell group. J. Neurosci. 8, 1286-1301.

Morrison, S.F., Reis, D.J., 1989c. Reticulospinal vasomotor neurons in the RVL mediate the somatosympathetic reflex. Am. J. Physiol. 256, R1084-R1097.

Natarajan, M., Morrison, S.F., 2000. Sympathoexcitatory CVLM neurons mediate responses to caudal pressor area stimulation. Am. J. Physiol. Regul. Integr. Comp. Physiol. 279, R364-R374.

Nauli, S.M., Pearce, W.J., Amer, A., Maher, T.J., Ally, A., 2001. Effects of nitric oxide and GABA interaction within ventrolateral medulla on cardiovascular responses during static muscle contraction. Brain Res. 922, 234-242.

Nicholas, A.P., Hancock, M.B., 1988. Immunocytochemical evidence for substance P and serotonin input to medullary bulbospinal adrenergic neurons. Synapse 2, 569-576.

Numao, Y., Koshiya, N., Gilbey, M.P., Spyer, K.M., 1987. Central respiratory drive-related activity in sympathetic nerves of the rat: the regional differences. Neurosci. Lett. 81, 279-284.

Ohta, A., Takagi, H., Matsui, T., Hamai, Y., Iida, S., Esumi, H., 1993. Localization of nitric oxide synthase immunoreactive neurons in the solitary nucleus and ventrolateral medulla-oblongata of the rat - their relation to catecholaminergic neurons. Neurosci. Lett. 158, 333-335.

Ootsuka, Y., Rong, W.F., Kishi, E., Koganezawa, T., Terui, N., 2002. Rhythmic activities of the sympatho-excitatory neurons in the medulla of rabbits, neurons controlling cutaneous vasomotion. Auton. Neurosci. 101, 48-59.

Ootsuka, Y., Terui, N., 1997. Functionally different neurons are organized topographically in the rostral ventrolateral medulla of rabbits. J. Auton. Nerv. Syst. 67, 67-78.

Oz, M., Renaud, L.P., 2002. Angiotensin AT(1)-receptors depolarize neonatal spinal motoneurons and other ventral horn neurons via two different conductances. J. Neurophys. 88, 2857-2863.

Paton, J.F., Boscan, P., Murphy, D., Kasparov, S., 2001. Unravelling mechanisms of action of angiotensin II on cardiorespiratory function using in vivo gene transfer. Acta Physiol. Scand. 173, 127-137.

Peyron, C., Tighe, D.K., Van den Pol, A.N., De Lecea, L., Heller, H.C., Sutcliffe, J.G., Kilduff, T.S., 1998. Neurons containing hypocretin (orexin) project to multiple neuronal systems. J. Neurosci. 18, 9996-10015.

Possas, O.S., Campos, R.R., Cravo, S.L., Lopes, O.U., Guertzenstein, P.G., 1994. A fall in arterial blood pressure produced by inhibition of the caudalmost ventrolateral medulla: The caudal pressure area. J. Auton. Nerv. Syst. 49, 235-246.

Reis, D.J., Granata, A.R., Joh, T.H., Ross, C.A., Ruggiero, D.A., Park, D.H., 1984. Brain stem catecholamine mechanisms in tonic and reflex control of blood pressure. Hypertension 6, II7-II15.

Reis, D.J., Piletz, J.E., 1997. The imidazoline receptor in control of blood pressure by clonidine and allied drugs. Am. J. Physiol.-Reg. Integr. Comp. Physiol. 273, R1569-R1571.

Reja, V., Goodchild, A.K., Pilowsky, P.M., 2002. Catecholamine-related gene expression correlates with blood pressures in SHR. Hypertension 40, 342-347.

Ritter, S., Bugarith, K., Dinh, T.T., 2001. Immunotoxic destruction of distinct catecholamine subgroups produces selective impairment of glucoregulatory responses and neuronal activation. J. Comp. Neurol. 432, 197-216.

Ritter, S., Llewellyn-Smith, I., Dinh, T.T., 1998. Subgroups of hindbrain catecholamine neurons are selectively activated by 2-deoxy-D-glucose induced metabolic challenge. Brain Res. 805, 41-54.

Ross, C.A., Armstrong, D.M., Ruggiero, D.A., Pickel, V.M., Joh, T.H., Reis, D.J., 1981. Adrenaline neurons in the rostral ventrolateral medulla innervate thoracic spinal cord: a combined immunocytochemical and retrograde transport demonstration. Neurosci. Lett. 25, 257-262.

Ross, C.A., Ruggiero, D.A., Park, D.H., Joh, T.H., Sved, A.F., Fernandez-Pardal, J., Saavedra, J.M., Reis, D.J., 1984. Tonic vasomotor control by the rostral ventrolateral medulla: Effect of electrical or chemical stimulation of the area containing C1 adrenaline neurons on arterial pressure, heart rate, and plasma catecholamines and vasopressin. J. Neurosci. 4, 474-494.

Saavedra, J.M., Grobecker, H., Axelrod, J., 1976. Adrenaline-forming enzyme in brainstem, elevation in genetic and experimental hypertension. Science 191, 483-484.

Sartor, D.M., Verberne, A.J., 2002. Cholecystokinin selectively affects presympathetic vasomotor neurons and sympathetic vasomotor outflow. Am. J. Physiol.- Reg. Integr. Comp. Physiol. 282, R1174-R1184.

Sato, M.A., Menani, J.V., Lopes, O.U., Colombari, E., 2001. Lesions of the commissural nucleus of the solitary tract reduce arterial pressure in spontaneously hypertensive rats. Hypertension 38, 560-564.

Sawchenko, P.E., Imaki, T., Potter, E., Kovacs, K., Imaki, J., Vale, W., 1993. The functional neuroanatomy of corticotropin-releasing factor. Ciba Foundation Symposium 172, 5-21.

Schreihofer, A.M., Guyenet, P.G., 1997. Identification of C1 presympathetic neurons in rat rostral ventrolateral medulla by juxtacellular labeling in vivo. J. Comp. Neurol. 387, 524-536.

Schreihofer, A.M., Guyenet, P.G., 2000. Sympathetic reflexes after depletion of bulbospinal catecholaminergic neurons with anti-D βH-saporin. Am. J. Physiol. - Regul. Integr. Comp. Physiol. 279, R729-R742.

Schreihofer, A.M., Guyenet, P.G., 2002. The baroreflex and beyond: Control of sympathetic vasomotor tone by GABAergic neurons in the ventrolateral medulla. Clin. Exp. Pharmacol. Physiol. 29, 514-521.

Schreihofer, A.M., Stornetta, R.L., Guyenet, P.G., 2000. Regulation of sympathetic tone and arterial pressure by rostral ventrolateral medulla after depletion of Cl cells in rat. J. Physiol. (Lond.) 529, 221-236.

Seyedabadi, M., Goodchild, A.K., Pilowsky, P.M., 2002. Site-specific effects of apelin-13 in the rat medulla oblongata on arterial pressure and respiration. Auton. Neurosci. 101, 32-38.

Sofroniew, M.V., Schrell, U., 1981. Evidence for a direct projection from oxytocin and vasopressin neurons in the hypothalamic paraventricular nucleus to the medulla oblongata, immunohistochemical visualization of both the horseradish peroxidase transported and the peptide produced by the same neurons. Neurosci. Lett. 22, 211-217.

Stornetta, R.L., Akey, P.J., Guyenet, P.G., 1999. Location and electrophysiological characterization of rostral medullary adrenergic neurons that contain neuropeptide Y mRNA in rat. J. Comp. Neurol. 415, 482-500.

Stornetta, R.L., Morrison, S.F., Ruggiero, D.A., Reis, D.J., 1989. Neurons of rostral ventrolateral medulla mediate somatic pressor reflex. Am. J. Physiol. 256, R448-R462.

Stornetta, R.L., Schreihofer, A.M., Pelaez, N.M., Sevigny, C.P., Guyenet, P.G., 2001. Preproenkephalin mRNA is expressed by C1 and non-C1 barosensitive bulbospinal neurons in the rostral ventrolateral medulla of the rat. J. Comp. Neurol. 435, 111-126.

Stornetta, R.L., Sevigny, C.P., Guyenet, P.G., 2002a. Vesicular glutamate transporter DNPI/VGLUT2 mRNA is present in C1 and several other groups of brainstem catecholaminergic neurons. J. Comp. Neurol. 444, 191-206.

Stornetta, R.L., Sevigny, C.P., Schreihofer, A.M., Rosin, D.L., Guyenet, P.G., 2002b. Vesicular glutamate transporter DNPI/GLUT2 is expressed by both C1 adrenergic and nonaminergic presympathetic vasomotor neurons of the rat medulla. J. Comp. Neurol. 444, 207-220.

Stornetta, R.L., Sevigny, C.P., Guyenet, P.G., 2003. Inspiratory augmenting bulbospinal neurons express both glutamatergic and enkephalinergic phenotypes. J. Comp. Neurol. 455, 113-124.

Strack, A.M., Loewy, A.D., 1990. Pseudorabies virus: a highly specific transneuronal cell body marker in the sympathetic nervous system. J. Neurosci. 10, 2139-2147.

Sun, M.K., 1995. Central neural organization and control of sympathetic nervous system in mammals. Prog. Neurobiol. 47, 157-233.

Sun, M.K., 1996. Pharmacology of reticulospinal vasomotor neurons in cardiovascular regulation. Pharm. Rev. 48, 465-494.

Sun, M.K., Guyenet, P.G., 1985. GABA-mediated baroreceptor inhibition of reticulospinal neurons. Am. J. Physiol. 249, R672-R680.

Sun, M.K., Guyenet, P.G., 1986a. Hypothalamic glutamatergic input to medullary sympathoexcitatory neurons in rats. Am. J. Physiol. 251, R798-R810.

Sun, M.K., Guyenet, P.G., 1986b. Medullospinal sympathoexcitatory neurons in normotensive and spontaneously hypertensive rats. Am. J. Physiol. 250, R910-R917.

Sun, M.K., Guyenet, P.G., 1987. Arterial baroreceptor and vagal inputs to sympathoexcitatory neurons in rat medulla. Am. J. Physiol. 252, R699-R709.

Sun, M.K., Guyenet, P.G., 1989. Effects of vasopressin and other neuropeptides on rostral medullary sympathoexcitatory neurons 'in vitro'. Brain Res. 492, 261-270.

Sun, M.K., Young, B.S., Hackett, J.T., Guyenet, P.G., 1988a. Reticulospinal pacemaker neurons of the rat rostral ventrolateral medulla with putative sympathoexcitatory function: an intracellular study in vitro. Brain Res. 442, 229-239.

Sun, M.K., Young, B.S., Hackett, J.T., Guyenet, P.G., 1988b. Rostral ventrolateral medullary neurons with intrinsic pacemaker properties are not catecholaminergic. Brain Res. 451, 345-349.

Sun, Q.J., Llewellyn-Smith, I., Minson, J., Arnolda, L., Chalmers, J., Pilowsky, P., 1996. Thyrotropin-releasing hormone immunoreactive boutons form close appositions with medullary expiratory neurons in the rat. Brain Res. 715, 136-144.

Sved, A.F., Ito, S., Yajima, Y., 2002. Role of excitatory amino acid inputs to the rostral ventrolateral medulla in cardiovascular regulation. Clin. Exp. Pharmacol. Physiol. 29, 503-506.

Sved, A.F., Mancini, D.L., Graham, J.C., Schreihofer, A.M., Hoffman, G.E., 1994. PNMT-containing neurons of the C1 cell group express c-fos in response to changes in baroreceptor input. Am. J. Physiol.- Regul. Integr. Comp. Physiol. 266, R361-R367.

Tagawa, T., Dampney, R.A., 1999. AT(1) receptors mediate excitatory inputs to rostral ventrolateral medulla pressor neurons from hypothalamus. Hypertension 34, 1301-1307.

Terui, N., Masuda, N., Saeki, Y., Kumada, M., 1990. Activity of barosensitive neurons in the caudal ventrolateral medulla that send axonal projections to the rostral ventrolateral medulla in rabbits. Neurosci. Lett. 118, 211-214.

Tsuruo, Y., Hökfelt, T., Visser, T., 1987. Thyrotropin releasing hormone (TRH)-immunoreactive cell groups in the rat central nervous system. Exp. Brain Res. 68, 213-217.

Urbanski, R.W., Murugaian, J., Krieger, A.J., Sapru, H.N., 1989. Cardiovascular effects of substance P receptor stimulation in the ventrolateral medullary pressor and depressor areas. Brain Res. 491, 383-389.

Verberne, A.J.M., Guyenet, P.G., 1992. Medullary pathway of the Bezold-Jarisch reflex in the rat. Am. J. Physiol. 263, R1195-R1202.

Wang, H., Germanson, T.P., Guyenet, P.G., 2002. Depressor and tachypneic responses to chemical stimulation of the ventral respiratory group are reduced by ablation of neurokinin-1 receptor-expressing neurons. J. Neurosci. 22, 3755-3764.

Wang, W.H., Lovick, T.A., 1992b. Excitatory 5-HT2-mediated effects on rostral ventrolateral medullary neurones in rats. Neurosci. Lett. 141, 89-92.

Wang, W.H., Lovick, T.A., 1992a. Inhibitory serotonergic effects on rostral ventrolateral medullary neurons. Pflugers Archiv. - Eur. J. Physiol. 422, 93-97.

Weiss, M.L., Kenney, M.J., Musch, T.I., Patel, K.P., 2003. Modifications to central neural circuitry during heart failure. Acta Physiol. Scand. 177, 57-67.

Yamamoto, H., Lee, C.E., Marcus, J.N., Williams, T.D., Overton, J.M., Lopez, M.E., Hollenberg, A.N., Baggio, L., Saper, C.B., Drucker, D.J., Elmquist, J.K., 2002. Glucagon-like peptide-1 receptor stimulation increases blood pressure and heart rate and activates autonomic regulatory neurons. J. Clin. Invest. 110, 43-52.

Zhang, K., Li, Y.F., Patel, K.P., 2002. Reduced endogenous GABA-mediated inhibition in the PVN on renal nerve discharge in rats with heart failure. Am. J. Physiol.-Regul. Integr. Comp. Physiol. 282, R1006-R1015.

Zhong, S., Huang, Z.S., Gebber, G.L., Barman, S.M., 1993. Role of the brainstem in generating the 2-6Hz oscillation in sympathetic nerve discharge. Am. J. Physiol. 265, R1026-R1035.

Chapter 10

SEROTONIN NEURONS IN THE BRAINSTEM AND SPINAL CORD: DIVERSE PROJECTIONS AND MULTIPLE FUNCTIONS

Paul M.Pilowsky
Departments of Physiology and Neurosurgery, University of Sydney and Royal North Shore Hospital, Sydney, Australia

Abstract: Serotonin neurons are located in the midline of the brainstem. The axons of these neurons branch profusely throughout the entire neuraxis providing an innervation to neurons that are involved in activities as diverse as the sleep/wake cycle, autonomic control and motor control. Here we review some of the work that has been done in relation to serotonin neurons in the brainstem with particular regard to the anatomy and co-localisation of other neurotransmitters. Particular attention is paid to the role of serotonin neurons in control of the cardiovascular system, but the role played by these neurons in modulating pain, chemosensation, thermoregulation and motor control as well as the sleep/wake cycle and sexual function is also dealt with. We conclude that despite considerable work in the past, much remains to be learned about the serotonin containing neurons and the other neurotransmitters that are released from these aminergic neurons.

Key words: Baroreceptors, RVMM, motor control, chemoreception, sleep wake

INTRODUCTION

The main source of input to sympathetic preganglionic neurons (SPN) arises from at least five major supramedullary cell groups (Strack et al., 1989; Smith et al., 1998) as well as possible intraspinal sources. The five supraspinal sites have been the subject of much careful anatomical study over the past decades and include:
- ➢ The rostral ventrolateral medulla (RVLM),
- ➢ The rostral ventromedial medulla (RVMM),

➢ The caudal midline raphe nuclei (CMR), including the raphe pallidus, raphe magnus and raphe obscurus,
➢ The pontine A5 group of noradrenaline containing neurons,
➢ And parts of the hypothalamic paraventricular nuclei.

Other sites have been reported by some authors, but they have not been so intensively investigated (Aicher et al., 1995).

The five main sites mentioned above have also been the focus of considerable investigation aimed at determining their functional role in normal activity and changes that occur following destruction or enhancement of their activity or after application of a vast range of pharmacological agents (Chalmers et al., 1985, 1988; Pilowsky et al., 1986a, 1995a, b; Jensen et al., 1995; Pilowsky and Goodchild, 2002). Of particular interest for many investigators is the RVLM (Fig. 1) since there is now overwhelming evidence that this site is key to the maintenance of sympathetic tone (Dampney, 1981; Ross et al., 1984; Willette et al., 1984; Ito and Sved, 1997; Schreihofer and Guyenet, 2002) and integration of reflexes (Granata et al., 1985; Terui et al., 1986, 1988; Gieroba et al., 1995; Miyawaki et al., 1996, 2001, 2002). Bilateral destruction of this nucleus normally causes blood pressure and sympathetic nerve activity to fall to very low levels (Guertzenstein and Silver, 1974; Pilowsky et al., 1985), although if the animal is permitted to recover from anaesthesia blood pressure is maintained by vasopressin, angiotensin and aldosterone release (Cochrane and Nathan, 1989, 1993, 1994). In fact it has been suggested that the RVLM is the final common pathway for much of the sympathetic premotor activity that regulates vasomotor pathways in the periphery, including visceral, renal, vascular smooth muscle and cardiac. In addition, premotor neurons regulating sympathetic outflow to adrenaline- and noradrenaline-secreting chromaffin cells are also located in the RVLM. As well as 'cardiovascular' neurons, the ventrolateral medulla also contains respiratory premotoneurons (Bötzinger cell group) and cranial motoneurons in the nucleus ambiguus pars compacta and externa (NAc and NAe) (Fig. 1).

Despite this very reasonable interest in the activity of neurons in the RVLM, many other pathways also seem to have an important role to play in controlling sympathetic activity. The principal difficulty for the physiologist in this era is to determine when these pathways are active and why. Of the well-described sites lying outside the RVLM, the midline or caudal midline raphe (obscurus, pallidus and magnus) and parapyramidal region as a whole is certainly one that has received considerable attention (Fig. 1). Here I will provide a brief summary of work that has been done on the anatomy, neurochemistry and function of neurons whose cell bodies lie in this region. Axons or other fibres of passage in the midline will not be discussed here.

10. SEROTONIN NEURONS IN THE BRAINSTEM AND SPINAL CORD: DIVERSE PROJECTIONS AND MULTIPLE FUNCTIONS

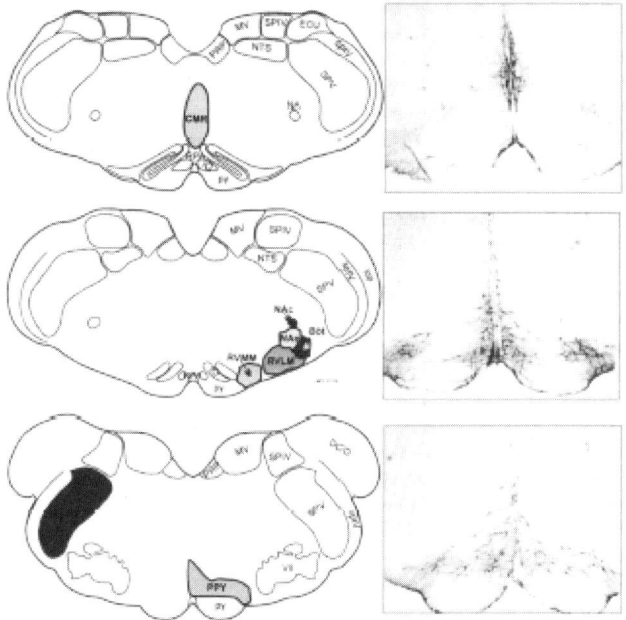

Figure 1. Location of nuclei in the rostral medulla immediately caudal to the facial nucleus (level 63, AP = -12.68; level 60, AP=-11.90; level 57, AP=-11.4; Swanson, 2000). The location of key medullary autonomic nuclei is shown on the left. The nucleus ambiguus pars compacta (NAc) contains the somata of oesophageal motoneurons. Ventral to it is the nucleus ambiguus pars externa (NAe), which contains glossopharyngeal and vagal motoneurons that subserve a variety of functions. The Bötzinger cell group (Böt) are a population of glycinergic expiratory-active respiratory neurons with very widespread intramedullary and spinal axon collaterals. The RVLM is found below and slightly intermingled with these nuclei and, in its rostral portion, contains a population of spinally-projecting neurons that are inhibited by baroreceptor activation. The RVMM is a region whose precise location is a subject of some debate, but which is commonly placed in the parapyramidal (PPY) region shown here. The raphe nuclei (caudal midline raphe, CMR; raphe pallidus, RPA) lie in the midline and have a very complex physiology, anatomy and neurochemistry. Photomicrographs on the right illustrate immunocytochemistry for PH8 (an antibody that detects serotonin neurons) at the corresponding level. Swanson, 1998, reproduced with permission.

The first issue to be addressed is that of the anatomical pathways involved. In this we are aided by two valuable approaches. First, standard tract-tracing with all of the various tracers that are now available (Clark and Proudfit, 1991; Kwiat and Basbaum, 1992; Minson et al., 1994; Jeske et al., 1995; Lan et al., 1997; Stornetta and Guyenet, 1999; Wang et al., 2001); and, second, viral tract-tracing that allows, when applied carefully, polysynaptic tracing of neuronal pathways (Loewy, 1998; Aston-Jones and Card, 2000; Lee et al., 2002; Martinov et al., 2002). When combined with immunohistochemistry (Minson et al., 1994; Stornetta and Guyenet, 1999; Phillips et al., 2001; Basura et al., 2001; Marsala et al., 2002) or *in situ* hybridisation histochemistry (Stornetta and Guyenet, 1999; Duffield et al., 1999; Basura et al., 2001) these approaches become extremely powerful tools for investigating connectivity in the autonomic and other nervous systems.

ANATOMY OF THE RVMM

To cardiovascular neuroscientists, the RVMM consists of a rather ill-defined region that lies close to the level of the facial nucleus in the ventral medulla. In the rat, it lies approximately between 0.5 and 1 mm lateral to the midline (0.0-0.5 defines the CMR). Rostrocaudally it starts at about the level of the facial nucleus and continues caudally for about 1.0 mm. These borders vary considerably depending on the basis upon which they are defined, as does the nomenclature used to describe this region. Names include the RVMM (Pilowsky et al., 1986a), the midline and the raphe nuclei. The region also traditionally refers to the parapyramidal region (Miura et al., 1996; Pelaez et al., 2002). I will use the term RVMM as a collective term for all of these sites unless specific differentiation is appropriate.

Because the RVMM contains such a heterogenous mix of neurons, the boundaries based on retrograde labelling from the spinal cord may differ somewhat from those that are defined on the basis of functional effects following activation or inhibition of neurons in these regions. Single unit recording from the region of the RVMM also reveals a multiplicity of behavioural types (Fields et al., 1983; Fornal et al., 1985; Pilowsky et al., 1995b; Ribeiro-do-Valle and Lucena, 2001). Anatomical studies using multiple tract tracers and immunohistochemistry (Allen and Cechetto, 1994) reveal that many neurons in the RVMM have highly collateralised axons that innervate many different regions of the brainstem and spinal cord. It seems likely that such a diverse arrangement of projections means that at least some of the neurons in the RVMM and parapyramidal region may serve a role as gain-setting neuromodulators (McCall and Aghajanian, 1979) that act to change the general level of excitability of large populations of neurons rather than acting in specific pathways.

The definition of the RVMM is complicated by the fact that authors in the pain literature also use this term to refer to the midline or to the raphe nuclei (Mason et al., 1990; Heinricher and Kaplan, 1991; Morgan and Fields, 1993; Potrebic et al., 1994; Thomas et al., 1995; Urban et al., 1999;

Odeh et al., 2003), while authors in the cardiovascular literature tend to refer to the midline as a region that is separate to the RVMM. In the cardiovascular literature RVMM commonly refers to a region immediately medial to the RVLM (Cox and Brody, 1988), or in the parapyramidal region (Pilowsky et al., 1986a; Miura et al., 1996; Pelaez et al., 2002).

In addition to problems with definition of nuclear borders, which are fuzzy at best in this region, it is also clear that there are many types of neurons in the RVMM with a range of different projections that express a variety of different neurochemicals. Neurons in the RVMM are commonly bulbospinal where they innervate all levels and segments of the spinal cord. Neurons in the dorsal and median raphe provide a dense innervation of supramedullary structures such as the hypothalamus, cortex, olfactory bulbs and so forth. This very widespread input from raphe neurons has led to considerable speculation about a role for these neurons in the sleep-wake cycle and arousal. Many other functions have been canvassed for the bulbospinal neurons including: regulation of penile erection (Marson et al., 1993; Giuliano et al., 1995; Sugaya et al., 1998); airways patency (Miura et al., 1996; Haxhiu et al., 1996; Lalley et al., 1997); modulation of pain transmission (Basbaum and Fields, 1984; Bouhassira et al., 1995; Urban et al., 1996; Mason, 2001); and thermoregulation by changes in blood flow, to the tail in rats (Smith et al., 1998; Nagashima et al., 2000; Tanaka et al., 2002), and to skin in hairless vertebrates such as man (Aoki et al., 2003).

NEUROCHEMISTRY OF THE RVMM

The neurochemistry of the RVMM is one of the most complex in the brain. Neurotransmitters that have been detected in this region include all of the known types from amino acids to peptides (Johansson et al., 1981; Hokfelt et al., 2000), but not gases (Leger et al., 2002). Similarly, a vast range of receptors, enzymes and other proteins are also found within neurons in this region. Colocalisation studies have revealed a further layer of complexity with numerous examples of substances being found together in combinations. Unfortunately very few of these studies are quantitative so that it is generally impossible to determine what the proportions of different populations are. Finally, some of the immunocytochemical studies have been combined with tract-tracing in order to reveal the complement of genes expressed in neurons that project to different sites. Again, this type of study has revealed a rich complexity of neurotransmitters and projections that is yet to be clearly understood in terms of function.

The neurotransmitter most commonly associated with the RVMM is serotonin (Proudfit et al., 1980; Ruda et al., 1981; Hunt and Lovick, 1982; Lovick and Hunt, 1983; Pilowsky et al., 1986a, 1995a; Chalmers et al., 1988; Dean et al., 1993; Zagon, 1993; Potrebic et al., 1994; Tanaka et al., 1994). This association may be for historical reasons since histochemical techniques that permitted the detection of neurotransmitters allowed detection of serotonin even before the advent of immunohistochemistry. However, it is always noted that almost all serotonin-synthesising neurons in this region seem to contain at least one, often several, other putative neurotransmitters. These include, amongst others, glutamate, enkephalin, GABA and substance P. Both serotonin containing (Fig. 1) and nonserotonergic (Leger et al, 2002) neurons arise from the medullary raphe and innervate all parts of the spinal grey matter (Pilowsky et al., 1990; Allen and Cechetto, 1994) (Fig. 2).

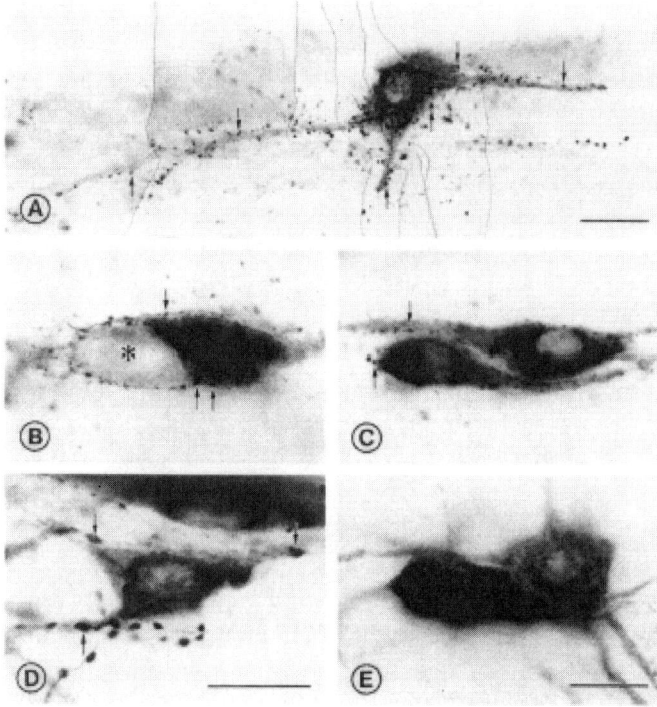

Figure 2. 5-HT immunoreactive nerve fibres and retrogradely labelled SPN whose cell bodies lie within the intermediolateral cell column. A-C: CTB retrogradely transported from the superior adrenal medulla. D-E: CTB retrogradely transported from the superior cervical ganglion. Many 5-HT immunoreactive varicosities are closely apposed to the SPN (arrows).

However, not all SPN receive a serotonin input (E). Jensen et al., 1995, reproduced with permission.

Ultimately, it is more important to determine the target, and type of synaptic connection, that the axons of these neurons form (pre- or post-synaptic) and the nature of the receptor upon which they act. Serotonin for example acts on at least 15 receptors one of which, the 5HT3 receptor, is ionotropic. The other serotonin receptors are metabotropic, some of which act to increase cellular activity, whilst others exert inhibitory effects (see Barnes and Sharp, 1999 for review). The complexity of actions of such a neurotransmitter, and in fact most others, suggests that we can no longer simplify and use the term 'serotonin neurons' as has been done parsimoniously in the past. Rather, we should take the actions of serotonin into account in the specific physiological context in which it is operating with regard to the activity of other neurons, the post-synaptic receptors present and co-transmitters released.

FUNCTIONS OF THE RVMM

A large number of functions have been ascribed to neurons in the midline/parapyramidal region. These include, amongst others, modulation of blood pressure, sensory afferent information, chemosensation, thermoregulation through modulation of cutaneous blood vessels (man) or tail blood vessels (rodents), innervation of brown adipose tissue, and the sleep-wake cycle. Given such a wide range of functions it seems unreasonable to refer to this region by one name simply on the basis of diffuse anatomical landmarks. Better perhaps is to refer to the functions that this region is responsible for and to define cell groups neurochemically as more information becomes available. It is therefore critical, in the absence of better information, to at least report precisely the sites from which recordings are made or cells are inhibited or stimulated. Fortunately, this now appears to be becoming the norm.

Cardiovascular

Activation of neurons in the midline (Chalmers et al., 1985; Pilowsky et al., 1986a; Coleman and Dampney, 1995; Bernard, 1998), medially in the medulla (Cox and Brody, 1989) or in the parapyramidal region (Chalmers et al., 1985; Pilowsky et al., 1986a) leads to variable effects on blood pressure and sympathetic nerve activity. Some authors have reported no effect, others increases, and others decreases (Adair et al., 1977; Coleman and Dampney, 1995; Bernard, 1998). The studies are further complicated by, amongst other things, the fact that they are carried out in several different species and using different anaesthetic agents and with different modes of stimulation. Early studies used electrical stimulation (Adair et al., 1977; Howe et al., 1983; Cox and Brody, 1989) but since then chemical stimulation (Goodchild et al., 1982), when used with care (Lipski et al., 1988), has become the accepted norm since it selectively activates somata and not fibres of passage. Moreover, with an increasing availability of specific agonists and antagonists a number of studies have investigated activation, not just of whole populations of neurons, with for example glutamate (Goodchild et al., 1982; Chalmers et al., 1985; Coleman and Dampney, 1995; Bernard, 1998), but also of specific populations with drugs that only act on receptor subtypes (Howe et al., 1983; Chalmers et al., 1985; Helke et al., 1993).

It has also become clear that very accurate localisation of microinjection sites is critical in order to understand the physiological responses elicited. One of the earliest studies to attempt a correlated anatomical and physiological study of this type was conducted by Adair et al. (1977) who used electrical stimulation in chloralose-anaesthetised cats. Adair et al. (1977) reported extremely variable changes in blood pressure with both pressor and depressor responses. On the other hand there were no consistent changes in heart rate. The key finding of this work is that the midline seems to be a rather heterogenous region with respect to control of blood pressure, a point that was also strongly made by Coote (1990). Experiments in rabbit with acute intracisternal administration of 5,7-dihydroxytryptamine (5,7-DHT) or 5,6-DHT revealed massive sympathoexcitation and pressor responses (Wing and Chalmers, 1974; Korner et al., 1984; Pilowsky et al., 1986b) (Fig. 3).

In a series of experiments aimed at demonstrating that neurons in the brainstem that contain serotonin are involved in the cardiovascular responses that can be elicited by stimulation of the RVMM, neurons that synthesise serotonin were destroyed with the neurotoxin 5,7- (or 5,6-) DHT prior to experimentation. Electrical or chemical stimulation of the RVMM was then performed to determine if 5,7-DHT pretreatment altered the response. In a study by Howe et al. (1983) there was a clear attenuation of the pressor response to electrical stimulation of the parapyramidal region coinciding with the lateral portion of the B1 and B3 cell groups. Activation of this region causes a pressor response and release of serotonin in the spinal cord (Pilowsky et al., 1986a) (Fig. 4). This region is known to contain bulbospinal serotonin neurons (Loewy and McKellar, 1981). Although some reports suggest that neurons in the midline that are involved in cardiovascular

10. SEROTONIN NEURONS IN THE BRAINSTEM AND SPINAL CORD: DIVERSE PROJECTIONS AND MULTIPLE FUNCTIONS

control are sympathoinhibitory (Barman and Gebber, 1988; Gebber and Barman, 1988), a sympathoexcitatory role for at least some neurons in the midline is supported by the finding of single neurons in this region that are pulse-modulated, spinally-projecting and sensitive to aortic nerve stimulation (Pilowsky et al., 1995b) (Fig. 5).

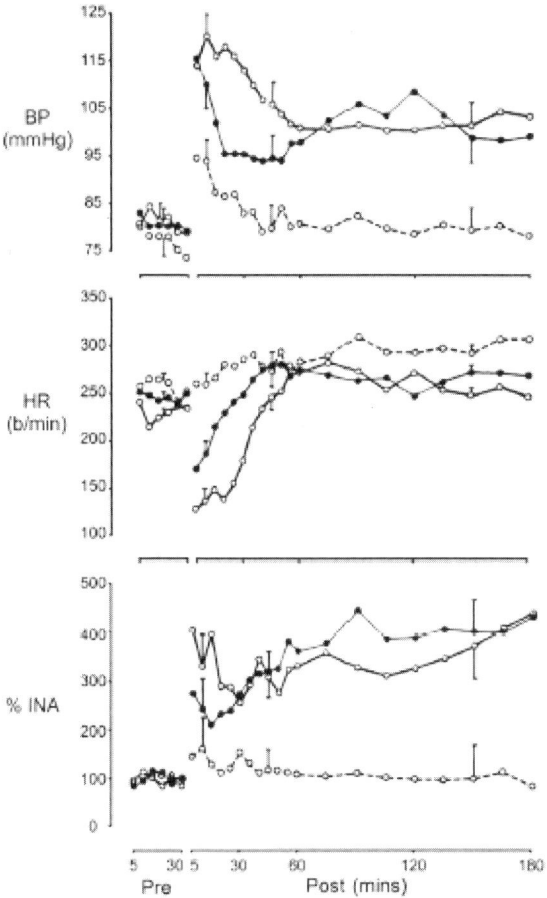

Figure 3. Effect of the serotonin neurotoxin (5,7-DHT) or catecholamine neurotoxin (6-OHDA) administered icv in conscious rabbits on blood pressure, heart rate and renal nerve activity. Mean arterial pressure (BP), heart rate (HR) and integrated renal nerve activity (INA) before and after intracisternal injections of 5,7-DHT (open circles, continuous line), 6-OHDA (closed circles, continuous line) or creatinine sulphate-vehicle (open circles, interrupted line).

Values are means, SEM indicated. The toxins cause an acute release of neurotransmitter from serotonin and noradrenaline terminals respectively, and an associated sympathoexcitation and pressor response. Pilowsky et al., 1986b, reproduced with permission.

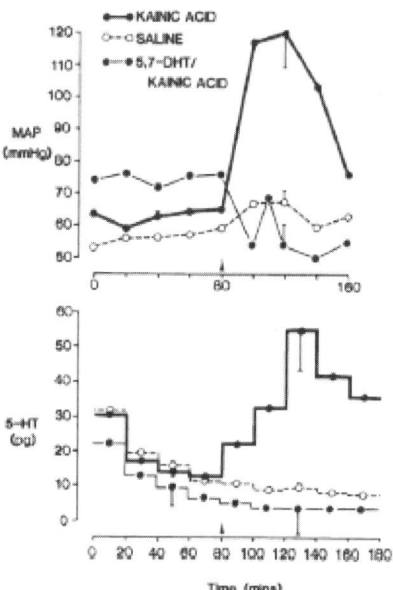

Figure 4. The effect on mean arterial pressure (MAP) (mmHg) and serotonin efflux (5-HT) (pg per 20 min collected by microdialysis) following bilateral microinjections of kainic acid into the region of the lateral B3 region (RVMM) (Fig. 1) of normal rats (black circles and thick black lines) and rats pre-treated with 5, 7-dihydroxytryptamine 2 weeks earlier (black circles and thin black lines). Vehicle-injected rats (open circles and dotted line). Microinjections were given at 80 minutes (arrowheads). Values are means, SEM indicated. Pilowsky et al., 1986a, reproduced with permission.

Other studies have provided data that suggests a depressor role for neurons in the midline raphe obscurus (Coleman and Dampney, 1995; Bernard, 1998) and in the raphe pallidus (Coleman and Dampney, 1995). However, it is possible that these depressor effects are mediated through a GABAergic projection to the RVLM (Coleman and Dampney, 1998), so it is still possible that intermingled bulbospinal sympathoexcitatory neurons exist in this area.

10. SEROTONIN NEURONS IN THE BRAINSTEM AND SPINAL CORD: DIVERSE PROJECTIONS AND MULTIPLE FUNCTIONS

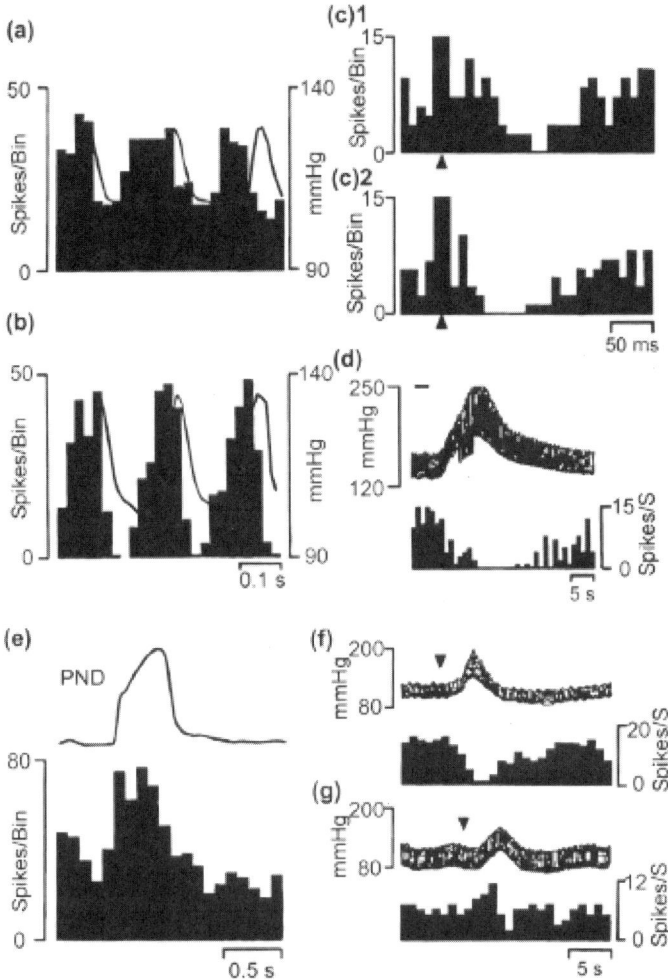

Figure 5. (a) Modulation of firing probability in a bulbospinal raphe unit as revealed by an R-wave triggered histogram. (b) In a rostral ventrolateral medulla unit at the same level of arterial blood pressure as in (a), a much stronger modulation was observed. (c) Electrical stimulation of the aortic nerve (arrowheads) caused a decrease in the firing probability of a bulbospinal raphe neuron after a delay of about 70 ms (plot 1). Aortic nerve stimulation had a more pronounced effect on bulbospinal neurons in the rostral ventrolateral medulla (plot 2). (d) Intravenous administration of 1.0 µg phenylephrine (bar) raised blood pressure and lowered the firing rate of a raphe spinal unit. (e) A midinspiratory increase in firing probability (the most common respiratory modulation observed) as revealed by a periphrenic histogram. Brief tail pinch (arrowheads) affected the firing rate of a bulbospinal barosensitive raphe neuron (f) and a rostral ventrolateral medulla neuron (g). Pilowsky et al., 1995b, reproduced with permission.

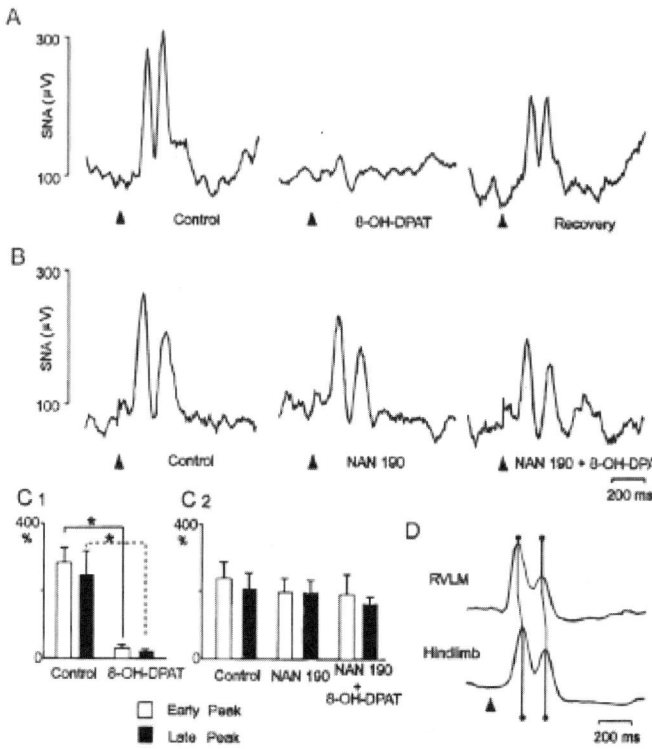

Figure 6. A: Two distinct excitatory responses in sympathetic nerve activity (SNA) were evoked by stimulation of the hindlimb (arrowheads). These were abolished following injection of 8-OH-DPAT (a 5HT1A receptor agonist) in the rostral ventrolateral medulla (RVLM). B: After preinjection of NAN-190 (a 5HT1A receptor antagonist) in the RVLM, 8-OH-DPAT in the RVLM did not abolish the somatosympathetic reflex. C: Summary of data in A and B. Injection of 8-OH-DPAT almost abolished the somatosympathetic reflex. (n=6, C1). Pretreatment with NAN-190 antagonised the inhibitory effect of 8-OH-DPAT (n=6, C2). * $P<0.01$. D: Electrical stimulation of the RVLM evoked 2 excitatory responses in SNA similar to those evoked by hindlimb stimulation. The interval between the two excitatory potentials evoked by RVLM and hindlimb stimulation is the same. Miyawaki et al., 2001, reproduced with permission.

A physiological role for the depressor responses that can be elicited from the caudal midline medulla (Coleman and Dampney, 1995; Bernard, 1998) may be the hypotensive response that occurs following haemorrhage

(Barcroft et al., 1944; Schadt and Ludbrook, 1991). It is now established that blockade of the midline eliminates the hypotensive response to haemorrhage (Henderson et al., 2000; Heslop et al., 2002), an effect that is apparently mediated by activation of 5HT1A receptors in the RVLM. A role for the subependymal serotonin neurons that become activated during haemorrhage remains to be established (Pelaez et al., 2002). A role for the midline and parapyramidal regions in mediating baroreceptor responses has been ruled out (McCall and Harris, 1987; Henderson et al., 2000), although cells in this region may well receive baroreceptor inputs (McCall and Clement, 1989) as noted above (Pilowsky et al., 1995b). Recently, a role for 5HT1A receptors in mediating somatosensory (sympathoexcitatory) effects on the RVLM has been demonstrated (Miyawaki et al., 2001) (Fig. 6).

Pain

The midline/parapyramidal region contains many neurons that are involved in pain pathways (Basbaum and Fields, 1984; Fields et al., 1991; Harris, 1996; Mason, 2001). Physiologically, these neurons have a characteristic pattern of activity such that stimulation of peripheral nociceptors by brief stimuli such as tail pinch will elicit an abrupt onset (ON cells) or cessation (OFF cells) of activity (Fields et al., 1983; Vanegas et al., 1984; Leung and Mason, 1999). Neurons in other parts of the nervous system including cardiovascular and respiratory neurons will also respond to nociceptive stimuli, but not with the same abrupt behaviour that can be observed in the ON and OFF cells. The ON/OFF cells project to the spinal cord (Vanegas et al., 1984; Fields and Heinricher, 1985; Fields et al., 1995; Fields, 2000) where they innervate neurons in the cervical dorsal horn (Fields et al., 1995). In the thoracic spinal cord, electrical stimulation of the raphe magnus has potent effects on the firing patterns of spinal neurons that respond to both visceral and somatic inputs (Cervero and Lumb, 1988). In the cat, projections within the medulla have also been demonstrated by intracellular recording and dye-filling (Mason and Fields, 1989). Mason and colleagues have investigated the neurochemistry of ON and OFF cells in an elegant series of experiments (Potrebic et al., 1994; Mason, 1997; Gao and Mason, 2000) using intracellular electrophysiology, dye-labelling and immunocytochemistry. Although this work has established that ON and OFF cells are non-serotonergic (Potrebic et al., 1994; Mason, 1997; Gao and Mason, 2000), there is still evidence that serotonin in the spinal cord plays a role in pain modulation (Mason, 2001; Ochi et al., 2002) so the issue is still controversial (Sorkin et al., 1993; Sorkin and McAdoo, 1993).

Chemosensation

Many areas of the brain contain neurons that are chemosensitive to some extent (Schlaefke, 1981; Jarolimek et al., 1990; Neubauer et al., 1991; Coates et al., 1993; Ritucci et al., 1998; Wang and Richerson, 1999; Nattie, 1999; Remmers et al., 2001; Ballantyne and Scheid, 2001; Jiang et al., 2001; Nattie et al., 2002; Filosa et al., 2002) and many neurons in the brainstem are especially so. This has been demonstrated in experiments in which recordings from neurons are made while the pH of the bath solution is changed by small increments (Richerson, 1995; Wang et al., 1998, 2001, 2002). In these studies, physiology was combined with dye-labelling and immunohistochemistry and it was revealed that the chemosensitive neurons commonly displayed a serotonergic phenotype (Wang et al., 1998, 2001; Richerson, 2001). This finding is also interesting, as it would suggest that the chemosensitive neurons, within normal physiological conditions, are not part of the pain pathways described above. These findings are also supported by the findings of others who have reported the appearance of Fos (a marker of cellular activation) in monoamine cells after hypercapnia (Haxhiu et al., 2001). Bernard et al. (1996) took a pharmacological approach and injected acetazolamide into the midline raphe in order to induce hypercapnia by carbonic anhydrase inhibition. This resulted in an increase in phrenic nerve discharge compared with injections of control agents. Alternatively, a solution of carbon dioxide has been dialysed into the ventral medullary raphe of rats, resulting in increases in ventilation, but only during sleep (Nattie and Li, 2001).

Thermoregulation

Thermoregulation is another system in which serotonin is now recognised to play a key role. Mice lacking the 5HT7 receptor do not become hypothermic in response to 5HT or the 5HT1/7 agonist 5-carboxamidotryptamine (Hedlund et al., 2003). Interestingly, oleamide, which acts allosterically at the 5HT7 receptor, still has a hypothermic effect even in mice that lack the 5HT7 receptor, indicating an independent site of action. The caudal midline is now recognised as a key nucleus in thermoregulation (Szelenyi and Hinckel, 1987). The key role played by the RVMM is as a relay nucleus from the hypothalamic preoptic region. Normally, prostaglandins, especially PGE2, act as pyrogens that increase the activity of neurons in the preoptic nucleus (Nakamura et al., 2002). Neurons in the preoptic nucleus then send axons to the RVMM, which in turn projects to sympathetic preganglionic neurons, resulting in activation of brown adipose tissue and tail vasoconstriction. Both are very effective means of increasing body temperature (Romanovsky et al., 1997; Nakamura et al., 2002). Application of muscimol (a GABA-A receptor agonist) into the RVMM effectively blocked the thermogenic effects of PGE2 injected into the preoptic region (Nakamura et al., 2002). Anatomical tracing studies with virus injections into the tail (Smith et al., 1998) confirm the existence of such a pathway – in addition to the other autonomic pathways that are revealed with this approach (Smith et al., 1998). Several recent physiological studies also confirm the importance of the RVMM as the source of premotor neurons in the regulation of body temperature (Nagashima et al., 2000; Morrison, 2001; Rathner et al., 2001; Tanaka et al., 2002). In humans, Johnson and colleagues have shown that thermoregulation via cutaneous vasodilation and vasoconstriction is caused both by alterations in sympathetically-induced vasoconstriction (Pergola et al., 1994; Stephens et al., 2001) and changes in active vasodilation (mediated by cotransmitters in cholinergic nerves) (Kellogg, Jr. et al., 1995, 1998).

Motor Control

There is now ample evidence that serotonin neurons are involved as premotoneurons throughout the neuraxis. Cranial motoneurons (Sun et al., 2002) (Fig. 7), hypoglossal motoneurons (Tallaksen-Greene et al., 1993) and phrenic motoneurons (Pilowsky et al., 1990) all have a rich serotonergic input. Physiologically, serotonin normally acts on motoneurons, through a variety of biophysical mechanisms to enhance excitability (Berger et al., 1992; Kubin et al., 1992; Schwarzacher et al., 2002; Perrier et al., 2003) and to facilitate the action of other neurotransmitters such as glutamate (McCall and Aghajanian, 1979).

Sleep Wake Cycle

Numerous studies now support a role for serotonin in the sleep/wake cycle (Kubin et al., 1996, 1998; Ursin, 2002; Wisor et al., 2003). Serotonin has also been implicated in the linkage between sleep and depression (Adrien, 2002). In the RVMM, recordings of unit activity in the midline of un-anaesthetised freely moving cats reveals many units whose activity is linked to the sleep wake cycle (Leung and Mason, 1999; Ribeiro-do-Valle and Lucena, 2001). Many of these cells are believed to be of a serotonergic phenotype (Leung and Mason, 1999).

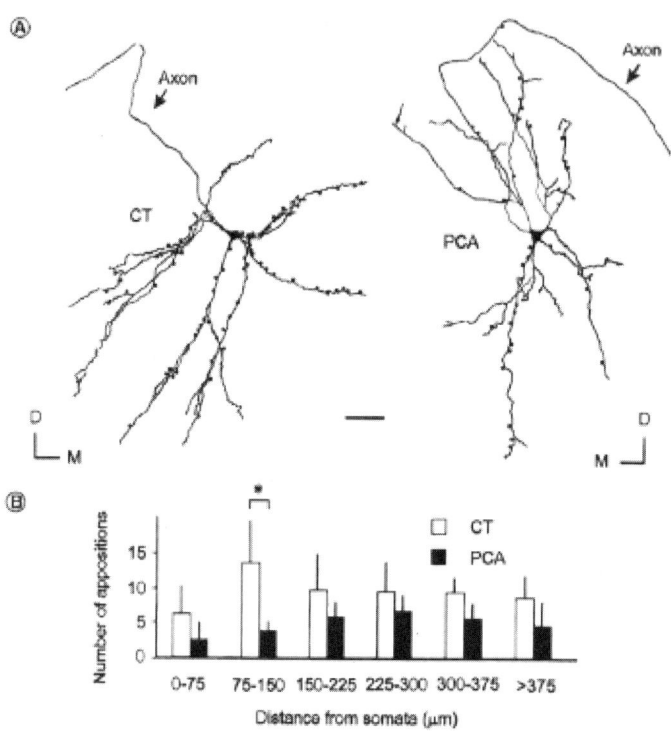

Figure 7. Reconstructions of rat cricoarytenoid (CT) and posterior cricoarytenoid (PCA) motoneurons (A) that were intracellularly labelled with neurobiotin in the rat. Immunohistochemistry for serotonin was then used to reveal the location of serotonin-containing varicosities opposed to the neurons. CT neurons received a larger input than PCA neurons (B) and this was significant 75-150 µm from the soma (P<0.001). Scale bar = 50 µm in A. Sun et al., 2002, reproduced with permission.

Sexual Function

Viral tracing studies also reveal a role for serotonin in control of the penis, with co-localisation of serotonin and tracer in the RVMM after injection of virus into the rat corpus cavernosum (Marson et al., 1993; Bancila et al., 2002). Serotonin may be facilatory or inhibitory depending on the receptor and pathway (Giuliano et al., 1995; Andersson, 2001). Pharmacological studies suggest that serotonin is important in sexual function in men and women (Allard and Giuliano, 2001; Meston and Frohlich, 2001; Popova and Amstislavskaya, 2002; Salonia et al., 2002).

CONCLUSION

The midline/parapyramidal region is a poorly understood region with a complex neurochemistry, a vast array of functions and neurons that project to almost every part of the nervous system. In light of this functional and chemical heterogeneity it is essential to have more studies to clarify the role of this clearly important, but still mysterious, brain region.

ACKNOWLEDGEMENTS

I am grateful to Miss N. Costin for her excellent assistance in the preparation of this manuscript and to Dr. Ann Goodchild for the production of Fig. 1. Work in the author's laboratory is supported by grants from the National Health and Medical Research Council (211023, 211196), National Heart Foundation of Australia (G00S0716), Garnett Passe and Rodney Williams Memorial Foundation ('03-'05), the North Shore Heart Research Foundation (14-00/01, 17-00/01) and Northern Sydney Health.

REFERENCES

Adair, J.R., Hamilton, B.L., Scappaticci, K.A., Helke, C.J., Gillis R.A., 1977. Cardiovascular responses to electrical stimulation of the medullary raphe area of the cat. Brain Res. 128, 141-145.

Adrien, J., 2002. Neurobiological bases for the relation between sleep and depression. Sleep Med. Rev. 6, 341-351.

Aicher, S.A., Reis, D.J., Nicolae, R., Milner, T.A., 1995. Monosynaptic projections from the medullary gigantocellular reticular formation to sympathetic preganglionic neurons in the thoracic spinal cord. J. Comp. Neurol. 363, 563-580.

Allard, J., Giuliano, F., 2001. Central nervous system agents in the treatment of erectile dysfunction: how do they work? Curr. Urol. Rep. 2, 488-494.

Allen, G.V., Cechetto, D.F., 1994. Serotoninergic and nonserotoninergic neurons in the medullary raphe system have axon collateral projections to autonomic and somatic cell groups in the medulla and spinal cord. J. Comp. Neurol. 350, 357-366.

Andersson, K.E., 2001. Pharmacology of penile erection. Pharmacol. Rev. 53, 417-450.

Aoki, K., Stephens, D.S.A., Johnson, J., Johnson, J.M., 2003. Cutaneous vasoconstrictor response to whole body skin cooling is altered by time of day. J. Appl. Physiol. 94(3), 930-934.

Aston-Jones, G., Card, J.P., 2000. Use of pseudorabies virus to delineate multisynaptic circuits in brain: opportunities and limitations. J. Neurosci. Methods 103, 51-61.

Ballantyne, D., Scheid, P., 2001. Central chemosensitivity of respiration: a brief overview. Respir. Physiol. 129, 5-12.

Bancila, M., Giuliano, F., Rampin, O., Mailly, P., Brisorgueil, M.J., Calas, A., Verge, D., 2002. Evidence for a direct projection from the paraventricular nucleus of the hypothalamus to putative serotoninergic neurons of the nucleus paragigantocellularis involved in the control of erection in rats. Eur. J. Neurosci. 16, 1240-1248.

Barcroft, H., McMichael, J., Edholm, O.G., Sharpey-Schafer, E.P., 1944. Posthaemorrhagic fainting: Study by cardiac output and forearm flow. Lancet i, 489-491.

Barman, S.M., Gebber, G.L., 1988. The axons of raphespinal sympathoinhibitory neurons branch in the cervical spinal cord. Brain Res. 441, 371-376.

Barnes, N.M., Sharp, T., 1999. A review of central 5-HT receptors and their function. Neuropharmacology 38, 1083-1152.

Basbaum, A.I., Fields, H.L., 1984. Endogenous pain control systems: brainstem spinal pathways and endorphin circuitry. Annu. Rev. Neurosci. 7, 309-338.

Basura, G.J., Zhou, S.Y., Walker, P.D., Goshgarian, H.G., 2001. Distribution of serotonin 2A and 2C receptor mRNA expression in the cervical ventral horn and phrenic motoneurons following spinal cord hemisection. Exp. Neurol. 169, 255-263.

Berger, A.J., Bayliss, D.A., Viana, F., 1992. Modulation of neonatal rat hypoglossal motoneuron excitability by serotonin. Neurosci. Lett. 143, 164-168.

Bernard, D.G., 1998. Cardiorespiratory responses to glutamate microinjected into the medullary raphe. Respir. Physiol. 113, 11-21.

Bernard, D.G., Li, A., Nattie, E.E., 1996. Evidence for central chemoreception in the midline raphe. J. Appl. Physiol. 80, 108-115.

Bouhassira, D., Chitour, D., Villanueva, L., Le Bars, D., 1995. The spinal transmission of nociceptive information: modulation by the caudal medulla. Neurosci. 69, 931-938.

Cervero, F., Lumb, B.M., 1988. Bilateral inputs and supraspinal control of viscerosomatic neurones in the lower thoracic spinal cord of the cat. J. Physiol. 403, 221-37.

10. SEROTONIN NEURONS IN THE BRAINSTEM AND SPINAL CORD: DIVERSE PROJECTIONS AND MULTIPLE FUNCTIONS

Chalmers, J.P., Kapoor, V., Macrae, I.M., Minson, J.B., Pilowsky, P., West, M.J., 1985. New approaches to the study of bulbospinal (b3) serotonergic neurons in the control of blood-pressure. J. Hypertension 3, S5-S9.

Chalmers, J.P., Pilowsky, P.M., Minson, J.B., Kapoor, V., Mills, E., West, M.J., 1988. Central serotonergic mechanisms in hypertension. Am. J. Hypertension 1, 79-83.

Clark, F.M., Proudfit, H.K., 1991. Projections of neurons in the ventromedial medulla to pontine catecholamine cell groups involved in the modulation of nociception. Brain Res. 540, 105-115.

Coates, E.L., Li, A., Nattie, E.E., 1993. Widespread sites of brain stem ventilatory chemoreceptors. J. Appl. Physiol. 75, 5-14.

Cochrane, K.L., Nathan, M.A., 1989. Normotension in conscious rats after placement of bilateral electrolytic lesions in the rostral ventrolateral medulla. J. Auton. Nerv. Sys. 26, 199-211.

Cochrane, K.L., Nathan, M.A., 1993. Cardiovascular effects of lesions of the rostral ventrolateral medulla and the nucleus-reticularis parvocellularis in rats. J. Auton. Nerv. Sys. 43, 69-82.

Cochrane, K.L., Nathan, M.A., 1994. Pressor systems involved in the maintenance of arterial-pressure after lesions of the rostral ventrolateral medulla. J. Auton. Nerv. Sys. 46, 9-18.

Coleman, M.J., Dampney, R.A., 1995. Powerful depressor and sympathoinhibitory effects evoked from neurons in the caudal raphe pallidus and obscurus. Am. J. Physiol. 268, R1295-R1302.

Coleman, M.J., Dampney, R.A., 1998. Sympathoinhibition evoked from caudal midline medulla is mediated by GABA receptors in rostral VLM. Am. J. Physiol. 274, R318-R323.

Coote, J.H., 1990. Bulbospinal serotonergic pathways in the control of blood pressure. J. Cardiovasc. Pharmacol. 15 Suppl 7, S35-S41.

Cox, B.F., Brody, M.J., 1988. Evidence for two functionally distinct vasomotor subregions of rostral ventral medulla. Clin. Exp. Hypertension A 10 Suppl 1, 11-18.

Cox, B.F., Brody, M.J., 1989. Subregions of rostral ventral medulla control arterial pressure and regional hemodynamics. Am. J. Physiol. 257, R635-R640.

Dampney, R.A., 1981. Brain stem mechanisms in the control of arterial pressure. Clin. Exp. Hypertension 3, 379-391.

Dean, C., Marson, L., Kampine, J.P., 1993. Distribution and co-localization of 5-hydroxytryptamine, thyrotropin-releasing hormone and substance P in the cat medulla. Neurosci. 57, 811-822.

Duffield, G.E., McNulty, S., Ebling, F.J.P., 1999. Anatomical and functional characterisation of a dopaminergic system in the suprachiasmatic nucleus of the neonatal Siberian hamster. J. Comp. Neurol. 408, 73-96.

Fields, H.L., 2000. Pain modulation: expectation, opioid analgesia and virtual pain. Prog. Brain Res. 122, 245-253.

Fields, H.L., Bry, J., Hentall, I., Zorman, G., 1983. The activity of neurons in the rostral medulla of the rat during withdrawal from noxious heat. J. Neurosci. 3, 2545-2552.

Fields, H.L., Heinricher, M.M., 1985. Anatomy and physiology of a nociceptive modulatory system. Philos. Trans. R. Soc. Lond. B. Biol. Sci. 308, 361-374.

Fields, H.L., Heinricher, M.M., Mason, P., 1991. Neurotransmitters in nociceptive modulatory circuits. Annu. Rev. Neurosci. 14, 219-245.

Fields, H.L., Malick, A., Burstein, R., 1995. Dorsal horn projection targets of ON and OFF cells in the rostral ventromedial medulla. J. Neurophysiol. 74, 1742-1759.

Filosa, J.A., Dean, J.B., Putnam, R.W., 2002. Role of intracellular and extracellular pH in the chemosensitive response of rat locus coeruleus neurones. J. Physiol. 541, 493-509.

Fornal, C., Auerbach, S., Jacobs, B.L., 1985. Activity of serotonin-containing neurons in nucleus raphe magnus in freely moving cats. Exp. Neurol. 88, 590-608.

Gao, K., Mason, P., 2000. Serotonergic raphe magnus cells that respond to noxious tail heat are not ON or OFF cells. J. Neurophysiol. 84, 1719-1725.

Gebber, G.L., Barman, S.M., 1988. Studies on the origin and generation of sympathetic nerve activity. Clin. Exp. Hypertension A 10 Suppl 1, 33-44.

Gieroba, Z.J., Messenger, J.P., Blessing, W.W., 1995. Abdominal vagal-stimulation excites bulbospinal barosensitive neurons in the rostral ventrolateral medulla. Neurosci. 65, 355-364.

Giuliano, F.A., Rampin, O., Benoit, G., Jardin, A., 1995. Neural control of penile erection. Urol. Clin. North Am. 22, 747-766.

Goodchild, A.K., Dampney, R.A., Bandler, R., 1982. A method for evoking physiological responses by stimulation of cell bodies, but not axons of passage, within localized regions of the central nervous system. J. Neurosci. Methods 6, 351-363.

Granata, A.R., Ruggiero, D.A., Park, D.H., Joh, T.H., Reis, D.J., 1985. Brain-stem area with c-1 epinephrine neurons mediates baroreflex vasodepressor responses. Am. J. Physiol. 248, H547-H567.

Guertzenstein, P.G., Silver, A., 1974. Fall in blood pressure produced from discrete regions of the ventral surface of the medulla by glycine and lesions. J. Physiol. 242, 489-503.

Harris, J.A., 1996. Descending antinociceptive mechanisms in the brainstem: their role in the animal's defensive system. J. Physiol. Paris 90, 15-25.

Haxhiu, M.A., Erokwu, B.O., Cherniack, N.S., 1996. The brainstem network involved in coordination of inspiratory activity and cholinergic outflow to the airways. J. Auton. Nerv. Sys. 61, 155-161.

Haxhiu, M.A., Tolentino-Silva, F., Pete, G., Kc, P., Mack, S.O., 2001. Monoaminergic neurons, chemosensation and arousal. Respir. Physiol. 129, 191-209.

Hedlund, P.B., Danielson, P.E., Thomas, E.A., Slanina, K., Carson, M.J., Sutcliffe, J.G., 2003. No hypothermic response to serotonin in 5-HT7 receptor knockout mice. Proc. Natl. Acad. Sci. U.S.A. 100, 1375-1380.

Heinricher, M.M., Kaplan, H.J., 1991. GABA-mediated inhibition in rostral ventromedial medulla - role in nociceptive modulation in the lightly anesthetized rat. Pain 47, 105-113.

Helke, C.J., McDonald, C.H., Phillips, E.T., 1993. Hypotensive effects of 5-HT1A receptor activation: ventral medullary sites and mechanisms of action in the rat. J. Auton. Nerv. Syst. 42, 177-188.

Henderson, L.A., Keay, K.A., Bandler, R., 2000. Caudal midline medulla mediates behaviourally-coupled but not baroreceptor-mediated vasodepression. Neurosci. 98, 779-792.

Heslop, D.J., Keay, K.A., Bandler, R., 2002. Haemorrhage-evoked compensation and decompensation are mediated by distinct caudal midline medullary regions in the urethane-anaesthetised rat. Neurosci. 113, 555-567.

Hokfelt, T., Arvidsson, U., Cullheim, S., Millhorn, D., Nicholas, A.P., Pieribone, V., Seroogy, K., Ulfhake, B., 2000. Multiple messengers in descending serotonin neurons: localization and functional implications. J. Chem. Neuroanat. 18, 75-86.

Howe, P.R., Kuhn, D.M., Minson, J.B., Stead, B.H., Chalmers, J.P., 1983. Evidence for a bulbospinal serotonergic pressor pathway in the rat brain. Brain Res. 270, 29-36.

Hunt, S.P., Lovick, T.A., 1982. The distribution of serotonin, met-enkephalin and beta-lipotropin-like immunoreactivity in neuronal perikarya of the cat brainstem. Neurosci. Lett. 30, 139-145.

10. SEROTONIN NEURONS IN THE BRAINSTEM AND SPINAL CORD: DIVERSE PROJECTIONS AND MULTIPLE FUNCTIONS

Ito, S. Sved, A.F., 1997. Tonic glutamate-mediated control of rostral ventrolateral medulla and sympathetic vasomotor tone. Am. J. Physiol. 273, R487-R494.

Jarolimek, W., Misgeld, U., Lux, H.D., 1990. Neurons sensitive to pH in slices of the rat ventral medulla oblongata. Pflugers Arch. 416, 247-253.

Jensen, I., Llewellyn-Smith, I.J., Pilowsky, P., Minson, J.B., Chalmers, J., 1995. Serotonin inputs to rabbit sympathetic preganglionic neurons projecting to the superior cervical-ganglion or adrenal-medulla. J. Comp. Neurol. 353, 427-438.

Jeske, I., Reis, D.J., Milner, T.A., 1995. Neurons in the barosensory area of the caudal ventrolateral medulla project monosynaptically on to sympathoexcitatory bulbospinal neurons in the rostral ventrolateral medulla. Neurosci. 65, 343-353.

Jiang, C., Xu, H., Cui, N., Wu, J., 2001. An alternative approach to the identification of respiratory central chemoreceptors in the brainstem. Respir. Physiol. 129, 141-157.

Johansson, O., Hokfelt, T., Pernow, B., Jeffcoate, S.L., White, N., Steinbusch, H.W., Verhofstad, A.A., Emson, P.C., Spindel, E., 1981. Immunohistochemical support for three putative transmitters in one neuron: coexistence of 5-hydroxytryptamine, substance P- and thyrotropin releasing hormone-like immunoreactivity in medullary neurons projecting to the spinal cord. Neurosci. 6, 1857-1881.

Kellogg, D.L., Jr., Crandall, C.G., Liu, Y., Charkoudian, N., Johnson, J.M., 1998. Nitric oxide and cutaneous active vasodilation during heat stress in humans. J. Appl. Physiol. 85, 824-829.

Kellogg, D.L., Jr., Pergola, P.E., Piest, K.L., Kosiba, W.A., Crandall, C.G., Grossmann, M., Johnson, J.M., 1995. Cutaneous active vasodilation in humans is mediated by cholinergic nerve cotransmission. Circ. Res. 77, 1222-1228.

Korner, P.I., Head, G.A., Bobik, A., Badoer, E., Aberdeen, J.A., 1984. Central and peripheral autonomic mechanisms involved in the circulatory actions of methyldopa. Hypertension 6, II63-II70.

Kubin, L., Davies, R.O., Pack, A.I., 1998. Control of Upper Airway Motoneurons During REM Sleep. News Physiol. Sci. 13, 91-97.

Kubin, L., Tojima, H., Davies, R.O., Pack, A.I., 1992. Serotonergic excitatory drive to hypoglossal motoneurons in the decerebrate cat. Neurosci. Lett. 139, 243-248.

Kubin, L., Tojima, H., Reignier, C., Pack, A.I., Davies, R.O., 1996. Interaction of serotonergic excitatory drive to hypoglossal motoneurons with carbachol-induced, REM sleep-like atonia . Sleep 19, 187-195.

Kwiat, G.C., Basbaum, A.I., 1992. The origin of brain-stem noradrenergic and serotonergic projections to the spinal-cord dorsal horn in the rat. Somatosensory and Motor Res. 9, 57-173.

Lalley, P.M., Benacka, R., Bischoff, A.M., Richter, D.W., 1997. Nucleus raphe obscurus evokes 5-HT-1A receptor-mediated modulation of respiratory neurons. Brain Res. 747, 156-159.

Lan, C.T., Wu, W.C., Ling, E.A., Chai, C.Y., 1997. Evidence of a direct projection from the cardiovascular-reactive dorsal medulla to the intermediolateral cell column of the spinal cord in cats as revealed by light and electron microscopy. Neurosci. 77, 521-533.

Lee, S., Miselis, R., Rivier, C., 2002. Anatomical and functional evidence for a neural hypothalamic-testicular pathway that is independent of the pituitary. Endocrinology 143, 4447-4454.

Leger, L., Gay, N., Cespuglio, R., 2002. Neurokinin NK1- and NK3-immunoreactive neurons in serotonergic cell groups in the rat brain. Neurosci. Lett. 323, 146-150.

Leung, C.G., Mason, P., 1999. Physiological properties of raphe magnus neurons during sleep and waking. J. Neurophysiol. 81, 584-595.

Lipski, J., Bellingham, M.C., West, M.J., Pilowsky, P., 1988. Limitations of the technique of pressure microinjection of excitatory amino-acids for evoking responses from localized regions of the cns. J. Neurosci. Methods 26, 169-179.

Loewy, A.D., 1998. Viruses as transneuronal tracers for defining neural circuits. Neurosci. Biobeh. Rev. 22, 679-684.

Loewy, A.D., McKellar, S., 1981. Serotonergic projections from the ventral medulla to the intermediolateral cell column in the rat. Brain Res. 211, 146-152.

Lovick, T.A., Hunt, S.P., 1983. Substance P-immunoreactive and serotonin-containing neurones in the ventral brainstem of the cat. Neurosci. Lett. 36, 223-228.

Marsala, J., Lukacova, N., Cizkova, D., Kafka, J., Katsube, N., Kucharova, K., Marsala, M., 2002. The case for the bulbospinal respiratory nitric oxide synthase- immunoreactive pathway in the dog. Exp. Neurol. 177, 115-132.

Marson, L., Platt, K.B., McKenna, K.E., 1993. Central nervous system innervation of the penis as revealed by the transneuronal transport of pseudorabies virus. Neurosci. 55, 263-280.

Martinov, V.N., Sefland, I., Walaas, S.I., Lomo, T., Nja, A., Hoover, F., 2002. Targeting functional subtypes of spinal motoneurons and skeletal muscle fibers in vivo by intramuscular injection of adenoviral and adeno-associated viral vectors. Anat. Embryol. 205, 215-221.

Mason, P., 1997. Physiological identification of pontomedullary serotonergic neurons in the rat. J. Neurophysiol. 77, 1087-1098.

Mason, P., 2001. Contributions of the medullary raphe and ventromedial reticular region to pain modulation and other homeostatic functions. Annu. Rev. Neurosci. 24, 737-77.

Mason, P., Fields, H.L., 1989. Axonal trajectories and terminations of on- and off-cells in the cat lower brainstem. J. Comp. Neurol. 288, 185-207.

Mason, P., Floeter, M.K., Fields, H.L., 1990. Somatodendritic morphology of on-cell and off-cells in the rostral ventromedial medulla. J. Comp. Neurol. 301, 23-43.

McCall, R.B., Aghajanian, G.K., 1979. Serotonergic facilitation of facial motoneuron excitation. Brain Res. 169, 11-27.

McCall, R.B., Clement, M.E., 1989. Identification of serotonergic and sympathetic neurons in medullary raphe nuclei. Brain Res. 477, 172-182.

McCall, R.B. Harris, L.T., 1987. Sympathetic alterations after midline medullary raphe lesions. Am. J. Physiol. 253, R91-R100.

Meston, C.M., Frohlich, P.F., 2001. Update on female sexual function. Curr. Opin. Urol. 11, 603-609.

Minson, J.B., Llewellyn-Smith, I.J., Arnolda, L.F., Pilowsky, P.M., Oliver, J.R., Chalmers, J.P., 1994. Disinhibition of the rostral ventral medulla increases blood-pressure and fos expression in bulbospinal neurons. Brain Res. 646, 44-52.

Miura, M., Okada, J., Takayama, K., 1996. Parapyramidal rostroventromedial medulla as a respiratory rhythm modulator. Neurosci. Lett. 203, 41-44.

Miyawaki, T., Goodchild, A.K., Pilowsky, P.M., 2001. Rostral ventral medulla 5-HT1A receptors selectively inhibit the somatosympathetic reflex. Am. J. Physiol-Reg. Int. Comp. Physiol. 280, R1261-R1268.

Miyawaki, T., Goodchild, A.K., Pilowsky, P.M., 2002. Activation of mu-opioid receptors in rat ventrolateral medulla selectively blocks baroreceptor reflexes while activation of delta opioid receptors blocks somato-sympathetic reflexes. Neurosci. 109, 133-144.

Miyawaki, T., Minson, J., Arnolda, L., Llewellyn-Smith, I., Chalmers, J., Pilowsky, P.M., 1996. AMPA/kainate receptors mediate sympathetic chemoreceptor reflex in the rostral ventrolateral medulla. Brain Res. 726, 64-68.

Morgan, M.M., Fields, H.L., 1993. Activity of nociceptive modulatory neurons in the rostral ventromedial medulla associated with volume expansion-induced antinociception. Pain 52, 1-9.

Morrison, S.F., 2001. Differential regulation of brown adipose and splanchnic sympathetic outflows in rat: Roles of raphe and rostral ventrolateral medulla neurons. Clin. Exp. Pharmacol. Physiol. 28, 138-143.

Nagashima, K., Nakai, S., Tanaka, M., Kanosue, K., 2000. Neuronal circuitries involved in thermoregulation. Auton Neurosci 20;85, 18-25.

Nakamura, K., Matsumura, K., Kaneko, T., Kobayashi, S., Katoh, H., Negishi, M., 2002. The rostral raphe pallidus nucleus mediates pyrogenic transmission from the preoptic area. J. Neurosci. 22, 4600-4610.

Nattie, E., 1999. CO_2, brainstem chemoreceptors and breathing. Prog. Neurobiol. 59, 299-331.

Nattie, E., Li, A., Meyerand, E., Dunn, J.F., 2002. Ventral medulla pHi measured in vivo by 31P NMR is not regulated during hypercapnia in anesthetized rat. Respir. Physiol. Neurobiol. 130, 139-149.

Nattie, E.E., Li, A., 2001. CO_2 dialysis in the medullary raphe of the rat increases ventilation in sleep. J. Appl. Physiol. 90, 1247-1257.

Neubauer, J.A., Gonsalves, S.F., Chou, W., Geller, H.M., Edelman, N.H., 1991. Chemosensitivity of medullary neurons in explant tissue cultures. Neurosci. 45, 701-708.

Ochi, T., Ohkubo, Y., Mutoh, S., 2002. FR143166 attenuates spinal pain transmission through activation of the serotonergic system. Eur. J. Pharmacol. 452, 319-324.

Odeh, F., Antal, M., Zagon, A., 2003. Heterogeneous synaptic inputs from the ventrolateral periaqueductal gray matter to neurons responding to somatosensory stimuli in the rostral ventromedial medulla of rats. Brain Res. 959, 287-294.

Pelaez, N.M., Schreihofer, A.M., Guyenet, P.G., 2002. Decompensated hemorrhage activates serotonergic neurons in the subependymal parapyramidal region of the rat medulla. Am. J. Physiol-Reg. Int. Comp. Physiol. 283, R688-R697.

Pergola, P.E., Kellogg, D.L., Jr., Johnson, J.M., Kosiba, W.A., 1994. Reflex control of active cutaneous vasodilation by skin temperature in humans. Am. J. Physiol. 266, H1979-H1984.

Perrier, J.F., Alaburda, A., Hounsgaard, J., 2003. 5-HT1A receptors increase excitability of spinal motoneurons by inhibiting a TASK-1-like K+ current in the adult turtle. J. Physiol. 548, 485-492.

Phillips, J.K., Goodchild, A.K., Dubey, R., Sesiashvili, E., Takeda, M., Chalmers, J., Pilowsky, P.M., Lipski, J., 2001. Differential expression of catecholamine biosynthetic enzymes in the rat ventrolateral medulla. J. Comp. Neurol. 432, 20-34.

Pilowsky, P.M., West, M., Chalmers, J., 1985. Renal sympathetic-nerve responses to stimulation, inhibition and destruction of the ventrolateral medulla in the rabbit. Neurosci. Lett. 60, 51-55.

Pilowsky, P.M., deCastro, D., Llewellyn-Smith, I., Lipski, J., Voss, M.D., 1990. Serotonin immunoreactive boutons make synapses with feline phrenic motoneurons. J. Neurosci. 10, 1091-1098.

Pilowsky, P.M., Goodchild, A.K., 2002. Baroreceptor reflex pathways and neurotransmitters: 10 years on. J. Hypertension 20, 1675-1688.

Pilowsky, P.M., Kapoor, V., Minson, J.B., West, M.J., Chalmers, J.P., 1986a. Spinal-cord serotonin release and raised blood-pressure after brain-stem kainic acid injection. Brain Res. 366, 354-357.

Pilowsky, P.M., Llewellyn-Smith, I.J., Minson, J.B., Arnolda, L.F., Chalmers, J.P., 1995a. Substance-P and serotonergic inputs to sympathetic preganglionic neurons. Clin. Exp. Hypertension 17, 335-344.

Pilowsky, P.M., Miyawaki, T., Minson, J.B., Sun, Q.J., Arnolda, L.F., Llewellyn-Smith, I.J., Chalmers, J.P., 1995b. Bulbospinal sympatho-excitatory neurons in the rat caudal raphe. J. Hypertension 13, 1618-1623.

Pilowsky, P.M., Morris, M.J., Kapoor, V., West, M.J., Chalmers, J.P., 1986b. Role of renal nerve activity, plasma-catecholamines and plasma vasopressin in cardiovascular-responses to intracisternal neurotoxins in the rabbit. J. Auton. Nerv. Sys. 17, 109-120.

Popova, N.K., Amstislavskaya, T.G., 2002. 5-HT2A and 5-HT2C serotonin receptors differentially modulate mouse sexual arousal and the hypothalamo-pituitary-testicular response to the presence of a female. Neuroendocrinol. 76, 28-34.

Potrebic, S.B., Fields, H.L., Mason, P., 1994. Serotonin immunoreactivity is contained in one physiological cell class in the rat rostral ventromedial medulla. J. Neurosci. 14, 1655-1665.

Proudfit, H.K., Larson, A.A., Anderson, E.G., 1980. The role of GABA and serotonin in the mediation of raphe-evoked spinal cord dorsal root potentials. Brain Res. 195, 149-165.

Rathner, J.A., Owens, N.C., McAllen, R.M., 2001. Cold-activated raphe-spinal neurons in rats. J. Physiol. 535, 841-854.

Remmers, J.E., Torgerson, C., Harris, M., Perry, S.F., Vasilakos, K., Wilson, R.J., 2001. Evolution of central respiratory chemoreception: a new twist on an old story. Respir. Physiol. 129, 211-217.

Ribeiro-do-Valle, L.E., Lucena, R.L., 2001. Behavioral correlates of the activity of serotonergic and non-serotonergic neurons in caudal raphe nuclei. Braz. J. Med. Biol. Res. 34, 919-937.

Richerson, G.B., 1995. Response to CO2 of neurons in the rostral ventral medulla in vitro. J. Neurophysiol. 73, 933-944.

Richerson, G.B., Wang, W., Tiwari, J., Bradley, S.R., 2001. Chemosensitivity of serotonergic neurons in the rostral ventral medulla. Respir. Physiol. 129, 175-189.

Ritucci, N.A., Chambers-Kersh, L., Dean, J.B., Putnam, R.W., 1998. Intracellular pH regulation in neurons from chemosensitive and nonchemosensitive areas of the medulla. Am. J. Physiol. 275, R1152-R1163.

Romanovsky, A.A., Kulchitsky, V.A., Simons, C.T., Sugimoto, N., Szekely, M., 1997. Cold defense mechanisms in vagotomized rats. Am. J. Physiol. 273, R784-R789.

Ross, C.A., Ruggiero, D.A., Park, D.H., Joh, T.H., Sved, A.F., Fernandez-Pardal, J., Saavedra, J.M., Reis, D.J., 1984. Tonic vasomotor control by the rostral ventrolateral medulla: effect of electrical or chemical stimulation of the area containing C1 adrenaline neurons on arterial pressure, heart rate, and plasma catecholamines and vasopressin. J. Neurosci. 4, 474-494.

Ruda, M., Allen, B., Gobel, S., 1981. Ultrastructure of descending serotoninergic axonal endings in layers I and II of the dorsal horn. J. Physiol. (Paris) 77, 205-209.

Salonia, A., Maga, T., Colombo, R., Scattoni, V., Briganti, A., Cestari, A., Guazzoni, G., Rigatti, P., Montorsi, F., 2002. A prospective study comparing paroxetine alone versus paroxetine plus sildenafil in patients with premature ejaculation. J. Urol. 168, 2486-2489.

Schadt, J.C., Ludbrook, J., 1991. Hemodynamic and neurohumoral responses to acute hypovolemia in conscious mammals. Am. J. Physiol. 260, H305-H318.

Schlaefke, M.E., 1981. Central chemosensitivity: a respiratory drive. Rev. Physiol. Biochem. Pharmacol. 90, 171-244.

10. SEROTONIN NEURONS IN THE BRAINSTEM AND SPINAL CORD: DIVERSE PROJECTIONS AND MULTIPLE FUNCTIONS

Schreihofer, A.M., Guyenet, P.G., 2002. The baroreflex and beyond: Control of sympathetic vasomotor tone by GABAergic neurons in the ventrolateral medulla. Clin. Exp. Pharmacol. Physiol. 29, 514-521.

Schwarzacher, S.W., Pestean, A., Gunther, S., Ballanyi, K., 2002. Serotonergic modulation of respiratory motoneurons and interneurons in brainstem slices of perinatal rats. Neurosci. 115, 1247-1259.

Smith, J.E., Jansen, A.S., Gilbey, M.P., Loewy, A.D., 1998. CNS cell groups projecting to sympathetic outflow of tail artery: neural circuits involved in heat loss in the rat. Brain Res. 786, 153-164.

Sorkin, L.S., McAdoo, D.J., 1993. Amino acids and serotonin are released into the lumbar spinal cord of the anesthetized cat following intradermal capsaicin injections. Brain Res. 607, 89-98.

Sorkin, L.S., McAdoo, D.J., Willis, W.D., 1993. Raphe magnus stimulation-induced antinociception in the cat is associated with release of amino acids as well as serotonin in the lumbar dorsal horn. Brain Res. 618, 95-108.

Stephens, D.P., Aoki, K., Kosiba, W.A., Johnson, J.M., 2001. Nonnoradrenergic mechanism of reflex cutaneous vasoconstriction in men. Am. J. Physiol. Heart Circ. Physiol. 280, H1496-H1504.

Stornetta, R.L., Guyenet, P.G., 1999. Distribution of glutamic acid decarboxylase mRNA-containing neurons in rat medulla projecting to thoracic spinal cord in relation to monoaminergic brainstem neurons. J. Comp. Neurol. 407, 367-380.

Strack, A.M., Sawyer, W.B., Hughes, J.H., Platt, K.B., Loewy, A.D., 1989. A general pattern of CNS innervation of the sympathetic outflow demonstrated by transneuronal pseudorabies viral infections. Brain Res. 491, 156-162.

Sugaya, K., Ogawa, Y., Hatano, T., Koyama, Y., Miyazato, T., Oda, M., 1998. Evidence for involvement of the subcoeruleus nucleus and nucleus raphe magnus in urine storage and penile erection in decerebrate rats. J. Urol. 159, 2172-2176.

Sun, Q.J., Berkowitz, R.G., Goodchild, A.K., Pilowsky, P.M., 2002. Serotonin inputs to inspiratory laryngeal motoneurons in the rat. J. Comp. Neurol. 451, 91-98.

Swanson, L.W., 1998. Brain Maps: Sturcture of the Rat Brain: A Laboratory Guide with Printed and Electronic Templates for Data, Models, and Schematics. Elsevier Science Ltd.

Szelenyi, Z., Hinckel, P., 1987. Changes in cold- and heat-defence following electrolytic lesions of raphe nuclei in the guinea-pig. Pflugers Arch. 409, 175-181.

Tallaksen-Greene, S.J., Elde, R., Wessendorf, M.W., 1993. Regional distribution of serotonin and substance P co-existing in nerve fibers and terminals in the brainstem of the rat. Neurosci. 53, 1127-1142.

Tanaka, M., Nagashima, K., McAllen, R.M., Kanosue, K., 2002. Role of the medullary raphe in thermoregulatory vasomotor control in rats. J. Physiol. 540, 657-664.

Tanaka, M., Okamura, H., Tamada, Y., Nagatsu, I., Tanaka, Y., Ibata, Y., 1994. Catecholaminergic input to spinally projecting serotonin neurons in the rostral ventromedial medulla oblongata of the rat. Brain Res. Bull. 35, 23-30.

Terui, N., Saeki, Y., Kumada, M., 1986. Barosensory neurons in the ventrolateral medulla in rabbits and their responses to various afferent inputs from peripheral and central sources. Japanese J. Physiol. 36, 1141-1164.

Terui, N., Saeki, Y., Kumada, M., 1988. Barosensory neurons in the rostral ventrolateral medulla mediate the renal-sympathetic reflex in rabbits. Clin. Exp. Hypertension A-Theory and Practice 10, 269-274.

Thomas, D.A., McGowan, M.K., Hammond, D.L., 1995. Microinjection of baclofen in the ventromedial medulla of rats - antinociception at low-doses and hyperalgesia at high-doses. J. Pharmacol. Exp. Ther. 275, 274-284.

Urban, M.O., Coutinho, S.V., Gebhart, G.F., 1999. Biphasic modulation of visceral nociception by neurotensin in rat rostral ventromedial medulla. J. Pharmacol. Exp. Ther. 290, 207-213.

Urban, M.O., Smith, D.J., Gebhart, G.F., 1996. Involvement of spinal cholecystokininB receptors in mediating neurotensin hyperalgesia from the medullary nucleus raphe magnus in the rat. J. Pharmacol. Exp. Ther. 278, 90-96.

Ursin, R., 2002. Serotonin and sleep . Sleep Med. Rev. 6, 55-69.

Vanegas, H., Barbaro, N.M., Fields, H.L., 1984. Tail-flick related activity in medullospinal neurons. Brain Res. 321, 135-141.

Wang, H., Stornetta, R.L., Rosin, D.L., Guyenet, P.G., 2001. Neurokinin-1 receptor-immunoreactive neurons of the ventral respiratory group in the rat. J. Comp. Neurol. 434, 128-146.

Wang, W., Bradley, S.R., Richerson, G.B., 2002. Quantification of the response of rat medullary raphe neurones to independent changes in pH(o) and P(CO2). J. Physiol. 540, 951-970.

Wang, W., Pizzonia, J.H., Richerson, G.B., 1998. Chemosensitivity of rat medullary raphe neurones in primary tissue culture. J. Physiol. 511, 433-450.

Wang, W., Richerson, G.B., 1999. Development of chemosensitivity of rat medullary raphe neurons. Neurosci. 90, 1001-1011.

Wang, W., Tiwari, J.K., Bradley, S.R., Zaykin, R.V., Richerson, G.B., 2001. Acidosis-stimulated neurons of the medullary raphe are serotonergic. J. Neurophysiol. 85, 2224-2235.

Willette, R.N., Punnen, S., Krieger, A.J., Sapru, H.N., 1984. Interdependence of rostral and caudal ventrolateral medullary areas in the control of blood pressure. Brain Res. 321, 169-174.

Wing, L.M., Chalmers, J.P., 1974. Participation of central serotonergic neurons in the control of the circulation of the unanesthetized rabbit. A study using 5,6-dihydroxytryptamine in experimental neurogenic and renal hypertension. Circ. Res. 35, 504-513.

Wisor, J.P., Wurts, S.W., Hall, F.S., Lesch, K.P., Murphy, D.L., Uhl, G.R., Edgar, D.M., 2003. Altered rapid eye movement sleep timing in serotonin transporter knockout mice. Neuroreport 14, 233-238.

Zagon, A., 1993. Innervation of serotonergic medullary raphe neurons from cells of the rostral ventrolateral medulla in rats. Neurosci. 55, 849-867.

Chapter 11

MEDULLARY RAPHE NEURONS IN AUTONOMIC REGULATION

Shaun F. Morrison
Neurological Sciences Institute, Oregon Health and Science University, Beaverton, OR 97006, USA

Abstract: The medullary raphe nuclei send both serotonergic and non-serotonergic projections to innervate spinal cord and brain stem targets, including sympathetic and parasympathetic preganglionic neurons in the spinal intermediolateral nucleus and dorsal motor nucleus, respectively. The recent discovery of retrogradely infected neurons within the medullary raphe following pseudorabies virus injections into a variety of autonomically-innervated tissues indicates the extensive influence of the medullary raphe on visceral function and homeostasis. This review summarizes our current understanding of the role of medullary raphe neurons in the regulation of arterial pressure and heart rate and in the control of body temperature, sexual function, insulin secretion and gastric motility and acid secretion. Many of these findings suggest that medullary raphe neurons function as autonomic premotor neurons, influencing the discharge of preganglionic neurons in functionally-specific reflex and central command pathways. Future research will provide insight into the integration of these medullary raphe neurons within the hierarchical control structure of specific autonomic functions and within the raphe-spinal pathways influencing somatic motor control, respiration and pain modulation.

Key words: cardiovascular, thermoregulation, sexual function, gastric function, insulin, heart rate, pseudorabies virus, sympathetic, parasympathetic

INTRODUCTION

The medullary raphe nuclei include the raphe obscurus, raphe pallidus and raphe magnus. These cell groups are located on the midline, but in the rostral medulla they also extend laterally over the pyramids. A narrow column of neurons located near the ventral medullary surface lateral to the pyramids and referred to as the parapyramidal cell group will also be considered as a component of the medullary raphe in this review due to the marked similarity between their neurochemical profiles and projection targets and those in the raphe pallidus at comparable rostrocaudal levels. While both medullary raphe nuclei and those in the pons and midbrain contain a high percentage of serotonergic neurons, medullary raphe nuclei are unique in their heavy projection pattern to sites in the spinal cord, including the intermediolateral nucleus containing sympathetic preganglionic neurons. Consistent with their spinal projections, medullary raphe neurons have been implicated in the modulation of sensory input, the control of autonomic outflow and the facilitation of somatic and respiratory motor outputs. Raphe projections to targets within the medulla, some of which include the branches of spinally-projecting axons, provide an additional substrate for influencing autonomic function through effects on populations of sympathetic premotor neurons and parasympathetic preganglionic neurons and on the processing of visceral sensory inputs. The role of medullary raphe neurons in the modulation of dorsal horn function (Mason, 2001) and motor control (Jacobs et al., 2002) have been recently reviewed. This chapter describes the effects on a variety of autonomic functions that have been attributed to medullary raphe neurons and delineates some of the results from the anatomical and physiological experiments that underlie our current appreciation of the diverse functions influenced by these medullary midline neurons.

MEDULLARY RAPHE NEURONS INFLUENCE CARDIOVASCULAR FUNCTION

RapheNeurons Participate in Medullary Sympathetic Networks Influencing Vasoconstriction

Early experiments demonstrating a variety of pressor and depressor responses during a simple mapping of the medullary raphe with electrical stimulation (Adair et al., 1977; Yen et al., 1983; McCall, 1984) suggested that the raphe contained multiple systems with differing effects on vasoconstrictor sympathetic outflows. Recordings of sympathetic nerve activities to the kidney, skin and skeletal muscle vasculature during electrical stimulation in the caudal medullary raphe provided direct evidence for heterogeneous sympathetic responses depending on stimulation site (Yusof and Coote, 1988). Evidence has been presented that the pressor and sympathoexcitatory responses evoked from electrical stimulation of raphe (McCall, 1984; Campos et al., 1993; Zhou and Gilbey, 1995) may be dependent on activation of neurons in the rostral ventrolateral medulla (Campos et al., 1993), while other data suggest they are dependent on activation of 5-HT receptors in the spinal cord (McCall, 1984). Although these data were obtained using a technique that does not distinguish activation of local cell bodies from stimulation of axonal pathways traversing the region, they do emphasize the potential for functional heterogeneity within the medullary raphe and the importance of localizing responsive sites within the raphe with minimal, effective quantities of cell body-selective excitatory or inhibitory agents. When this approach was used, glutamate-evoked activation of midline medullary raphe neurons in the rat evoked depressor responses that were attenuated by pretreatment with a serotonergic neurotoxin, while similar activation of parapyramidal neurons evoked pressor responses that were eliminated by the neurotoxin treatment, but not by lesions of the rostral ventrolateral medulla (Minson et al., 1987).

Potential cellular substrates for the electrically-evoked cardiovascular responses were suggested by recordings of cat medullary raphe-spinal neurons with activity synchronized to the spontaneous bursts in inferior cardiac sympathetic nerve activity. Populations of neurons, primarily in raphe pallidus, were inhibited while others were excited by activation of the baroreceptor reflex (Morrison and Gebber, 1984; Morrison and Gebber, 1985), suggesting that these groups of raphe neurons participated in medullospinal pathways mediating sympathoexcitatory and sympathoinhibitory effects, respectively. Raphespinal neurons inhibited by baroreceptor reflex activation were also identified in rat (Pilowsky et al., 1995). Subsequent studies suggested that sympathoinhibitory raphespinal neurons receive a synaptic drive from sympathoinhibitory neurons in the lateral tegmental field (Barman and Gebber, 1989). Additionally, anatomical evidence indicates the existence of GABAergic, raphespinal neurons in the caudal medulla (Stornetta and Guyenet, 1999). Raphespinal neurons exhibit activity correlated to both the 10 Hz and to the cardiac-related rhythms in the cat (Barman and Gebber, 1997) and some raphe neurons with 10 Hz-related activity receive a synaptic drive from neurons in the caudal ventrolateral medulla and send an axonal branch back to the same area of the caudal ventrolateral medulla (Barman et al., 1995). These studies are consistent with a model in which raphe-spinal neurons in the pallidus region of the caudal medulla function to inhibit sympathetic nerve activity, but also receive inputs from medullary neuronal networks involved in the elaboration of distinct rhythms in sympathetic nerve discharge. Interestingly, electrical stimulation in this raphe region produced a differential response consisting of an inhibition of renal and cardiac sympathetic nerve activities, but an increase in that on the vertebral nerve (Larsen et al., 2000). While electrolytic lesions of the medullary raphe region suggest an involvement of local neurons in a tonic, non-baroreceptor-mediated inhibition of sympathetic nerve activity and in the depressor responses from the anterior hypothalamus and the spinal trigeminal tract (McCall and Harris, 1987), the physiological roles of these responses and the pathways and neurotransmitters through which they are mediated, remain to be determined.

Caudal Medullary Raphe Neurons and the Decompensatory Response to Hemorrhage

In addition to a potential for sympathoinhibition mediated by spinally-projecting neurons in the caudal medullary raphe, neurons in this region also project to the rostral ventrolateral medulla to inhibit the discharge of cardiovascular sympathetic premotor neurons located there. Chemical excitation of cells in the caudal raphe pallidus and obscurus of the rat and rabbit markedly inhibits sympathetic nerve activity (Fig. 1) and reduces arterial pressure and heart rate (Coleman and Dampney, 1995; Henderson et al., 1998b; Verberne et al., 1999), a response similar to that seen during the sympathoinhibitory (decompensatory) phase of the response to hemorrhage (Henderson et al., 1998a; Schadt, 2003). Caudal medullary raphe neurons may mediate the vasodepressor response to activation of the ventrolateral periaqueductal gray (Wang and Lovick, 1993; Henderson et al., 1998b), an area proposed to play a key role in the response to hemorrhage (Cavun and Millington, 2001). Stimulation in the caudal medullary raphe elicits an inhibition of sympathetic premotor neurons in the rostral ventrolateral medulla that is blocked by application of the $GABA_A$ receptor antagonist, bicuculline, onto these rostral ventrolateral medullary neurons (McCall, 1988; Coleman and Dampney, 1998; Verberne et al., 1999). This caudal raphe-evoked sympathoinhibition does not involve the glutamate-mediated activation of GABAergic neurons in the caudal ventrolateral medulla that is an essential element of the baroreceptor reflex pathway (Potas and Dampney, 2003). The 5-HT1A receptor agonist, 8-OH DPAT, elicits a centrally-mediated reduction in sympathetic nerve activity which may be mediated in part through activation of post-synaptic 5-HT1A or alpha 2A adrenergic receptors in the rostral ventrolateral medulla (Nosjean and Guyenet, 1991), rather than by autoreceptor-mediated inhibition of serotonergic, sympathoexcitatory neurons (McCall et al., 1989). Further, sympathoinhibitory responses to activation of the ventrolateral

11. MEDULLARY RAPHE NEURONS IN AUTONOMIC REGULATION

periaqueductal gray and to hemorrhage arise from activation of 5-HT1A receptors in the rostral ventrolateral medulla (Bago and Dean, 2001; Dean and Bago, 2002). Together, these results suggest an important role for caudal medullary raphe neurons in the sympathoinhibitory responses to hypovolemia (see Fig. 1) and perhaps other behavioral responses (Henderson et al., 2000), that may involve an interplay of serotonergic and GABAergic receptor activation in the rostral ventrolateral medulla that remains to be elucidated.

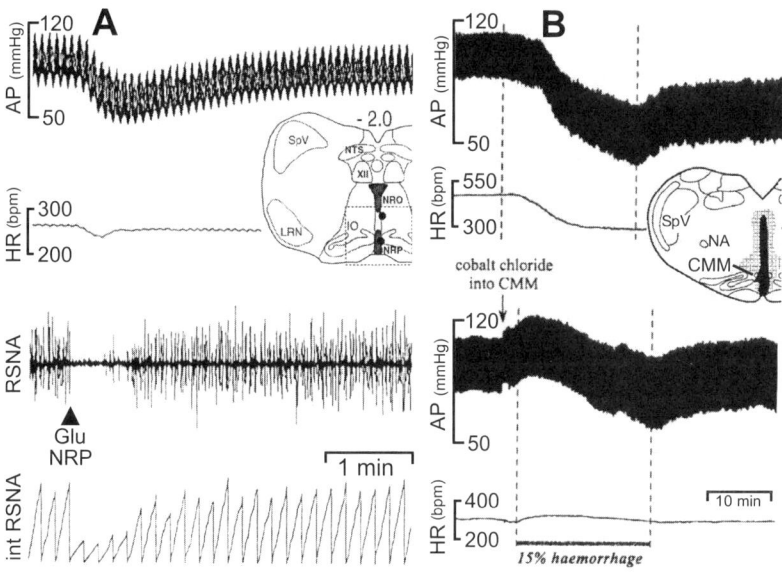

Figure 1. Excitation of caudal medullary raphe neurons evokes reductions in arterial pressure (AP) and heart rate (HR) and may contribute to the hypothension following acute hypovolemia. A: microinjection of glutamate into the caudal raphe pallidus (NRP) of the rabbit causes an inhibition of renal sympathetic nerve activity (RSNA) and integrated RSNA, a depressor response and a bradycardia. Inset: drawing of the rabbit medulla 2 mm caudal to the obex showing the injection site in NRP. SpV is spinal trigeminal nucleus, LRN is lateral reticular nuicleus, NRO is raphe obscurus, NTS is nucleus of the tractus solitarius, XII is the hypoglossal nucleus. Adapted from (Coleman and Dampney, 1995). B: Upper panel shows the fall in AP and HR in response to a 15% hemorrhage (between dotted lines) in a rat under control conditions. Lower panel shows that after blockade of synaptic transmission in the caudal midline medulla (CMM), the depressor response to hemorrhage is reduced and the bradycardia is eliminated. Inset: drawing of the rat medulla 0.5 mm rostral to the obex, indicating the spread of the cobalt chloride. NA is nucleus ambiguus. Adapted from (Henderson et al., 1998a).

ROSTRAL MEDULLARY RAPHE AND THERMOREGULATION

As environmental temperature deviates from thermoneutral, thermoregulatory responses are initiated to defend body temperature. These responses are triggered initially by activation of cutaneous temperature receptors and subsequently through hypothalamic sensation of alterations in body core temperature and in blood temperature. Two important autonomic effector components of the thermoregulatory response to defend body temperature are cutaneous vasoconstriction, which determines the degree of heat loss to the environment, and non-shivering thermogenesis in brown adipose tissue which consumes metabolic fuel to provide heat during exposure to a cold environment. The raphe nuclei in the rostral ventromedial medulla, particularly the rostral raphe pallidus, the caudal parts of raphe magnus and the parapyramidal cell group, contain neurons that play a critical role, possibly as sympathetic premotor neurons, in the elaboration of thermoregulatory responses involving cutaneous vasoconstriction and sympathetically-mediated thermogenesis in brown adipose tissue. The model developed from extensive studies of the functional organization of the central pathways regulating the cardiovascular system emphasizes the importance of sympathetic premotor neurons in determining the spontaneous and evoked levels of sympathetic outflow. Sympathetic premotor neurons are those supraspinal neurons that directly innervate the sympathetic preganglionic neurons. As with the sympathetic preganglionic neurons, considerable evidence supports the existence of functionally-specific sympathetic premotor neurons which, by virtue of their projection targets, regulate the activity of a single target tissue (Morrison, 2001).

Anatomical studies, beginning with early retrograde (Loewy, 1981) and anterograde (Bacon et al., 1990) tracer studies have long supported a direct innervation of sympathetic preganglionic neurons by neurons in the rostral medullary raphe. Using the recently-developed technique of transynaptic transport of pseudorabies virus, inoculations of brown adipose tissue (Bamshad et al., 1999; Oldfield et al., 2002; Cano et al., 2003), the stellate ganglion (Farkas et al., 1998) which is the principal site of ganglion cells innervating the brown adipose tissue (Morrison et al., 2000; Oldfield et al., 2002) or the heart, or the rat tail artery (Smith et al., 1998) result in infected neurons in the rostral ventromedial medulla, including serotonergic and non-serotonergic neurons. From the timing of these raphe infections, we can conclude that the rostral medullary raphe contains sympathetic premotor neurons that innervate sympathetic preganglionic neurons controlling thermoregulatory target tissues. This same region of the rostral ventromedial medulla is also the target of descending projections from rostral cell groups thought to be involved in thermoregulatory pathways, including the medial preoptic (Hermann et al., 1997; Nakamura et al., 2002), the paraventricular and the dorsomedial (Hosoya et al., 1989; Hermann et al., 1997) hypothalamus and the periaqueductal gray (Hermann et al., 1997; Farkas et al., 1998).

The role of rostral medullary raphe neurons in cutaneous vasoconstriction has been studied with regard to the tail circulation in the rat and the ear circulation in the rabbit. The data indicate that cutaneous vasoconstriction is regulated independently from the tone in other circulatory beds such as the renal and mesenteric that contributes significantly to the determination of arterial pressure. Inhibition of raphe pallidus or parapyramidal neurons in rabbit increased ear blood flow, while focal electrical stimulation reduced ear blood flow and neither treatment affected the mesenteric blood flow (Blessing et al., 1999). Inhibition of neuronal activity in these raphe regions eliminated the skin vasoconstrictions evoked by hypothalamic and amygdala stimulation (Nalivaiko and Blessing, 2001) or by trigeminal tract or noxious cutaneous stimulation (Blessing and Nalivaiko, 2000), without affecting the responses in the mesenteric bed. Similarly, activation of neurons in the rostral medullary raphe pallidus and magnus regions increased sympathetic outflow to the rat tail artery (Rathner and McAllen, 1999) and reduced rat tail blood flow (Blessing and Nalivaiko, 2001), with little effect on renal or mesenteric circulations. Activation of neurons in this region also blocked the vasodilation in rat tail normally elicited by warming the preoptic hypothalamus (Tanaka et al., 2002). Raphespinal, cold-activated neurons have been identified in the rat (Rathner et al., 2001) and these may function as sympathetic premotor neurons controlling cutaneous vasoconstriction. Nociceptive, amygdala-responsive, spinally-projecting neurons have been recorded in rabbit medullary raphe; however, their responses to thermal stimuli have not been described.

Activation of the rostral medullary raphe pallidus and caudal raphe magnus also increases the sympathetic outflow to brown adipose tissue (Fig. 2) (Morrison et al., 1999), without a comparable change in the vasoconstrictor tone in the splanchnic sympathetic nerve (Morrison, 1999). Inhibition of rostral raphe pallidus neurons in conscious rats evoked a large fall in body temperature (Zaretsky et al., 2003), suggesting that neurons in this region are tonically-active and contributing significantly to the excitatory drive necessary to sustain brown adipose tissue thermogenesis and cutaneous vasoconstriction at a level sufficient to maintain a homeostatic body temperature. Activation of brown adipose tissue thermogenesis and cutaneous vasoconstriction contribute to the increase in body temperature during the febrile component of the inflammatory response. Inhibition of the activity in rostral raphe pallidus neurons blocked the thermogenesis in brown adipose tissue and the increase in body temperature evoked by central administration of the fever-stimulating prostaglandin E2 (Nakamura et al., 2002). Together, these data strongly support the role of rostral medullary raphe neurons as sympathetic premotor neurons controlling target tissues involved in thermoregulatory responses to changes in environmental temperature or to immune or stressful stimuli that produce fever.

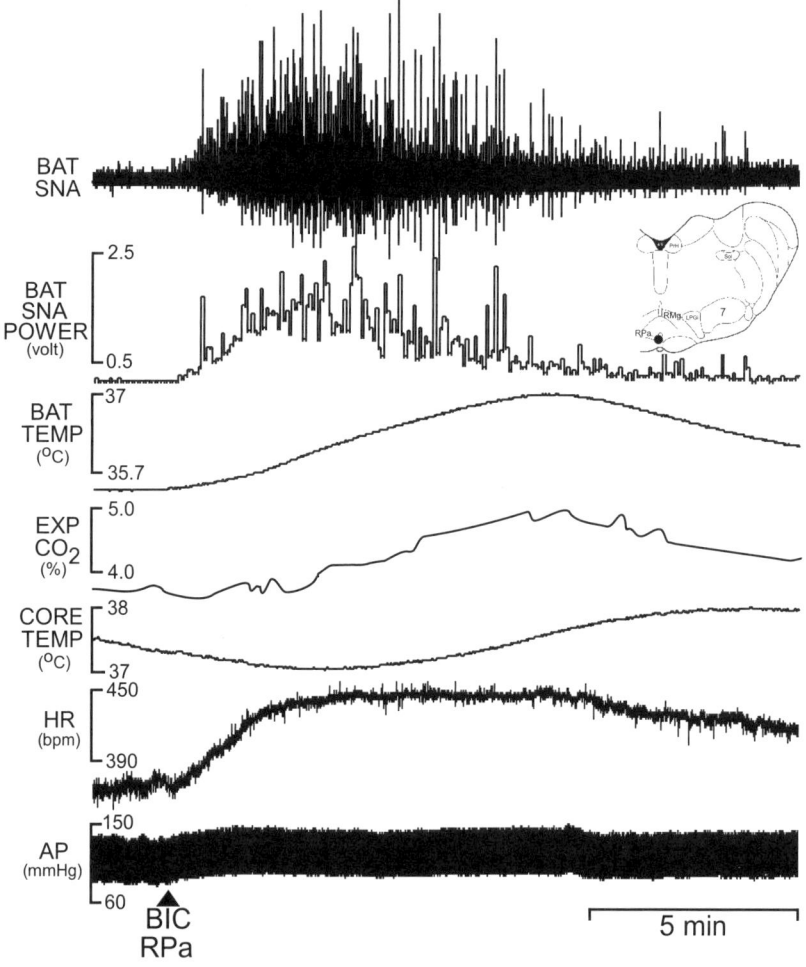

Figure 2. Stimulation of thermogenesis in brown adipose tissue (BAT) by activation of rostral raphe pallidus (RPa) neurons. Traces are BAT sympathetic nerve activity (SNA), cumulative

power/4 seconds in BAT SNA, BAT temperature, expired CO2, core temperature, heart rate (HR), arterial pressure (AP). Activation of RPa neurons with a microinjection of the GABAA receptor antagonist, bicuculline, increases BAT SNA, BAT temperature, expired CO2 and HR. Inset : bicuculline microinjection site in RPa. 7: facial nerve nucleus, RMg: raphe magnus, LPGi: lateral paragigantocellular nucleus, Sol: solitary nucleus, PrH: prepositus hypoglossi. Data similar to those in (Morrison et al., 1999).

MEDULLARY RAPHE NEURONS AND THE CONTROL OF PARASYMPATHETIC VAGAL OUTFLOW

Physiological studies have indicated that neurons in the medullary raphe can exert a potent influence on the parasympathetic vagal outflows to several target tissues. Anatomical support for some of these findings has been provided by experiments in which injections of the transynaptic, retrograde tracer, pseudorabies virus, were made into the pancreas (Loewy et al., 1994) or the distal airways (Hadziefendic and Haxhiu, 1999) of rats that had been spinally transected at C8 to prevent retrograde infections of supraspinal neurons that synapse on infected sympathetic preganglionic neurons. In both cases, infected neurons were found in the medullary raphe nuclei at survival times longer than those that labeled vagal preganglionic neurons in the dorsal motor nucleus of the vagus or the region of the nucleus ambiguus. In contrast, when pseudorabies virus was injected into the heart, the extensive retrograde labeling seen in the raphe nuclei of intact animals was largely eliminated in animals receiving a prior T1 transection (Ter Horst et al., 1996), indicating that medullary raphe neurons predominantly influence the sympathetic rather than the vagal inputs to the heart.

Activation of the vagal parasympathetic input to the pancreas induces insulin release from β cells. Similarly, chemical stimulation of neurons in the dorsal vagal complex (DVC) also increases circulating insulin levels. Neurons in the caudal medullary raphe pallidus, raphe obscurus and parapyramidal region, including those synthesizing thyrotropin-releasing hormone (TRH), project to the DVC (Lynn et al., 1991). One of the functions of the raphe neurons that constitute this projection has been suggested by the finding in anesthetized rats that microinjection of the excitatory amino acid, kainic acid, into the caudal medullary raphe nuclei produces a marked stimulation of insulin secretion (Krowicki and Hornby, 1995; Yang et al., 2002) that is mimicked by microinjection of a TRH analogue into the DVC and attenuated by microinjection of an anti-TRH IgG into the DVC (Yang et al., 2002). Bilateral vagotomy abolished these centrally-evoked increases in plasma insulin. These data indicate that a population of neurons in the caudal medullary raphe nuclei can increase the activity of vagal preganglionic neurons with projections to the pancreas to induce a marked increase in insulin secretion. This pathway presumably plays a significant role in the central regulation of metabolic substrate availability to insulin-sensitive tissues. The inputs that might normally be responsible for increasing the discharge of these insulin-controlling raphe neurons and the metabolic regulatory pathways in which they participate remain unknown.

Neurons in the caudal medullary raphe nuclei, including those that synthesize TRH, can also exert a significant effect on gastric function, including gastric motility, gastric secretion of acid, pepsin and mucus and gastric mucosal blood flow. These effects are mediated through the vagal parasympathetic input to the stomach. The dense innervation of the DVC by TRH-immunoreactive terminals arises from the caudal raphe nuclei (Palkovits et al., 1986; Lynn et al., 1991) and, within the DVC, gastric motor neurons are one of their primary postsynaptic targets (Rinaman and Miselis, 1990). TRH receptors are located in the DVC (Manaker and Rizio, 1989) overlaping the region containing vagal preganglionic neurons projecting to the stomach. Although no attempt was made to restrict uptake to vagal inputs, neurons in the caudal raphe were retrogradely infected following stomach injections of pseudorabies virus (Card et al., 1990). Together, these results provide an anatomical foundation for the role of raphe neurons in the central vagal regulation of gastric function.

Physiological studies in rats and cats have indicated that activation of caudal raphe neurons produces an increase in gastric motility and gastric secretion of acid (Fig. 3) and pepsin (Hornby et al., 1990; White et al., 1991; Yang et al., 1993; Garrick et al., 1994). In the example in Figure 3, activation of caudal raphe pallidus neurons with a microinjection of glutamate is shown to produce a 3-fold increase in gastric acid secretion (Fig. 3A) in control, but not in vagotomized rats. This effect is greatly reduced in rats receiving a prior injection of TRH antibody into the DVC (Fig. 3A). Similar activation of caudal raphe pallidus neurons evokes a 5-fold increase in the force of stomach contractions (Fig. 3B) that is also markedly reduced by application of TRH antibody at the site of vagal preganglionic neurons in DVC. Prolonged and intense activation of the pathway from the caudal raphe nuclei to the DVC can result in erosion of the gastric mucosa in fasted rats (Okumura et al., 1993; Kaneko and Taché, 1995). In addition to the anatomical data described above, functional studies substantialize the role of TRH release in the DVC in mediating the gastric effects following activation of caudal medullary raphe neurons (Fig. 3). Application of TRH into the DVC mimicks the effects of raphe stimulation on gastric motility, gastric acid and pepsin release, and gastric erosion production (Tache and Yang, 1993) and this vagal activation may arise from the excitatory effect of TRH on vagal preganglionic neurons (McCann et al., 1989). Serotonin exerts a potentiation of the TRH-evoked effects on DVC neurons that control gastric function. Serotonin is colocalized in some TRH-containing axon terminals of raphe neurons that project to the DVC (Wessendorf et al., 1990). Although serotonin evokes little effect when applied to DVC alone, it enhanced the increase in gastric motility and acid secretion to TRH microinjected into the DVC (McCann et al., 1988; Yoneda and Tache, 1995). Together, these studies have implications for the central neuronal pathways and transmitter systems that play a role in formation of gastric erosions and ulcers. Cold stress produces experimental gastric erosions in fasted animals. When either TRH antibodies or antisense oligodeoxynucleotides against the TRH-R1 receptor were added to the medullary cerebrospinal fluid, the cold-induced stimulation of gastric acid secretion, emptying and erosions was prevented (Niida et al., 1991; Yang et al., 1994; Martinez et al., 1998). Additionally, acute cold exposure increases TRH mRNA in the caudal medullary raphe (Yang et al., 1994). These data provide strong evidence for a key role for caudal raphe neurons in providing the elevated vagal drive that is postulated to result in damage to the gastric mucosa. Important steps in elucidating the central pathways involved in gastric ulcer formation will include the identification of raphe neurons with inputs to the DVC neurons controlling gastric function, the characterization of the inputs and reflex stimuli that determine the discharge of these raphe neurons and the determination of the mechanisms through which stressful stimuli can lead to exaggerated responses in the raphe-DVC gastric control network.

MEDULLARY RAPHE NEURONS INFLUENCE PELVIC ORGAN FUNCTIONS

Pseudorabies viral retrograde tracing has indicated that medullary raphe neurons are within pathways that can influence the function of pelvic organs involved in reproduction and micturition. Injections of the viral retrograde tracer into male and female sex organs (Marson et al., 1993; Marson, 1995; Marson and McKenna, 1996; Orr and Marson, 1998) or into the bladder (Marson, 1997; Nadelhaft and Vera, 2001) consistently label neurons in raphe magnus, in raphe pallidus and in the parapyramidal region, including portions of the paragigantocellular reticular nucleus that lie dorsolateral to the pyramids. This extensive labeling in the medullary raphe nuclei, as well as other brainstem and hypothalamic regions, is a reflection of the complexity of the neural circuits necessary to coordinate the autonomic, somatic, and respiratory motor components of sexual behavior and of micturition and to integrate the sensory and emotional information that plays a significant role in the initiation of these behaviors.

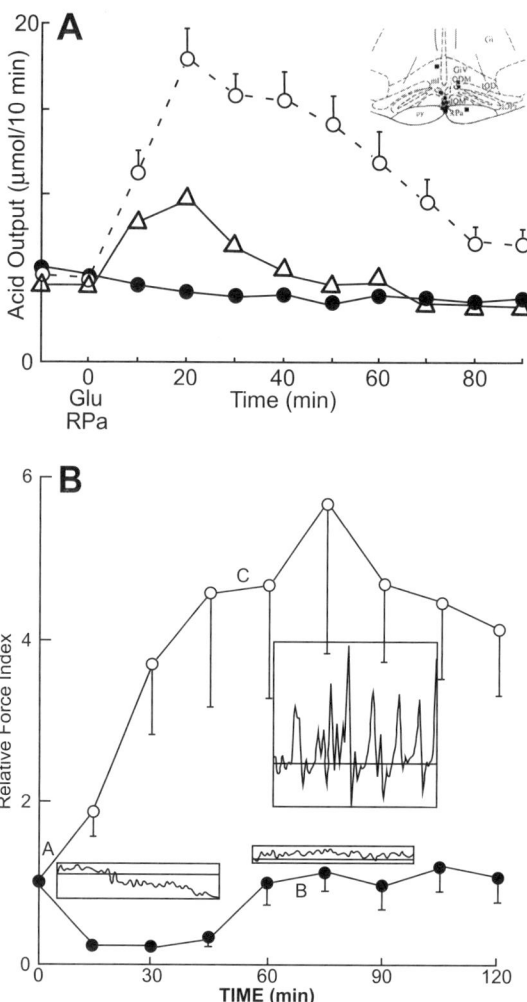

Figure 3. Increases gastric acid secretion and gastric contractility evoked by activation of caudal raphe pallidus (RPa) neurons are dependent on the vagus nerve and TRH receptor activation in the dorsal vagal complex. A: microinjection of glutamate (Glu) into RPa increases gastric acid secretion in control animals (open circles), but not in vagotomized animals (filled circles). Acid secretion response to activation of RPa neurons is markedly reduced following microinjection of TRH antibody into the dorsal vagal complex (open

triangles). Adapted from (Garrick et al., 1994). B: Gastric contractility is low under control conditions (point A, trace of force during 5 minute period of basal contractility), but is greatly enhanced (point C, trace of force during 5 minute period of increased contractility) following microinjection of glutamate into RPa prior to time 0 (open circles). The effect of activating RPa neurons (prior to time 0) is blocked by microinjection of TRH antibody into the dorsal vagal complex at time 0 (filled circles). Adapted from (Yang et al., 1993).

Although no functional information is available on the role that neurons in the medullary raphe might play in the control of micturition, a restricted region of the medullary raphe has been identified that contributes significantly to the tonic, descending inhibitory restraint on the spinal circuits mediating sexual reflex behaviors. Sexual responses are not normally elicited in anesthetized rats by urethral stimuli, but can be evoked following either transection of the neuraxis at the level of the inferior olive or placement of cytotoxic lesions in the rostral pole of the paragigantocellular reticular nucleus (Fig. 4) which contains a population of serotonergic neurons projecting to the lumbar spinal cord (Marson and McKenna, 1990; Yells et al., 1992). In the example in Figure 4, release of urethral occlusion in the intact rat (Fig. 4A) fails to elicit either bursts of electrical activity in the bulbospongiosus muscle or rhythmic pressure pulses in the urethra, whereas after kainic acid lesion of cells in the paragigantocellular reticular nucleus, similar release of urethral pressure results in large ejaculatory bursts in muscle activity and urethral pressure (Fig.4B). Neurotoxic lesion of the serotonergic system with 5,7 dihydroxytryptamine results in a similar loss of reflexive erections (Marson and McKenna, 1994) and intrathecal administration of serotonin inhibits these reflex responses (Marson and McKenna, 1992), suggesting a role for serotonin in mediating the descending inhibition from the paragigantocellular reticular nucleus. Projections from the medial preoptic area to the rostral medullary raphe (Simerly and Swanson, 1988; Murphy et al., 1999), including the paragigantocellular reticular nucleus, have been identified and this projection may contain a significant GABAergic component (Nakamura et al., 2002). These findings provide a potential pathway for the removal of the tonic, medullary serotonergic inhibition of spinal copulatory circuits during sexual behaviors evoked by hormonal or sexual stimuli integrated in the medial preoptic area.

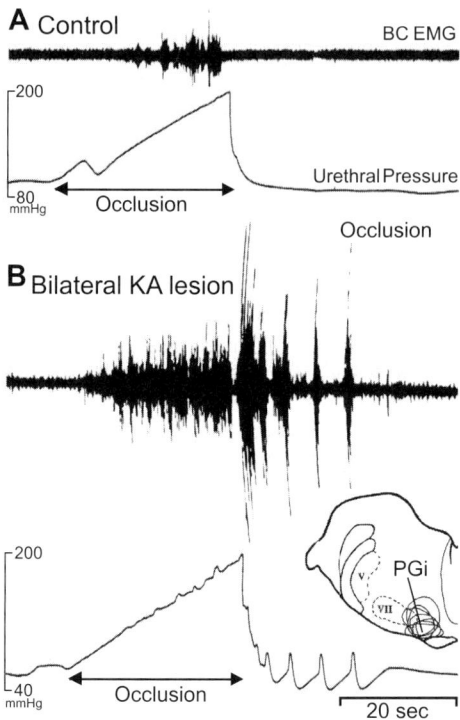

Figure 4. Effect of bilateral lesion of the paragigantocellular reticular nucleus (PGi) on the amplitude of the coitus reflex in a male rat. A: electromyographic activity of the bulbospongiosus muscle (BC EMG) and pressure within the urethra under control conditions. Distension of the urethra was accomplished by infusion of saline and occlusion (during the 30s period between the arrowheads) of the urethral meatus. Note the absence of rhythmic bursting of the muscle (coitus reflex) following release of the occlusion in control. B: coitus reflex 8 minutes following bilateral kainic acid (KA) lesion of cell bodies in the PGi. Note the ejaculatory bursting of the muscle following release of the urethral occlusion in the lesioned rat. Inset: effective lesion sites in PGi. V is spinal nucleus of the trigeminal nerve, VII is the facial nerve nucleus. Adapted from (Marson and McKenna, 1990).

CONCLUSIONS

In summary, the medullary raphe nuclei contain anatomically and functionally diverse populations of neurons. Although there are a large number of spinally-projecting serotonergic neurons in these nuclei, there are significant populations of neurons that do not contain serotonin and there are several populations in which multiple peptide and amino acid neurotransmitters have been colocalized. Identification of neurotransmitter phenotype with specific functions has not been accomplished. Based on retrograde labeling with pseudorabies virus, the medullary raphe nuclei are expected to influence function in an extensive array of autonomic tissues, both through their sympathetic and parasympathetic innervations. This has been the case in each of the key areas studied to date: cardiovascular regulation, control of body temperature, gastric function and energy substrate utilization and expression of mating and sexual behavior. These functional studies have suggested that medullary raphe neurons function as autonomic premotor neurons, influencing the discharge of preganglionic neurons in functionally-specific reflex and central command pathways. Future investigations on the medullary raphe hold the promise of a vertical integration of the current findings within the hierarchical control structure of specific autonomic functions and a horizontal integration of the raphe effects on these autonomic efferent pathways with those on somatic motor control, respiration and pain modulation.

REFERENCES

Adair, J.R., Hamilton, B.L., Scappaticci, K.A., Helke, C.J., Gillis, R.A., 1977. Cardiovascular responses to electrical stimulation of the medullary raphe area of the cat. Brain Res. 128, 141-145.

Bacon, S.J., Zagon, A., Smith, A.D., 1990. Electron microscopic evidence of a monosynaptic pathway between cells in the caudal raphe nuclei and sympathetic preganglionic neurons in the rat spinal cord. Exp. Brain Res. 79, 589-602.

Bago, M., Dean, C., 2001. Sympathoinhibition from ventrolateral periaqueductal gray mediated by 5-HT(1A) receptors in the RVLM. Am. J. Physiol. Regul. Integr. Comp. Physiol. 280, R976-R984.

Bamshad, M., Song, C.K., Bartness, T.J., 1999. CNS origins of the sympathetic nervous system outflow to brown adipose tissue. Am. J. Physiol. 276, R1569-R1578.

Barman, S.M., Gebber, G.L., 1989. Lateral tegmental field neurons of cat medulla: a source of basal activity of raphespinal sympathoinhibitory neurons. J. Neurophysiol. 61, 1011-1024.

Barman, S.M., Gebber, G.L., 1997. Subgroups of rostral ventrolateral medullary and caudal medullary raphe neurons based on patterns of relationship to sympathetic nerve discharge and axonal projections. J. Neurophysiol. 77, 65-75.

Barman, S.M., Orer, H.S., Gebber, G.L., 1995. Axonal projections of caudal ventrolateral medullary and medullary raphe neurons with activity correlated to the 10-Hz rhythm in sympathetic nerve discharge. J. Neurophysiol. 74, 2295-2308.

Blessing, W.W., Nalivaiko, E., 2000. Regional blood flow and nociceptive stimuli in rabbits: patterning by medullary raphe, not ventrolateral medulla. J. Physiol. 524(Pt 1), 279-292.

Blessing, W.W., Nalivaiko, E., 2001. Raphe magnus/pallidus neurons regulate tail but not mesenteric arterial blood flow in rats. Neurosci. 105, 923-929.

Blessing, W.W., Yu, Y.H., Nalivaiko, E., 1999. Raphe pallidus and parapyramidal neurons regulate ear pinna vascular conductance in the rabbit. Neurosci. Lett. 270, 33-36.

Campos, R.R., Futuro-Neto, H.A., Guertzenstein, P.G., 1993. Role of the rostral ventrolateral medulla in the pressor response to stimulation of the nucleus raphe obscurus. Brazilian J. Med. Biol. Res. 26, 1-9.

Cano, G., Passerin, A.M., Schiltz, J.C., Card, J.P., Morrison, S.F., Sved, A.F., 2003. Anatomical substrates for the central control of sympathetic outflow to interscapular adipose tissue during cold exposure. J. Comp. Neurol. In Press.

Card, J.P., Rinaman, L., Schwaber, J.S., Miselis, R.R., Whealy, M.E., Robbins, A.K., Enquist, L.W., 1990. Neurotropic properties of pseudorabies virus: uptake and transneuronal passage in the rat central nervous system. J. Neurosci. 10, 1974-1994.

Cavun, S., Millington, W.R., 2001. Evidence that hemorrhagic hypotension is mediated by the ventrolateral periaqueductal gray region. Am. J. Physiol. Regul. Integr. Comp. Physiol. 281, R747-R752.

Coleman, M.J., Dampney, R.A., 1995. Powerful depressor and sympathoinhibitory effects evoked from neurons in the caudal raphe pallidus and obscurus. Am. J. Physiol. 268, R1295-R1302.

Coleman, M.J., Dampney, R.A., 1998. Sympathoinhibition evoked from caudal midline medulla is mediated by GABA receptors in rostral VLM. Am. J. Physiol. 274, R318-R323.

Dean, C., Bago, M., 2002. Renal sympathoinhibition mediated by 5-HT(1A) receptors in the RVLM during severe hemorrhage in rats. Am. J. Physiol. Regul. Integr. Comp. Physiol. 282, R122-R130.

Farkas, E., Jansen, A.S., Loewy, A.D., 1998. Periaqueductal gray matter input to cardiac-related sympathetic premotor neurons. Brain Res. 792, 179-192.

Garrick, T., Prince, M., Yang, H., Ohning, G., Taché, Y., 1994. Raphe pallidus stimulation increases gastric contractility via TRH projections to the dorsal vagal complex in rats. Brain Res. 636, 343-347.

Hadziefendic, S., Haxhiu, M.A., 1999. CNS innervation of vagal preganglionic neurons controlling peripheral airways: a transneuronal labeling study using pseudorabies virus. J. Auton. Nerv. Syst. 76, 135-145.

Henderson, L.A., Keay, K.A., Bandler, R., 1998a. Hypotension following acute hypovolaemia depends on the caudal midline medulla. Neuroreport 9, 1839-1844.

Henderson, L.A., Keay, K.A., Bandler, R., 1998b. The ventrolateral periaqueductal gray projects to caudal brainstem depressor regions: a functional-anatomical and physiological study. Neurosci. 82, 201-221.

Henderson, L.A., Keay, K.A., Bandler, R., 2000. Caudal midline medulla mediates behaviourally-coupled but not baroreceptor-mediated vasodepression. Neurosci. 98, 779-792.

Hermann, D.M., Luppi, P.H., Peyron, C., Hinckel, P., Jouvet, M., 1997. Afferent projections to the rat nuclei raphe magnus, raphe pallidus and reticularis gigantocellularis pars alpha demonstrated by iontophoretic application of choleratoxin (subunit b). J. Chem. Neuroanat. 13, 1-21.

Hornby, P.J., Rossiter, C.D., White, R.L., Norman, W.P., Kuhn, D.H., Gillis, R.A., 1990. Medullary raphe: a new site for vagally mediated stimulation of gastric motility in cats. Am. J. Physiol. 258, G637-G647.

Hosoya, Y., Sugiura, Y., Zhang, F.Z., Ito, R., Kohno, K., 1989. Direct projection from the dorsal hypothalamic area to the nucleus raphe pallidus: a study using anterograde transport with Phaseolus vulgaris leucoagglutinin in the rat. Exp. Brain Res. 75, 40-46.

Jacobs, B.L., Martin-Cora, F.J., Fornal, C.A., 2002. Activity of medullary serotonergic neurons in freely moving animals. Brain Res. Brain Res. Rev. 40, 45-52.

Kaneko, H., Taché, Y., 1995. TRH in the dorsal motor nucleus of vagus is involved in gastric erosion induced by excitation of raphe pallidus in rats. Brain Res. 699, 97-102.

Krowicki, Z.K., Hornby, P.J., 1995. The nucleus raphe obscurus controls pancreatic hormone secretion in the rat. Am. J. Physiol. Endocrinol. Metab. 268, E1128-E1134.

Larsen, P.D., Zhong, S., Gebber, G.L., Barman, S.M., 2000. Differential pattern of spinal sympathetic outflow in response to stimulation of the caudal medullary raphe. Am. J. Physiol. Regul. Integr. Comp. Physiol. 279, R210-R221.

Loewy, A.D., 1981. Raphe pallidus and raphe obscurus projections to the intermediolateral cell column in the rat. Brain Res. 222, 129-133.

Loewy, A.D., Franklin, M.F., Haxhiu, M.A., 1994. CNS monoamine cell groups projecting to pancreatic vagal motor neurons: a transneuronal labeling study using pseudorabies virus. Brain Res. 638, 248-260.

Lynn, R.B., Kreider, M.S., Miselis, R.R., 1991. Thyrotropin-releasing hormone-immunoreactive projections to the dorsal motor nucleus and the nucleus of the solitary tract of the rat. J. Comp. Neurol. 311, 271-288.

Manaker, S., Rizio, G., 1989. Autoradiographic localization of thyrotropin-releasing hormone and substance P receptors in the rat dorsal vagal complex. J. Comp. Neurol. 290, 516-526.

Marson, L., 1995. Central nervous system neurons identified after injection of pseudorabies virus into the rat clitoris. Neurosci. Lett. 190, 41-44.

Marson, L., 1997. Identification of central nervous system neurons that innervate the bladder body, bladder base, or external urethral sphincter of female rats: a transneuronal tracing study using pseudorabies virus. J. Comp. Neurol. 389, 584-602.

Marson, L., McKenna, K.E., 1990. The identification of a brainstem site controlling spinal sexual reflexes in male rats. Brain Res. 515, 303-308.

Marson, L., McKenna, K.E., 1992. A role for 5-hydroxytryptamine in descending inhibition of spinal sexual reflexes. Exp. Brain Res. 88, 313-320.

Marson, L., McKenna, K.E., 1994. Serotonergic neurotoxic lesions facilitate male sexual reflexes. Pharmacol. Biochem. Behav. 47, 883-888.

Marson, L., McKenna, K.E., 1996. CNS cell groups involved in the control of the ischiocavernosus and bulbospongiosus muscles: a transneuronal tracing study using pseudorabies virus. J. Comp. Neurol. 374, 161-179.

Marson, L., Platt, K.B., McKenna, K.E., 1993. Central nervous system innervation of the penis as revealed by the transneuronal transport of pseudorabies virus. Neurosci. 55, 263-280.

Martinez, V., Wu, S.V., Tache, Y., 1998. Intracisternal antisense oligodeoxynucleotides to the thyrotropin-releasing hormone receptor blocked vagal-dependent stimulation of gastric emptying induced by acute cold in rats. Endocrinology 139, 3730-3735.

Mason, P., 2001. Contributions of the medullary raphe and ventromedial reticular region to pain modulation and other homeostatic functions. Annu. Rev. Neurosci. 24, 737-777.

McCall, R.B., 1984. Evidence for a serotonergically mediated sympathoexcitatory response to stimulation of medullary raphe nuclei. Brain Res. 311, 131-139.

McCall, R.B., 1988. GABA-mediated inhibition of sympathoexcitatory neurons by midline medullary stimulation. Am. J. Physiol. 255, R605-R615.

McCall, R.B., Clement, M.E., Harris, L.T., 1989. Studies on the mechanism of the sympatholytic effect of 8-OH DPAT: lack of correlation between inhibition of serotonin neuronal firing and sympathetic activity. Brain Res. 501, 73-83.

McCall, R.B., Harris, L.T., 1987. Sympathetic alterations after midline medullary raphe lesions. Am. J. Physiol. 253, R91-R100.

McCann, M.J., Hermann, G.E., Rogers, R.C., 1988. Dorsal medullary serotonin and gastric motility: enhancement of effects by thyrotropin-releasing hormone. J. Auton. Nerv. Syst. 25, 35-40.

McCann, M.J., Hermann, G.E., Rogers, R.C., 1989. Thyrotropin-releasing hormone: effects on identified neurons of the dorsal vagal complex. J. Auton. Nerv. Syst. 26, 107-112.

Minson, J.B., Chalmers, J.P., Caon, A.C., Renaud, B., 1987. Separate areas of rat medulla oblongata with populations of serotonin- and adrenaline-containing neurons alter blood pressure after L-glutamate stimulation. J. Auton. Nerv. Syst. 19, 39-50.

Morrison, S.F., 1999. RVLM and raphe differentially regulate sympathetic outflows to splanchnic and brown adipose tissue. Am. J. Physiol. 276, R962-R973.

Morrison, S.F., 2001. Differential control of sympathetic outflow. Am. J. Physiol. Regul. Integr. Comp. Physiol. 281, R683-R698.

Morrison, S.F., Gebber, G.L., 1984. Raphe neurons with sympathetic-related activity: baroreceptor responses and spinal connections. Am. J. Physiol. 246, R338-R348.

Morrison, S.F., Gebber, G.L., 1985. Axonal branching patterns and funicular trajectories of raphespinal sympathoinhibitory neurons. J. Neurophysiol. 53, 759-772.

Morrison, S.F., Ramamurthy, S., Young, J.B., 2000. Reduced rearing temperature augments responses in sympathetic outflow to brown adipose tissue. J. Neurosci. 20, 9264-9271.

Morrison, S.F., Sved, A.F., Passerin, A.M., 1999. GABA-mediated inhibition of raphe pallidus neurons regulates sympathetic outflow to brown adipose tissue. Am. J. Physiol. 276, R290-R297.

Murphy, A.Z., Rizvi, T.A., Ennis, M., Shipley, M.T., 1999. The organization of preoptic-medullary circuits in the male rat: evidence for interconnectivity of neural structures involved in reproductive behavior, antinociception and cardiovascular regulation. Neurosci. 91, 1103-1116.

Nadelhaft, I., Vera, P.L., 2001. Separate urinary bladder and external urethral sphincter neurons in the central nervous system of the rat: simultaneous labeling with two immunohistochemically distinguishable pseudorabies viruses. Brain Res. 903, 33-44.

Nakamura, K., Matsumura, K., Kaneko, T., Kobayashi, S., Katoh, H., Negishi, M., 2002. The rostral raphe pallidus nucleus mediates pyrogenic transmission from the preoptic area. J. Neurosci. 22, 4600-4610.

Nalivaiko, E., Blessing, W.W., 2001. Raphe region mediates changes in cutaneous vascular tone elicited by stimulation of amygdala and hypothalamus in rabbits. Brain Res. 891, 130-137.

Niida, H., Takeuchi, K., Okabe, S., 1991. Role of thyrotropin-releasing hormone in acid secretory response induced by lowering of body temperature in the rat. Eur. J. Pharmacol. 198, 137-142.

Nosjean, A., Guyenet, P.G., 1991. Role of ventrolateral medulla in sympatholytic effect of 8-OHDPAT in rats. Am. J. Physiol. 260, R600-R609.

Okumura, T., Uehara, A., Taniguchi, Y., Watanabe, Y., Tsuji, K., Kitamori, S., Namiki, M., 1993. Kainic acid injection into medullary raphe produces gastric lesions through the vagal system in rats. Am. J. Physiol. 264, G655-G658.

Oldfield, B.J., Giles, M.E., Watson, A., Anderson, C., Colvill, L.M., McKinley, M.J., 2002. The neurochemical characterisation of hypothalamic pathways projecting polysynaptically to brown adipose tissue in the rat. Neurosci. 110, 515-526.

Orr, R., Marson, L., 1998. Identification of CNS neurons innervating the rat prostate: a transneuronal tracing study using pseudorabies virus. J. Auton. Nerv. Syst. 72, 4-15.

Palkovits, M., Mezey, E., Eskay, R.L., Brownstein, M.J., 1986. Innervation of the nucleus of the solitary tract and the dorsal vagal nucleus by thyrotropin-releasing hormone-containing raphe neurons. Brain Res. 373, 246-251.

Pilowsky, P.M., Miyawaki, T., Minson, J.B., Sun, Q.J., Arnolda, L.F., Llewellyn-Smith, I.J., Chalmers, J.P., 1995. Bulbospinal sympatho-excitatory neurons in the rat caudal raphe. J. Hypertension 13, 1618-1623.

Potas, J.R., Dampney, R.A., 2003. Sympathoinhibitory pathway from caudal midline medulla to RVLM is independent of baroreceptor reflex pathway. Am. J. Physiol. Regul. Integr. Comp. Physiol. 284, R1071-R1078.

Rathner, J.A., McAllen, R.M., 1999. Differential control of sympathetic drive to the rat tail artery and kidney by medullary premotor cell groups. Brain Res. 834, 196-199.

Rathner, J.A., Owens, N.C., McAllen, R.M., 2001. Cold-activated raphe-spinal neurons in rats. J. Physiol. 535, 841-854.

Rinaman, L., Miselis, R.R., 1990. Thyrotropin-releasing hormone-immunoreactive nerve terminals synapse on the dendrites of gastric vagal motoneurons in the rat. J. Comp. Neurol. 294, 235-251.

Schadt, J.C., 2003. What is the role of serotonin during hemorrhage in conscious animals? Am. J. Physiol. Regul. Integr. Comp. Physiol. 284, R780-R781.

Simerly, R.B., Swanson, L.W., 1988. Projections of the medial preoptic nucleus: a Phaseolus vulgaris leucoagglutinin anterograde tract-tracing study in the rat. J. Comp. Neurol. 270, 209-242.

Smith, J.E., Jansen, A.S., Gilbey, M.P., Loewy, A.D., 1998. CNS cell groups projecting to sympathetic outflow of tail artery: neural circuits involved in heat loss in the rat. Brain Res. 786, 153-164.

Stornetta, R.L., Guyenet, P.G., 1999. Distribution of glutamic acid decarboxylase mRNA-containing neurons in rat medulla projecting to thoracic spinal cord in relation to monoaminergic brainstem neurons. J. Comp. Neurol. 407, 367-380.

Tache, Y., Yang, H., 1993. Role of medullary TRH in the vagal regulation of gastric function. In: Tache Y, Wingate DL, Burks TF (eds) Innervation of the gut: Pathophysiological implications. CRC Press, Boca Raton, pp 67-80.

Tanaka, M., Nagashima, K., McAllen, R.M., Kanosue, K., 2002. Role of the medullary raphe in thermoregulatory vasomotor control in rats. J. Physiol. 540, 657-664.

Ter Horst, G.J., Hautvast, R.W., De Jongste, M.J., Korf, J., 1996. Neuroanatomy of cardiac activity-regulating circuitry: a transneuronal retrograde viral labelling study in the rat. Eur. J. Neurosci. 8, 2029-2041.

Verberne, A.J., Sartor, D.M., Berke, A., 1999. Midline medullary depressor responses are mediated by inhibition of RVLM sympathoexcitatory neurons in rats. Am. J. Physiol. 276, R1054-R1062.

Wang, W.H., Lovick, T.A., 1993. The inhibitory effect of the ventrolateral periaqueductal grey matter on neurones in the rostral ventrolateral medulla involves a relay in the medullary raphe nuclei. Exp. Brain Res. 94, 295-300.

Wessendorf, M.W., Appel, N.M., Molitor, T.W., Elde, R.P., 1990. A method for immunofluorescent demonstration of three coexisting neurotransmitters in rat brain and spinal cord, using the fluorophores fluorescein, lissamine rhodamine, and 7-amino-4-methylcoumarin-3-acetic acid. J. Histochem. Cytochem. 38, 1859-1877.

White, R.L., Jr., Rossiter, C.D., Hornby, P.J., Harmon, J.W., Kasbekar, D.K., Gillis, R.A., 1991. Excitation of neurons in the medullary raphe increases gastric acid and pepsin production in cats. Am. J. Physiol. 260, G91-G96.

Yang, H., Ohning, G., Tache, Y., 1993. TRH in dorsal vagal complex mediates acid response to excitation of raphe pallidus neurons in rats. Am. J. Physiol. 265, G880-G886.

Yang, H., Tache, Y., Ohning, G., Go, V.L., 2002. Activation of raphe pallidus neurons increases insulin through medullary thyrotropin-releasing hormone (TRH)-vagal pathways. Pancreas 25, 301-307.

Yang, H., Wu, S.V., Ishikawa, T., Tache, Y., 1994. Cold exposure elevates thyrotropin-releasing hormone gene expression in medullary raphe nuclei: relationship with vagally mediated gastric erosions. Neurosci. 61, 655-663.

Yells, D.P., Hendricks, S.E., Prendergast, M.A., 1992. Lesions of the nucleus paragigantocellularis: effects on mating behavior in male rats. Brain Res. 596, 73-79.

Yen, C.T., Blum, P.S., Spath, J.A., 1983. Control of cardiovascular function by electrical stimulation within the medullary raphe region of the cat. Exp.Neurol. 79, 666-679.

Yoneda, M., Tache, Y., 1995. Serotonin enhances gastric acid response to TRH analogue in dorsal vagal complex through 5-HT2 receptors in rats. Am. J. Physiol. 269, R1-R6.

Yusof, A.P., Coote, J.H., 1988. Patterns of activity in sympathetic postganglionic nerves to skeletal muscle, skin and kidney during stimulation of the medullary raphe area of the rat. J. Auton. Nerv. Syst. 24, 71-79.

Zaretsky, D.V., Zaretskaia, M.V., DiMicco, J.A., 2003. Stimulation and blockade of GABAA receptors in the raphe pallidus: Effects on body temperature, heart rate and blood pressure in conscious rats. Am. J. Physiol. Regul. Integr. Comp. Physiol. In Press.

Zhou, S.Y., Gilbey, M.P., 1995. Sympathoexcitatory influence of a fast conducting raphe-spinal pathway in the rat. Am. J. Physiol. 268, R1230-R1235.

Chapter 12

INTERNEURONAL INPUTS TO SYMPATHETIC PREGANGLIONIC NEURONS: EVIDENCE FROM TRANSECTED SPINAL CORD

Ida J. Llewellyn-Smith[1] and Lynne C. Weaver[2]
[1]*Cardiovascular Neuroscience Group, Cardiovascular Medicine & Centre for Neuroscience, Flinders University of South Australia, Bedford Park, South Australia 5042 AUSTRALIA and* [2]*Laboratory of Spinal Cord Injury, BioTherapeutics Research Group, Robarts Research Institute, 100 Perth Drive, P.O. Box 5015, London Ontario N6A 5K8 CANADA*

Abstract: Sympathetic preganglionic neurons (SPN) in the spinal cord are important for controlling the heart, blood vessels and release of catecholamines from the adrenal medulla. The activity of SPN is predominantly controlled by nerve pathways originating in the brain; but SPN also receive information from interneurons, which lie near them in the spinal cord and relay peripheral sensory or other central nervous input. Very little is known about the chemical phenotypes of the interneurons that synapse on SPN. Substance P (SP)- and enkephalin (ENK)-immunoreactive axons persist after spinal cord injury and small, neuropeptide Y (NPY)-expressing neurons occur in the spinal cord laminae containing interneurons antecedent to SPN. We therefore examined choline acetyltransferase (ChAT)-immunoreactive SPN caudal to a complete transection of the upper thoracic cord done 7 days earlier for evidence of synapses by nerve fibers containing these 3 neuropeptides. Seven days was chosen because virtually all input to SPN from rostral to the transection has degenerated by this time, leaving only input from intraspinal neurons. We found that all three types of nerve fibers formed synapses on the cell bodies and dendrites of ChAT-positive SPN. These results indicate that spinal interneurons containing SP, ENK and NPY provide synaptic input to SPN. Whether these neuropeptides occur in the same or different populations of interneurons and whether their synapses occur on barosensitive SPN remain to be determined.

Key words: Blood pressure, Sympathetic preganglionic neurons, Interneurons, Spinal cord, Substance P, Enkephalin, Neuropeptide Y, Immunocytochemistry, Electron Microscopy

INTERNEURONS AND THE CONTROL OF SYMPATHETIC OUTFLOW

Through nerve pathways in the periphery, sympathetic preganglionic neurons (SPN) regulate the activity of autonomic effector organs, including the heart, blood vessels and the adrenal medulla. SPN, with cell bodies in the intermediolateral cell column (IML) in the thoracic (T) and upper lumbar (L) segments of the spinal cord, integrate information arising from many regions within the central nervous system. Neurons in the medulla, pons and higher centres have been acknowledged to play important roles in controlling the output of SPN (Dampney, 1994) and our recent ultrastructural studies indicate that SPN in spinal cord segments which have a significant involvement in cardiovascular control receive most of their synaptic input from supraspinal sources (Llewellyn-Smith and Weaver, 2001). However, there is also long-standing evidence that neurons with somata in the spinal cord also convey information to SPN (Gebber and McCall, 1976; McCall et al., 1977). Nevertheless, relatively little is known about how these spinal interneurons influence the activity of SPN or about their neurochemical phenotypes. Electrophysiological experiments have shed the most light on the role of spinal interneurons in blood pressure control and a few studies have provided anatomical information about this neuronal type. In general, current evidence suggests that:

- spinal interneurons convey information to SPN from central and peripheral sources (Fig. 1).
- both inhibitory and excitatory spinal interneurons synapse on SPN.
- some spinal interneurons respond to changes in blood pressure and could be in circuits that include cardiovascular SPN.
- many of the spinal interneurons synapsing on SPN are local and occur in spinal cord laminae V and VII as well as within the IML.
- spinal interneurons contain a diverse array of neurotransmitter-related markers.
- interneurons are both electrophysiologically and morphologically distinct from SPN.

Physiological Evidence for Interneuronal Circuits Contributing to the Activity of SPN

Physiological studies suggest that intraspinal interneurons are likely to play at least three roles in conveying information to SPN.

12. INTERNEURONAL INPUTS TO SYMPATHETIC PREGANGLIONIC NEURONS: EVIDENCE FROM TRANSECTED SPINAL CORD

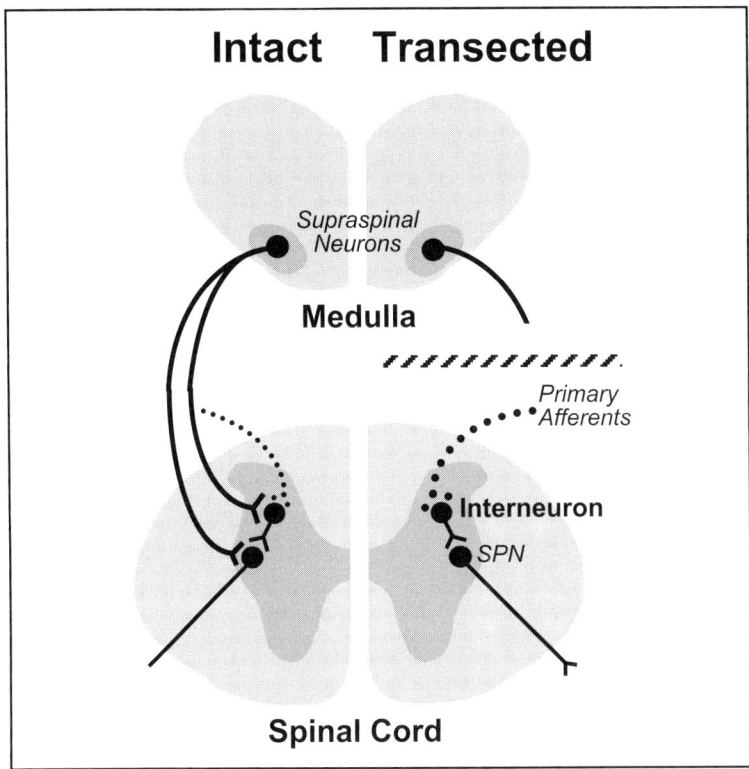

Figure 1. Central circuits that include SPN and spinal interneurons. **Left:** In intact spinal cord, SPN receive inputs from supraspinal neurons and from intraspinal neurons. The interneurons convey both sensory information from primary afferent neurons and information from the brain. Since supraspinal axons and intraspinal axons contain many of the same neurotransmitters and related markers, it is not possible to distinguish supraspinal from intraspinal synapses on SPN in intact spinal cord. **Right:** Caudal to a complete transection (diagonal lines) made 7 days earlier, virtually all supraspinal input to SPN has degenerated. Since synapses from intraspinal neurons are the only ones that remain after transection, they can be conclusively identified. By 14 days, primary afferent axons have sprouted to create a denser arbor in the deeper layers of the dorsal horn.

Processing input from the brain to SPN: In the 1970's, experiments in the cat showed that both excitatory and inhibitory spinal interneurons responded to baroreflex stimulation (Gebber and McCall, 1976; McCall et al., 1977). Excitatory interneurons in bulbospinal circuits to SPN were subsequently confirmed by intracellular studies in cat, where activation of descending fibers evoked polysynaptic excitatory post-synaptic potentials (PSP's) in SPN (Dembowsky et al., 1985) The neonatal rat brainstem-spinal cord preparation has provided evidence that inhibitory interneurons also convey input from the brain to SPN (Deuchars et al., 1997). In this preparation, stimulation of the rostral ventrolateral medulla (RVLM), an area crucial for the tonic and reflex control of blood pressure, has been shown to activate at least two pathways. Bulbospinal RVLM neurons use a monosynaptic excitatory pathway to communicate directly with SPN; and a direct projection has been confirmed ultrastructurally (Zagon and Smith, 1993). A second polysynaptic bulbospinal pathway involves inhibitory spinal interneurons. SPN show inhibitory PSP's when this bulbospinal pathway is activated.

Conveying primary afferent information to SPN: Circuits relaying sensory input to SPN involve both excitatory and inhibitory interneurons. In either rat spinal cord slices, neonatal brainstem-spinal cord preparations or adult cat spinal cord *in vivo*, stimulation of dorsal roots or the dorsal horn produces polysynaptic excitatory and inhibitory PSP's in SPN (Dembowsky et al., 1985; Shen et al., 1990; Deuchars et al., 1995). Furthermore, excitation is produced in both intracellularly impaled SPN and sympathetic nerves by stimulation of visceral or somatic afferents (Coote and Downman, 1966; Dembowsky et al., 1985), although whether these responses are due to activation of spinal or supraspinal pathways or both was not determined. Nevertheless, the involvement of spinal interneurons in sympathetic responses to sensory input has recently been confirmed in adult rat spinal cord after acute cervical transection (Chau et al., 1997; Chau et al., 2000). Extracellular recording was used to define populations of interneurons in the dorsal and lateral horns with ongoing activity positively correlated with renal sympathetic nerve activity despite the lack of supraspinal input. These interneurons were shown to respond to noxious somatic or visceral sensory stimuli. Interestingly, the neurons in the lateral and dorsal horns differed in their responses, with the former being excited by colorectal distension and the latter being inhibited (Chau et al., 2000). These results suggest that the physiological outcome of a sensory stimulus may depend upon the population of spinal interneurons that is involved in relaying the stimulus to SPN.

Participation in short-range local circuits involving SPN: Spontaneous PSP's occur in SPN in spinal cord slices, suggesting that spinal interneurons have short range interactions with SPN (Spanswick et al., 1994). In confirmation of such short range pathways within autonomic control circuitry, whole cell patch clamp has revealed that parasympathetic preganglionic neurons receive synapses from extracellularly-stimulated interneurons that are located less than 100μm away (Araki and de Groat, 1996).

12. INTERNEURONAL INPUTS TO SYMPATHETIC PREGANGLIONIC NEURONS: EVIDENCE FROM TRANSECTED SPINAL CORD

Anatomical Evidence for Interneurons in Neuronal Circuits Controlling SPN

Both transynaptic tracing with herpes simplex virus and transneuronal transport of wheat germ agglutinin have shown that interneurons antecedent to SPN are located in spinal cord laminae V and VII (Cabot et al., 1994; Joshi et al., 1995; Clarke et al., 1998). At least some of these interneurons are likely to relay sensory information to SPN since it is well documented that primary afferent fibers carrying sensory input from the periphery terminate in these same laminae. The occurrence of neurons with ongoing activity that correlates with renal sympathetic nerve activity in dorsal horn laminae III-V supports an interneuronal link between primary afferents and sympathetic outflow (Chau et al., 1997; Chau et al., 2000). There is also anatomical support for the presence of interneurons in some circuits from the brain. For example, oxytocin and vasopressin fibers, which arise from the hypothalamus, appear to specifically target spinal interneurons rather than SPN (Strack et al., 1989). Finally, many of the interneurons that synapse on SPN are likely to be located within a few hundred micrometers of their targets since responses mediated by interneurons (i.e., spontaneous PSP's) can be detected electrophysiologically in spinal cord slices that are 300 µm thick or less.

Neurochemistry of Interneurons

Amino Acids: Interneurons antecedent to SPN are likely to use these small molecules as their primary fast-acting neurotransmitters. In intact spinal cord, where synapses of both supraspinal and intraspinal origin are present, virtually all of the nerve fibers that synapse on SPN contain either the inhibitory amino acid, γ-aminobutyric acid (GABA), or the excitatory amino acid, glutamate (Llewellyn-Smith et al., 1992a; Llewellyn-Smith et al., 1998). Furthermore, glutamate and GABA synapses occur on SPN deprived of supraspinal input by a spinal cord transection (Llewellyn-Smith et al., 1997; Llewellyn-Smith and Weaver, 2001). Consequently, these transmitters must also occur in all spinal interneurons that communicate with SPN. Interneurons probably also use the inhibitory amino acid, glycine, as a transmitter. Glycine-immunoreactive nerve fibers form synapses on SPN (Cabot et al., 1992) and glycine-positive interneurons occur in lamina V and VII (Todd and Sullivan, 1998). Glycine interneurons are inhibitory since SPN hyperpolarize in response to glycine and the glycine antagonist strychnine blocks spontaneous PSP's in SPN and the production of inhibitory PSP's evoked by dorsal horn stimulation (Spanswick et al., 1994). Furthermore, the indirect effects of glutamate acting through N-methyl-D-aspartate receptors on SPN are likely to be mediated by inhibitory spinal interneurons (Dun and Mo, 1989).

Neuropeptides: Barosensitive spinal interneurons may also contain and release this type of neurotransmitter. Varicose nerve fibers immunoreactive for substance P (SP), enkephalin (ENK) and neuropeptide Y (NPY) are still present in the IML below a spinal cord transection (Davis and Cabot, 1984; Romagnano et al., 1987; Cassam et al., 1997), suggesting that these peptides also occur in spinal interneurons. We have evidence for the presence of NPY mRNA in spinal neurons in laminae V and VII of the thoracic cord (Minson et al., 2001), where interneurons antecedent to SPN are located; but whether or not these NPY interneurons are barosensitive and synapse on SPN is unknown. When released from either interneuronal or supraspinal inputs, any or all of these peptides could have direct post-synaptic effects on SPN. For example, SPN respond directly to substance P with slow depolarizations (e.g., Dun and Mo, 1988; Cammack and Logan, 1996). Alternatively, these peptides could presynaptically affect the release of amino acids from their own or adjacent terminals, as happens with opioid agonists in the RVLM (Hayar and Guyenet, 1998).

Other neuronal markers: Spinal autonomic interneurons contain other chemical markers, some of which are also found in SPN. Electrophysiological and immunohistochemical evidence indicates that the potassium channel subunit, Kv3.1, occurs in interneurons but is completely absent from SPN (Deuchars et al., 2001). These Kv3.1-positive interneurons are part of local circuitry involving SPN since they are transynaptically labelled by a GFP-expressing herpes virus injected into the adrenal medulla (Brooke et al., 2002). In contrast, the calcium binding protein, calbindin 28DK, and nitric oxide synthase (NOS) probably occur in both SPN and some interneurons. Calbindin- or NOS-containing SPN identified by retrogradely labelling after intraperitoneal injection of Fluorogold are intermixed with other Fluorogold-negative, calbindin- or NOS-immunoreactive neurons, which could be interneurons (Anderson, 1992; Grkovic and Anderson, 1997). We have demonstrated calbindin-immunoreactive synapses in the IML of intact cord and have found that these inputs can be divided into two groups, i.e., inputs containing calbindin + GABA and inputs containing calbindin + glutamate (Llewellyn-Smith et al., 2002). Since calbindin occurs in some bulbospinal neurons in the RVLM (Goodchild et al., 2000) and all of these neurons appear to be glutamatergic (Stornetta et al., 2002), at least some of the calbindin + glutamate terminals are likely to have a supraspinal origin. However, it

12. INTERNEURONAL INPUTS TO SYMPATHETIC PREGANGLIONIC NEURONS: EVIDENCE FROM TRANSECTED SPINAL CORD

remains to be established whether both types of calbindin synapses arise from spinal interneurons and whether calbindin-positive interneurons synapse on SPN that respond to changes in blood pressure. As well as all SPN, the acetylcholine synthesizing enzyme, choline acetyltransferase (ChAT) also occurs in some interneurons. At electron microscope level, we found that ChAT-positive axon terminals synapsed on ChAT-positive SPN in both intact and transected spinal cord (Llewellyn-Smith and Weaver, 2001). Some of these cholinergic synapses on SPN co-contain glutamate, whereas, others contain GABA (Llewellyn-Smith and Weaver, unpublished observations) so that at least two populations of spinal cholinergic neurons innervate SPN.

Electrophysiological and Morphological Characterization of Interneurons in and Near the IML

Using patch clamp techniques in slices of neonatal rat spinal cord, Deuchars et al. (2001) have shown that interneurons can be distinguished from SPN on the basis of 4 electrophysiological characteristics. Interneurons with cell bodies in or immediately adjacent to the IML were shown to have significantly shorter-lasting action potentials than SPN (4.16 $\pm$ 0.1 ms compared to 9.4 + 0.25 ms), mainly because they repolarize more rapidly. Interneurons also show significantly higher instantaneous and steady state firing rates. The differences in these two properties relate to the fact that the potassium channel subunit, Kv3.1, occurs in interneurons but not in SPN. The other electrophysiological features distinguishing interneurons from SPN are that interneurons have afterhyperpolarizations that are smaller in amplitude than those of SPN and that interneurons exhibit both I_H and I_A currents whereas SPN have only I_A.

As the neurons in the Deuchars study (Deuchars et al., 2001) were filled with Neurobiotin while they were being characterized pharmacologically and electrophysiologically, their morphology could be assessed after histological processing. Neurons that were classified as interneurons had cell bodies within or immediately adjacent to the IML and their axons ramified extensively within the spinal grey matter, including the IML and the central autonomic area. In contrast, SPN axons travelled into the ventral horn without branching. These data provide strong anatomical evidence that local circuit interneurons could indeed innervate SPN.

STUDIES ON TRANSECTED SPINAL CORD: ILLUMINATING SPINAL AUTONOMIC INTERNEURONS?

Since supraspinal axons and axons of intraspinal origin contain many of the same neurotransmitters and related markers, it is not possible to distinguish spinal interneuronal synapses on SPN from supraspinal ones in intact spinal cord. However, animals with complete spinal cord transections have provided a wealth of data that have significantly increased our understanding of how spinal interneurons might interact with SPN to affect sympathetic outflow (see Sections **Physiological Evidence for Interneuronal Circuits Contributing to the Activity of SPN** and **Neurochemistry of Interneurons**). In fact, some of the earliest indications that spinal interneurons participate in cardiovascular control came from studies of this type (e.g., Mannard and Polosa, 1973). Nevertheless, when trying to glean information about the neurochemistry of spinal autonomic interneurons, it is important to realize that many plastic changes occur in the spinal cord after injury and that some of these involve SPN, primary afferents and interneurons. These changes might confuse the interpretation of anatomical data on the neurotransmitter phenotypes of interneurons if experiments are done long enough after injury for significant reorganization to have taken place.

Injury-induced changes in SPN: Soon after a spinal cord injury, SPN retract and then regrow their dendrites and the size of their somata decreases and then returns to control (Krassioukov and Weaver, 1996; Krenz and Weaver, 1998a; Krassioukov et al., 1999; Llewellyn-Smith and Weaver, 2001; Weaver et al., 2001). The time course of this response probably correlates with the death and removal of severed presynaptic axons because, in rats, dendrites are maximally retracted at 3 days after a complete transection when the largest number of degenerating axon terminals are found in the IML (Weaver et al., 1997).

Injury-induced changes in primary afferents: By 14 days after transection, primary afferents have undergone significant sprouting so that, in the deeper layers of the dorsal horn, the axonal arbors of calcitonin gene-related peptide (CGRP)-immunoreactive afferents have approximately doubled in area without a change in the number of CGRP-positive dorsal root ganglion neurons (Krenz and Weaver, 1998b). Large diameter myelinated afferents sprout as well as the unmyelinated or lightly myelinated CGRP-immunoreactive ones (Wong et al., 2000). Interestingly, blocking the sprouting of CGRP-immunoreactive sensory neurons by intrathecal administration of a neutralizing antibody to nerve growth factor (NGF) reduces by about half the rise in blood pressure caused by colorectal

12. INTERNEURONAL INPUTS TO SYMPATHETIC PREGANGLIONIC NEURONS: EVIDENCE FROM TRANSECTED SPINAL CORD

distension in rats with complete spinal cord transections (Krenz et al., 1999). Eliminating NGF activity with an intrathecally-delivered fusion protein composed of the extracellular portion of the trkA receptor linked to the Fc portion of human immunoglobulin has a similar effect on rats with clip compression spinal cord injuries (Weaver et al., 2002; Marsh et al., 2002).

Injury-induced changes in interneurons: Immunostaining for growth associated protein 43 (GAP-43), which occurs in developing and regenerating neurons, indicates that the intraspinal axons of neurons with cell bodies caudal to a transection also begin to sprout and form new connections shortly after injury (Cassam et al., 1995; Weaver et al., 1997). Some of these new connections are made with SPN (Weaver et al., 1997). Since GAP-43-immunoreactive growth cones have been found in the cord at all post-injury times that have been examined (Cassam et al., 1995; Weaver et al., 1997), reorganization of spinal circuits is likely to be an ongoing and long-term process. This continuing synaptic reorganization may be the reason for the onset and increasing severity of maladaptive, injury-related physiological sequelae, such as autonomic dysreflexia (Mathias and Frankel, 1992; Krassioukov and Weaver, 1995; Maiorov et al., 1997; Teasell et al., 2000). After a transection, interneurons probably also turn on genes that are not normally expressed in intact cord. For example, the catecholamine synthetic enzyme, dopamine β-hydroxylase, appears to be upregulated by transection since 6 times as many spinal neurons are immunoreactive for this enzyme in 14-day injured compared to intact cord (Cassam et al., 1997). Interneurons that turn on the synthesis of catecholamine enzymes are likely to be the source of the tyrosine hydroxylase immunoreactive synapses that we have found on sympathoadrenal neurons caudal to a 14-day transection (Llewellyn-Smith et al., 1995). Whether these neurons actually synthesize and release a catecholamine is unknown. The physiological responses of interneurons also show changes in the chronic phase of spinal cord injury. In rats with one-month transections, many more sympathetically-correlated interneurons were excited by cutaneous sensory stimulation and from wider areas of the body surface than in rats in which cords had been cut only hours earlier (Krassioukov et al., 2002). These alterations in responsiveness may possibly reflect increased direct or indirect innervation of interneurons by the larger terminal fields of sprouted primary afferents discussed above.

INTERNEURONS PROVIDE NEUROPEPTIDE-IMMUNOREACTIVE INPUTS TO SPN

The considerations, detailed in **STUDIES ON TRANSECTED SPINAL CORD: ILLUMINATING SPINAL AUTONOMIC INTRNEURONS?** indicate that it is critical to examine the spinal cord shortly after a transection, before significant plastic reorganization has occurred, for insight to be gained into what sorts of inputs interneurons might provide to SPN under normal circumstances. Bearing this in mind, we have been examining tissue from rats with complete transections done 7 days before perfusion. This time point was chosen for two reasons. First, severed axons from above the lesion have been almost completely cleared from the cord by this time (Weaver et al., 1997). Second, reinnervation of SPN by sprouting interneurons is unlikely to be significant at 7 days post-operatively since SPN are still profoundly denervated at 14 days (Llewellyn-Smith and Weaver, 2001). Our results, reported below, are strongly suggestive of a direct innervation of SPN in intact cord by interneurons with axons that contain, and possibly release, SP, NPY and ENK.

Male Sprague-Dawley rats had their spinal cords completely transected at T4/5 using our standard procedures (Krassioukov and Weaver, 1995), as approved by the Animal Welfare Committee of Flinders University. Seven days later, the rats were perfused for electron microscopy (EM) with a fixative that contained formaldehyde and glutaraldehyde and Vibratome sections of fixed lower thoracic cord (T7-T11) were permeabilized by washing with 50% ethanol (Llewellyn-Smith and Minson, 1992). Double labelling avidin-biotin-peroxidase immunocytochemistry was used to localize ChAT with a tetramethylbenzidine tungstate reaction (Llewellyn-Smith et al., 1993) and a neuropeptide (SP, ENK or NPY) with a diaminobenzidine reaction intensified with imidazole (Llewellyn-Smith et al., 1992b). The stained sections were osmicated, dehydrated, infiltrated and embedded in resin. Ultrathin sections containing ChAT-immunoreactive SPN were stained with lead citrate and examined electron microscopically. Tissue from other rats with 7-day transections was reacted for light microscopic visualization of the peptides using avidin-biotin-peroxidase and a diaminobenzidine reaction intensified with nickel (Llewellyn-Smith et al., 1993).

Substance P: Light microscopy of 7-day transected cord revealed that many varicose SP-immunoreactive axons were present in the IML caudal to a transection (Fig. 2A), confirming our previous observations (Cassam et al., 1997). At the EM level, these axons were found to make synapses on the cell bodies and dendrites (Fig. 2B) of ChAT-immunoreactive SPN. Many

SP synapses also were found on ChAT-negative somata and dendrites near the ChAT-positive SPN.

Neuropeptide Y: As we have shown previously (Cassam et al., 1997), varicose NPY-immunoreactive fibers are also present in the lower thoracic IML of rats with transections at T4/5 done 7 days before perfusion although at reduced density compared to intact cord (Fig. 3A). Small NPY-immunoreactive neurons also occur near the IML (Fig. 3B). When examined electron microscopically, the NPY-immunoreactive axons were observed to form synapses on the cell bodies (Fig. 3C and 3D) and dendrites of ChAT-positive SPN but such synapses were difficult to find.

Enkephalin: Evidence from light microscopy suggests that, at 1 week post-operatively, ENK-immunoreactive innervation persists in the IML caudal to a spinal cord transection but the fibers are less numerous than in intact cord (Romagnano et al., 1987). We have confirmed those observations here (Fig. 4A) and have also observed ENK-immunoreactive neurons dorsal to the IML (Fig. 4B) in lamina V, where interneurons antecedent to SPN are located. We also found ultrastructurally that many ENK-positive boutons formed synapses in the lower thoracic IML of 7-day transected cord. Although some of these synapses occurred on SPN dendrites (Fig. 4C) and rare ones on SPN cell bodies, most ENK boutons innervated ChAT-negative IML neurons.

CONCLUSION

Despite some advances in recent years, our understanding of spinal interneurons that might contribute to blood pressure regulation is still only very rudimentary. Most importantly, it is still unknown whether a particular subset or subsets of spinal interneurons innervate vasomotor SPN preferentially instead of SPN controlling other targets. Increasing evidence suggests that central autonomic circuitry is organized into discrete channels, each of which has a function-specific output (Morrison, 2001); and it seems likely that this organizational pattern also extends to spinal autonomic interneurons. This functional specificity is likely to be manifest not only physiologically but also in anatomical differences between interneurons in their neurochemistry and the types of inputs that they receive. Studies employing immunocytochemistry, *in situ* hybridization, tract tracing and physiological stimuli to activate Fos should be particularly revealing. We need to learn much more about the interneurons that innervate SPN if we are to understand their role in regulating cardiovascular and other types of sympathetic outflow.

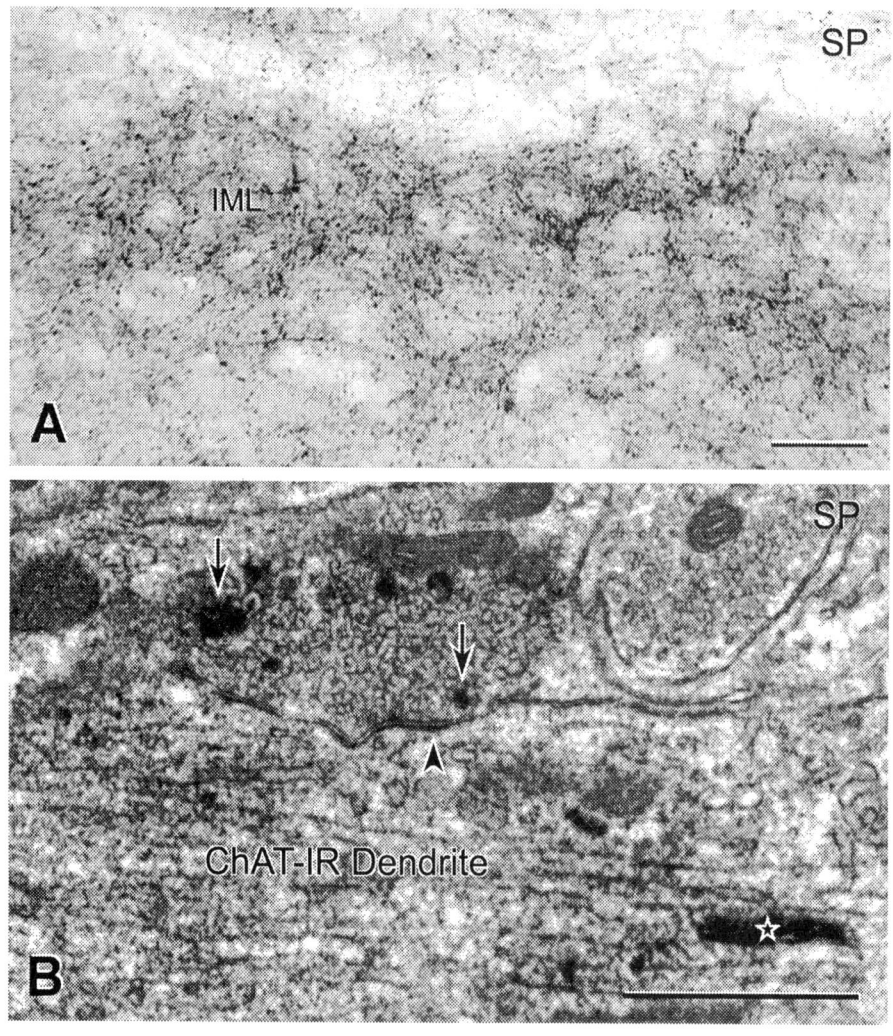

Figure 2. **A:** Light micrograph showing innervation of the IML by SP-immunoreactive axons caudal to a 7-day transection. Parasagittal section of T10. Dorsal is to the top. Scale bar, 100 µm. **B:** Electron micrograph showing that caudal to a 7-day transection, an SP-immunoreactive axon terminal synapses (arrowhead) onto a ChAT-immunoreactive (IR) dendrite of an SPN in the IML. Arrows, SP-immunoreactive large granular vesicles; star, TMB-tungstate crystal indicating ChAT-immunoreactivity. Scale bar, 1 µm.

12. INTERNEURONAL INPUTS TO SYMPATHETIC PREGANGLIONIC NEURONS: EVIDENCE FROM TRANSECTED SPINAL CORD

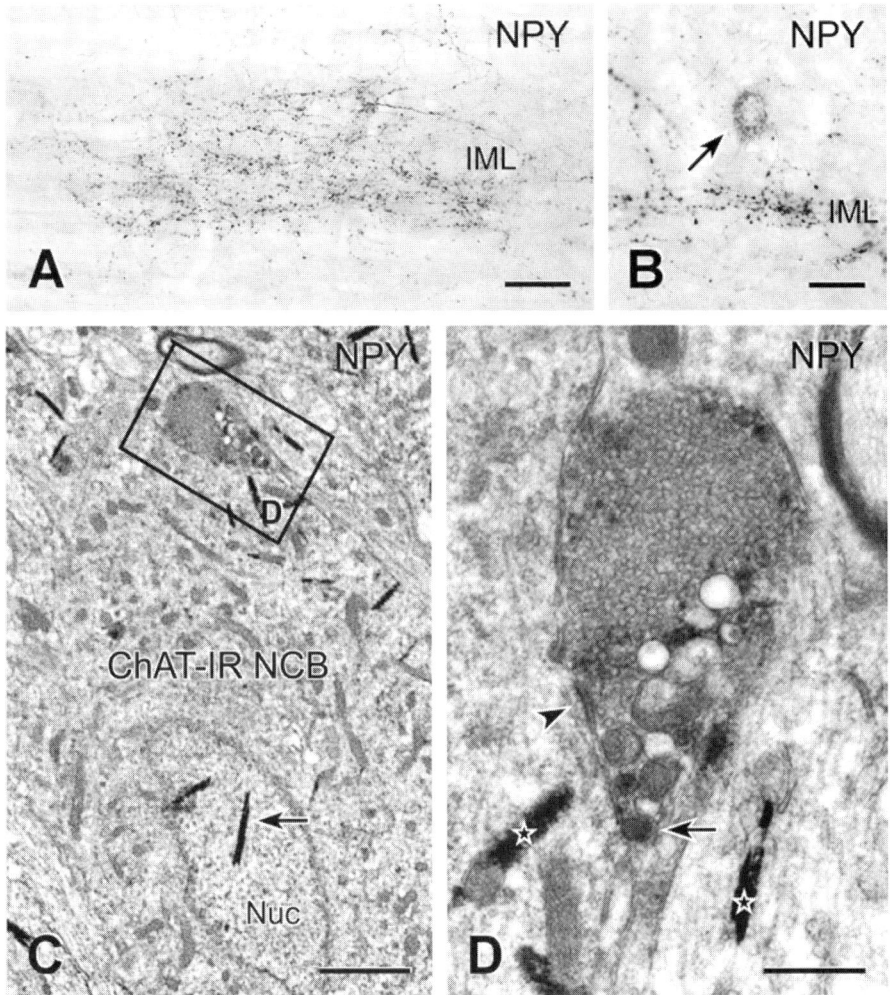

Figure 3. **A:** Light micrograph showing innervation of the IML by NPY-immunoreactive axons caudal to a 7 day transection. Parasagittal section of T7. Dorsal is to the top. Scale bar, 100 µm. **B:** Light micrograph of an NPY-immunoreactive neuron (arrow) lying just dorsal to the IML in T8. Scale bar, 50 µm. **C:** Electron micrograph showing the ChAT-immunoreactive nerve cell body of an SPN in the IML caudal to a 7-day transection. An NPY-immunoreactive terminal (box D) abuts the cell body. Arrow, TMB-tungstate crystal. Scale bar, 2 µm. **D:** Electron micrograph showing the NPY-immunoreactive terminal in box D in Figure 3C at higher magnification. The terminal forms a synapse (arrowhead) on the ChAT-immunoreactive cell body. Arrow, NPY-immunoreactive large granular vesicle; stars, TMB-tungstate crystals indicating ChAT-immunoreactivity. Scale bar, 500 nm.

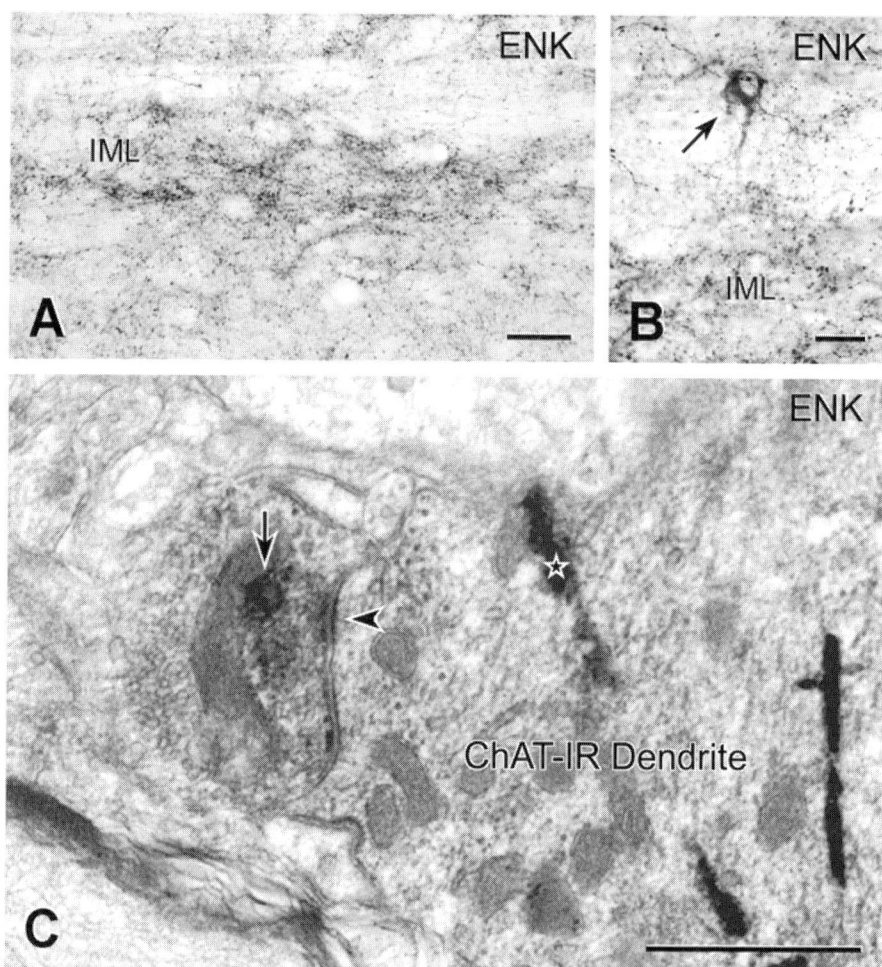

Figure 4. **A:** Light micrograph showing innervation of the IML by ENK-immunoreactive axons caudal to a 7-day transection. Parasagittal section of T11. Dorsal is to the top. Scale bar, 100 μm. **B:** Light micrograph of an ENK-immunoreactive neuron (arrow) lying just dorsal to the IML in T9. Parasagittal section. Scale bar, 50 μm. **C:** Electron micrograph showing an ENK-immunoreactive axon terminal synapsing (arrowhead) onto a ChAT-immunoreactive (IR) dendrite of an SPN caudal to a 7-day transection. Arrow, an ENK-immunoreactive large granular vesicle; star, TMB-tungstate crystal indicating ChAT-immunoreactivity. Scale bar, 1 μm.

ACKNOWLEDGEMENTS

This work was supported by grants from the National Health and Medical Research Council of Australia, National Heart Foundation of Australia, Canadian Institutes of Health Research and Heart and Stroke Foundation of Ontario, Canada. Carolyn Martin, Lee Travis and Barbara Atkinson provided expert technical assistance.

REFERENCES

Anderson, C.R., 1992. NADPH diaphorase-positive neurons in the rat spinal cord include a subpopulation of autonomic preganglionic neurons. Neurosci. Lett. 139, 280-284.

Araki, I., de Groat, W.C., 1996. Unitary excitatory synaptic currents in preganglionic neurons mediated by two distinct groups of interneurons in neonatal rat sacral parasympathetic nucleus. J. Neurophysiol. 76, 215-226.

Brooke, R.E., Pyner, S., McLeish, P., Buchan, S., Deuchars, J., Deuchars, S.A., 2002. Spinal cord interneurones labelled transneuronally from the adrenal gland by a GFP-herpes virus construct contain the potassium channel subunit Kv3.1b. Auton. Neurosci. 98, 45-50.

Cabot, J.B., Alessi, V., Bushnell, A., 1992. Glycine-like immunoreactive input to sympathetic preganglionic neurons. Brain Res. 571, 1-18.

Cabot, J.B., Alessi, V., Carroll, J., Ligorio, M., 1994. Spinal cord lamina V and lamina VII interneuronal projections to sympathetic preganglionic neurons. J. Comp. Neurol. 347, 515-530.

Cammack, C., Logan, S.D., 1996. Excitation of rat sympathetic preganglionic neurones by selective activation of the NK1 receptor. J. Auton. Nerv. Syst. 57, 87-92.

Cassam, A.K., Krassioukov, A.V., Weaver, L.C., 1995. Growth associated protein-43 in spinal autonomic nuclei in rats seven to thirty days after cord injury. J. Neurotrauma 12, 415-415.

Cassam, A.K., Llewellyn-Smith, I.J., Weaver, L.C., 1997. Catecholamine enzymes and neuropeptides are expressed in fibres and somata in the intermediate gray matter in chronic spinal rats. Neuroscience 78, 829-841.

Chau, D., Johns, D.G., Schramm, L.P., 2000. Ongoing and stimulus-evoked activity of sympathetically correlated neurons in the intermediate zone and dorsal horn of acutely spinalized rats. J. Neurophysiol. 83, 2699-2707.

Chau, D., Kim, N., Schramm, L.P., 1997. Sympathetically correlated activity of dorsal horn neurons in spinally transected rats. J. Neurophysiol. 77, 2966-2974.

Clarke, H.A., Dekaban, G.A., Weaver, L.C., 1998. Identification of lamina V and VII interneurons presynaptic to adrenal sympathetic preganglionic neurons in rats using a recombinant herpes simplex virus type 1. Neuroscience 85, 863-872.

Coote, J.H., Downman, C.B.B., 1966. Central pathways of some autonomic reflex discharges. J. Physiol. 183, 714-729.

Dampney, R.A.L., 1994. Functional organization of central pathways regulating the cardiovascular system. Physiol. Rev. 74, 323-364.

Davis, B.M., Cabot, J.B., 1984. Substance P-containing pathways to avian sympathetic preganglionic neurons: evidence for major spinal-spinal circuitry. J. Neurosci. 4, 2145-2159.

Dembowsky, K., Czachurski, J., Seller, H., 1985. An intracellular study of the synaptic input to sympathetic preganglionic neurones of the third thoracic segment of the cat. J. Auton. Nerv. Syst. 13, 201-244.

Deuchars, S.A., Brooke, R.E., Frater, B., Deuchars, J., 2001. Properties of interneurones in the intermediolateral cell column of the rat spinal cord: role of the potassium channel subunit Kv3.1. Neuroscience 106, 433-446.

Deuchars, S.A., Spyer, K.M., Brooks, P.A., Gilbey, M.P., 1995. A study of sympathetic preganglionic neuronal activity in a neonatal rat brainstem-spinal cord preparation. J. Auton. Nerv. Syst. 52, 51-63.

12. INTERNEURONAL INPUTS TO SYMPATHETIC PREGANGLIONIC NEURONS: EVIDENCE FROM TRANSECTED SPINAL CORD

Deuchars, S.A., Spyer, K.M., Gilbey, M.P., 1997. Stimulation within the rostral ventrolateral medulla can evoke monosynaptic GABAergic IPSPs in sympathetic preganglionic neurons in vitro. J. Neurophysiol. 77, 229-235.

Dun, N.J., Mo, N., 1989. Inhibitory postsynaptic potentials in neonatal rat sympathetic preganglionic neurones in vitro. J. Physiol. 410, 267-281.

Gebber, G.L., McCall, R.B., 1976. Identification and discharge patterns of spinal sympathetic interneurons. Am. J. Physiol. 231, 722-733.

Goodchild, A.K., Llewellyn-Smith, I.J., Sun, Q.-J., Chalmers, J., Cunningham, A.M., Pilowsky, P.M., 2000. Calbindin-immunoreactive neurons in the reticular formation of the rat brainstem: Catecholamine content and spinal projections. J. Comp. Neurol. 424, 547-562.

Grkovic, I., Anderson, C.R., 1997. Calbindin D28K-immunoreactivity identifies distinct subpopulations of sympathetic pre- and postganglionic neurons in the rat. J. Comp. Neurol. 386, 245-259.

Hayar, A., Guyenet, P.G., 1998. Pre- and postsynaptic inhibitory actions of methionine-enkephalin on identified bulbospinal neurons of the rat RVL. J. Neurophysiol. 80, 2003-2014.

Joshi, S., LeVatte, M.A., Dekaban, G A., Weaver, L.C., 1995. Identification of spinal interneurons antecedent to adrenal sympathetic preganglionic neurons using transsynaptic transport of herpes simplex virus type-1. Neuroscience 65, 893-903.

Krassioukov, A., Johns, D.G., Schramm, L.P., 2002. Sensitivity of sympathetically correlated spinal interneurons, renal sympathetic nerve activity, and arterial pressure to somatic and visceral stimuli after chronic spinal cord injury. J. Neurotrauma 19, 1521-1530.

Krassioukov, A.V., Bunge, R.P., Pucket, W.R., Bygrave, M.A., 1999. The changes in human spinal sympathetic preganglionic neurons after spinal cord injury. Spinal Cord. 37, 6-13.

Krassioukov, A.V., Weaver, L.C., 1995. Episodic hypertension due to autonomic dysreflexia in acute and chronic spinal cord-injured rats. Am. J. Physiol. 268, H2077-H2083.

Krassioukov, A.V., Weaver, L.C., 1996. Morphological changes in sympathetic preganglionic neurons after spinal cord injury in rats. Neuroscience 70, 211-225.

Krenz, N.R., Meakin, S.O., Krassioukov, A.V., Weaver, L.C., 1999. Neutralizing intraspinal nerve growth factor blocks autonomic dysreflexia caused by spinal cord injury. J. Neurosci. 19, 7405-7414.

Krenz, N.R., Weaver, L.C., 1998a. Changes in the morphology of sympathetic preganglionic neurons parallel the development of autonomic dysreflexia after spinal cord injury in rats. Neurosci. Lett. 243, 61-64.

Krenz, N.R., Weaver, L.C., 1998b. Sprouting of primary afferent fibers after spinal cord transection in the rat. Neuroscience 85, 443-458.

Llewellyn-Smith, I.J., Arnolda, L.F., Pilowsky, P.M., Chalmers, J.P., Minson, J.B., 1998. GABA- and glutamate-immunoreactive synapses on sympathetic preganglionic neurons projecting to the superior cervical ganglion. J. Auton. Nerv. Syst. 71, 96-110.

Llewellyn-Smith, I.J., Cassam, A.K., Krenz, N.R., Krassioukov, A.V., Weaver, L.C., 1997. Glutamate- and GABA-immunoreactive synapses on sympathetic preganglionic neurons caudal to a spinal cord transection in rats. Neuroscience 80, 1225-1235.

Llewellyn-Smith, I.J., Martin, C.L., Minson, J.B., 2002. Glutamate and GABA content of calbindin-immunoreactive nerve terminals in the rat intermediolateral cell column. Auton. Neurosci. 98, 7-11.

Llewellyn-Smith, I.J., Minson, J.B., 1992. Complete penetration of antibodies into vibratome sections after glutaraldehyde fixation and ethanol treatment: Light and electron microscopy for neuropeptides. J. Histochem. Cytochem. 40, 1741-1749.

Llewellyn-Smith, I.J., Minson, J.B., Pilowsky, P.M., 1992b. Retrograde tracing with cholera toxin B-gold or with immunocytochemically detected cholera toxin B in the central nervous system. In: P.M.Conn (Ed.), Methods in neuroscience. Academic Press, New York, pp. 180-201.

Llewellyn-Smith, I.J., Phend, K.D., Minson, J.B., Pilowsky, P.M., Chalmers, J.P., 1992a. Glutamate immunoreactive synapses on retrogradely labelled sympathetic neurons in rat thoracic spinal cord. Brain Res. 581, 67-80.

Llewellyn-Smith, I.J., Pilowsky, P., Minson, J.B., 1993. The tungstate-stabilized tetramethylbenzidine reaction for light and electron microscopic immunocytochemistry and for revealing biocytin-filled neurons. J. Neurosci. Methods 46, 27-40.

Llewellyn-Smith, I.J., Weaver, L.C., 2001. Changes in synaptic inputs to sympathetic preganglionic neurons after spinal cord injury. J. Comp. Neurol. 435, 226-240.

Maiorov, D.N., Krenz, N.R., Krassioukov, A.V., Weaver, L.C., 1997. Role of spinal NMDA and AMPA receptors in episodic hypertension in conscious spinal rats. Am. J. Physiol. 273, H1266-H1274.

Mannard, A., Polosa, C., 1973. Analysis of background firing of single sympathetic preganglionic neurons of cat cervical nerve. J. Neurophysiol. 36, 398-408.

Marsh, D.R., Wong, S.T., Meakin, S.O., MacDonald, J.I S., Hamilton, E.F., Weaver, L.C., 2002. Neutralizing intraspinal nerve growth factor with a trkA-IgG fusion protein blocks the development of autonomic dysreflexia in a clip-compression model of spinal cord injury. J. Neurotrauma 19, 1531-1543.

Mathias, C.J. Frankel, H.L., 1992. The cardiovascular system in tetraplegia and paraplegia. In: H.L.Frankel (Ed.), Handbook of Clinical Neurology. Elsevier Science Publishers pp. 435-456.

McCall, R.B., Gebber, G.L., Barman, S.M., 1977. Spinal interneurons in the baroreceptor reflex arc. Am. J. Physiol. 232, H657-H665.

Minson, J.B., Llewellyn-Smith, I.J., Arnolda, L.F., 2001. Neuropeptide Y mRNA expression in interneurons in rat spinal cord. Auton. Neurosci. 93, 14-20.

Morrison, S.F., 2001. Differential regulation of sympathetic outflow to vasoconstrictor and thermoregulatory effects. Ann. N.Y. Acad. Sci. 940, 286-298.

Romagnano, M.A., Braiman, J., Loomis, M., Hamill, R.W., 1987. Enkephalin fibers in autonomic nuclear regions: Intraspinal vs. supraspinal origin. J. Comp. Neurol. 266, 319-331.

Shen, E., Mo, N., Dun, N.J., 1990. APV-sensitive dorsal root afferent transmission to neonate rat sympathetic preganglionic neurons in vitro. J. Neurophysiol. 64, 991-999.

Spanswick, D., Pickering, A.E., Gibson, I.C., Logan, S.D., 1994. Inhibition of sympathetic preganglionic neurons by spinal glycinergic interneurons. Neuroscience 62, 205-216.

Stornetta, R.L., Sevigny, C.P., Schreihofer, A.M., Rosin, D.L., Guyenet, P.G., 2002. Vesicular glutamate transporter DNPI/VGLUT2 is expressed by both C1 adrenergic and nonaminergic presympathetic vasomotor neurons of the rat medulla. J. Comp. Neurol. 444, 207-220.

Strack, A.M., Sawyer, W.B., Platt, K.B., Loewy, A.D., 1989. CNS cell groups regulating the sympathetic outflow to adrenal gland as revealed by transneuronal cell body labeling with pseudorabies virus. Brain Res. 491, 274-296.

Teasell, R.W., Arnold, J.M., Krassioukov, A., Delaney, G.A., 2000. Cardiovascular consequences of loss of supraspinal control of the sympathetic nervous system after spinal cord injury. Arch. Phys. Med. Rehabil. 81, 506-516.

12. INTERNEURONAL INPUTS TO SYMPATHETIC PREGANGLIONIC NEURONS: EVIDENCE FROM TRANSECTED SPINAL CORD

Todd, A.J., Sullivan, A.C., 1998. Light microscopic study of the co-existence of GABA-like and glycine-like immunoreactivities in the spinal cord of the rat. J. Comp. Neurol. 296, 496-505.

Weaver, L.C., Cassam, A.K., Krassioukov, A.V., Llewellyn-Smith, I.J., 1997. Changes in immunoreactivity for growth associated protein-43 suggest reorganization of synapses on spinal sympathetic neurons after cord transection. Neuroscience 81, 535-551.

Weaver, L.C., Marsh, D.R., Gris, D., Meakin, S.O., Dekaban, G.A., 2002. Central mechanisms for autonomic dysreflexia after spinal cord injury. Prog. Brain Res. 137, 83-95.

Weaver, L.C., Verghese, P., Bruce, J.C., Fehlings, M.G., Krenz, N.R., Marsh, D.R., 2001. Autonomic dysreflexia and primary afferent sprouting after clip-compression injury of the rat spinal cord. J. Neurotrauma 18, 1107-1119.

Wong, S.T., Atkinson, B.A., Weaver, L.C., 2000. Confocal microscopic analysis reveals sprouting of primary afferent fibres in rat dorsal horn after spinal cord injury. Neurosci. Lett. 296, 65-68.

Zagon, A., Smith, A.D., 1993. Monosynaptic projections from the rostral ventrolateral medulla oblongata to identified sympathetic preganglionic neurons. Neuroscience 54, 729-743.

Chapter 13

SYMPATHETIC PREGANGLIONIC NEURONS: ELECTRICAL PROPERTIES AND RESPONSE TO NEUROTRANSMITTERS

G. Cristina Brailoiu and Nae J. Dun
Department of Pharmacology, James H. Quillen College of Medicine, East Tennessee State University, Johnson City, TN 37614, USA

Abstract: In response to external and internal signals, the sympathetic nervous system executes complex and highly differentiated commands to multiple target organs including the heart and blood vessels. For example, humans as well as animals, when presented with a threatening stimulus, display a series of physiological responses that may include increased cardiac output, pupillary dilation, constriction of cutaneous and splanchnic vasculatures, dilation of skeletal muscle vasculatures, and increased glucose utilization. The completion of this complex, yet organized, task is dependent on the temporal and spatial coordination of hundreds or thousands of synaptic events at different levels of the neuraxis. The efferent portion of the sympathetic nervous system consists of two sets of serially connected neurons: the preganglionic neurons whose cell bodies reside in the thoracolumbar spinal cord and the postganglionic neurons that aggregate in para- or pre-vertebral ganglia and adrenal medulla. Viewed in this context, sympathetic preganglionic neurons (SPN) in the spinal cord are the final common site where supraspinal and segmental inputs related to the sympathetic nerve activity are integrated and processed prior to transmission to the peripheral sympathetic ganglia. The frequency and/or pattern of discharges encoded by SPN is an important determinant of spinal sympathetic outflow to the sympathetic ganglia, which is in turn dictated by the intrinsic membrane properties of SPN and interplay between synaptically released transmitters/modulators and circulating neurohumorals. This chapter represents a selective overview of membrane properties and actions of putative transmitters and modulators, which are derived principally from studies conducted on SPN in spinal cord slice preparations.

Key words: glutamate, intermediolateral cell column, glycine, gaba, neuropeptides

INTRODUCTION

In the rat, where most of the current knowledge concerning the electrical behavior and synaptic transmission is derived, a population of SPN in the neonate is intrinsically active (Spanwick & Logan, 1990a; Shen et al., 1994). Consequently the activity of these neurons may be intrinsically programmed. Putative transmitters including 5-HT and norepinephrine can effectively modify the rate and pattern of spontaneous discharges of these neurons or enable a quiescent neuron to discharge rhythmically by suppressing various K^+ currents, suggesting that the electrical behavior of intrinsically active or quiescent neurons may be subject to modification by putative transmitters (Shen et al., 1994). Evidence is reasonably strong that excitatory and inhibitory amino acids are the principal transmitters mediating excitatory and inhibitory synaptic transmission to the SPN. Immunohistochemical studies reveal a plethora of putative transmitters/modulators including 5-HT, epinephrine, dopamine, enkephalin, nociceptin, substance P, pituitary adenylate cyclase activating peptide (PACAP), oxytocin, and vasopressin, that are expressed in nerve fibers projecting into the spinal sympathetic nuclei. Nitric oxide synthase, which converts L-arginine into nitric oxide (NO), is present in a subset of SPN (Dun et al., 1993). Multiple transmitters impinging upon SPN may represent chemical coding of different descending and/or segmental pathways converging onto SPN. Thus, a specific stimulus may activate one or more pathways impinging upon SPN, which in turn carry out the final adjustment of central sympathetic outflow to target tissues including vasculatures and myocardium.

LOCATION

SPN, situated exclusively in the spinal cord between cervical segment C8 and lumbar segment L2, are topographically organized in four distinct nuclei: intermediolateral cell column (the nucleus intermediolateralis thoraco-lumbalis pars principalis, IML), lateral funicular area (the nucleus intermediolateralis thoraco-lumbalis pars funicularis, LF), intercalated cell group (the nucleus intercalatus spinalis, IC) and central autonomic nucleus (the nucleus intercalatus pars paraependymalis, CA) (Petras and Cummings, 1972). A horizontal section of rat thoraco-lumbar spinal cord labeled with neuronal nitric oxide synthase (NOS) antiserum, revealing the cytoarchitecture of the four sympathetic preganglionic nuclei, is illustrated in Fig. 1A. The majority of SPN are located in the IML, which occupies the lateral horn of the spinal gray matter. SPN are arranged in target-specific horizontal columns in the spinal cord, projecting to the superior cervical ganglion, stellate ganglion, paravertebral and prevertebral ganglia and to the adrenal medulla; some SPN, however, project to both the superior cervical and stellate ganglion (Pyner and Coote, 1994).

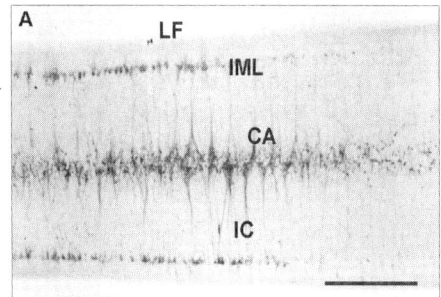 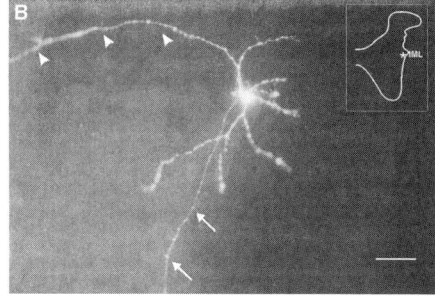

Figure 1. Anatomic location of rat sympathetic preganglionic neurons. A. Sympathetic preganglionic neurons revealed by nitric oxide synthase-immunoreactivity in a horizontal section of rat thoraco-lumbar spinal cord; SPN are arranged in a ladder-like cytoarchitecture, with four distinct nuclei, LF, IML, CA and IC. B. Location and morphology of a single SPN intracellularly labeled with Lucifer Yellow. Arrowheads and arrows indicate a medially projecting dendrite and ventrally projecting axon, respectively. Calibration scale: 250 µm in A and 50 µm in B.

DEVELOPMENT

SPN and spinal somatic motor neurons from the thoracic spinal cord develop synchronously on the embryonic days 11-12 (E11-12) (Phelps et al., 1993) and subsequently migrate from the ventricular zone to form a common primitive motor column by E13. SPN translocate dorsally and medially from their original position, and are probably guided, by circumferential axons and glial elements to reach their final location between E14-18 (Markham and Vaughn, 1991; Phelps et al., 1993; Schober and Unsicker, 2001). During the pre- and postnatal development, cell death, which appears to be regulated by postsynaptic targets and local factors, occurs within a specific time window for a distinct neuronal population (Burek and Oppenheim, 1996). The developmental cell death process appears to differ between somatic and autonomic neurons in that SPN are not dependent on target-derived trophic factors during embryonic and postnatal development (E12-P22) (Wetts and Vaughn, 1998).

MORPHOLOGY

A number of studies utilizing retrograde labeling and/or histochemical techniques including horseradish peroxidase, DiI, Fluorogold, Fast Blue, cholera toxin B subunit, pseudorabies virus, recombinant herpes simplex virus type 1, choline acetyltransferase (ChAT) immunohistochemistry, acetylcholinesterase histochemistry or intracellular labeling techniques with Lucifer Yellow or biocytin, have provided useful information relative to the topography and morphology of SPN in mammals (Chung et al., 1975; Barber et al., 1984; Dembowsky et al., 1985a; Forehand, 1990; Shen and Dun, 1990; Strack et al., 1989; Hosoya et al., 1994; Anderson and Edwards, 1994; Forehand et al., 1994; Pyner and Coote, 1994; Sah and McLachlan, 1995; Clarke et al., 1998; Krassioukov et al., 1998; Deuchars et al., 2001b). The morphology of SPN appears to vary among neurons of the four nuclei. For example, SPN in the IML appear mostly fusiform, 20-25 µm in diameter; a few are round and small or multipolar and large (Barber et al., 1984; Dembowsky et al., 1985a; Shen and Dun, 1990; Forehand, 1990). The dendritic tree is composed of 4-8 primary dendrites with spines, oriented principally in the rostro-caudal direction; while some dendrites project mediolaterally toward the central canal. The axon of IML neurons arises from the cell body or as a direct extension of a first- or second-order dendrite and projects medially and ventrally along the border of gray matter, then turns laterally in the ventral horn, to exit via the ventral root (Fig. 1B). The medially oriented dendrites observed in the neonatal rat SPN have been suggested to be a developmental feature such that the majority of these dendrites undergo a re-orientation to a longitudinal direction in mature animals (Schramm et al., 1976). SPN from the CA have round or triangular perikarya, 20-30 µm in diameter; and their dendrites form a loose, longitudinal plexus throughout the nucleus (Hosoya et al., 1994). In the LF, SPN appear medium fusiform or large triangular, with dendrites extending laterally or medially toward the IML (Barber et al., 1984). The IC is composed of fusiform or oval neurons positioned between the IML and CA (Barber et al., 1984). Interneurons, which occur in close proximity to or between SPN, are smaller cells (15 µm in diameter), solitary or in clusters, with a more variable morphology. They are generally multipolar with varicose dendrites extending in many directions (Shen and Dun, 1990; Sah and McLachlan, 1995; Clarke et al., 1998; Deuchars et al., 2001b).

PHENOTYPES

SPN are cholinergic, as the acetylcholine synthetizing enzyme ChAT is detectable in these neurons (Barber et al., 1984). Subpopulations of SPN express immunoreactivity to one or more of the following neuropeptides: enkephalin, neurotensin, somatostatin, substance P, corticotropin-releasing factor (Krukoff et al., 1985; Krukoff, 1986, 1987), vasoactive intestinal peptide, peptide histidine isoleucine amide (Baldwin et al., 1991), calcitonin gene-related peptide (Yamamoto et al., 1989; Gibbins, 1992), calretinin (Grkovic and Anderson, 1995), secretoneurin (Dun et al., 1997a), nociceptin (Dun et al., 1997b), and cocaine and amphetamine-regulated transcript (CART) (Dun et al., 2000). *In situ* hybridization studies reveal the presence of PACAP mRNA in some of the rat SPN (Beaudet et al., 1998). On the other hand, immunohistochemical studies failed to detect PACAP-immunoreactivity in SPN nor in the paravertebral ganglia of the rats (Dun et al., 1996a, b). In addition, the majority of SPN contain the enzyme NOS (Dun et al., 1993). The expression of diverse neuropeptides and enzymes in subsets of SPN suggests that they are chemically coded and that they are arranged in a target-specific manner.

SYNAPTIC INPUTS

SPN receive synaptic inputs arising from a multitude of supraspinal sites as well as segmental afferents. Anatomical and/or neurophysiological studies identify five major supraspinal cell groups that innervate SPN; these include the rostral ventrolateral medulla (RVLM), rostral ventromedial medulla (RVMM), caudal raphe nuclei, pontine A5 cell group, and hypothalamic paraventricular nucleus. Although less attention has been paid to the intraspinal pathways, it is clear that SPNs receive excitatory and inhibitory inputs from spinal interneurons (Dun and Mo, 1989). Ultrastructural studies have shown synaptic contacts between retrogradely labeled SPN and nerve terminals containing monoamines and neuropeptides; e.g., substance P, neuropeptide Y, somatostatin, enkephalin, neurotensin, VIP, oxytocin, and cholecystokinin (Chiba and Masuko, 1987; Bacon and Smith, 1988; Milner et al., 1988; Chiba, 1989; Pilowsky et al., 1992; Hosoya et al., 1995; Jensen et al., 1995), or amino acids; e.g., glutamate (Llewellyn-Smith et al., 1992, 1998), γ-aminobutyric (GABA) (Bogan et al., 1989), and glycine (Cabot et al., 1992).

ELECTROPHYSIOLOGY

Membrane Properties

Intracellular or whole-cell patch clamp recordings have been made from SPN of neonatal rats, adult cats and guinea pigs in vitro and in vivo (Dembowsky et al., 1986, Yoshimura et al., 1986b; Dun and Mo, 1988; Spanswick and Logan, 1990b; Dun et al., 1992; Inokuchi et al., 1993; Miyazaki et al., 1996). SPN of immature rats have a resting membrane potential between -50 and -75 mV; the action potential amplitude and duration varies between 60 and 85 mV and 2-4.5 ms, respectively. The input membrane resistance and time constant obtained in intracellular recordings are lower (70-150 MΩ and 4-12 ms) than those (200-1200 MΩ, 25-65 ms) reported in the whole-cell patch studies (Shen et al., 1990; Spanswick and Logan, 1990b; Wu and Dun, 1993, 1995, 1996; Spanswick et al., 1995; Miyazaki et al., 1996). Higher input resistance and time constant are likely to be due to less damage to the membrane integrity in the whole-cell configuration. A notch or shoulder, which is attributed to an influx of Ca^{2+}, is noted on the falling phase of the action potential of some of the SPN (Dembowsky et al., 1986). The action potential is followed by a prolonged after-hyperpolarization (AHP), lasting up to a few seconds. In the case of cat SPN, the AHP consists of two components: a fast and slow AHP, which may be caused, respectfully, by an increase of K^+ and Ca^{2+}-activated K^+ conductance (Yoshimura et al., 1986a). Functionally, the amplitude of synaptic responses is attenuated during the course of AHP, suggesting that the later plays an important role in influencing the input and output ratio of SPN (Yoshimura et al., 1986a).

An interesting feature that has been noted in all the studies involving neonatal rats is the occurrence of spontaneous firing in a population of SPN. Spontaneous firing is strictly defined as non synaptically driven action potentials that occur in a regular or irregular fashion (Spanswick and Logan, 1990a; Shen et al., 1994). The cellular or ionic basis underlying spontaneous discharges in SPN has yet to be fully defined. Electrical synapses have been documented in neonatal rat SPN (Nolan et al., 1999). The electrotonic coupling could explain the earlier findings regarding spontaneous rhythmic activity observed in a population of SPN (Spanswick and Logan, 1990a; Shen et al., 1994) and the dye-coupled cells that were revealed when SPN were filed with Lucifer Yellow (Shen and Dun, 1990). Alternatively, divalent ions, such as barium and cesium that block potassium channels, can convert a silent to a rhythmically firing SPN in vitro (Shen et al., 1994). Putative transmitters including monoamines and peptides are known to block various potassium channels. It is possible that endogenously released amines and peptides, by suppressing potassium conductance, may induce rhythmic discharge in an otherwise silent SPN. An interesting question that remains to be addressed is whether or not rhythmically firing SPN exhibit a specific transmitter phenotype(s) and/or innervate a specific sympathetic ganglion.

Owing to the difficulty in making satisfactory intracellular or whole-cell patch recording from SPN of adult rat spinal cord, information on the membrane properties of adult rat SPN is limited. In particular, the issue whether or not spontaneous rhythmic discharge occurs in a population of adult rat SPN has not been satisfactorily resolved.

Similarly, information relative to interneurons antecedent to SPN in the spinal cord is limited. Injection of herpes simplex virus type I into the adrenal gland of hamsters labeled a group of small diameter cells in lamina V and VII (Joshi et al., 1995). These cells are morphologically distinct from SPN. In the neonatal rat spinal cord slices, a small number of neurons intracellularly labeled with Lucifer Yellow are not activated antidromically and, are therefore, tentatively classified as interneurons (Shen and Dun, 1990).

Membrane Currents

Several types of voltage-dependent K$^+$ currents have been identified in SPNs: delayed rectifier current (I_K), Ca^{2+}-dependent transient current (I_C), Ca^{2+}-dependent sustained current (I_{AHP}), A-current (I_A), M-current (I_M), inward rectifying current (I_{ir}) and quinine-sensitive outward rectifier (Brown and Selyanko, 1985; McFarlane and Cooper, 1992; Bordey et al., 1995; Sah and McLachlan, 1995; Miyazaki et al., 1996, Wilson et al., 2002). A more recent study shows that interneurons in the vicinity of the IML, but not SPN, possess a K$^+$ current sensitive to low concentrations of 4-aminopyridine and tetraethylammonium (Deuchars et al., 2001b). The presence of this voltage-gated K$^+$ channel expressing the Kv3.1 subunits confers the interneuron a fast repolarization phase, consequently a shorter duration of action potential, as compared to the SPN (Deuchars et al., 2001b).

SPNs possess low threshold voltage activated, T types, and high threshold Ca^{2+}currents, which can be modulated by putative neurotransmitters (Wilson et al., 2002).

Synaptic Transmission

On-going sub-threshold excitatory postsynaptic potentials (EPSPs) and supra-threshold discharges are regular features of intracellular recordings from SPN in vivo (Dembowsky et al., 1985b; McLachlan and Hirst, 1980), indicating that neurons presynaptic to SPN are tonically active. On-going synaptic activities sensitive to the non-NMDA glutamate receptor antagonist can also be recorded in SPN of transverse rat spinal cord slices (Shen et al., 1994). Four types of synaptic responses are reported in SPN: a fast and slow EPSP and a fast and slow inhibitory postsynaptic potential (IPSP) (Dembowsky et al., 1985b; Dun and Mo, 1989; Shen et al., 1990; Inokuchi et al., 1992a, b). The fast EPSP can be recorded from almost every SPN, whereas other types of synaptic responses are not elicited in all the SPNs tested. Particularly, slow EPSPs and slow IPSPs are infrequently encountered (Yoshimura et al., 1987a, b).

EPSPs

Electrical stimulation of the dorsal root, lateral funiculus or focal stimulation evokes fast EPSPs, referred to herein as EPSPs, in IML neurons from rat and/or cat mediated by excitatory amino acids acting on AMPA/kainate or NMDA receptors (Mo and Dun, 1987; Shen et al., 1990; Inokuschi et al., 1992b; Krupp and Feltz, 1995). Importantly, EPSPs evoked by stimulation of dorsal roots are electrophysiologically and pharmacologically distinct from those elicited by stimulation of lateral funiculus (Mo and Dun, 1987; Shen et al., 1990; Krupp and Feltz, 1995). Accordingly, the synaptic latency of EPSPs elicited by dorsal root stimulation is long, viz. 5 to 15 ms, implying a di- or polysynaptic pathway. Moreover, dorsal root-evoked EPSPs exhibit a relatively slow rising and falling phase (Fig. 2A) as compared to EPSPs elicited by lateral funiculus stimulation (Fig. 2B). Further, dorsal root-evoked EPSPs are enhanced in Mg^{2+}-free solution, and show negative slope conductance, which are characteristics of NMDA receptor-mediated responses (Mo and Dun, 1987). A significant pharmacological distinction is that EPSPs elicited by dorsal root stimulation are largely eliminated by NMDA receptor antagonists (Fig. 2A) and those by lateral funiculus stimulation are largely blocked by non-NMDA receptor antagonists (Fig. 2B). It is estimated that approximately 75% and 29% of the total charge of excitatory postsynaptic currents (EPSCs) induced by dorsal root or lateral funiculus stimulation is carried by, respectively, NMDA and non-NMDA components (Krupp and Feltz, 1995). In a rat brainstem-spinal cord preparation, stimulation of rostral ventrolateral

13. SYMPATHETIC PREGANGLIONIC NEURONS: ELECTRICAL PROPERTIES AND RESPONSE TO NEUROTRANSMITTERS

medulla elicits a short latency EPSP in the SPN that is blocked by non-NMDA receptor antagonist or by a combination of NMDA and non-NMDA receptor antagonists (Deuchars et al., 1995). Collectively, these findings suggest that SPN receive two distinct glutamatergic inputs of supraspinal and sensory (or segmental) origin and that NMDA and AMPA/kainate receptors may be topographically distributed on SPN.

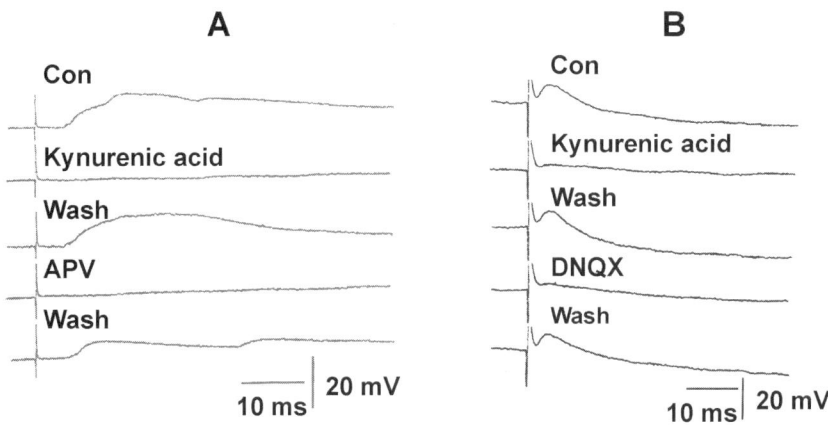

Figure 2. Pharmacology of EPSPs evoked by stimulation of dorsal root or lateral funiculus. A. EPSPs evoked by dorsal root stimulation show a long synaptic latency, i.e., 5ms, slowly rising and falling phase, and are eliminated by the non-selective glutamate receptor antagonist kynurenic acid or the NMDA receptor antagonist APV. B. EPSPs evoked by lateral funiculus stimulation show a short latency, fast rising and falling phase, and are blocked by the non-selective glutamate receptor antagonist kynurenic acid or the AMPA/kainate receptor antagonist DNQX.

In addition to the fast EPSP, which is the primary transmission pathway, repetitive nerve stimulation elicits a slow developing, longer lasting (seconds to minutes) depolarization, termed slow EPSP, in a population of SPN. The slow EPSP has the following characteristics: 1) the amplitude and duration are graded with stimulus intensity and summated with increasing number of pulses; 2) the response is increased with depolarization, decreased with hyperpolarization and has a reversal around -90mV; 3) the response is accompanied by increased membrane excitability (Dun et al., 1992). In contrast to the fast EPSP where glutamate or a closely related amino acid is the likely transmitter, the transmitter(s) mediating the slow EPSP appears to be diverse. Catecholamines and neuropeptides are potential candidates responsible for generating the slow EPSP, because the action of these putative transmitters mimics the slow EPSP. For example, norepinephrine acting on adrenergic α_1-receptors may mediate the slow EPSP evoked in the cat SPN (Yoshimura et al., 1987a). In neonate rats, the slow EPSP evoked in a population of SPN may be mediated by substance P (Dun and Mo, 1988). Vasopressin is another possible candidate in mediating slow EPSPs (Ma and Dun, 1985). The physiological role of slow EPSPs remains a topic of interest. As the membrane excitability is invariably increased during the course of slow EPSPs, the latter may serve to prime the target neuron to incoming sub-threshold EPSPs.

IPSPs

Stimulation of the dorsal roots evoked a fast, strychnine-sensitive IPSP in approximately 50% of rat SPN (Dun and Mo, 1989). Fast IPSPs evoked by focal stimulation in cat SPN are mediated by glycine or GABA or glycine and GABA (Inokuchi et al., 1992a). IPSPs can be induced in SPN in spinal cord slices by the application of excitatory amino acid receptor agonists, indicating that glycine and/or GABA is released from inhibitory interneurons bearing excitatory amino acid receptors (Dun and Mo, 1989; Inokuchi et al., 1992a; Spanswick et al., 1994). The location of inhibitory interneurons in relationship to SPN is not clear.

A hyperpolarization lasting for seconds, termed slow IPSP, has been reported in cat SPN (Yoshimura et al., 1987b, 1997b). The response is thought to be mediated by norepinephrine acting on α_2-receptors (Yoshimura et al., 1987b, 1997b). A corresponding response has yet to be demonstrated in the rat SPN.

Modulation of Synaptic Transmission

Nerve fibers immunoreactive to a large number of putative transmitters including monoamines and neuropeptides are noted in the spinal sympathetic nuclei. Their physiological function is largely unknown. Pharmacological studies show that a number of peptides and amines when applied to SPN either depolarize or hyperpolarize the membrane. In many instances, the membrane depolarization is associated with increased membrane resistance due to closure of certain K^+ channels, leading to increased membrane excitability and enhanced synaptic responses. Few putative transmitters have been shown to hyperpolarize SPN, for example, catecholamines acting on α_2-receptors and the opioid-like peptide nociceptin hyperpolarize SPN (Yoshimura et al., 1987b; Miyazaki et al., 1989; Lai et al., 2000). In addition to membrane effects, several putative transmitters increase or decrease the amplitude of EPSPs or IPSPs, probably by a presynaptic action. For example, glutamate acting on metabotropic receptors (Wu and Dun, 1993), GABA acting on $GABA_B$ receptors (Wu and Dun, 1992), norepinephrine (Miyazaki et al., 1998), nociceptin (Lai et al., 2000) and adenosine (Deuchars et al., 2001a) suppress glutamate release, thereby attenuating excitatory or inhibitory transmission.

The presence of PACAP in nerve fibers projecting to the lateral horn area was first demonstrated by immunohistochemical techniques (Dun et al., 1996a, b). Electrophysiological studies have revealed some unique features of the action of PACAP on SPN. PACAP at higher concentrations depolarizes a subset of SPN (Lai et al., 1997). At low concentrations, the peptide selectively increases the NMDA receptor-mediated depolarization in a cyclic AMP-dependent manner, but not the AMPA receptor-mediated response (Wu and Dun, 1997). As a corollary, PACAP may play a role in synaptic plasticity.

Nitric Oxide and Synaptic Modulation

SPN express neuronal NOS in rodent and primate spinal cords (Dun et al., 1993). NO has been implicated as a messenger molecule mediating a variety of cellular processes including long-term potentiation or depression of synaptic transmission (Ito and Karachot, 1990; Bohme et al., 1991; O'Dell et al., 1991; Schuman and Madison, 1991). The presence of NOS in SPN raised the possibility that NO may be released from these neurons and that it may influence excitatory and/or inhibitory synaptic transmission. In spinal cord slice preparations, repetitive discharge of SPN induced by a train of depolarizing current pulses was followed by a long-lasting increase of the amplitude of EPSPs, which could be prevented by pretreating the slices with the NOS inhibitors or bovine hemoglobin. More importantly, inclusion of the Ca^{2+} chelator BAPTA in the patch electrodes nullified the increase of EPSPs after repetitive discharge of SPN. These observations suggest that Ca^{2+} influx via voltage-gated channels, which activates NOS of SPN, resulting in a release of NO and potentiation of EPSPs (Wu and Dun, 1995). In addition to demonstrating a potentiation of EPSPs by NO (Wu et al., 1997), NO released from SPNs also enhances IPSPs (Wu and Dun, 1996). On the basis of these findings, it is hypothesized that NO may have a dual effect on synaptic transmission to SPN. During high frequency discharge of SPN, NO released from these neurons may function as a positive retrograde messenger to ensure adequate excitatory transmitter release. Concurrently,

NO may protect the neuron from excessive stimulation by increasing the liberation of inhibitory transmitters. The spatial relationship of the site of NO release to the excitatory and inhibitory synapse may determine the concentration of NO reaching the respective synapses, and thereby, the strength of increase of inhibitory or excitatory synaptic responses.

CONCLUSIONS

SPNs in the spinal cord are the final common site where supraspinal and segmental inputs related to sympathetic nerve activity are integrated and processed prior to transmitting to the sympathetic postganglionic neurons. The strategic importance of SPN in cardiovascular control is evidenced by an abrupt fall in blood pressure in an anesthetized animal or human upon intrathecal infusion of local anesthetics; e.g., lidocaine. As SPNs are the only link between the spinal cord and peripheral sympathetic ganglia, they must provide a tonic control over sympathetic postganglionic neurons whose axons in turn innervate vascular smooth muscles and cardiac muscles. SPN express a multitude of membrane conductances, particularly K^+, which confer SPN the ability to generate different and variable output patterns in response to various physiological and environmental stimuli.

In parallel, SPNs receive multiple inputs expressing diverse groups of neurotransmitters/ neuromodulators, which probably represent chemical coding of different descending and/or segmental pathways converging onto the SPN. For example, oxytocin and vasopressin-containing fibers may arise from parvocellular neurons in the parventricular nucleus, catecholaminergic fibers from neurons in the RVLM, 5-HT fibers from neurons in the caudal raphe nuclei, and substance P and enkephalin may arise from neurons in the ventral medulla (Dampney, 1994). NO may be released directly from SPN in an activity-dependent manner to facilitate the release of excitatory or inhibitory transmitters from incoming nerve fibers (Wu and Dun, 1996). Thus, a specific stimulus may activate one or more pathways that impinge upon SPN, which in turn carry out the final adjustment of central sympathetic outflow that dictates the activity of target tissues including vasculatures and myocardium.

ACKNOWLEDGEMENTS

This work was supported by the National Institutes of Health grants NS18710 and HL51314.

REFERENCES

Anderson, C.R., Edwards, S.L., 1994. Intraperitoneal injections of Fluorogold reliably labels all sympathetic preganglionic neurons in the rat. J. Neurosci. Methods 53, 137-141.

Bacon, S.J., Smith, A.D., 1988. Preganglionic sympathetic neurones innervating the rat adrenal medulla: immunocytochemical evidence of synaptic input from nerve terminals containing substance P, GABA or 5-hydroxytryptamine. J. Auton. Nerv. Syst. 24, 97-122.

Baldwin, C., Sasek, C.A., Zigmond, R.E., 1991. Evidence that some preganglionic sympathetic neurons in the rat contain vasoactive intestinal peptide- or peptide histidine isoleucine amide-like immunoreactivities. Neurosci. 40, 175-184.

Barber, R.P., Phelps, P.E., Houser, C.R., Crawford, G.D., Salvaterra, P.M., Vaughn, J.E., 1984. The morphology and distribution of neurons containing choline acetyltransferase in the adult rat spinal cord: an immunocytochemical study. J. Comp Neurol. 229, 329-346.

Beaudet, M.M., Braas, K.M., May, V., 1998. Pituitary adenylate cyclase activating polypeptide (PACAP) expression in sympathetic preganglionic projection neurons to the superior cervical ganglion. J. Neurobiol. 36, 325-336.

Bogan, N., Mennone, A., Cabot, J.B., 1989. Light microscopic and ultrastructural localization of GABA-like immunoreactive input to retrogradely labeled sympathetic preganglionic neurons. Brain Res. 505, 257-270.

Bohme, G.A., Bon, C., Stutzmann, J.M., Doble, A., Blanchard J.C., 1991. Possible involvement of nitric oxide in long-term potentiation. Eur. J. Pharmacol. 199, 379-381.

Bordey, A., Feltz, P., Trouslard, J., 1995. Kinetics of A-currents in sympathetic preganglionic neurones and glial cells. Neuroreport 7, 37-40.

Brown, D.A., Selyanko, A.A., 1985. Membrane currents underlying the cholinergic slow excitatory post-synaptic potential in the rat sympathetic ganglion. J. Physiol. 365, 365-387.

Burek, M.J., Oppenheim, R.W., 1996. Programmed cell death in the developing nervous system. Brain Pathol. 6, 427-446.

Cabot, J.B., Alessi, V., Bushnell, A., 1992. Glycine-like immunoreactive input to sympathetic preganglionic neurons. Brain Res. 571, 1-18.

Chiba, T., 1989. Direct synaptic contacts of 5-hydroxytryptamine-, neuropeptide Y-, and somatostatin-immunoreactive nerve terminals on the preganglionic sympathetic neurons in the guinea pig. Neurosci. Lett. 105, 281-286.

Chiba, T., Masuko, S., 1987. Synaptic structure of the monoamine and peptide nerve terminals in the intermediolateral nucleus of the guinea pig thoracic spinal cord. J. Comp. Neurol. 262, 242-255.

Chung, J.M., Chung, K., Wurster, R.D., 1975. Sympathetic preganglionic neurons of the cat spinal cord: horseradish peroxidase study. Brain Res. 91, 126-131.

Clarke, H.A., Dekaban, G.A., Weaver, L.C., 1998. Identification of lamina V and VII interneurons presynaptic to adrenal sympathetic preganglionic neurons in rats using a recombinant herpes simplex virus type 1. Neurosci. 85, 863-872.

Dampney, R.A., 1994. Functional organization of central pathways regulating the cardiovascular system. Physiol. Rev. 74, 323-364.

Dembowsky, K., Czachurski, J., Seller, H., 1985a. Morphology of sympathetic preganglionic neurons in the thoracic spinal cord of the cat: an intracellular horseradish peroxidase study. J. Comp. Neurol. 238, 453-465.

Dembowsky, K., Czachurski, J., Seller, H., 1985b. An intracellular study of the synaptic input to sympathetic preganglionic neurones of the third thoracic segment of the cat. J. Auton. Nerv. Syst. 13, 201-244.

Dembowsky, K., Czachurski, J., Seller, H., 1986. Three types of sympathetic preganglionic neurones with different electrophysiological properties are identified by intracellular recordings in the cat. Pflugers Arch. 406, 112-120.
Deuchars, S.A., Morrison, S.F., Gilbey, M.P., 1995. Medullary-evoked EPSPs in neonatal rat sympathetic preganglionic neurones in vitro. J. Physiol. 487, 453-463.
Deuchars, S.A., Brooke, R.E., Deuchars, J. 2001a. Adenosine A1 receptors reduce release from excitatory but not inhibitory synaptic inputs onto lateral horn neurons. J. Neurosci. 21, 6308-6320.
Deuchars, S.A., Brooke, R.E., Frater, B., Deuchars, J., 2001b. Properties of interneurones in the intermediolateral cell column of the rat spinal cord: role of the potassium channel subunit Kv3.1. Neurosci. 106, 433-446.
Dun, N.J., Dun, S.L., Lin, H.H., Hwang, L.L., Saria, A., Fischer-Colbrie, R., 1997a. Secretoneurin-like immunoreactivity in rat sympathetic, enteric and sensory ganglia. Brain Res. 760, 8-16.
Dun, N.J., Dun, S.L., Hwang, L.L., 1997b. Nociceptin-like immunoreactivity in autonomic nuclei of the rat spinal cord. Neurosci. Lett. 234, 95-98.
Dun, N.J., Dun, S.L., Kwok, E.H., Yang, J., Chang, J., 2000. Cocaine- and amphetamine-regulated transcript-immunoreactivity in the rat sympatho-adrenal axis. Neurosci. Lett. 283, 97-100.
Dun, N.J., Dun, S.L., Wu, S.Y., Forstermann, U., Schmidt, H.H., Tseng, L.F., 1993. Nitric oxide synthase immunoreactivity in the rat, mouse, cat and squirrel monkey spinal cord. Neurosci. 54, 845-857.
Dun, N.J., Miyazaki, T., Tang, H., Dun, E.C., 1996a. Pituitary adenylate cyclase activating polypeptide immunoreactivity in the rat spinal cord and medulla: implication of sensory and autonomic functions. Neurosci. 73, 677-686.
Dun, N.J., Mo, N., 1988. In vitro effects of substance P on neonatal rat sympathetic preganglionic neurons. J. Physiol. 399, 321-333.
Dun, N.J., Mo, N., 1989. Inhibitory postsynaptic potentials in neonatal rat sympathetic preganglionic neurones in vitro. J. Physiol. 410, 267-281.
Dun, N.J., Tang, H., Dun, S.L., Huang, R., Dun E.C., Wakade A.R., 1996b. Pituitary adenylate cyclase activating polypeptide-immunoreactive sensory neurons innervate rat adrenal medulla. Brain Res. 716, 11-21.
Dun, N.J., Wu, S.Y., Shen, E., Miyazaki, T., Dun, S.L., Ren, C., 1992. Synaptic mechanisms in sympathetic preganglionic neurons. Can. J. Physiol. Pharmacol. 70, S86-S91.
Forehand, C.J., 1990. Morphology of sympathetic preganglionic neurons in the neonatal rat spinal cord: an intracellular horseradish peroxidase study. J. Comp. Neurol. 298, 334-342.
Forehand, C.J., Ezerman, E.B., Rubin, E., Glover, J.C., 1994. Segmental patterning of rat and chicken sympathetic preganglionic neurons: correlation between soma position and axon projection pathway. J. Neurosci. 14, 231-241.
Gibbins, I.L., 1992. Vasoconstrictor, vasodilator and pilomotor pathways in sympathetic ganglia of guinea-pigs. Neurosci. 47, 657-672.
Grkovic, I., Anderson, C.R., 1995. Calretinin-containing preganglionic nerve terminals in the rat superior cervical ganglion surround neurons projecting to the submandibular salivary gland. Brain Res. 684, 127-135.
Hosoya, Y., Matsukawa, Okado, N., Sugiura, Y., Kohno, K., 1995. Oxytocinergic innervation to the upper thoracic sympathetic preganglionic neurons in the rat. A light and electron microscopical study using a combined retrograde transport and immunocytochemical technique. Exp. Brain Res. 107, 9-16.
Hosoya, Y., Nadelhaft, I., Wang, D., Kohno, K., 1994. Thoracolumbar sympathetic preganglionic neurons in the dorsal commissural nucleus of the male rat: an

immunohistochemical study using retrograde labeling of cholera toxin subunit B. Exp. Brain Res. 98, 21-30.
Inokuchi, H., Masuko, S., Chiba, T., Yoshimura, M., Polosa, C., Nishi S., 1993. Membrane properties and dendritic arborization of the intermediolateral nucleus neurons in the guinea-pig thoracic spinal cord in vitro. J. Auton. Nerv. Syst. 43, 97-106.
Inokuchi, H., Yoshimura, M., Trzebski, A., Polosa, C., Nishi, S., 1992a. Fast inhibitory postsynaptic potentials and responses to inhibitory amino acids of sympathetic preganglionic neurons in the adult cat. J. Auton. Nerv. Syst. 41, 53-59.
Inokuchi, H., Yoshimura, M., Yamada, S., Polosa, C., Nishi, S., 1992b. Fast excitatory postsynaptic potentials and the responses to excitant amino acids of sympathetic preganglionic neurons in the slice of the cat spinal cord. Neurosci. 46, 657-667.
Ito, M., Karachot, L., 1990. Messengers mediating long-term desensitization in cerebellar Purkinje cells. Neuroreport 1, 129-132.
Jensen, I., Llewellyn-Smith, I.J., Pilowsky, P., Minson, J.B., Chalmers, J., 1995. Serotonin inputs to rabbit sympathetic preganglionic neurons projecting to the superior cervical ganglion or adrenal medulla. J. Comp. Neurol. 353, 427-438.
Joshi, S., Levatte, A., Dekaban, G.A., Weaver, L.C., 1995. Identification of spinal interneurons antecedent to adrenal sympathetic preganglionic neurons using trans-synaptic transport of herpes simplex virus type 1. Neurosci. 65, 893-903.
Krassioukov, A.V., Bygrave, M.A., Puckett, W.R., Bunge, R.P., Rogers, K.A., 1998. Human sympathetic preganglionic neurons and motoneurons retrogradely labeled with DiI. J. Auton. Nerv. Syst. 70, 123-128.
Krukoff, T.L., 1986. Segmental distribution of corticotropin-releasing factor-like and vasoactive intestinal peptide-like immunoreactivities in presumptive sympathetic preganglionic neurons of the cat. Brain Res. 382, 153-157.
Krukoff, T.L., 1987. Coexistence of neuropeptides in sympathetic preganglionic neurons of the cat. Peptides 8, 109-112.
Krukoff, T.L., Ciriello, J., Calaresu, F.R., 1985. Segmental distribution of peptide-like immunoreactivity in cell bodies of the thoracolumbar sympathetic nuclei of the cat. J. Comp. Neurol. 240, 90-102.
Krupp, J., Feltz, P., 1995. Excitatory postsynaptic currents and glutamate receptors in neonatal rat sympathetic preganglionic neurons in vitro. J. Neurophysiol. 73, 1503-1512.
Lai, C.C., Wu, S.Y., Chen, C.T., Dun, N.J., 2000. Nociceptin inhibits rat sympathetic preganglionic neurons in situ and in vitro. Am. J. Physiol. Regul. Integr. Comp. Physiol. 278, R592-R597.
Lai, C.C., Wu, S.Y., Lin, H.H., Dun N.J., 1997. Excitatory action of pituitary adenylate cyclase activating polypeptide on rat sympathetic preganglionic neurons in vivo and in vitro. Brain Res. 748, 189-194.
Llewellyn-Smith, I.J., Arnolda, L.F., Pilowsky, P.M., Chalmers ,J.P., Minson, J.B., 1998. GABA- and glutamate-immunoreactive synapses on sympathetic preganglionic neurons projecting to the superior cervical ganglion. J. Auton. Nerv. Syst. 71, 96-110.
Llewellyn-Smith, I.J., Phend, K.D., Minson, J.B., Pilowsky, P.M., Chalmers, J.P., 1992. Glutamate-immunoreactive synapses on retrogradely-labelled sympathetic preganglionic neurons in rat thoracic spinal cord. Brain Res. 581, 67-80.
Ma, R.C., Dun, N.J., 1985. Vasopressin depolarizes lateral horn cells of the neonatal rat spinal cord in vitro. Brain Res. 348, 36-43.
Markham, J.A., Vaughn, J.E., 1991. Migration patterns of sympathetic preganglionic neurons in embryonic rat spinal cord. J. Neurobiol. 22, 811-822.

McFarlane, S., Cooper, E., 1992. Postnatal development of voltage-gated K currents on rat sympathetic neurons. J. Neurophysiol. 67, 1291-1300.
McLachlan, E.M., Hirst, G.D., 1980. Some properties of preganglionic neurons in upper thoracic spinal cord of the cat. J. Neurophysiol. 43, 1251-1265.
Milner, T.A., Morrison, S.F., Abate, C., Reis, D.J., 1988. Phenylethanolamine N-methyltransferase-containing terminals synapse directly on sympathetic preganglionic neurons. Brain Res. 448, 205-222.
Miyazaki, T., Coote, J.H., Dun N.J., 1989. Excitatory and inhibitory effects of epinephrine on neonatal rat sympathetic preganglionic neurons in vitro. Brain Res. 497, 108-116.
Miyazaki, T., Dun, N.J., Kobayashi, H., Tosaka, T., 1996. Voltage-dependent potassium currents of sympathetic preganglionic neurons in neonatal rat spinal cord thin slices. Brain Res. 743, 1-10.
Miyazaki, T., Kobayashi, H., Tosaka, T., 1998. Presynaptic inhibition by noradrenaline of the EPSC evoked in neonatal rat sympathetic preganglionic neurons. Brain Res. 790, 170-177.
Mo, N., Dun, N.J., 1987. Excitatory postsynaptic potentials in neonatal rat sympathetic preganglionic neurons: possible mediation by NMDA receptors. Neurosci. Lett. 77, 327-332.
Nolan, M.F., Logan, S.D., Spanswick, D., 1999. Electrophysiological properties of electrical synapses between rat sympathetic preganglionic neurones in vitro. J. Physiol. 519, 753-764.
O'Dell, T.J., Hawkins, R.D., Kandel, E.R., Arancio O., 1991. Tests of the roles of two diffusible substances in long-term potentiation: evidence for nitric oxide as a possible early retrograde messenger. Proc. Natl. Acad. Sci. U.S.A. 88, 11285-11289.
Petras, J.M., Cummings, J.F., 1972. Autonomic neurons in the spinal cord of the Rhesus monkey: a correlation of the findings of cytoarchitectonics and sympathectomy with fiber degeneration following dorsal rhizotomy. J. Comp. Neurol. 146, 189-218.
Phelps, P.E., Barber, R.P., Vaughn, J.E., 1993. Embryonic development of rat sympathetic preganglionic neurons: possible migratory substrates. J. Comp. Neurol. 330, 1-14.
Pilowsky, P., Llewellyn-Smith, I.J., Lipski, J., Chalmers, J., 1992. Substance P immunoreactive boutons form synapses with feline sympathetic preganglionic neurons. J. Comp. Neurol. 320, 121-135.
Pyner, S., Coote, J.H., 1994. Evidence that sympathetic preganglionic neurones are arranged in target-specific columns in the thoracic spinal cord of the rat. J. Comp. Neurol. 342, 153-166.
Sah, P., McLachlan, E.M., 1995. Membrane properties and synaptic potentials in rat sympathetic preganglionic neurons studied in horizontal spinal cord slices in vitro. J. Auton. Nerv. Syst. 53, 1-15.
Schober, A., Unsicker, K., 2001. Growth and neurotrophic factors regulating development and maintenance of sympathetic preganglionic neurons. Int. Rev. Cytol. 205, 37-76.
Schramm, L.P, Stribling, J.M., Adair, J.R., 1976. Developmental reorientation of sympathetic preganglionic neurons in the rat. Brain Res. 106, 166-171.
Schuman, E.M., Madison, D.V., 1991. A requirement for the intercellular messenger nitric oxide in long-term potentiation. Science 254, 1503-1506.
Shen, E., Dun, N.J., 1990. Neonate rat sympathetic preganglionic neurons intracellularly labelled with Lucifer Yellow in thin spinal cord slices. J. Auton. Nerv. Syst. 29, 247-254.
Shen, E., Mo, N., Dun, N.J., 1990. APV-sensitive dorsal root afferent transmission to neonate rat sympathetic preganglionic neurons in vitro. J. Neurophysiol. 64, 991-999.
Shen, E., Wu, S.Y., Dun, N.J., 1994. Spontaneous and transmitter-induced rhythmic activity in neonatal rat sympathetic preganglionic neurons in vitro. J. Neurophysiol. 71, 1197-1205.

Spanswick, D., Logan, S.D., 1990a. Spontaneous rhythmic activity in the intermediolateral cell nucleus of the neonate rat thoracolumbar spinal cord in vitro. Neurosci. 39, 395-403.

Spanswick, D., Logan, S.D., 1990b. Sympathetic preganglionic neurones in neonatal rat spinal cord in vitro: electrophysiological characteristics and the effects of selective excitatory amino acid receptor agonists. Brain Res. 525, 181-188.

Spanswick, D., Pickering, A.E., Gibson, I.C., Logan, S.D., 1994. Inhibition of sympathetic preganglionic neurons by spinal glycinergic interneurons. Neurosci. 62, 205-216.

Spanswick, D., Pickering, A.E., Gibson, I.C., Logan, S.D., 1995. Excitation of sympathetic preganglionic neurons via metabotropic excitatory amino acid receptors. Neurosci. 68, 1247-1261.

Strack, A.M., Sawyer, W.B., Hughes, J.H., Platt, K.B., Loewy, A.D., 1989. A general pattern of CNS innervation of the sympathetic outflow demonstrated by transneuronal pseudorabies viral infections. Brain Res. 491, 156-162.

Wetts, R., Vaughn, J.E., 1998. Differences in developmental cell death between somatic and autonomic motor neurons of rat spinal cord, J. Comp. Neurol. 396, 483-492.

Wilson, J.M., Coderre, E., Renaud, L.P., Spanswick, D., 2002. Active and passive membrane properties of rat sympathetic preganglionic neurones innervating the adrenal medulla. J. Physiol. 545, 945-960.

Wu, S.Y., Dun, N.J., 1992. Presynaptic GABAB receptor activation attenuates synaptic transmission to rat sympathetic preganglionic neurons in vitro. Brain Res. 572, 94-102.

Wu, S.Y., Dun, N.J., 1993. Excitatory amino acids depress synaptic currents in neonate rat sympathetic preganglionic neurons. J. Neurophysiol. 69, 2030-2038.

Wu, S.Y., Dun, N.J., 1995. Calcium-activated release of nitric oxide potentiates excitatory synaptic potentials in immature rat sympathetic preganglionic neurons. J. Neurophysiol. 74, 2600-2603.

Wu, S.Y., Dun, N.J., 1996. Potentiation of IPSCs by nitric oxide in immature rat sympathetic preganglionic neurons in vitro. J. Physiol. 495, 479-490.

Wu, S.Y., Dun, N.J., 1997. Potentiation of NMDA currents by pituitary adenylate cyclase activating polypeptide in neonatal rat sympathetic preganglionic neurons. J. Neurophysiol. 78, 1175-1179.

Wu, S.Y., Dun, S.L., Förstermann, U., Dun, N.J., 1997. Nitric oxide and excitatory postsynaptic currents in immature rat sympathetic preganglionic neurons *in vitro*. Neurosci. 79, 237-245.

Yamamoto, K., Senba, E., Matsunaga, T., Tohyama, M., 1989. Calcitonin gene-related peptide containing sympathetic preganglionic and sensory neurons projecting to the superior cervical ganglion of the rat. Brain Res. 487, 158-164.

Yoshimura, M., Polosa, C., Nishi, S., 1986a. Afterhyperpolarization mechanisms in cat sympathetic preganglionic neuron in vitro. J. Neurophysiol. 55, 1234-1246.

Yoshimura, M., Polosa, C., Nishi, S., 1986b. Electrophysiological properties of sympathetic preganglionic neurons in the cat spinal cord in vitro. Pflugers Arch. 406, 91-98.

Yoshimura, M., Polosa, C., Nishi, S., 1987a. Slow EPSP and the depolarizing action of noradrenaline on sympathetic preganglionic neurons. Brain Res. 414, 138-142.

Yoshimura, M., Polosa, C., Nishi, S., 1987b. Slow IPSP and the noradrenaline-induced inhibition of the cat sympathetic preganglionic neuron in vitro. Brain Res. 419, 383-386.

Chapter 14

NEUROCHEMICAL HETEROGENEITY IN SYMPATHETIC GANGLIA AND ITS IMPLICATIONS FOR CARDIOVASCULAR REGULATION

Miguel A. Morales[1], John C. Hancock,[2] and Donald B. Hoover[2]

[1]*Department of Cell Biology & Physiology, Biomedical Research Institute, UNAM, Mexico DF; and* [2]*Department of Pharmacology, East Tennessee State University, Johnson City, TN 37614, USA*

Abstract: Sympathetic ganglia are important components of the neural circuitry for regulation of vascular tone and cardiac dynamics and may contribute to the pathophysiology of hypertension and heart failure. For a long time, these ganglia were viewed as simple relay stations where acetylcholine mediated the transfer of signals from the central nervous system to peripheral effector neurons. However, new information gathered over the past few decades has established that sympathetic ganglia are structurally and functionally more complex than previously envisioned. Numerous peptides have been identified in subpopulations of preganglionic nerve fibers that supply sympathetic ganglia. These peptides can be co-transmitters in cholinergic nerve fibers or transmitters in preganglionic nerve fibers that are non-cholinergic. Additionally, other classical transmitters besides acetylcholine have been identified in some preganglionic fibers. Several preganglionic fibers with different transmitters can surround individual postganglionic neurons. Postganglionic neurons can also receive input by collateral processes of sensory neurons, providing a mechanism for modulation of ganglion function by axon reflexes. Lastly, there is evidence that ganglionic transmission can be influenced through interneurons, presynaptic receptors on preganglionic nerve fibers, and by the actions of hormones and paracrine mediators. Collectively, these findings indicate that postganglionic neurons function as integrators. Most postganglionic sympathetic neurons are noradrenergic but these cells also express peptides (e.g., neuropeptide Y) as co-transmitters. Topographical distribution of chemically coded preganglionic fibers and target-dependent localization of postganglionic neurons have been identified in some sympathetic ganglia. As our understanding of the neurochemical, anatomical and functional complexities of sympathetic ganglia evolve, they may become attractive targets for the development of novel pharmaceuticals to treat cardiovascular diseases.

Key words: sympathetic ganglia, noradrenaline, acetylcholine, nitric oxide, hormones, topography, co-transmitters, blood pressure, neuropeptides, substance P, transmitters, chemical coding, ganglionic transmission, cholinergic, classical transmitters, neuromodulation, ganglionic LTP

INTRODUCTION

Early Concepts of Sympathetic Ganglion Function (i.e., relay station)

Early studies of the autonomic nervous system, considered ganglia as simple, peripheral relay stations of the visceral motor efferent outflow (Langley, 1921; Dale, 1935), comprising a monosynaptic pathway with preganglionic axon fibers, which contain acetylcholine (ACh), and noradrenergic (sympathetic) or cholinergic (parasympathetic) ganglionic neurons (Elfvin et al., 1993). From the origin of these concepts, during the first decades of twentieth century, to the sixties, it was generally accepted that ganglionic synapses are monosynaptic with a unique chemical mediator. During that decade, some reports appeared of the presence and storage of other chemical mediators in the same axon fibers or ganglionic neurons. These mediators, named co-transmitters, were proposed as neuromodulators of the action of the principal or classical transmitter (Burnstock, 1976). Subsequently, several chemical messengers have been described, including many peptides (Klimaschewski et al., 1996). The presence of more than one mediator in a synaptic pathway, required a reconsideration of ganglionic pharmacology; they were not a pathway with a single transmitter anymore. In fact, they contained at least two transmitters involved in ganglionic transmission. Later on it was determined that preganglionic axon fibers are highly specific, contain a particular combination of messengers, display a specific target-dependent topographic distribution along ganglia, and contact specific neurons containing a precise mix of transmitters. These properties lead to the concept of chemical coding of ganglionic transmission (Costa et al., 1986). Several reports showing different types of matching between preganglionic fibers and ganglionic neurons have been presented for diverse peptides in different autonomic ganglia (Flet and Bell, 1991; Heym et al., 1991; Carbó et al., 1997). Moreover, an important population of ganglionic neurons in sympathetic ganglia was found to store ACh instead of containing noradrenaline (NA) (Lindh et al., 1986; Morales et al., 1995). Furthermore, neurons exist that can express both classical transmitters, NA and ACh (Jimenez et al., 2002).

Current View that Sympathetic Ganglia Exhibit Anatomical and Functional Complexity

At present it is known that ganglia, in particular ganglionic synapses, show anatomical and functional complexity. A series of examples support this property. According to their chemical content, various populations of axon boutons exist. Each set of axons follows a distinctive pathway to reach ganglionic neurons. Preganglionic cholinergic fibers, although distributed almost homogeneously within a ganglion, exhibit an approximation to ganglionic neurons that varies from several fibers surrounding single neurons in a glomerulus–like form, to neurons approximated only by single nerve fibers (Jiménez et al., 2002). Not all preganglionic populations are cholinergic; many preganglionic fibers contain peptides without expressing ACh. Preganglionic enkephalinergic fibers often show a spiral form around some ganglionic cholinergic neurons. Neurotensin that facilitates and enkephalin that inhibits ganglionic transmission are stored in separate preganglionic nerve fibers, which show a complementary distribution along the ganglia (Jiménez et al., 2002).

Integration of Input from Multiple Sources (CNS, Afferents, Humoral Substances)

Upon exiting the ventral roots and white rami, preganglionic fibers branch into several segmental ganglia; within the ganglia they branch further to contact a number of postganglionic neurons (Njå and Puves, 1977). In addition, postganglionic neurons receive convergent inputs from several levels of the spinal cord (Szurszewski, 1981). This circuitry for divergence and convergence potentially yields amplification of the vasoconstrictor signal. The cholinergic nature of these preganglionic axons is generally accepted. Sympathetic ganglia also receive innervation from preganglionic fibers containing various peptides and nitric oxide and sensory innervation from dorsal root ganglia (Gibbins et al., 2000). Although these ganglia are often considered as simple relay stations for autonomic motor activity generated from the central nervous system, they more likely serve as stations for integration of information from multiple convergent inputs (Gibbins et al., 2000).

CHEMICAL CODING IN SYMPATHETIC GANGLIA

Preganglionic Transmitters and Co-transmitters

It is generally accepted that all preganglionic sympathetic neurons are cholinergic (Koelle, 1955; Fredricksson and Sjöqvist, 1962; Lindh et al., 1986). Beside acetylcholine (ACh), many studies have shown the existence of several neuropeptides stored and released from various populations of preganglionic neurons. These peptides include neurotensin (NT) (Lundberg et al., 1982a; Morales et al., 1993), calcitonin gene-related peptide (CGRP) (Heym et al., 1993; Morales et al., 1995), vasoactive intestinal peptide (VIP) (Hökfelt et al., 1977b; Morales et al., 1995), enkephalin (ENK) (Schultzberg et al., 1979; Morales et al., 1995; Jiménez et al., 2002), somatostatin (SOM) (Bachoo et al., 1987; Schultzberg and Lindh 1988), substance P (SP) (Hökfelt et al., 1977c; Nozdrachev et al., 2002), atrial natriuretic factor (ANF) (Inagaki et al., 1986), bombesin/gastrin-releasing peptide (BOM/GRP) (Schultzberg 1983), cholecystokinin (CCK) (Larsson and Rehfeld, 1979), dynorphin (DYN) (Heym et al., 1990) and secretoneurin (Dun et al, 1997). It had been assumed, but rarely demonstrated (Kondo et al., 1985), that the preganglionic nerves containing peptides are cholinergic. Recently, we demonstrated with simultaneous double immunodetection of choline acetyltransferase (ChAT, the ACh synthetic enzyme) and the peptides NT or methionine ENK (met-ENK), the existence of a large population of preganglionic nerves that express NT or met-ENK and not ChAT (Fig. 1) (Jímenez et al., 2002). The presence of such a large population of peptidergic non-cholinergic preganglionic nerves opens the possibility that other classical transmitters occur in these nerves. Various studies have reported the presence of other classical transmitters, including GABA (Dobó et al., 1989; Párducz et al., 1992; Nozdrachev et al., 2002), glutamate (Morrison et al., 1989), dopamine (Heym et al., 1984) and serotonin (Ma et al., 1985; Karhula, 1995), or ATP (Silinsky and Gerzanich, 1993). There is also the possibility that some preganglionic nerves express only peptides without containing a classical transmitter. In this case, peptides released from these nerves may act as classical transmitters (Kow and Pfaff, 1988) or as modulators of the classical transmitter released from other preganglionic nerve fibers. In fact, some convergence of preganglionic

14. NEUROCHEMICAL HETEROGENEITY IN SYMPATHETIC GANGLIA AND ITS IMPLICATIONS FOR CARDIOVASCULAR REGULATION

nerves containing different transmitters over single ganglionic neurons has been reported (Jiménez et al., 2002).

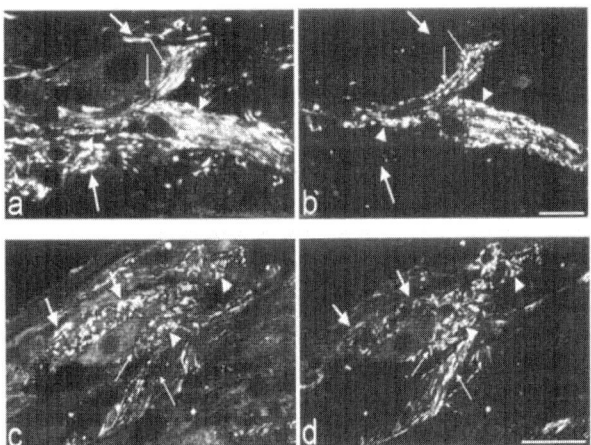

Figure 1. Confocal microscopic images of cat stellate ganglion sections simultaneously reacted against ChAT and Met-ENK (a-b) and ChAT and NT (c-d). Optical sections of 1 µm showing the occurrence and co-occurrence of these transmitters. There were many ChAT-IR fibers and boutons negative to Met-ENK or NT (arrows). Likewise, several fibers and boutons, more evident for ChAT-NT, reactive to the peptides but negative to ChAT were detected (thin arrows). Few sites of co-occurrence of ChAT and peptide immunoreactivity were also detected (arrow heads). Calibration bar: a-b 15 µm, c-d 25 µm. Reprinted from Jiménez et al., (2002) by permission of Wiley-Liss Inc., a subsidiary of John Wiley & Sons, Inc.

Studies on topographic distribution of peptidergic preganglionic nerves along the ganglia have shown specific patterns of distribution, which are often correlated with their modulatory actions (Elfvin et al., 1993). In this matter, by combining retrograde axonal tracing and immunodetection, we have shown in the cat stellate ganglia a particular pattern of distribution of NT and met-ENK. NT is preferentially located in the caudal and dorsomedial regions, where most of the ganglionic neurons with axons that project through the vertebral nerve are located, while met-ENK is mainly located in the ventromedial region, a site that has fewer NT nerves and where most of the sympathetic neurons that innervate the heart are found (Fig.2) (Carbó et al., 1997; Jiménez et al., 2002).

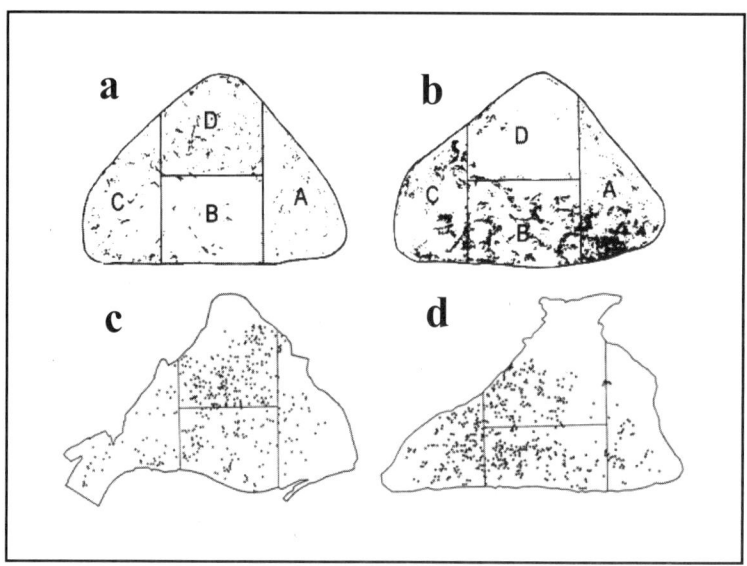

Figure 2. Image processing drawings showing the patterns of distribution of the Met-ENK- (a) and NT-IR (b) preganglionic axon boutons and the site of the cardiac (c) and vertebral (d) neurons retrograde labeling. Each drawing represents a longitudinal 14 μm ganglion section and black dots immunofluorescent or HRP labeled sites, axon boutons (a-b) and cell bodies (c-d). Ganglia were divide in four regions: caudal (A), dorsomedial (B), cranial (C), and ventromedial (D). While Met-ENK fibers are preferably detected in D region where the majority of the cardiac neurons were found, NT-IR was found more abundant in A and B, matching the localization of both cardiac and vertebral neurons; on the other hand D region was almost lacking NT-IR. Reprinted modified from Carbó et al., (1997) and Jiménez et al., (2002) by permission of Wiley-Liss, Inc., a subsidiary of John Wiley & Sons, Inc.

Postganglionic Transmitters and Co-transmitters

Although the majority of principal neurons in sympathetic ganglia are noradrenergic and show immunoreactivity to tyrosine hrydoxylase (TH) (rate-limiting enzyme for NA synthesis), some principal ganglionic neurons in the cat superior cervical ganglion (SCG) were suggested to be dopaminergic (Heym et al., 1984). Also, a considerable population of neurons (10-15%) is reactive to ChAT and negative to TH (Lindh et al., 1986; Morales et al., 1995; Jiménez et al., 2002); these neurons innervate sweat glands in the foot pad. Within this population of ChAT-positive neurons, some cells co-express TH (Fig. 3). We have found that denervation of SCG by sectioning of preganglionic nerves increases the number of neurons only reactive to ChAT or co-expressing ChAT and TH compared to

14. NEUROCHEMICAL HETEROGENEITY IN SYMPATHETIC GANGLIA AND ITS IMPLICATIONS FOR CARDIOVASCULAR REGULATION

those expressing only TH (Jiménez et al., 2002). This finding indicates the importance of presynaptic cholinergic activity for transmitter induction and expression in postsynaptic neurons. A similar example of transynaptic influence has been found in rat SCG where denervation decreases TH activity of principal ganglionic neurons (Hendry et al., 1973). Some factors with the property of switching noradrenergic neurons to cholinergic have been identified in cocultures of sympathocytes with cardiomyocytes (Rao and Landis, 1990). Our data suggest the presence of a factor that inhibits the switching to the cholinergic phenotype. This factor would be removed after denervation.

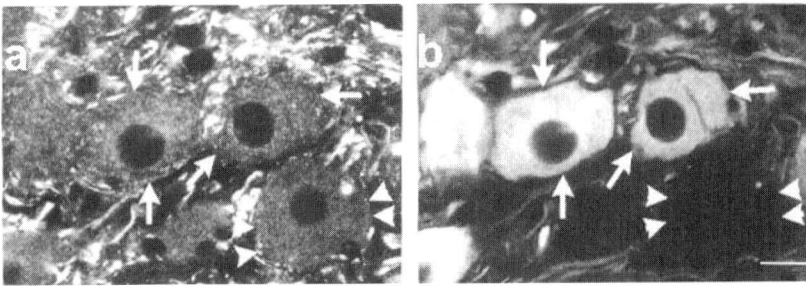

Figure 3. Double-stained immunofluorescence micrographs of the cat stellate ganglion sections reacted with anti-ChAT (a) and anti-TH (b) antibodies. ChAT-IR cell bodies were detected, most of which were TH-negative (arrowheads). Two types of ChAT-IR cell bodies were found, TH-positive (arrows) and TH-negative (arrowheads). Calibration bar: 25 µm. Reprinted from Jiménez et al., (2002) by permission of Wiley-Liss Inc., a subsidiary of John Wiley & Sons, Inc.

In comparison with preganglionic neurons, less peptides have been detected in principal ganglionic neurons; these include SOM (Hökfelt et al., 1977a; Leranth et al., 1980), ENK (Schultzberg et al., 1979; Jiménez et al., 2002), vasopressin (VP) (Hanley et al., 1984), neuropeptide Y (NPY), which has been found in a large population of ganglion cells (Lundberg and Tatemoto, 1982; Gibbins, 1991), galanin (GAL) (Kummer, 1987; Lindh et al., 1989), DYN (Heym et al., 1990) and secretoneurin (Dun et al., 1997). These peptides are detected in noradrenergic neurons, which means they are co-localized with TH. In cholinergic non-noradrenergic cells, other peptides such as VIP (Hökfelt et al., 1977b; Morales et al., 1995) and CGRP (Kummer and Heym, 1988; Morales et al., 1995) are present. Apart from NA, ACh, and neuropeptides, some ganglionic neurons contain GABA

(Häppölä et al., 1987; Nozdrachev et al., 2002). We have detected that the number of neurons expressing met-ENK and TH increases after denervation and expression of NT, which is not regularly found in ganglionic neurons, emerges in some neurons. A possible explanation of this increase in peptide expression, shown for ENK in other ganglia (Henion and Landis, 1993), would be that axons of these neurons project out of the ganglia through the preganglionic nerve, and transection induces an up-regulation of peptide synthesis. We explored this possibility by using a retrograde axon tracer combined with immunodetection of the peptides. Although we found several neurons labeled with the tracer applied in the proximal end of the nerve, they did not correspond to the cells with induced peptide immunoreactivity. Therefore, the mechanism by which this peptide expression increases after denervation remains to be elucidated.

MECHANISMS UNDERLYING AND MODULATING GANGLIONIC TRANSMISSION

Presynaptic Modulation

It is well established that nerve terminals possess a diversity of receptors, which influence the release of neurotransmitters. Autoreceptors and heteroreceptors are present at preganglionic sympathetic nerve endings and exert either positive or negative effects on ganglionic neurotransmission.

Liang and Vizi have used the rat SCG as a model system to study the role of presynaptic cholinergic receptors in modulating release of ACh from preganglionic sympathetic nerve endings (Liang and Vizi, 1997). They found that the release of ACh evoked by preganglionic nerve stimulation was augmented by treatment with nicotinic agonists and attenuated by exposure to nicotinic antagonists. These observations suggest that release of endogenous ACh is amplified through activation of presynaptic nicotinic receptors and that ganglionic neurotransmission is thereby facilitated. Other experiments in this study provided evidence that preganglionic nerve fibers in the SCG also contain muscarinic receptors, which have an inhibitory influence on ACh release. Evoked release of ACh was augmented during treatment with atropine, but this effect was only observed in the absence of nicotinic receptor blockers (Liang and Vizi, 1997). Accordingly, it appears that the inhibitory influence of muscarinic receptors is restricted to the portion of ACh release that results from positive modulation via presynaptic nicotinic receptors.

14. NEUROCHEMICAL HETEROGENEITY IN SYMPATHETIC GANGLIA AND ITS IMPLICATIONS FOR CARDIOVASCULAR REGULATION

Sympathetic ganglia can be exposed to catecholamines that are released from the adrenal medulla and arrive via the circulation. It is also possible that catecholamines are released locally from dendritic processes of postganglionic neurons (Zaidi and Matthews, 1997) and from SIF cells. These catecholamines could modulate the release of ACh by binding to presynaptic adrenergic receptors. Evidence from studies of rat SCG and dog stellate ganglia demonstrates the presence of presynaptic α_2-adrenergic receptors that are inhibitory and presynaptic β-adrenergic receptors that are facilitatory (Medgett, 1983; McCallum et al., 1998). The inhibitory adrenergic receptors are normally dominant, and their influence is dependant on the frequency of preganglionic nerve stimulation.

Several paracrine mediators can affect ganglionic neurotransmission through binding to presynaptic receptors. The strongest evidence has been provided for adenosine and histamine. It has been reported that 5'-nucleotidase, an ectoenzyme that produces adenosine, is present at sites within sympathetic ganglia (Nacimiento et al., 1991), so it is feasible that adenosine could be produced locally under appropriate conditions. Adenosine might also occur within the synaptic regions as a metabolite of ATP released from nerve fibers. Studies of rat SCG have demonstrated that stimulation evoked release of ACh is increased by treatment with an A_1 adenosine receptor antagonist, which supports the concept that endogenous adenosine has a tonic inhibitory effect on ganglionic transmission in this species (Liang and Vizi, 1997). Although inhibitory A_1 adenosine receptors are also present in guinea pig SCG, they are not tonically activated in this species (Borasio et al., 1995). Accordingly, the precise role of these receptors in modulating transmission is unresolved. In contrast, there is very compelling evidence that histamine functions as an immunomodulator of ganglionic neurotransmission primarily by acting at presynaptic receptors. Weinreich and coworkers have performed sophisticated electrophysiological studies of guinea pig SCG to investigate this topic (Christian and Weinreich, 1992). Their work has established that preganglionic cholinergic nerves contain H_1 receptors that augment release of ACh and, less commonly, H_3 receptors that attenuate release. This conclusion is supported by quantal analysis of the effects of histamine on synaptically evoked responses. Furthermore, these investigators demonstrated that mast cells were present in the SCG (Undem et al., 1990) and that histamine was released from ganglia of ovalbumin-sensitized animals when the tissue was exposed to the sensitizing agent (Undem et al., 1991). Importantly, they established that antigen challenge caused effects on ganglionic neurotransmission that were mediated in large part by endogenous histamine. Accordingly, presynaptic histamine receptors are clearly sites where the immune system is able to influence sympathetic function.

Postsynaptic Effects

Cholinergic Mechanisms

Two distinctive excitatory cholinoceptive sites and one inhibitory cholinoceptive site have been described in sympathetic ganglia (Eccles and Libet, 1961; Nishi and Koketsu, 1968; Libet and Tosaka, 1969; Weight et al., 1979). The first excitatory site is sensitive to activation by nicotine or related drugs; this receptor is termed nicotinic. There are at least 10 types of nicotinic receptors which are named $\alpha 1$ through $\alpha 10$ (Alexander et al., 2001; Taylor, 2001). The receptor on sympathetic ganglia, termed $\alpha 3$, is selectively blocked by hexamethonium-like drugs. This receptor has a pentameric structure consisting of five subunits that form a central channel. Multiple combinations of subunits or their mRNAs have been identified in ganglia with $\alpha 3$ and $\beta 2$ being in abundance. Nicotinic receptors are ligand-gated ion channels whose activation causes an increase in the permeability to Na^+, depolarization of the cell membrane, and excitation. The depolarization has a latency of less than 1 msec and decays in 10 to 50 msec. From the results of several studies in the early literature, it is reasonable to conclude that the primary pathway for ganglionic transmission occurs via nicotinic receptors.

The second type of excitatory cholinoceptive site and the inhibitory cholinoceptive site are muscarinic receptors. These receptors mediate delayed-onset slow depolarizations and hyperpolarizations of the cell membrane. M_1 muscarinic receptors mediate a secondary depolarization of the postganglionic cell membrane termed the slow EPSP. The slow EPSP follows the initial fast depolarization caused by nerve stimulation or iontophoretic application of ACh and occurs after a delay of about 50 msec. M_1 muscarinic receptors couple to the G-protein $G_{q/11}$ and cause activation of phospholipase C; this results in the hydrolysis of polyphosphoinositides and mobilization of intracellular Ca^{+2}. The slow EPSP results from a decreased K^+ conductance, which is called *M current* (Weight et al., 1979; Adams et al., 1982).

M_2 muscarinic receptors mediate a secondary hyperpolarization termed the slow IPSP. The slow IPSP follows the initial depolarization of the cell membrane caused by nerve stimulation or iontrophoretic application of ACh and precedes the slow EPSP. M_2 muscarinic receptors activate the G-proteins termed G_o and G_i with inhibition of adenylyl cyclase, activation of

14. NEUROCHEMICAL HETEROGENEITY IN SYMPATHETIC GANGLIA AND ITS IMPLICATIONS FOR CARDIOVASCULAR REGULATION

K^+ channels through the $\beta\gamma$-subunits of G_i, and suppression of voltage-gated L-type calcium channels. Increased K^+ permeability leads to hyperpolarization. It is relevant that the atropine-sensitive cholinoceptive sites can be activated directly by ACh liberated from preganglionic nerve endings (Hilton, 1961; Long and Eckstein, 1961; Volle, 1962). Substantial evidence has accumulated to suggest that M_2 sites can also be activated indirectly by dopamine and NA. These catecholamines cause hyperpolarization of ganglionic neurons, and both synaptically evoked and catecholamine induced hyperpolarization are blocked by α-adrenoceptor antagonists or atropine. ACh that is released from the preganglionic terminal may thus act on an interneuron that releases dopamine or NA such that the catecholamine causes hyperpolarization of the ganglion cell. These dopamine and NA containing interneurons are probably SIF cells (Eränkö et al., 1980; Taylor, 2001).

Postsynaptic Actions of Some Non-cholinergic Transmitters

Neurotensin

The tridecapeptide NT (Carraway & Leeman, 1973; 1975) has been demonstrated by immunocytochemistry in the perikarya of sympathetic preganglionic neurons (Krukoff et al., 1985) as well as in nerve fibers within sympathetic ganglia (Lundberg et al., 1982a; Bachoo et al., 1987; Morales et al., 1993; Carbó et al., 1997; Zetina et al., 1999; Jiménez et al., 2002). We have also demonstrated its restricted location in large dense core vesicles (LDCV, Morales et al., 1993) as well as the heterogeneous distribution of nerve fibers immunoreactive to NT within the cat stellate ganglia (Carbó et al., 1997; Jiménez et al., 2002). A frequency-dependent depletion of NT after prolonged high frequency stimulation has likewise been reported (Caverson et al., 1989; Maher et al., 1991; Zetina et al., 1999). This depletion is followed by a receptor-mediated internalization, which has been proposed as part of the signaling mechanism of this peptide (Zetina et al., 1999). In general, the modulatory effects of NT in sympathetic ganglia are a facilitation of ganglionic transmission. NT increases membrane excitability and efficiency of nicotinic synaptic transmission in a large proportion of neurons of the guinea-pig prevertebral sympathetic ganglia, two actions that are mediated by the presence of functional NT receptors in both the postsynaptic membrane and presynaptic non-cholinergic nerve terminals

(Stapelfeldt and Szurszewski, 1989a). It also causes a transient depolarization and facilitation of electrically evoked non-cholinergic slow EPSPs in neurons of guinea pig inferior mesenteric ganglion, changes that involve presynaptic release of SP from collateral nerve terminals of primary afferent nerve fibers (Stapelfeldt and Szurszewski, 1989b). Close arterial injection of NT into the cat stellate ganglion increases heart rate by direct activation of ganglionic neurons and produces a transient facilitation of the heart rate response to preganglionic nerve stimulation (Bachoo and Polosa, 1988).

Opioids

Enkephalins have been proposed as inhibitors of ganglionic transmission since the pioneering works of Konishi and collaborators and North and Williams (Konishi et al., 1979; North and Williams, 1976). Such inhibition can be induced by the release of endogenous opioids after high frequency stimulation of preganglionic nerves to the SCG (Zhang et al., 1991). Furthermore, it is antagonized by naloxone and enhanced by peptidase inhibitors. Inhibition is also induced by the injection of exogenous ENK or opioid agonist (Zhang et al., 1991; 1993). By immunohistochemical detection, different patterns of distribution of ENK-IR fibers have been described in sympathetic ganglia. While in guinea pig stellate ganglion ENK fibers are found in the ventral region preferentially surrounding ganglionic neurons that project to the lung (Heym et al., 1991), in the dog, ENK-IR fibers have been detected mainly in the caudal two thirds of the stellate ganglion (Darvesh et al., 1987), and we have found met-ENK-IR fibers predominantly in the ventromedial region of the cat stellate ganglion where most of the cardiac neurons lie (Jiménez et al., 2002). In the same ganglion, direct local application of opiates inhibited tachycardia induced by preganglionic stimulation or chemical stimulation with cholinergic agonist, while tachycardia induced by postganglionic stimulation was not affected (Prosdocimi et al., 1986).

Nitric oxide

The presence of neuronal nitric oxide synthase (NOS) in some preganglionic sympathetic nerve fibers suggests that endogenous nitric oxide (NO) could modulate neurotransmission at sympathetic ganglia (Anderson et al., 1993; Dun et al., 1993). Cyclic GMP, the second messenger often utilized by NO, has also been identified in sympathetic ganglia where it is localized to a variable extent in nerve fibers and in postganglionic neurons (Holmberg et al., 2001). This observation implies that NO could exhibit pre- and postsynaptic actions mediated by cGMP. Early studies established that stimulation of preganglionic nerves to the rat SCG increased cGMP levels in the tissue and that at least a portion of this cGMP was localized to postganglionic neurons (see Briggs, 1992). Experiments with isolated rat SCG later demonstrated that treatment with a membrane permeable cGMP analog or a NO donor potentiated ganglionic transmission (Briggs, 1992). Patch-clamp studies of isolated neurons from the same ganglion showed that NO donors acted directly on postganglionic cells to increase the amplitude of Ca^{2+} currents (Chen and Schofield, 1993). More recently, endogenous NO has been implicated in the maintenance of long-term potentiation of nicotinic transmission at rat SCG, and it has been proposed that this effect may occur through a presynaptic action to enhance release of ACh (Altememi and Alkadhi, 1999).

A more extensive array of effects has been attributed to NO at prevertebral ganglia. NO donors cause hyperpolarization of most neurons in mouse superior mesenteric ganglia, however, this response is followed by fast EPSPs in a significant number of cells, with some generating action potentials (Mazet et al., 1996). It was proposed that endogenous NO attenuates a late slow EPSP evoked by repetitive stimulation of preganglionic nerves. In contrast, depolarization and action potential generation could reflect presynaptic effects of NO to enhance release of ACh. In studies of rabbit celiac ganglion, the influence of endogenous NO varied between neurons, causing augmentation, attenuation, or dual effects on the number of action potentials generated by trains of supramaximal preganglionic stimuli (Quinson et al., 1998). Lastly, patch-clamp studies of guinea pig celiac ganglion neurons have provided evidence that NO can exert opposing, state-dependent effects on membrane currents. These actions could differentially influence nicotinic and peptidergic transmission (Browning et al., 1998).

Long-Term Potentiation

Sympathetic ganglia possess a plasticity that allows them to modify synaptic transmission under certain conditions of stimulation, resulting in facilitation, augmentation, potentiation (Zengel et al., 1980), and even in the long-lasting effect of potentiation known as long-term potentiation (LTP) (Brown and McAfee, 1982). As in hippocampus and other regions of the

CNS, LTP in sympathetic ganglia consists of a long-lasting increase in synaptic efficacy that can be elicited *in vitro* and *in vivo* after application of a brief (few seconds) stimulus train of high frequency (20-40 Hz). This capacity, first described by Dunant and Dolivo (1968), was named LTP in view of its similarity to the phenomenon found in hippocampus (Bliss and Lomo, 1973). The cellular mechanisms underlying LTP remain unclear but pre- and postsynaptic components have been proposed (Briggs et al., 1985; Collier, 1996). These components might be an increase in ACh release or an enhancement of nicotinic responsiveness at the postsynaptic membrane. Recently, supporting the postsynaptic proposal, a morphological study demonstrated a structural remodeling of shaft synapses associated with LTP in cat SCG (Kadota and Kadota, 2002). A plausible mechanism for LTP that involves both pre- and post-synaptic sites is the release of a modulatory substance (e.g., peptide) from presynaptic terminals during the tetanic stimulation that may serve to enhance postsynaptic efficacy (Morales et al., 1994; Collier, 1996).

Hippocampal LTP was proposed as a model of learning and memory, since its feature of long-lasting changes elicited by brief stimuli resembles enduring modifications presented in learning and memory. However, LTP in sympathetic ganglia must represent a different adaptive process, not related to learning and memory. In spite of the clear evidence for LTP with tetanic stimulation, there has been no clear demonstration of physiological responses that can induce ganglionic LTP (Briggs, 1995). The observation that hypoxia as well as high frequency stimulation causes an increase in the ACh content of SCG (Birks, 1978) could be a clue to the possible role of LTP in the phenomenon of compensatory hypoxic firing. However, it has not been demonstrated that LTP correlates with an increase in ganglionic ACh content (Briggs et al., 1995). Another possible physiological role of ganglionic LTP, might be to maintain high levels of autonomic tone such that high and lasting ganglionic activity can be elicited and sustained by repetitive brief high frequency trains of stimuli (Briggs and McAfee, 1988). Lastly, it has been suggested that altered synaptic plasticity expressed as LTP in SCG may contribute to the induction and maintenance of high blood pressure in conditions where elevated circulating levels of ouabain have been described (Aileru et al., 2001).

14. NEUROCHEMICAL HETEROGENEITY IN SYMPATHETIC GANGLIA AND ITS IMPLICATIONS FOR CARDIOVASCULAR REGULATION

HUMORAL INFLUENCES

Angiotensin

Angiotensin II (ANG II) is a peptide that plays an important role in the regulation of blood pressure and in the pathophysiology of hypertension and heart failure. Inhibitors of angiotensin converting enzyme (ACE) and drugs that block the AT_1 subtype of ANG II receptor have become vital tools in the control of the latter diseases. Many of the adverse effects attributed to ANG II occur because this peptide increases sympathetic drive to blood vessels and the heart (Reid, 1992). Sites where ANG II acts to increase sympathetic drive include the brainstem, sympathetic ganglia, the adrenal medulla, and sympathetic nerve endings (Reid, 1992).

It has been known for several decades that ANG II can stimulate noradrenergic neurons in sympathetic ganglia. This was first shown in functional studies where close arterial injection of the peptide evoked tissue responses that were inhibited by adrenergic blockers (Reid, 1992). Electrophysiological studies subsequently showed that ANG II causes slow depolarization of postganglionic sympathetic neurons (Dun et al., 1978) and can increase the efficacy of ganglionic neurotransmission (Sullivan and Bolter, 1997). It is also clear that ANG II can activate postganglionic neurons independent of central input. This was convincingly demonstrated in two recent studies, which showed that ANG II caused a dose-dependent increase in renal nerve activity in mice and rats during ganglion blockade with hexamethonium (Ma et al., 2001; Dendorfer et al., 2002). Both studies further demonstrated that AT_1 receptors mediated the ganglion stimulating action of ANG II.

It has been suggested that the concentration of ANG II in the blood may not be sufficient to influence sympathetic ganglia under physiological conditions (Sullivan and Bolter, 1997). Since ANG II levels are elevated in cardiac hypertrophy, effects of circulating ANG II on ganglionic transmission are more likely to be expressed under such pathophysiological conditions. Furthermore, recent work has provided evidence for a local renin-angiotensin system (RAS) in sympathetic ganglia (Kushiku et al., 2001). Basal expression of the angiotensinogen gene occurs in dog stellate ganglia, suggesting the potential for local production of ANG II. This thesis is supported by the observations that 1) isolated stellate ganglia release ANG II into their incubation media and 2) stimulation of the preganglionic nerve increases the amount of ANG II released into the media. Intense nerve stimulation also increased the expression of angiotensinogen mRNA. Evoked release of ANG II was attenuated by treatment with captopril, which suggests that ACE is likewise present in the stellate ganglia. The cellular localization of RAS components in sympathetic ganglia is presently unknown.

Atrial Natriuretic Peptide

Atrial natriuretic peptide (ANP) is another hormone postulated to affect neurotransmission at sympathetic ganglia. In this case, an inhibitory influence has been proposed. Studies of normal men have shown that infusions of ANP produced a much smaller reflex increase in muscle sympathetic nerve activity (MSNA) than occurred with infusion of sodium nitroprusside as a hemodynamic control (Floras, 1990). It was concluded that ANP blunted the normal reflex responses to decreased diastolic and central venous pressures by an inhibitory action at sympathetic ganglia or an effect that was mediated through the CNS. Subsequent experiments demonstrated that MSNA was increased by edrophonium, a short-acting cholinesterase inhibitor that does not enter the CNS, and that ANP attenuated the effect of edrophonium on MSNA (Floras, 1995). Since edrophonium is a quaternary amine, it may increase MSNA, at least in part, by direct stimulation of nicotinic receptors on postganglionic sympathetic neurons. This mechanism for edrophonium would support the conclusion that ANP exerts its inhibitory influence at sympathetic ganglia. However, if ANP caused a centrally mediated decrease in preganglionic sympathetic tone, this would result in less release of ACh into the synaptic cleft. The presence of less ACh would be expected to reduce the impact of edrophonium if it was acting primarily as a cholinesterase inhibitor.

Animal studies have demonstrated that ANP receptors are present in cultured rat thoracic sympathetic chain ganglia where they are localized to subpopulations of fibroblasts and glial cells but not to neurons (James et al., 1990). These findings are consistent with a report that ANP has no effect on neurons in the rat SCG (Pant and Smith, 1989). Nevertheless, *in vitro* studies have established that treatment with ANP increases cGMP accumulation and attenuates carbachol-stimulated synthesis of catecholamines from radiolabeled tyrosine in the SCG (Debinski et al., 1987). It has also been reported that atriopeptin III depresses the excitability of sympathetic neurons in guinea pigs (Cheung, 1988).

Collectively, these observations support the concept that natriuretic peptides could function as inhibitory modulators of neurotransmission at sympathetic ganglia. Since circulating levels of natriuetic peptides are elevated in many cardiovascular diseases including hypertension, such effects on the sympathetic nervous system deserve further study.

SENSORY INNERVATION OF SYMPATHETIC GANGLIA

In addition to the efferent-motor pathways that reach or traverse sympathetic ganglia, several evidences are available that afferent-sensory connections are also present in these ganglia (Elfvin et al., 1993; Klimaschewski et al., 1996). Such connections have been revealed either by anterograde axonal tracer applied to spinal ganglia, which labeled axon fibers in sympathetic ganglia, or by retrograde axonal tracer injected into the sympathetic ganglia, which labeled cell bodies in spinal ganglia (Quigg et al., 1990; Nozdrachev et al., 2002). The application to spinal ganglia of the tracer cholera toxin subunit B-conjugated to HRP, which can also be transported transsynaptically, revealed not only fibers in sympathetic ganglia but cell bodies of ganglionic neurons, too. This labeling of ganglionic neurons was interpreted to mean that sensory fibers, on their way to the peripheral target, pass through the sympathetic ganglia and emit some collaterals that synapse on ganglionic neurons (Quigg et al., 1990). Besides the probable sensory function, these fibers could play a role of transporting trophic molecules from peripheral targets to the ganglionic neurons, as has been suggested in other systems (Purves et al., 1988). However, we have found that injection of HRP alone, which cannot diffuse transsynaptically, also labels cell bodies of sympathetic ganglionic neurons. An alternative explanation for these labeled ganglionic neurons is that they are efferent neurons that innervate blood vessels of spinal dura or sensory neurons that pass through spinal ganglia on their way to dorsal radices (Nozdrachev et al., 2002). Many primary sensory fibers projecting into the sympathetic ganglia surround ganglionic neurons and contain SP and CGRP (Gamse et al., 1980; Gibbins et al., 1985). However, SP and CGRP can occur in separate nerve fibers, which are not sensory (Heym et al., 1993). Thus, CGRP immunoreactive fibers can be of two kinds: thick fiber bundles with non-varicose appearance and thinner varicose fibers. While the first might correspond to sensory fibers, the others could be either sensory collaterals that synapse ganglionic neurons or efferent preganglionic axon fibers (Yamamoto et al., 1989).

SYMPATHETIC GANGLIA INFLUENCING BLOOD PRESSURE AND CARDIAC FUNCTION

Postganglionic vasoconstrictor and cardioaccelerator neurons receive their innervation from preganglionic neurons located in the intermediolateral cell columns of the thoracolumbar spinal column extending into the lateral funiculi. The vasoconstrictor neurons are readily identifiable *in vivo* by the entrainment of their firing to the cardiac cycle (Jänig, 1988). Preganglionic fibers project from these neurons to associated paravertebral and prevertebral ganglia where they make synaptic contact with postganglionic sympathetic neurons.

Postganglionic sympathetic neurons influencing blood pressure are distributed throughout all sympathetic ganglia. Ganglia at certain levels of the spinal column project to specific vascular beds where they control local blood flow through changes in regional vascular resistance. Since blood pressure reflects the sum of resistances in the peripheral vasculature, no specific sympathetic ganglia can be identified as having a predominant or exclusive role in the control of blood pressure. Sympathetic postganglionic neurons that innervate blood vessels are located throughout the ganglia and are thus not readily identifiable. There is some evidence, however, that vasomotor neurons are smaller than those lying in secretomotor pathways (Gibbins et al., 2000). Postganglionic neurons in paravertebral ganglia send fibers to resistance vessels in the skin, kidneys, thorax and somatic vasculature. They also send fibers to blood vessels of the brain, abdominal and pelvic viscera and other organs for the local regulation of blood flow. Other central preganglionic vasomotor fibers pass through paravertebral ganglia to synapse in prevertebral ganglia that provide the primary innervation to the vasculature of internal organs.

Sympathetic efferent postganglionic neurons innervating the heart are found primarily in the cranial poles of the stellate ganglia bilaterally (Armour, 1994). Some postganglionic sympathetic fibers also originate from neurons located throughout the middle cervical and mediastinal ganglia. A few neurons providing innervation to the heart are found in the SCG (Armour, 1994). Fibers passing to these ganglia originate ipsilaterally from the spinal cord at the levels T1 through T5. Postganglionic fibers from these ganglia provide innervation to all regions of the heart. There is also evidence from functional studies for efferent postganglionic sympathetic neurons in the intrinsic cardiac nervous system (Armour, 1994).

14. NEUROCHEMICAL HETEROGENEITY IN SYMPATHETIC GANGLIA AND ITS IMPLICATIONS FOR CARDIOVASCULAR REGULATION

SYMPATHETIC GANGLIA IN HYPERTENSION AND HEART FAILURE

Hypertension

Studies on spontaneously hypertensive rats (SHR) have demonstrated elevated firing in postganglionic sympathetic nerves (Fig. 4) (Judy et al., 1976; Schramm and Chornoboy, 1982; Hancock and Lindsay, 2000). Nicotinic cholinergic receptor blockade decreases both the basal level of nerve firing and bursting to similar levels in these hypertensive rats (Hancock and Lindsay, 2000) and in normotensive rats suggesting that the firing originates largely in the central nervous system.

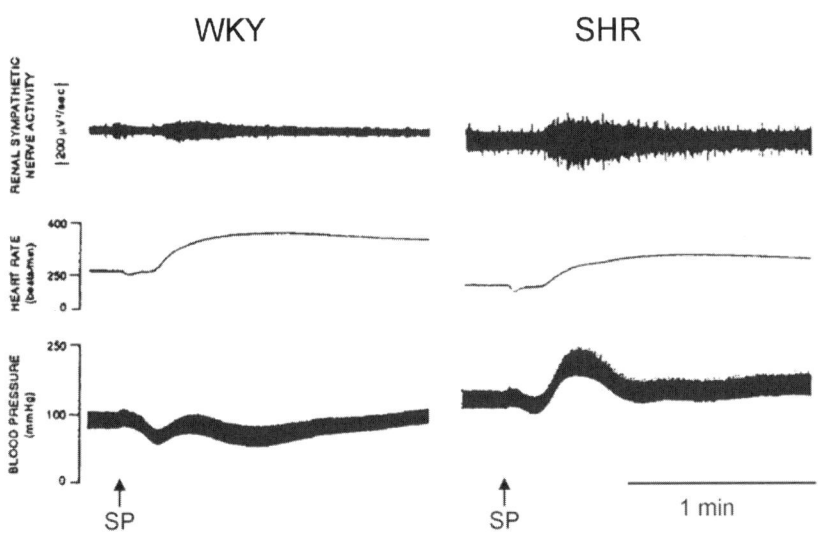

Figure 4. Recorder tracings showing the effects of 32 nmol/kg of substance P (SP) on renal sympathetic nerve activity, blood pressure and heart rate in SHR and WKY rats. Rats were treated with chlorisondamine (10.5 μmol/kg). Arrows indicate the time SP was injected i.v. Reprinted modified from Hancock and Lindsay (2000) by permission of Elsevier Science Inc.

Hypertension is also associated with remodeling and changes in sympathetic ganglia that facilitate ganglionic transmission and may contribute to the elevated postganglionic sympathetic nerve activity and blood pressure. Recordings of compound action potentials from SCG showed that ganglia of SHRs have an enhanced synaptic transmission with physiological frequencies of stimulation (Magee and Schofield, 1992). Intracellular recordings showed that preganglionic nerve stimulation elicited larger amplitude EPSPs in neurons of the SCG of SHRs compared to those of normotensive rats. Analysis of variance of both compound and unitary EPSP amplitude suggested that an increase in transmitter release was responsible for the elevated EPSP amplitude in these neurons (Magee and Schofield, 1994). There may also be intrinsic membrane alterations of postganglionic sympathetic neurons. For example, spike accommodation has been reported to be decreased in SHR SCG neurons and this increased neuronal excitability has been suggested to be involved in the elevated sympathetic nerve activity in these rats (Yarowsky and Weinreich, 1985; Jubelin and Kannan, 1990). Using similar techniques, we did not find a difference in the ability of SHR and normotensive rats to follow repetitive stimulation (Tompkins and Hancock, 2002). In addition, the nicotinic receptor stimulant DMPP caused similar increases in postganglionic firing in SHRs and WKY rats (Hancock and Lindsay, 2000). These results suggested that differences in the active and passive neuronal membrane properties of SHRs alone do not account for the increased sympathetic nerve activity observed in this strain.

In addition to cholinergic innervation from the spinal cord, sympathetic ganglia receive input from nerves containing Met-ENK, NT, CGRP (Gibbins et al., 2000; Jimenez et al., 2002), and NOS. We do not know if there are changes in these transmitter systems in hypertension. Ganglia also receive innervation by tachykinins and other sensory peptides that have their origin outside of the CNS. SP is a primary tachykinin in sympathetic ganglia along with the tachykinin, neurokinin A (NKA); both peptides are synthesized in dorsal root ganglia and transported in caudally directed sensory fibers to sympathetic ganglia (Hökfelt et al., 1975; Dalsgaard et al., 1983; Cuello, 1987; Matthews et al., 1987; Elfvin et al., 1993). SP is released in sympathetic ganglia where it functions as an efferent transmitter to cause depolarization and to evoke firing in postganglionic sympathetic nerve fibers (Dun and Jiang, 1982; Konishi et al., 1985; Cuello, 1987). SP is also found in a population of preganglionic neurons. These neurons appear to be primarily in lumbar and pelvic ganglia and make synapses selectively with cholinergic, VIP-containing vasodilator neurons (Gibbins et al., 2000). Another population of SP-containing fibers project from the gastrointestinal tract to prevertebral ganglia. SP's effect on sympathetic ganglia is mostly via NK_1 receptors with some actions mediated by NK_3 receptors. We have shown that intravenous injection of SP into the femoral artery or close arterial injection of SP into the circulation of the SCG evokes increases in postganglionic sympathetic nerve firing and blood pressure (Hancock and

14. NEUROCHEMICAL HETEROGENEITY IN SYMPATHETIC GANGLIA AND ITS IMPLICATIONS FOR CARDIOVASCULAR REGULATION

Lindsay, 2000; Tompkins and Hancock, 2002). The pressor response to SP is totally blocked by the NK_1 selective agonist GR82334 indicating that it is mediated via NK_1 receptors (Schoborg et al., 2000).

Several lines of evidence suggest that there is an up regulation of the tachykinergic system in SHRs and that the ganglion responsiveness to tachykinins is correlated with blood pressure (Gursinge and Bell, 1989; Hancock and Lindsay, 2000; Schoborg et al., 2000). SCG of SHRs are more densely innervated by SP-immunoreactive nerves than ganglia of Wistar Kyoto (WKY) rats, which serve as the normotensive control (Gurusinghe and Bell, 1989). We have shown that there is a three-fold increase in the expression of the NK_1 receptor message and NK_1 receptor protein in SCG of SHRs compared to in WKY rats (Fig. 5) (Schoborg et al., 2000). Further, we have shown that intravenous injection of SP or an NK_1 agonist elicits enhanced postganglionic sympathetic nerve activity, tachycardia and a pressor response in SHRs (Fig. 4). Conversely, SP has a primary effect to stimulate NK_1 receptors on vascular endothelial cells to cause vasodilation in WKY rats and has minimal effects on sympathetic nerve activity in this strain (Zawadzki et al., 1981; Maggio, 1988; Hancock and Lindsay, 2000). Our *in vitro* studies with intracellular recording from the SCG have shown that SHRs are more responsive to ganglion stimulation by SP and NK_1 agonists because of a greater number of responsive neurons within the SCG rather than an enhanced responsiveness of individual neurons (Tompkins and Hancock, 2002).

These observations suggest that the genes responsible for increased NK_1 receptor expression in SHRs are cosegregrated with the genes for hypertension and may contribute to hypertension by enhancing sympathetic nerve activity. Our unpublished studies have shown that sympathetic nerve and blood pressure responses to SP and NK_1 receptor agonists are not upregulated in a nongenetic model of hypertension in which rats were made hypertenisive by deoxycorticosterone acetate (DOCA)-salt treatments. In contrast, sympathetic nerve and blood pressure responses to SP in SHRs treated with captopril from age 6 weeks to prevent the development of hypertension were similar to those of untreated hypertensive SHR. These observations reduce the possibility that factors associated with elevated blood pressure cause the upregulation of NK_1 receptor genes in SHRs. Overall, these observations suggest plasticity in ganglia such that intrinsic ganglionic function can be modified in disease states.

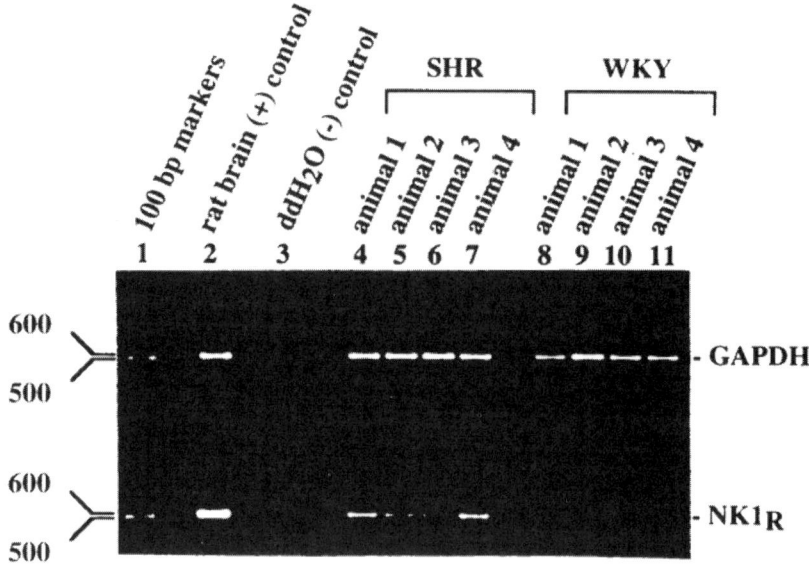

Figure 5. RT-PCR analysis of NK1 receptor mRNA expression in SHRs and WKYs superior cervical ganglion. RT(+) cDNA reactions were PCR amplified and electrophoresed. GAPDH specific amplification reaction products were loaded in the upper set of lanes; the NK1 receptor specific amplification products were loaded in the lower set. Lane 1 contains a 100 base pair molecular weight ladder (Promega). Lane 2 contains PCR amplified rat brain (+) control cDNA. Lane 3 contains an identical PCR reaction in which double distilled water was substituted for template DNA. Lanes 4-7 contain PCR amplified cDNA from the superior cervical ganglion of four SHRs. Lanes 8-11 contain amplified superior cervical ganglion cDNA from four WKYs. The positions of the GAPDH (510 bp) and NK1 receptor (537 bp) amplified products are indicated to the right. The sizes of the pertinent molecular weight marker DNAs are indicated to the left. Reprinted modified from Schoborg et al., (2000) by permission of The American Physiological Society.

Heart Failure

Little is known about functional and structural alterations of cardiac ganglia in heart failure. As in hypertension, there is a general activation of the sympathetic nervous system in both experimental and human heart failure (Ferguson and Mark, 1994). This has been the explanation for the tachycardia that occurs in heart failure and is supported by several lines of evidence. In contrast, control of the heart by the parasympathetic nervous system is attenuated in heart failure (Eckberg et al., 1971). The anatomic sites responsible for the increase in sympathetic nervous system activity is not completely understood but appears to be cardiovascular centers within the central nervous system (Ferguson and Mark, 1994).

14. NEUROCHEMICAL HETEROGENEITY IN SYMPATHETIC GANGLIA AND ITS IMPLICATIONS FOR CARDIOVASCULAR REGULATION

PERSPECTIVES

Sympathetic ganglia can no longer be considered mere relay stations for sympathetic efferent outflow. As we have shown in this chapter, there is a high degree of complexity in organization and function of these ganglia. The following examples illustrate this concept. A variety of peptides found in both preganglionic axon fibers and principal ganglionic neurons exert modulatory effects on cholinergic transmission that can be detected in target organs such as heart and blood vessels. In addition, other mediators, different from the known classical ganglionic transmitters, ACh and NA, have been proposed to participate as principal transmitters in sympathetic ganglia; these include dopamine, GABA, glutamate, ATP, NO and probably some peptides. The observation of preganglionic axon fibers innervating specific principal ganglionic neurons based on their specific transmitters has led to the concept of chemical coding. Furthermore, ganglionic activity may be modified, either inhibited or enhanced, by changes in the signal pattern of input activity. As a final point, in addition to the motor efferent outflow, sensory afferent pathways from peripheral organs including the heart and blood vessels make synaptic contact with principal ganglionic neurons and probably provide efferent sensory information to these neurons. This neural circuit between ganglia and peripheral organs can be the basis of a short-loop reflex. Further understanding of the complexity of information processing within sympathetic ganglia and its influence in cardiovascular regulation could contribute to the development of novel therapeutic tools for treating hypertension and cardiac diseases.

ACKNOWLEDGEMENTS:

Research support to the authors has been provide by CONACYT grant 32685-N (MAM), a Grant-in-Aid from the Southeastern Affiliate of the American Heart Association (JCH) and by NIH grants HL54268 (JCH) and HL54633 (DBH).

REFERENCES

Adams, P.R., Brown, D.A., Constanti, A., 1982. Pharmacological inhibition of the M-current. J. Physiol. 332, 223-262.

Aileru, A.A., De Albuquerque, A., Hamlyn, J.M., Manunta, P., Shah, J.R., Hamilton, M.J., Weinreich, D., 2001 Synaptic plasticity in sympathetic ganglia from acquired and inherited forms of ouabain-dependent hypertension. Am. J. Physiol. Regul. Integr. Comp. Physiol. 281, R635-644.

Alexander, A., Mathie, A., Peters, J.A., 2001. Acetylcholine receptors. TiPS Receptor Nomenclature Supplement 6-11.

Altememi, G.F., Alkadhi, K.A., 1999. Nitric oxide is required for the maintenance but not initiation of ganglionic long-term potentiation. Neuroscience 94, 897-902.

Anderson, C.R., Edwards, S.L., Furness, J.B., Bredt, D.S., Snyder, S.H., 1993. The distribution of nitric oxide synthase-containing autonomic preganglionic terminals in the rat. Brain Res. 614, 78-85.

Armour, J.A., 1994. Peripheral autonomic neuronal interactions in cardiac regulation. In, Neurocardiology. (Armour, J.A. and Ardell, J.L. eds.) Oxford Press, New York, pp. 219-244.

Bachoo, M., Ciriello, J., Polosa, C., 1987. Effect of preganglionic stimulation on neuropeptide-like immunoreactivity in the stellate ganglion of the cat. Brain Res. 400, 377-382.

Bachoo, M., Polosa, C., 1988. Cardioacceleration produced by close intra-arterial injection of neurotensin into the stellate ganglion of the cat. Can. J. Physiol. Pharmacol. 66, 408-412.

Birks, R.I., 1978. Regulation by patterned preganglionic neural activity of transmitter stores in a sympathetic ganglion. J. Physiol. 280, 559-572.

Bliss, T.V., Lomo, T., 1973. Long-lasting potentiation of synaptic transmission in the dentate area of the anaesthetized rabbit following stimulation of the perforant path. J. Physiol. 232, 331-356.

Borasio, P.G., Pavan, B., Fabbri, E., Ginanni-Corradini, F., Arcelli, D., Poli, A., 1995. Adenosine analogs inhibit acetylcholine release and cyclic AMP synthesis in the guinea-pig superior cervical ganglion. Neurosci. Lett. 184, 97-100.

Briggs, C. A., 1995. Long-term potentiation of synaptic transmission in the sympathetic ganglion: Multiple types and mechanisms. In Autonomic Ganglia (ed McLachlan E.M.), Vol. 6, pp. 297-339. Harwood Academic Publishers, Australia.

Briggs, C.A., 1992. Potentiation of nicotinic transmission in the rat superior cervical sympathetic ganglion: effects of cyclic GMP and nitric oxide generators. Brain Res. 573, 139-146.

Briggs, C.A., McAfee, D.A., 1988. Long-term potentiation at nicotinic synapses in the rat superior cervical ganglion. J. Physiol. 404, 129-144.

Briggs, C.A., McAfee, D.A., McCaman, R., 1985. Long-term potentiation of synaptic acetylcholine release in the superior cervical ganglion of the rat. J. Physiol. 363,181-190.

Brown, T.H.., McAfee, D.A., 1982. Long-term potentiation in the superior cervical ganglion. Science 215, 1411-1413.

Browning, K.N., Zheng, Z.L., Kreulen, D.L., Travagli, R.A., 1998. Effects of nitric oxide in cultured prevertebral sympathetic ganglion neurons. J. Pharmacol. Exp. Ther. 286, 1086-1093.

Burnstock, G., 1976. Do some nerve cells release more than one transmitter? Neuroscie. 1, 239-248.

14. NEUROCHEMICAL HETEROGENEITY IN SYMPATHETIC GANGLIA AND ITS IMPLICATIONS FOR CARDIOVASCULAR REGULATION

Carbó, R., Zetina, M.E., Corkidi, G., Morales, M.A., 1997. Topographic relationship of neurotensin-containing axon terminals with cardiac and non-cardiac principal ganglion cells in the stellate ganglia of the cat. Synapse 25, 277-284.

Carraway, R., Leeman, S.E., 1973. The isolation of a new hypotensive peptide, neurotensin, from bovine hypothalami. J. Biol. Chem. 248, 6854-6861.

Caverson, M.N., Bachoo, M., Ciriello, J., Polosa, C., 1989. Effect of preganglionic stimulation of chronic decentralization on neurotensin-like immunoreactivity in sympathetic ganglia of the cat. Brain Res. 482, 365-370.

Chen, C., Schofield, G.G., 1993. Nitric oxide modulates Ca^{2+} channel currents in rat sympathetic neurons. Eur. J. Pharmacol. 243, 83-86.

Cheung, D.W., 1988. Atriopeptin III depresses the excitability of sympathetic neurones. Biochem. Biophys. Res. Commun. 154, 411-416.

Christian, E.P., Weinreich, D., 1992. Presynaptic histamine H_1 and H_3 receptors modulate sympathetic ganglionic synaptic transmission in the guinea-pig. J. Physiol. 457, 407-430.

Costa, M., Furness, J.B., Gibbins, I.L., 1986. Chemical coding of enteric neurons. Prog. Brain Res. 68,217-239.

Collier, B., 1996. Activity related modulation of cholinergic transmission. Prog. Brain Res. 109, 243-249.

Cuello, A.C., 1987. Peptides as neuromodulators in primary sensory neurons. Neuropharmacology 26, 971-979.

Dale, H., 1935. Pharmachology and nerve endings. Proc. Roy. Soc. Med. 28, 319-332.

Dalsgaard, C.J., Jonsson C.E., Hökfelt T., Cuello A.C., 1983. Localization of substance P-immunoreactive nerve fibers in the human digital skin. Experientia 39, 1018-1020.

Darvesh, S., Nance, D.M., Hopkins, D.A., Armour, J.A., 1987. Distribution of neuropeptide-like immunoreactivity in intact and chronically decentralized middle cervical and stellate ganglia of dogs. J. Auton. Nerv. Syst. 21, 67-80.

Debinski, W., Kuchel, O., Buu, N.T., Cantin, M., Genest, J., 1987. Atrial natriuretic factor partially inhibits the stimulated catecholamine synthesis in superior cervical ganglia of the rat. Neurosci. Lett. 77, 92-96.

Dendorfer, A., Thornagel, A., Raasch, W., Grisk, O., Tempel, K., Dominiak, P., 2002. Angiotensin II induces catecholamine release by direct ganglionic excitation. Hypertension 40, 348-354.

Dobó, E., Kása, P., Wenthold, R.J., Joó, F., Wolf, J.R., 1989. Evidence for GABA-ergic fibers entering the superior cervical ganglion of rat from the preganglionic nerve trunk. Histochemistry 92, 133-136.

Dun, N.J., Dun, S.L., Lin, H.H., Hwang, L.L., Saria, A., Fischer-Colbrie, R., 1997. Secretoneurin-like immunoreactivity in rat sympathetic, enteric and sensory ganglia. Brain Res. 760, 8-16

Dun, N.J., Dun, S.L., Wu, S.Y., Forstermann, U., 1993. Nitric oxide synthase immunoreactivity in rat superior cervical ganglia and adrenal glands. Neurosci. Lett. 158, 51-54.

Dun N.J., Jiang Z.G., 1982. Non-cholinergic excitatory transmission in inferior mesenteric ganglia of the guinea-pig: possible mediation by substance P. J. Physiol. (Lond.) 325, 145-159.

Dun, N.J., Nishi, S., Karczmar, A.G., 1978. An analysis of the effect of angiotensin II on mammalian ganglion cells. J. Pharmacol. Exp. Ther. 204, 669-675.

Dunant, Y., Dolivo, M., 1968. Plasticity of synaptic functions in the excised sympathetic ganglion of the rat. Brain Res. 10, 271-273.

Eccles, R.M., Libet, B., 1961. Origin and Blockade of the synaptic responses of curarized sympathetic ganglia. J. Physiol. 157, 484-503.
Eckberg, D.L., Drabinsky, M., Braunwald, E., 1971. Defective cardiac parasympathetic control in patients with heart disease. N. Engl. J. Med. 285, 877-883.
Elfvin, L., Lindh, B., Hökfelt, T., 1993. The chemical neuroanatomy of sympathetic ganglia. Annu. Rev. Neurosci. 16, 471-507.
Eränkö, O., Sonila, S., Pävärinta, H., eds. *Histochemistry and Cell Biology of Autonomic Neurons, SIF cells and Paraneurons*. Ravin Press, New York 1980.
Ferguson, D., Mark, A.L., 1994. Clinical Neurocardiology: Role of the autonomic nervous system in clinical heart failure. In, *Neurocardiology*. (Armour, J.A. and Ardell, J.L. eds.) Oxford Press, New York, pp. 397-423.
Flet, D.L., Bell, C., 1991. Topography of functional subpopulations of neurons in the superior cervical ganglion of the rat. J. Anat. 177, 55-66.
Floras, J.S., 1990. Sympathoinhibitory effects of atrial natriuretic factor in normal humans. Circulation 81, 1860-1873.
Floras, J.S., 1995. Inhibitory effect of atrial natriuretic factor on sympathetic ganglionic neurotransmission in humans. Am. J. Physiol. 269, R406-R412.
Fredricksson, B., Sjöqvist, F. 1962. A cytomorphological study of cholinesterase in sympathetic ganglia of the cat. Acta Morphol. Neerl. Scand. 5, 140-166.
Gamse, R., Holzer, P., Lembeck, F., 1980. Decrease of substance P in primary afferent neurons and impairment of neurogenic plasma extravasation by capsaicin. Br. J. Pharmacol. 68, 207-213
Gibbins, I.L., 1991. Vasomotor, pilomotor and secretomotor neurons distinguished by size and neuropeptide content in superior cervical ganglia of mice. J. Auton. Nerv. Syst. 34, 171-183.
Gibbins, I.L., Furness, J.B., Costa, M., McTntyre, I., Hillyard, C.J., Girgis S., 1985. Co-localization of calcitonin gene-related peptide-like immunoreactivity with substance P in cutaneous vascular and visceral sensory neurons of guinea pigs. Neuroscie. Lett. 57, 125-130.
Gibbins, I.L., Jobling, P., Messenger, J.P., Teo, E.H., Morris, J.L., 2000. Neuronal morphology and synaptic organization of sympathetic ganglia. J. Auton. Nerv. Syst. 81, 104-109.
Gurusinghe, C.J., Bell C., 1989. Different patterns of immunolocalization of calcitonin gene-related peptide and substance P in sympathetic ganglia of normotensive and genetically hypertensive rats. Neurosci. Lett. 106, 89-94.
Hancock, J.C., Lindsay G.W., 2000. Enhanced ganglionic responses to substance P in spontaneously hypertensive rats. Peptides 21, 535-541.
Hanley, M.R., Benton, H.P., Lightman, S.L., Todd, K., Bone, E.A., Fretten, P., Palmer, S., Kirk, C.J., Michell, RH., 1984. A vasopressin-like peptide in the mammalian sympathetic nervous system. Nature 309, 258-261.
Häppölä, O., Päivärinta, H., Soinila, S., Wu, J.Y., Panula, P., 1987. Localization of L-glutamate decarboxylase and GABA transaminase immunoreactivity in the sympathetic ganglia of the rat. Neuroscience 21, 271-281.
Hendry, I.A., Iversen, L.L., Black, I.B., 1973. A comparison of the neural regulation of tyrosine hydroxylase activity in sympathetic ganglia of adult mice and rats. J. Neurochem. 20,1683-1689.
Henion, P.D., Landis, S.C., 1993. Modulation of the enkephalinergic phenotype of rat sympathetic neurons by hormonal and transsynaptic mechanisms. J. Neurobiol. 24, 1243-1251.

14. NEUROCHEMICAL HETEROGENEITY IN SYMPATHETIC GANGLIA AND ITS IMPLICATIONS FOR CARDIOVASCULAR REGULATION

Heym, C., Kummer, W., Gleich, A., Asmar, R., Liu, N., 1991. The guinea pig stellate ganglion: neurochemical and somatotopic organization. J. Auton. Nerv. Syst. 33, 104-105.

Heym, C., Liu, N., Gleich, A., Oberst, P., Kummer, W., 1993. Immunohistochemical evidence for different pathways immunoreactive to substance P and calcitonin gene-related peptide (CGRP) in the guinea-pig stellate ganglion. Cell Tissue Res. 272, 563-574.

Heym, C.H., Reinecke, M., Weihe, E., Forssmann, W.G., 1984. Dopamine-β-hydroxylase-, neurotensin-, substance P-, vasoactive intestinal polypeptide- and enkephalin-immunohistochemestry of paravertebral and prevertebral ganglia in the cat. Cell. Tiss. Res. 235, 411-418.

Heym, C., Webber, R., Horn, M., Kummer, W., 1990. Neuronal pathways in the guinea-pig lumbar sympathetic ganglia as revealed by immunohistochemistry. Histochemistry 93, 547-557.

Hilton, J.G., 1961. The pressor response to neostigmine after ganglionic blockade. J. Pharmacol. Exp. Ther. 132, 23-28.

Hökfelt, T., Elfvin, L.G., Elde, R., Schultzberg M., Goldstein, M., Luft R., 1977a. Occurrence of somatostatin-like immunoreactivity in some peripheral sympathetic noradrenergic neurons. Proc. Natl. Acad. Sci. USA. 74, 3587-3591.

Hökfelt, T., Elfvin, L.G., Schultzberg, M., Fuxe, K., Said, S.I., Goldstein, M., 1977b. Immunohistochemical evidence of vasoactive intestinal polypeptide-containing neurons and nerve fibers in sympathetic ganglia. Neurosci. 2, 885-896.

Hökfelt, T., Elfvin, L.G., Schultzberg, M., Goldstein, M., Nilsson, G., 1977c. On the occurrence of substance P-containing fibers in sympathetic ganglia: immunohistochemical evidence. Brain Res. 132, 29-41.

Hökfelt T., Kellerth J.O., Nilsson G., Pernow B., 1975. Substance P, localization in the central nervous system and in some primary sensory neurons. Science 190, 889-890.

Holmberg, K., Steinbusch, H.M., de Vente, J., Hökfelt, T., 2001. Distribution of cGMP in guinea pig autonomic ganglia after stimulation with sodium nitroprusside. Auton.Neurosci. 89, 7-15.

Inagaki, S., Kubota, Y., Kito, S., Kangawa, K., Matsuo, H., 1986. Immunoreactive atrial natriuretic polypeptides in the adrenal medulla and sympathetic ganglia. Regul. Pept. 15, 249-260.

James, S., Hassall, C.J., Polak, J.M., Burnstock, G., 1990. Visualisation of specific binding sites for atrial natriuretic peptide on non-neuronal cells of cultured rat sympathetic ganglia. Cell Tissue Res. 259, 129-137.

Jänig, W., 1988. Pre- and postganglionic vasoconstrictor neurons: differentiation, types and discharge properties. Ann. Rev. Physiol. 50, 525-539.

Jiménez, B., Mora-Valladares, E., Zetina, M.E., Morales, M.A., 2002. Occurrence, co-occurrence and topographic distribution of choline acetyltransferase, met-enkephalin and neurotensin in the stellate ganglion of the cat. Synapse 43,163-174.

Jubelin, B.C., Kannan, M.S., 1990. Neurons from neonatal hypertensive rats exhibit abnormal membrane properties in vitro. Am. J. Physiol. 259, C389-C396.

Judy, W.V., Watanabe, A.M., Henry, D.P., Besch, H.R., Jr., Murphy, W.R., Hockel, G.M., 1976. Sympathetic nerve activity, role in regulation of blood pressure in the spontaneously hypertensive rat. Circ. Res. 38, 21-29.

Kadota, T., Kadota, K., 2002. Rapid structural remodeling of shaft synapses associated with long-term potentiation in the cat superior cervical ganglion in situ. Neuroscie. Res. 43, 135-146.

Karhula, T., 1995. Comparison of immunohistochemical localization of [Met5]enkephalin-Arg6-Gly7-Leu8, [Met5]enkephalin, neuropeptide Y and vasoactive intestinal polypeptide in the superior cervical ganglion of the rat. J. Auton. Nerv. Syst. 51, 9-18.

Klimaschewski, L., Kummer, W., Heym, C., 1996. Localization, regulation and functions of neurotransmitters and neuromodulators in cervical sympathetic ganglia. Microsc. Res. Tech. 35, 44-68.

Koelle, G.B., 1955. The histochemical identification of acetylcholinesterase in cholinergic, adrenergic and sensory neurons. J. Pharmacol. Exp. Ther. 114, 167-184.

Kondo, H., Kuramoto, H., Wainer, B.H., Yanaihara, N., 1985. Evidence for the coexistence of acetylcholine and enkephalin in the sympathetic preganglionic neurons of rats. Brain Res. 335, 309-314.

Konishi, S., Okamoto, K., Otsuka, M., 1985. Substance P as a neurotransmitter released from peripheral branches of primary afferent neurons producing slow synaptic excitation in autonomic ganglion cells. In: Jordon C.C., Oehme P. (Eds), Substance P: Metabolism and Biological Actions. Taylor and Francis, Philadelphia, pp. 121-136.

Konishi, S., Tsunoo, A., Otsuka, M., 1979. Enkephalins presynaptically inhibit cholinergic transmission in sympathetic ganglia. Nature 282, 515–516.

Kow, L.M., Pfaff, D.W., 1988. Neuromodulatory actions of peptides. Annu. Rev. Pharmacol. Toxicol. 28, 163-188.

Krukoff, T., Cirielo, J., Calaresu, F., 1985. Segmental distribution of peptide-like immunoreactivity in cell bodies of the thoracolumbar sympathetic nuclei of the cat. J. Comp. Neurol. 240, 90-102.

Kummer, W., 1987. Galanin- and neuropeptide Y-like immunoreactivities coexist in paravertebral sympathetic neurons of the cat. Neurosci. Lett. 78, 127-131.

Kummer, W., Heym, C., 1988. Neuropeptide distribution in the cervico-thoracic paravertebral ganglia of the cat with particular reference to calcitonin gene-related peptide immunoreactivity. Cell Tissue Res. 252, 463-471.

Kushiku, K., Yamada, H., Shibata, K., Tokunaga, R., Katsuragi, T., Furukawa, T., 2001. Upregulation of immunoreactive angiotensin II release and angiotensinogen mRNA expression by high-frequency preganglionic stimulation at the canine cardiac sympathetic ganglia. Circ. Res. 88, 110-116.

Langley, J.N., 1921. The Autonomic Nervous System. Heffer & Sons, Cambridge.

Larsson, L.I., Rehfeld, J.F., 1979. Localization and molecular heterogeneity of cholecystokinin in the central and peripheral nervous system. Brain Res. 165, 201-218.

Leranth, C., Williams, T.H., Jew, J.Y., Arimura, A., 1980. Immuno-electron microscopic identification of somatostatin in cells and axons of sympathetic ganglia in the guinea pig. Cell Tissue Res. 212, 83-89.

Liang, S.D., Vizi, E.S., 1997. Positive feedback modulation of acetylcholine release from isolated rat superior cervical ganglion. J. Pharmacol. Exp. Ther. 280, 650-655.

Libet, B., Tosaka, T., 1969. Slow inhibitory and excitatory postsynaptic responses in single cells of mammalian sympathetic ganglia. J. Neurophysiol. 32, 43-50.

Lindh, B., Lundberg, J.M., Hökfelt, T., 1989. NPY-, galanin-, VIP/PHI-, CGRP- and substance P-immunoreactive neuronal subpopulations in cat autonomic and sensory ganglia and their projections. Cell Tissue Res. 256, 259-273.

Lindh, B., Staines, W., Hökfelt, T., Terenius, L., Salvaterra, P.M., 1986. Immunohistochemical demostration of choline acetyltransferase-immunoreactive

preganglionic nerve fibers in guinea pig autonomic ganglia. Proc. Natl. Acad. Sci. USA. 83, 5316-5320.
Long , J.P., Eckstein, J.W., 1961. Gamgionic actions of neostigmine methylsulphate. J. Pharmacol. Exp. Ther. 133, 216-222.
Lundberg, J.M., Rokaeus, A., Hokfelt, T., Rosell, S., Brown, M., Goldstein, M., 1982a. Neurotensin-like immunoreactivity in the preganglionic sympathetic nerves and in the adrenal medulla of the cat. Acta Physiol. Scand. 14, 153-155.
Lundberg, J.M., Tatemoto, K., 1982. Pancreatic polypeptide family (APP, BPP, NPY and PYY) in relation to sympathetic vasoconstriction resistant to alpha-adrenoceptor blockade. Acta Physiol. Scand. 116, 393-402.
Ma, X., Abboud, F.M., Chapleau, M.W., 2001. A novel effect of angiotensin on renal sympathetic nerve activity in mice. J. Hypertension 19, 609-618.
Ma, R.C., Horwitz, J., Kiraly, M., Perlman, R.L., Dun, N.J., 1985. Immunohistochemical and biochemical detection of serotonin in the guinea pig celiac-superior mesenteric plexus. Neurosci. Lett. 56, 107-112.
Magee, J.C., Schofield, G.G., 1992. Neurotransmission through sympathetic ganglia of spontaneously hypertensive rats. Hypertension 20, 367-373.
Magee, J.C., Schofield, G.G., 1994. Alterations of synaptic transmission in sympathetic ganglia of spontaneously hypertensive rats. Am. J. Physiol. 267, R1397-R1407.
Maggio, J.E., 1988. Tachykinins. Ann. Rev. Neurosci. 11, 13-28.
Maher, E., Bachoo, M., Cernacek, P., Polosa, C., 1991. Dynamics of neurotensin stores in the stellate ganglion of the cat. Brain Res. 562, 258-264.
Matthews, M.R., Connaughton M., Cuello A.C., 1987. Ultrastructure and distribution of substance P-immunoreactive sensory collaterals in the guinea pig prevertebral sympathetic ganglia. J. Comp. Neurol. 258, 28-51.
Mazet, B., Miller, S.M., Szurszewski, J.H., 1996. Electrophysiological effects of nitric oxide in mouse superior mesenteric ganglion. Am. J. Physiol. 270, G324-G331.
McCallum, J.B., Boban, N., Hogan, Q., Schmeling, W.T., Kampine, J.P., Bosnjak, Z.J., 1998. The mechanism of α_2-adrenergic inhibition of sympathetic ganglionic transmission. Anesth. Analg. 87, 503-510.
Medgett, I.C., 1983. Modulation of transmission in rat sympathetic ganglia by activation of presynaptic α- and β-adrenoceptors. Br. J. Pharmacol. 78, 17-27.
Morales, M.A., Bachoo, M., Beaudet, A., Collier, B., Polosa, C., 1993. Ultrastructural localization of neurotensin immunoreactivity in the stellate ganglion of the cat. J. Neurocitol. 22, 1017-1021.
Morales, M.A., Bachoo, M., Polosa, C., 1994. Pre-and postsynaptic components of nicotinic long-term potentiation in the superior cervical ganglion of the cat. J. Neurophysiol. 72, 819-824.
Morales, M.A., Holmberg, K., Xu, Z.Q., Cozzaris, C., Hartmans, B.K., Emson, P., Goldstein, M., Elfvin L.C., Hökfelt, T., 1995. Localization of choline acetyltransferase in rat peripheral sympathetic neurons and its coexistence with nitric oxide synthase and neuropeptides. Proc. Natl. Acad. Sci. USA. 92,11819-11823.
Morrison, S.F., Callaway, J., Milner, T.A., Reis, D.J., 1989. Glutamate in the spinal sympathetic intermediolateral nucleus: localization by light and electron microscopy. Brain Res. 503, 5-15.
Nacimiento, W., Schoen, S.W., Nacimiento, A.C., Kreutzberg, G.W., 1991. Cytochemistry of 5'-nucleotidase in the superior cervical ganglion of cat and guinea pig. Brain Res. 567, 283-289.

Nishi, S., Koketsu, K., 1968. Early and late after-discharges of amphibian sympathetic ganglion cells. J. Neurophysiol. 31, 109-121.
Njä, A., Purves, D., 1977. Specific innervation of guinea pig superior cervical ganglion cells by preganglionic fibers arising from different levels of the spinal cord. J. Physiol. 264, 565-583.
North, R.A., Williams, J.T., 1976. Enkephalin inhibits firing of myenteric neurons. Nature 264, 460-461.
Nozdrachev, A.D., Jiménez, B., Morales, M.A. Fateev, M.M., 2002. Neuronal organization and cell interactions of the cat stellate ganglion. Autonom. Neuroscience 95, 43-56.
Pant, K.K., Smith, P.A., 1989. Atrial natriuretic factor suppresses M-current in frog but not in rat sympathetic neurones. Neurosci. Lett. 100, 243-248.
Párducz, A., Dobo, E., Joo, F., Wolff, J.R., 1992. Termination pattern and fine structural characteristics of GABA- and [Met]enkephalin-containing nerve fibers and synapses in the superior cervical ganglion of adult rat. Neuroscience 49, 963-971.
Prosdocimi, M., Finesso, M., Gorio, A., 1986. Enkephalin modulation of neural transmission in the cat stellate ganglion: pharmacological actions of exogenous opiates. J. Auton. Nerv. Syst. 17, 217-30.
Purves, D., Snider, W.D., Voyvodic, J.T., 1988. Trophic regulation of nerve cell morphology and innervation in the autonomic nervous system. Nature 336, 123-128.
Quigg, M., Elfvin, L.G., Aldskogius, H., 1990. Anterograde transsynaptic transport of WGA-HRP from spinal afferents to postganglionic sympathetic cells of the stellate ganglion of the guinea pig. Brain Res. 518, 173-178.
Quinson, N., Catalin, D., Miolan, J.P., Niel, J.P., 1998. Nerve-induced release of nitric oxide exerts dual effects on nicotinic transmission within the coeliac ganglion in the rabbit. Neuroscience 84, 229-240.
Rao, M.S., Landis, S.C., 1990. Characterization of a target-derived neuronal cholinergic differentiation factor. Neuron. 5, 899-910.
Reid, I.A., 1992. Interactions between ANG II, sympathetic nervous system, and baroreceptor reflexes in regulation of blood pressure. Am. J. Physiol. 262, E763-E778.
Schramm, L.P., Chornoboy, E.S., 1982. Sympathetic activity in spontaneously hypertensive rats after spinal transection. Am. J. Physiol. 243, R506-R511.
Schoborg, R.V., Hoover, D.B., Tompkins, J.D., Hancock J.C., 2000. Increased ganglionic responses to substance P in hypertensive rats due to upregulation of NK_1 receptors. Am. J. Physiol. Regul. Integr. Comp. Physiol. 279, R1685-R1694.
Schultzberg, M., 1983. Bombesin-like immunoreactivity in sympathetic ganglia. Neuroscience 8, 363-374.
Schultzberg, M., Hökfelt, T., Terenius, L., Elfvin, L.G., Lundberg, J.M., Brandt, J., Elde, R.P., Goldstein, M., 1979. Enkephalin immunoreactive nerve fibres and cell bodies in sympathetic ganglia of the guinea-pig and rat. Neuroscie. 4, 249-270.
Schultzberg, M., Lindh, B., 1988. Transmitters and peptides in autonomic ganglia. In: Handbook of Chemical Neuroanatomy. The Peripheral Nervous System 6, 297-326.
Silinsky, E.M., Gerzanich, V., 1993. On the excitatory effects of ATP and its role as a neurotransmitter in celiac neurons of the guinea-pig. J. Physiol. 464, 197-212.
Stapelfeldt, W.H., Szurszewski, J.H., 1989a. The electrophysiology effects of neurotensin on neurons of guinea-pig prevertebral sympathetic ganglia. J. Physiol. 411, 301-323.
Stapelfeldt, W.H., Szurszewski, J.H., 1989b. Neurotensin facilitates release of substance P in the guinea-pig inferior mesenteric ganglion. J. Physiol. 411, 325-343.
Sullivan, C.J., Bolter, C.P., 1997. Influence of angiotensin II on transmission in the superior cervical ganglion. Arch. Physiol. Biochem. 105, 577-582.

Szurszewski, J.H., 1981. Physiology of mammalian prevertebral ganglia. Annu. Rev. Physiol. 81;43, 53-68.

Taylor, P., 2001. Agents acting at the neuromuscular junction and autonomic ganglia. In, *Goodman and Gilman's The Pharmacological Basis of Therapeutics.* Edition 10. (Hardman J.G., Limbird, L.E. and Goodman Gilman. A. eds.) McGraw-Hill, New York, pp. 193-213.

Tompkins, J.D., Hancock, J.C., 2002. Electrophysiological effects of tachykinin agonists on sympathetic ganglia of spontaneously hypertensive rats. Auton. Neurosci. 97, 26-34.

Undem, B.J., Hubbard, W.C., Christian, E.P., Weinreich, D., 1990. Mast cells in the guinea pig superior cervical ganglion: a functional and histological assessment. J. Auton. Nerv. Syst. 30, 75-87.

Undem, B.J., Myers, A.C., Weinreich, D., 1991. Antigen-induced modulation of autonomic and sensory neurons in vitro. Int. Arch. Allergy Appl. Immunol. 94, 319-324.

Volle, R.L., 1962. The actions of several ganglion blocking agents on the postganglionic discharge induced by diisopropyl phosphorofurodate (DFP) in sympathetic ganglia. J. Pharmacol. Exp. Ther. 135, 45-53.

Weight, F.F., Schulman, J.A., Smith, P.A., Busis, N.A., 1979. Long-lasting synaptic potentials and the modulation of synaptic transmission. Fed. Proc. 38, 2084-2094.

Yamamoto, K., Senba, E., Matsunaga, T., Tohyama, M., 1989. Calcitonin gene-related peptide containing sympathetic preganglionic and sensory neurons projecting to the superior cervical ganglion of the rat. Brain Res. 487, 158-164.

Yarowsky, P., Weinreich D., 1985. Loss of accommodation in sympathetic neurons from spontaneously hypertensive rats. Hypertension 7, 268-276.

Zaidi, Z.F., Matthews, M.R., 1997. Exocytotic release from neuronal cell bodies, dendrites and nerve terminals in sympathetic ganglia of the rat, and its differential regulation. Neuroscience 80, 861-891.

Zawadzki, J.V., Furchgott, R.F., Cherry, P., 1981. The obligatory role of endothelial cells in the relaxation of arterial smooth muscle by substance P. Fed. Proc. 40,689.

Zengel, J.E., Magleby, K.L., Horn, J.P., McAfee, D.A., Yarowsky, P.J., 1980. Facilitation, augmentation, and potentiation of synaptic transmission at the superior cervical ganglion of the rabbit. J. Gen. Physiol. 76, 213-31.

Zetina, M.E., Jiménez, B., Diaz-Luna, F., Mora-Valladares, E., Morales, M.A., 1999. Release-depletion and receptor-mediated neuronal internalization of endogenous neurotensin in the stellate ganglion of the cat. Neuroscience 92, 655-664.

Zhang, C., Bachoo, M., Polosa, C., 1991. Naloxone-sensitive inhibition of nicotinic transmission in the superior cervical ganglion of the cat. Brain Res. 548, 29-34.

Zhang, C., Bachoo, M., Polosa, C., 1993. The receptors activated by exogenous and endogenous opioids in the superior cervical ganglion of the cat. Brain Res. 622, 211-214.

Chapter 15

MAMMALIAN CARDIAC GANGLIA AS LOCAL INTEGRATION CENTERS: HISTOCHEMICAL AND ELECTROPHYSIOLOGICAL EVIDENCE

Rodney L. Parsons
University of Vermont College of Medicine, Burlington, VT 05405, USA

Abstract: The mammalian parasympathetic cardiac ganglia form a complex intrinsic cardiac nervous system composed of parasympathetic postganglionic projection neurons, local interneurons and possibly afferent neurons with central projections. In addition to the preganglionic parasympathetic cholinergic excitatory innervation, the cardiac ganglia are innervated by sympathetic postganglionic fibers and afferent fibers derived from both the spinal dorsal root ganglia and the vagal sensory ganglia. Thus, multiple neurotransmitters can potentially modulate activity of the cardiac neurons. Results discussed here identify the neurochemically-coded fiber inputs and summarize the neuronal response to some of these transmitters in a model system, the guinea pig cardiac ganglia. The diverse functional and neurochemical properties of cardiac neurons and their strategic location position them to mediate local reflex mechanisms initiated by changes in cardiac performance or by changes in the local environment within the myocardium. Thus, it is critical to better understand the integrative mechanisms within the cardiac ganglia.

Key words: Parasympathetic neuron, cholinergic neurons, neuropeptides, neurotransmitters

INTRODUCTION

Both parasympathetic inhibitory and sympathetic excitatory signals to the heart originate within the central nervous system and reach cardiac targets via two-neuron pathways. In the parasympathetic pathway, depending on the species, intervening cardiac ganglia are located in fat pads, on the epicardium and, in some cases, within the myocardium. Historically, the parasympathetic cardiac ganglia were considered simple relay stations;

however, it is now thought that the cardiac ganglia form an integrative intrinsic cardiac nervous system intimately involved in local control of cardiac function (Parsons et al., 1987; Randall and Wurster, 1994; Armour, 1999; Ardell, 2001).

The cardiac ganglia contain two cell types: cardiac neurons and small intensely fluorescent (SIF) cells. Both cell types can be innervated by fibers containing a variety of neurotransmitters/neuromodulators. In addition to the preganglionic parasympathetic cholinergic excitatory innervation, cardiac ganglia can be innervated by sympathetic postganglionic fibers and afferent fibers derived from both the spinal dorsal root ganglia (DRG) and the vagal sensory ganglia. The specific neuropeptide or neuromodulator expressed in different nerve fibers can be species-specific. Thus, comparison of histochemical and pharmacological results must be made with caution when different animal model preparations are used. However, in all instances, spike activity of neurons in the cardiac ganglia is determined by their electrical personality and the coordinated action of the diverse neurochemical signals released from fibers innervating them.

Cardiac neurons can exhibit diverse functional and neurochemical properties and are strategically located to provide neuroregulatory responses to changes in cardiac performance and changes in the local environment within the myocardium. Neurons within cardiac ganglia may have different functions; some are the classical parasympathetic postganglionic projection neurons whereas others may be afferent neurons with central projections or local interneurons (Randall and Wurster, 1994; Edwards et al., 1995; Cheng et al., 1997; Armour, 1999; Ardell, 2001). Although the concept of an integrative function for the cardiac ganglia appears common across mammalian species, the complexity of the intrinsic cardiac nervous system likely increases when compared in primates, cats and dogs to that in rodents. In all species, distinct cardiac ganglia are positioned appropriately to regulate heart rate, conductivity and contractility (Ardell and Randall, 1986; Billman et al., 1989; Randall et al., 1991; Gatti et al., 1995). These distinct cardiac ganglia receive preferential innervation from neurons in cardioregulatory brainstem nuclei (Gatti et al., 1997; Cheng et al., 1999; Cheng and Powley, 2000).

ELECTROPHYSIOLOGICAL PROPERTIES OF CARDIAC NEURONS

The electrophysiological properties of mammalian cardiac neurons have been analyzed using cells in intact whole mount ganglia preparations and dissociated neurons of many species including rat, guinea pig, pig and dog

15. MAMMALIAN CARDIAC GANGLIA AS LOCAL INTEGRATION CENTERS: HISTOCHEMICAL AND ELECTROPHYSIOLOGICAL EVIDENCE

(Seabrook et al., 1990; Xu and Adams, 1992a,b; Selyanko, 1992; Selyanko and Skok, 1992b; Smith et al., 1992; Xi-Moy et al., 1993; Xi et al., 1994; Adams and Harper, 1995; Edwards et al., 1995; Hardwick et al., 1995; Xi-Moy and Dun, 1995; Smith, 1999). Commonly, the cardiac ganglia contain two general classes of excitable neurons: phasic and tonic cells. The predominant phasic cells generate a maximum of one or two action potentials (APs) in response to long suprathreshold depolarizing current pulses, whereas tonic cells fire multiple APs with the number of APs roughly proportional to the stimulus strength. Both sodium (Na^+) and calcium (Ca^{2+}) influx contribute to the depolarizing phase of the AP and repolarization is due to the activation of Ca^{2+}- and voltage-activated potassium (K^+) conductances (Xu and Adams, 1992a,b; Adams and Harper, 1995). An afterhyperpolarization (AHP) follows spike repolarization (Adams and Harper, 1995). Cardiac neuron excitability is regulated in part by the non-inactivating voltage-dependent K^+ current I_M and the hyperpolarization-activated cationic conductance I_h (Edwards et al., 1995; Xi-Moy and Dun, 1995; Cuevas et al., 1997). Modulation of both currents can influence the level of the resting membrane potential and both can be modulated by neurotransmitters. Neuronal excitability also can be increased following a decrease in the AHP that occurs after an AP (Adams and Harper, 1995). Commonly, the AHP results from the activation of Ca^{2+}-activated K^+ conductances by Ca^{2+} entering neurons through Ca^{2+}-selective membrane channels. Reduction of Ca^{2+} influx or blockade of Ca^{2+}-activated K^+ channels can decrease the amplitude and duration of the AHP, leading to repetitive firing.

Both fast excitatory postsynaptic potentials (fEPSPs) and slow excitatory postsynaptic potentials (sEPSPs) can be elicited by stimulation of vagal trunks or by stimulation of interganglionic nerve bundles (Konishi et al., 1985; Seabrook et al., 1990; Selyanko and Skok, 1992b; Smith et al., 1992; Xi-Moy et al., 1993; Edwards et al., 1995; Hardwick et al., 1995; Smith, 1999). Fast EPSPs are generated by acetylcholine (ACh), which is released from preganglionic nerves and acts on nicotinic receptors that gate a nonselective cationic conductance (Fieber and Adams, 1991a; Selyanko and Skok, 1992a; Xi-Moy et al., 1993). The response to exogenous ACh and the fEPSP are blocked by ganglionic nicotinic antagonists.

Adams and colleagues have shown that rat cardiac neurons also express ATP-gated P2X and P2Y receptors (Fieber and Adams, 1991b). ATP activation of P2X receptors gates a nonselective cationic conductance distinct from that activated by ACh (Fieber and Adams, 1991b; Liu and Adams, 2001) whereas activation of P2Y receptors mobilizes intracellular

Ca^{2+} through the phospholipase C/inositol 1,4,5-triphosphate signaling pathway (Liu et al., 2000b). Iontophoretic application of ATP to guinea pig cardiac neurons produced no depolarization; thus these neurons apparently do not express P2X receptors (Inokuchi and McLachlan, 1995).

Slow EPSPs are generated following repetitive neural stimulation and could be generated by ACh released from preganglionic neurons acting on muscarinic receptors (Seabrook et al., 1990; Selyanko and Skok, 1992a,b; Xi-Moy et al., 1993). Alternatively, for the guinea pig, the sEPSP recorded from cardiac neurons could be generated by substance P (SP), which is released from afferent fibers and activates NK_3 tachykinin receptors (Konishi et al., 1985; Hardwick et al., 1995, 1997). This view is supported by the observations that capsaicin application, which stimulates the release of tachykinins from afferent nerves, initiates a depolarization similar to that of the sEPSP and that sEPSPs can be elicited by fiber tract stimulation in the guinea pig when muscarinic receptors are blocked by atropine (Hardwick et al., 1995).

Mammalian intracardiac neurons express multiple voltage-dependent Ca^{2+} channels (VDCCs) (Adams and Harper, 1995). An initial voltage clamp analysis of Ca^{2+} currents (I_{Ca}), done on dissociated neonatal rat cardiac neurons maintained in cell culture, indicated that there were at least three pharmacologically distinct types of Ca^{2+} channels (Xu and Adams, 1992b). More recent studies done on acutely dissociated adult rat cardiac neurons confirm that the major component of I_{Ca} is current carried through N type channels, with smaller contributions to the total through L-type, Q-type and R-type channels (Jeong and Wurster, 1997a). I_{Ca} in rat cardiac neurons is inhibited by activation of a number of G-protein coupled receptors via pertussis toxin sensitive and voltage dependent pathways (Xu and Adams, 1993; Cuevas and Adams, 1997; Jeong and Wurster, 1997b; Adams and Trequattrini, 1998; Jeong et al., 1999; Zhang and Cuevas, 2002).

HISTOCHEMICAL ORGANIZATION OF THE GUINEA PIG CARDIAC GANGLIA

Our laboratory has used the guinea pig atrial whole mount preparation to establish the histochemical organization of mammalian cardiac ganglia. The cardiac ganglia are considered integration centers; thus, it was of interest to identify or confirm the class of fibers containing different neurotransmitters or neuropeptides in order to establish the action of these neurochemical-signaling molecules on individual cardiac neurons. The guinea pig atrial whole mount preparation was chosen as a model system because in this species the cardiac ganglia form a network of interconnected neuron clusters

15. MAMMALIAN CARDIAC GANGLIA AS LOCAL INTEGRATION CENTERS: HISTOCHEMICAL AND ELECTROPHYSIOLOGICAL EVIDENCE

that are essentially cell monolayers on the atrial epicardium, allowing immunohistochemical localization of bioactive substances within nerve fibers and cells, and intracellular recording from intact, innervated neurons (Hardwick et al., 1995; Mawe et al., 1996; Braas et al., 1998; Kennedy et al., 1998; Leger et al., 1999; Zhang et al., 2001).

Chemical Phenotype of Neurons in the Guinea Pig Cardiac Ganglia

Two cell types are easily discernable in guinea pig cardiac ganglia: large (20-40 µm) relatively simple cardiac neurons and much smaller SIF cells (<15 µm). Approximately 1500 cardiac neurons are present in the guinea pig intrinsic cardiac nervous system (Leger et al., 1999). Guinea pig cardiac neurons are readily labeled with an anti-serum directed against the acetylcholine synthetic enzyme choline acetyltransferase (ChAT)(Fig. 1A) (Mawe et al., 1996; Horackova et al., 1999; Leger et al., 1999; Calupca et al., 2000a,b; Calupca et al., 2001), an anti-serum directed against the somal cytoskeletal protein microtubule associated protein 2 (MAP-2) (Mawe et al., 1996) and by an anti-serum directed against protein gene product 9.5 (PGP 9.5) (Leger et al., 1999; Horackova et al., 1999). The SIF cells exhibit immunoreactivity to the catecholamine synthetic enzyme tyrosine hydroxylase (TH) (Mawe et al., 1996; Leger et al., 1999). Some SIF cells also exhibit ChAT immunoreactivity (Mawe et al., 1996).

The guinea pig cardiac neurons have the capability to express multiple intercellular signaling molecules. A quantitative analysis of the chemical coding of guinea pig cardiac neurons completed by Steele et al. (1994) indicates that subpopulations of cardiac neurons express a variety of neuropeptides including somatostatin, neuropeptide Y (NPY), vasoactive intestinal polypeptide (VIP), dynorphin, SP, and the nitric oxide (NO) synthesizing enzyme, nitric oxide synthase (NOS), with many cells expressing multiple putative neurotransmitters. Other studies also indicate that subsets of guinea pig cardiac neurons exhibit immunoreactivity to NPY or NOS (Hassall et al., 1992; Klimaschewski et al., 1992; Mawe et al., 1996; Sosunov et al., 1997; Kennedy et al., 1998; Horackova et al., 1999). We have confirmed in doubly-labeled, freshly fixed cardiac ganglia whole mounts that subpopulations of cardiac neurons exhibit immunoreactivity for somatostatin, NOS and NPY, but have not observed SP-immunoreactive (IR) cardiac neurons. In freshly fixed whole mount preparations, the percentage of cardiac neurons exhibiting somatostatin-IR is ~60% (Table 1). Fewer cardiac neurons exhibit immunoreactivity to NPY or NOS (Table 1) and the percentage of cells exhibiting immunoreactivity for NPY or NOS is quite variable between preparations (Calupca and Parsons, unpublished observations). A few (<5 %) neurons in freshly fixed cardiac ganglia exhibit immunoreactivity to pituitary adenylate cyclase-activating polypeptide (PACAP) (Braas et al., 1998; Calupca et al., 2000a) or cocaine- and amphetamine-regulated transcript peptide (CARTp) (Calupca et al., 2001). The CARTp-IR intracardiac neurons also exhibit NOS immunoreactivity.

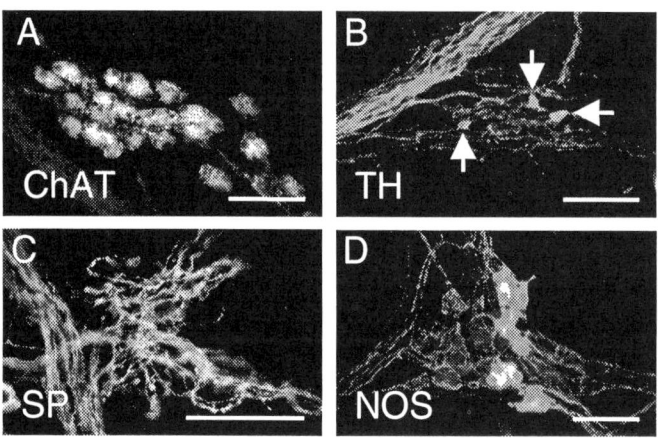

Figure 1. Cholinergic cardiac neurons are innervated by nerve fibers immunoreactive for tyrosine hydroxylase (TH), substance P (SP) or nitric oxide synthase (NOS). A: ChAT-IR neurons in a cardiac ganglion. B: A network of TH-IR fibers within a cardiac ganglion. Although no cardiac neurons exhibit TH immunoreactivity, SIF cells intermingled amongst the neurons are TH immunoreactive (indicated by arrows). C: SP-IR fibers surrounding neurons in a cardiac ganglion. D: NOS-IR cells and fibers within a cardiac ganglion. Although NOS cells are present in this ganglion, most of the NOS-IR fibers are derived from sources outside of the heart (Calupca et al., 2000b). The calibration bar equals 100 µm. Panel A is reprinted from the Journal of Comparative Neurology, Vol. 426, Calupca, M.A., Vizzard, M.A. and Parsons, R.L., Origin of Neuronal Nitric Oxide Synthase (NOS)-Immunoreactive Fibers in Guinea Pig Parasympathetic Cardiac Ganglia, pp. 493-504, 2000, with permission from Wiley-Liss, Inc.

Using reverse transcription–PCR, we determined that pro-PACAP, pro-NPY and pro-somatostatin mRNA are expressed within guinea pig cardiac ganglia preparations (Braas et al., 1998; Kennedy et al., 1998; Parsons et al., 2000). These observations are consistent with the presence of PACAP-IR, NPY-IR and somatostatin-IR cells and confirm that guinea pig cardiac neurons have the synthetic capability for PACAP peptide, NPY and somatostatin production in addition to ACh.

We also have shown for guinea pig cardiac neurons that the expression of neuropeptides by the cardiac neurons can exhibit considerable plasticity. Either maintenance in explant culture alone or in combination with colchicine treatment affects the percentage of ChAT-IR or MAP-2-IR cardiac neurons that exhibit immunoreactivity for somatostatin, NPY, PACAP or CARTp (Table 1) (Lynch et al., 1999; Calupca et al., 2000a; Parsons et al., 2000; Calupca et al., 2001). However, no change in the percentage of NOS-IR neurons was evident following explant culture for 72 hours (Calupca et al., 2000b). A few SP-IR neurons were noted in 72 hour explanted preparations (Hardwick and Parsons, unpublished observations), an observation consistent with Steele et al. (1994), who analyzed chemical coding of cardiac neurons in explanted cardiac ganglia preparations treated with colchicine. Thus, different experimental conditions very likely contribute to differences in published results related to cardiac neuron chemical coding. The mechanism(s) responsible for the change in neuropeptide expression in explanted cardiac neurons remain(s) to be determined, but it could be a response to axotomy or loss of trophic factors from extrinsically originating nerve fibers that degenerate in culture.

Table -1. Plasticity of neurochemical expression in guinea pig cardiac ganglia

	\multicolumn{4}{c}{% of ChAT-IR or MAP-2-IR neurons that co-express a specific neuropeptide/synthetic enzyme}		
	controls	n	72 hour explants	n	Reference
NPY*	4 ± 0.3^1	7	16 ± 2	7	Lynch et al., 1999
PACAP♦	4 ± 1	3	34 ± 2	3	Calupca et al., 2000a
Som°	60 ± 1	4	42 ± 2	3	Parsons et al., 2000
nNOS•	13 ± 3	5	13 ± 2	8	Calupca et al., 2000b
CARTp+	<4	13	33 ± 4	4	Calupca et al., 2001

[1]mean ± SEM, * NeuropeptideY, ♦ Pituitary adenylate cyclase activating polypeptide, °Somatostatin, • neuronal nitric oxide synthase, + Cocaine- and amphetamine-regulated transcript peptide, n Number of cardiac ganglia whole mount preparations examined.

Chemical Coding of Extrinsic Neural Inputs to Neurons Within the Guinea Pig Intracardiac Ganglia

Guinea pig cardiac neurons are surrounded by ChAT-IR fibers, most of which are parasympathetic preganglionic fibers (Fig. 2A) (Mawe et al.,

1996; Calupca et al., 2000a,b). Cardiac neurons also can be innervated by TH-IR sympathetic postganglionic nerve fibers (Fig. 1B) (Leger et al., 1999; Calupca et al., 2000a,b) and in double-labeled whole mount ganglia preparations, all of the TH-IR fibers exhibit immunoreactivity to NPY (Sternini and Brecha, 1985; Kennedy et al., 1998). In addition, the guinea pig cardiac ganglia are innervated by SP-IR fibers (Fig. 1C) and all the SP-IR fibers exhibit calcitonin gene-related peptide (CGRP) immunoreactivity (Wharton et al., 1981; Urban and Papka, 1985; Dalsgaard et al., 1986; Gerstheimer and Metz, 1986; Gibbins et al., 1987; Calupca et al., 2000a,b). The SP-/CGRP-IR fibers are afferent fibers derived primarily from DRG (Urban and Papka, 1985).

As would be expected for fibers derived from sources extrinsic to the heart, most of the ChAT-IR fibers surrounding individual cardiac neurons (Calupca et al., 2000a,b; 2001) and all of the TH-/NPY-IR fibers and SP-/CGRP-IR are absent in cardiac ganglia whole mounts maintained in culture for 72 hours (Lynch et al., 1999; Calupca et al., 2000a).

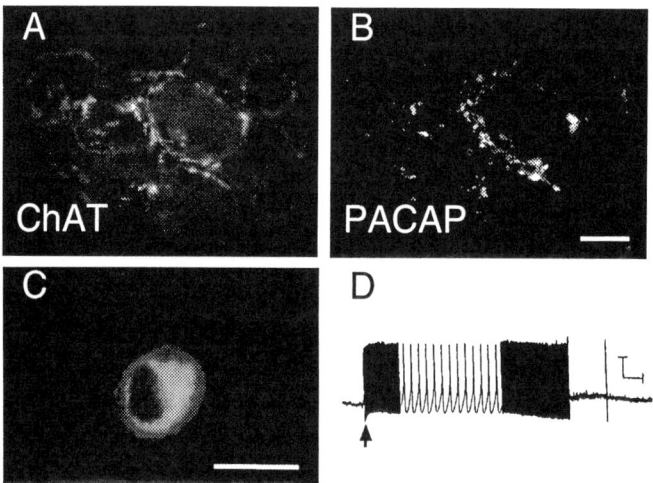

Figure 2. ChAT and PACAP are co-localized in fibers surrounding cardiac neurons. A,B: confocal optical slices (approximately 0.8 µm thick) of ChAT fibers (A) and PACAP fibers (B) around neurons in a double-labeled cardiac ganglia preparation. Note in A that the cardiac neurons are ChAT immunoreactive and that ChAT and PACAP are co-localized in fibers surrounding the cells. C: confocal image of a dissociated cardiac neuron immunocytochemically-labeled against MAP-2 and against the PAC_1 receptor. The staining of the cell interior surrounding the nucleus indicates MAP-2 immunoreactivity, whereas the portion of the neuron near the plasma membrane exhibits PAC_1 receptor immunoreactivity. The calibration bar equals 20 µm for A and B and 25 µm for C. D: puffer application of $PACAP_{27}$ (1 sec pulse of 50 µM, onset indicated by arrow) to an enzymatically dissociated

15. MAMMALIAN CARDIAC GANGLIA AS LOCAL INTEGRATION CENTERS: HISTOCHEMICAL AND ELECTROPHYSIOLOGICAL EVIDENCE

cardiac neuron with a resting membrane potential of -50 mV caused depolarization and a burst of action potentials. Calibration: y-axis, 20 mV; x-axis, 2 sec except for an expanded recording period of 200 msec. Panels C and D are reprinted from the Annals of the New York Academy of Sciences, Vol. 921, Parsons, R.L., Rossignol, T.M., Calupca, M.A., Hardwick, J.C., and Braas, K.M., PACAP Peptides Modulate Guinea Pig Cardiac Neuron Membrane Excitability and Neuropeptide Expression, pp. 202-210, 2000, with permission from the New York Academy of Sciences, U.S.A.

Many cardiac ganglia in the guinea pig atrial whole mount preparation are innervated by NOS-IR fibers (Fig. 1D) and within many ganglia, NOS IR fibers with prominent varicosities surround individual cardiac neurons (Tanaka et al., 1993; Calupca et al., 2000b). Even though some cardiac neurons exhibit NOS immunoreactivity, the majority of the NOS-IR fibers within the ganglia appear to be derived from sources outside the heart as they degenerate after 72 hours in culture (Calupca et al., 2000b). In addition, fibers derived from NOS-IR cardiac neurons primarily enter interganglionic fiber bundles without making any pericellular complexes around adjacent neurons. This is particularly evident in the 72 hour cultured whole mount ganglia preparations devoid of extrinsic NOS fibers (Calupca et al., 2000b). In the guinea pig cardiac ganglia, the NOS-IR fibers do not exhibit immunoreactivity for SP, CGRP, NPY or ChAT (Calupca et al., 2000b). Tanaka and Chiba (1998) demonstrated the NOS fibers were vagal in origin. Their observations coupled with our result that NOS and ChAT were not co-localized (Calupca et al., 2000b) lead us to propose that the extrinsic NOS-IR fibers innervating the guinea pig cardiac ganglia are primarily afferent fibers derived from vagal sensory ganglia. This conclusion is consistent with the presence of numerous NOS immunoreactive cells in the nodose ganglia (Calupca et al., 2000b).

CARTp-IR fibers also surround neurons within many cardiac ganglia (Calupca et al., 2001). These CARTp-IR fibers are not observed in 72-hour explant cultured preparations. The CARTp-IR fibers also exhibit NOS immunoreactivity and the NOS-/CARTp-IR fibers are hypothesized to be afferent fibers derived from the vagal sensory ganglia (Calupca et al., 2001).

In the guinea pig, PACAP-IR fibers surround neurons within the cardiac ganglia in freshly fixed preparations (Braas et al., 1998; Calupca et al., 2000a). In contrast, no PACAP-IR fibers innervate cardiac neurons in 72 hour cultured whole mount cardiac ganglia preparations even though the number of cardiac neurons exhibiting PACAP immunoreactivity increases approximately 8 fold (Table 1) (Calupca et al., 2000a). In double-labeled whole mount cardiac ganglia preparations, there is no co-localization of

PACAP in fibers immunoreactive for NPY, SP or NOS, indicating that the PACAP-IR fibers are not sympathetic postganglionic axons or DRG or vagal afferent fibers. However, PACAP-IR is co-localized with ChAT-IR within fibers innervating cardiac neurons (Fig 2B); thus, the PACAP-IR fibers are preganglionic parasympathetic axons (Calupca et al., 2000a).

NEUROCHEMICAL MODULATION OF GUINEA PIG CARDIAC NEURONS

Pharmacological studies indicate that NPY, SP, NO and PACAP modify the parasympathetic inhibitory drive to the heart (Pardini et al., 1992; Seebeck et al., 1996; Conlon and Kidd, 1999; Tompkins et al., 1999; Chang et al., 2000). Our initial studies investigated the effects of NPY and SP on cardiac neuron activity. More recent studies focus on the modulation of cardiac neurons by PACAP.

Other investigators have shown that NOS inhibitors reduce vagal bradycardia and that treatment with a NO donor can enhance responses to vagal stimulation. Thus, NO facilitates the negative chronotropic effects of vagal stimulation (Conlon and Kidd, 1999), possibly by acting within the cardiac ganglia (Armour et al., 1995) and/or by enhancing release of ACh from postganglionic nerves (Herring et al., 2000; Herring and Paterson, 2001).

NPY and Norepinephrine (NE) Have Multiple Effects on Guinea Pig Intracardiac Neurons

Our recent observations demonstrate that NPY hyperpolarizes cardiac neurons, decreases the amplitude and duration of the AHP that follows an AP, and decreases the amplitude of fEPSPs elicited by stimulation of interganglionic nerve bundles (Fig. 3A-C) (Kennedy et al., 1998).

15. MAMMALIAN CARDIAC GANGLIA AS LOCAL INTEGRATION CENTERS: HISTOCHEMICAL AND ELECTROPHYSIOLOGICAL EVIDENCE

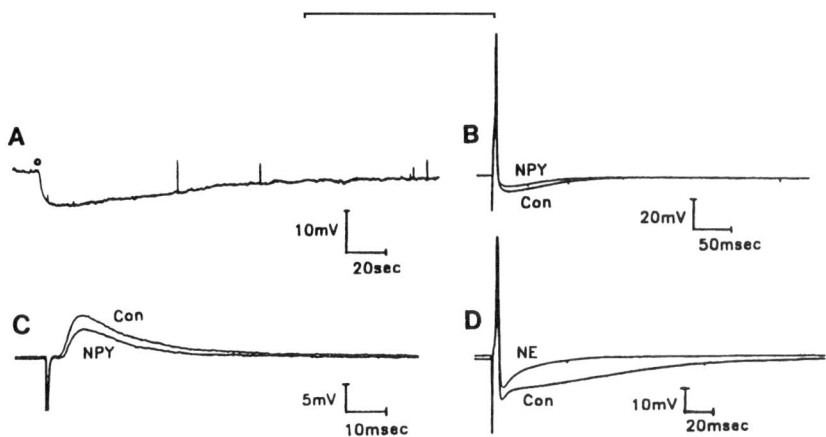

Figure 3. Effects of neuropeptide Y (NPY) and norepinephrine (NE) on guinea pig intracardiac neurons and ganglionic transmission. NPY (100 µM) and NE (100 µM) were applied by pressure ejection (1 sec duration, indicated by open circle in A. A: a NPY-induced membrane hyperpolarization of 9 mV in a cell with resting membrane potential of -59 mV. B: NPY decreased the action potential after hyperpolarization (AHP). Action potentials were elicited by 50 msec depolarizing current pulses. Each trace is the average of 5 action potentials recorded prior to and after NPY application from a cell with resting membrane potential of -60 mV. C: NPY decreased the subthreshold fEPSPs elicited by interganglionic fiber bundle stimulation in a cell with resting membrane potential of -53 mV. D: NE decreased the action potential AHP in a cell with resting membrane potential of -44 mV. Each trace is the average of 5 action potentials recorded prior to (con) and after NE application. Panels A-C are reprinted from the Journal of the Autonomic Nervous System, Vol. 71, Kennedy, A.L., Harakall, S.A., Lynch, S.W., Braas, K.M., Hardwick, J.C., Mawe, G.M. and Parsons, R.L., Expression and physiological actions of neuropeptide Y in guinea pig parasympathetic cardiac ganglia, pp. 190-195, 1998, with permission from Elsevier Science.

Norepinephrine also can affect cardiac neurons (Kennedy, Hardwick and Parsons, unpublished observations). NE had mixed effects in 30% of 35 cells tested; 7 cells were depolarized, 3 cells were hyperpolarized and 25 cells were unaffected. When noted, the change in membrane potential was only a few mV. NE also decreased the amplitude and duration of the directly elicited spike AHP in 25% of the neurons studied (Fig. 3D). These observations suggest that a subset of guinea pig cardiac neurons express adrenergic receptors as reported for rat cardiac neurons (Xu and Adams, 1993).

Substance P Depolarizes Guinea Pig Cardiac Neurons Through Activation of NK_3 Tachykinin Receptors

Konishi et al. (1985) first reported that SP produces a slow depolarization in guinea pig cardiac neurons that is similar in amplitude and duration to the sEPSP initiated by fiber tract stimulation. We showed that application of capsaicin, an agent known to release SP from sensory fibers, elicits a membrane depolarization very similar to that produced by SP, NKA, and NKB (Fig 4A) (Hardwick et al., 1995). The SP-induced slow depolarization is often accompanied by a decrease in membrane resistance. In addition, AP activity produced by a long depolarizing current pulse or following a long hyperpolarizing current pulse can be increased during SP or NKB application. These effects are not due to the tachykinin-induced membrane depolarization, as a comparable depolarization elicited electrotonically does not change either membrane resistance or excitability (Hardwick et al., 1995).

The SP-induced depolarization is due to NK_3 tachykinin receptor mediated activation of a nonselective cationic conductance (Hardwick et al., 1997). The rank order potency NKB=senktide>NKA>SP indicates a primary role of NK_3 receptors mediating the depolarization. Additional support for this conclusion is obtained using selective tachykinin receptor antagonists. Only the NK_3 selective antagonist SR 142801 and not SR 140333, a NK_1 selective antagonist, or SR 48986, a NK_2 selective antagonist, inhibits the SP-induced depolarization (Fig 4B).

Evidence for activation of a nonselective conductance being primarily responsible for generation of the SP-induced depolarization was obtained from ion substitution studies and voltage clamp recordings. Reduction of extracellular Na^+ markedly reduces the amplitude of the SP-induced depolarization measured at the resting membrane potential (~-50 mV) indicating involvement of Na^+ influx in generation of the depolarization. Furthermore, in voltage clamped cells, the amplitude of the SP-induced inward current increases as the membrane potential is clamped at hyperpolarized levels and decreases as the membrane potential is clamped at depolarized levels, with the reversal potential for the SP-induced current being -2 mV (Hardwick et al., 1997).

15. MAMMALIAN CARDIAC GANGLIA AS LOCAL INTEGRATION CENTERS: HISTOCHEMICAL AND ELECTROPHYSIOLOGICAL EVIDENCE

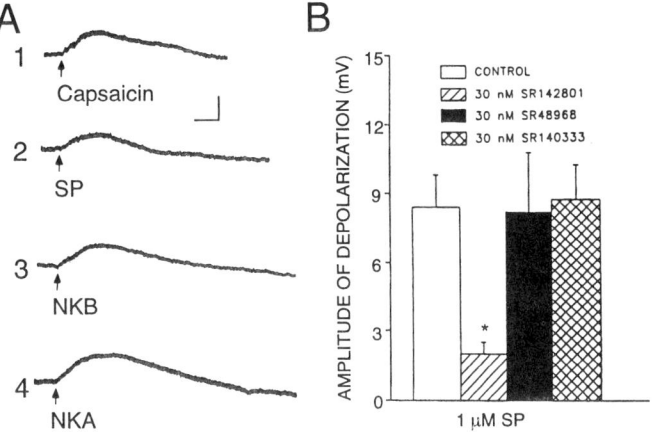

Figure 4. Tachykinins and capsaicin depolarize cardiac neurons. A: the long-lasting depolarization produced by 1 sec pressure ejection (indicated by the arrow) of capsaicin (1 mM) and substance P (SP), neurokinin B (NKB) and neurokinin A (NKA) (all at 100 µM) to intracardiac neurons. The resting membrane potential in each cell was -50 mV in A_1, -50 mV in A_2, -46 in A_3 and -48 in A_4. Calibration: y-axis, 10 mV; x-axis, 10 sec. B: the SP-induced depolarization is significantly inhibited by a NK_3 tachykinin receptor antagonist SR14280, but not by SR140333 a NK_1 receptor antagonist or SR48968 a NK_2 receptor antagonist. Panel A is reprinted from the Journal of the Autonomic Nervous System, Vol. 53, Hardwick, J.C., Mawe, G.M. and Parsons, R.L., Evidence for afferent fiber innervation of parasympathetic neurons of the guinea-pig cardiac ganglion, pp. 166-174, 1995, with permission from Elsevier Science. Panel B is reprinted from the Journal of Physiology, Vol. 504.1, Hardwick, J.C., Mawe, G.M. and Parsons, R.L., Tachykinin-induced activation of non-specific cation conductance via NK_3 neurokinin receptors in guinea pig intracardiac neurones, pp. 65-74, 1997, with permission from the Physiological Society.

It should be noted that some effects of SP on cardiac neurons are species-dependent. SP depolarizes and increases excitability of guinea pig cardiac neurons, but does not directly affect excitability of rat cardiac neurons (Cuevas and Adams, 2000). However, SP modulates the response to ACh in both rat and guinea pig intracardiac neurons (Cuevas and Adams, 2000, Zhang et al., 2001).

PACAP Peptides Depolarize and Increase Membrane Excitability of Guinea Pig Cardiac Neurons

As indicated above, the cardiac neurons are innervated by PACAP-IR nerve fibers (Braas et al., 1998; Calupca et al., 2000a). Also, the cardiac neurons express PAC_1 selective receptors (Fig. 2C). Characterization of the alternative splice variants of the PACAP-selective receptor indicate that the predominant form of the PAC_1 receptor is the *very short* variant containing neither the HIP nor HOP cassettes in the third cytoplasmic loop (Braas et al., 1998). Consistent with the expression of PAC_1 receptors, $PACAP_{27}$ depolarizes the guinea pig cardiac neurons (Fig. 5A) (Braas et al., 1998). Since the depolarization can be elicited in acutely dissociated neurons (Fig. 2D), this is a direct effect of the peptide. $PACAP_{27}$ also produces a concentration-dependent increase in membrane excitability established by determining the number of action potentials produced in response to 500 msec suprathreshold depolarizing current pulses prior to and following peptide application (Fig. 5B) (Braas et al., 1998). Although $PACAP_{27}$ increases excitability markedly, VIP had little or no effect on excitability; an observation consistent with PAC_1 receptor-mediated actions. Surprisingly, $PACAP_{27}$ is noticeably more effective than $PACAP_{38}$; the rank order of peptide-induced increase in membrane excitability is $PACAP_{27} > PACAP_{38} > VIP$.

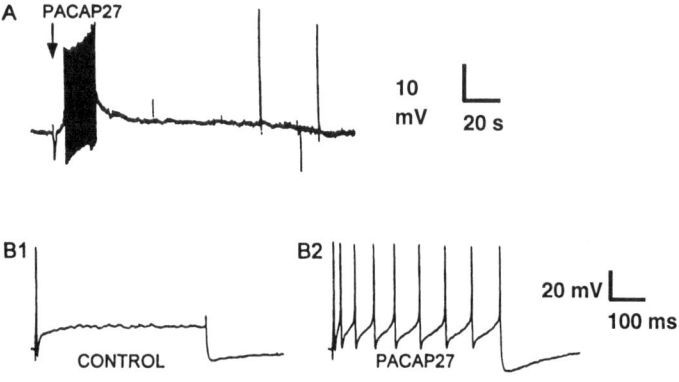

Figure 5. $PACAP_{27}$ depolarizes and increases excitability of cardiac neurons. A: a phasic cell depolarized by local pressure application (1 sec, 50 µM) of $PACAP_{27}$. B_1: a 500 msec depolarizing current pulse (0.3 nA) elicited a single action potential prior to peptide application, but B_2: the same current pulse elicited multiple APs during superfusion with 100 nM $PACAP_{27}$. Modified and reprinted from The Journal of Neuroscience, Vol. 18(23), Brass, K.M., May, V., Harakall, S.A., Hardwick, J.C. and Parsons, R.L., Pituitary Adenylate Cyclase-Activating Polypeptide Expression and Modulation of Neuronal Excitability in Guinea Pig Cardiac Ganglia, pp. 9766-9779, 1998, with permission from the Society for Neuroscience.

15. MAMMALIAN CARDIAC GANGLIA AS LOCAL INTEGRATION CENTERS: HISTOCHEMICAL AND ELECTROPHYSIOLOGICAL EVIDENCE

The PACAP$_{27}$-induced change in excitability is not associated with either a change in membrane resistance or action potential configuration (Braas et al., 1998). Furthermore, treatment of cardiac neurons with 1 mM barium to inhibit I_M does not eliminate the PACAP-induced increase in excitability (Parsons et al., 2000). Thus, the potent excitatory actions of PACAP$_{27}$ do not result simply from an alteration in action potential properties or from inhibition of the potassium current, I_M, which commonly regulates membrane excitability in cardiac neurons (Xi-Moy and Dun, 1995; Cuevas et al., 1997).

PACAP peptides acting through PAC$_1$ receptors can stimulate phospholipase C (PLC) and adenylate cyclase signaling cascades (Braas and May, 1999). Recent studies suggest that a PACAP$_{27}$-induced activation of PLC is not a prerequisite for either the depolarization or increase in excitability. During exposure to the potent PLC inhibitor U73122 (10 µM), 100 nM PACAP$_{27}$ still depolarized and increased membrane excitability of the cardiac neurons (Fig. 6).

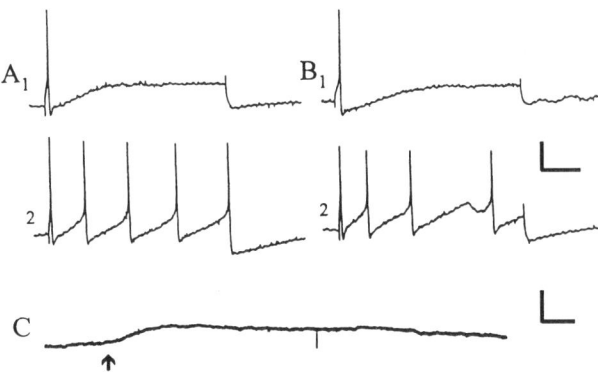

Figure 6. The PACAP$_{27}$-induced increase in excitability and depolarization is not eliminated during exposure to a phospholipase (PLC) inhibitor. A: local pressure application of PACAP$_{27}$ (0.2 sec, 100 µM) increased excitability as determined by the change in action potential generation produced by a 500 msec depolarizing current pulse (0.4 sec). Prior to PACAP$_{27}$ application, the depolarizing current pulse typically elicited a single action potential (A$_1$, B$_1$). Pressure application of PACAP$_{27}$ increased membrane excitability under the same depolarizing conditions before (A$_2$) and following treatment with the PLC inhibitor U73122 (10 µM) (B$_2$). C: the membrane depolarization produced by local pressure application of PACAP$_{27}$ to another cell exposed to 10 µM U73122. The resting membrane potential was -47 mV. Calibration: y-axis, 20 mV; x-axis, 100 sec in A, B and y-axis, 25 mV; x-axis, 10 sec in C.

Ongoing studies investigate whether a PACAP-induced modulation of the hyperpolarization-activated inward current, I_h might contribute to the PACAP-induced increase in excitability. I_h is modulated by c-AMP and as PACAP peptides can activate adenylate cyclase leading to increased c-AMP, we have initiated an analysis of $PACAP_{27}$ effects on I_h using perforated patch techniques on dissociated guinea pig intracardiac neurons (Merriam and Parsons, 2002). Initial results demonstrate that I_h is evident in essentially all dissociated guinea pig cardiac neurons. Furthermore, $PACAP_{27}$ increases I_h by shifting the activation curve, an effect consistent with the presence of a PACAP-facilitated inward current that could contribute to an increase in excitability and the shift from a phasic to apparent tonic firing behavior (Braas et al., 1998).

PACAP peptides also can affect the membrane potential and increase excitability of rat cardiac neurons (De Haven and Cuevas, 2002). In addition, PACAP peptides affect ACh responses in rat intracardiac neurons (Liu et al., 2000a), an effect shared with VIP (Cuevas and Adams, 1996). To date, this effect of PACAP has not been reported for guinea pig cardiac neurons

CONCLUSIONS

Results obtained over the past 15 years by many investigators have demonstrated that the intrinsic cardiac nervous system is much more complex than originally proposed. Neurons within a cardiac ganglion can be innervated by a variety of neurochemically-coded fiber types and the extensive repertoire of neurochemical signals provide the basis for the potential participation of the mammalian intrinsic cardiac nervous system in local, integrative regulatory mechanisms modulating cardiac function. Given that heart disease remains the primary cause of mortality in the United States and that considerable evidence suggests that neural mechanisms could play a prominent role in the pathophysiology of several cardiac disorders, it is critical that we understand the neurochemistry and function of neural pathways that control the heart, and in particular, within cardiac ganglia.

ACKNOWLEDGEMENTS

Many colleagues contributed to the work from the author's laboratory that is included in this chapter. I specifically thank Dr. Jean Harwick for assistance in the experiments shown in figure 6 and Ms. Laura Merriam and Dr. Cynthia Forehand for critical review of the manuscript. Work in the author's laboratory has been supported by NIH grants NS-23978 and HL-65481.

REFERENCES

Adams, D.J., Harper, A.A., 1995. Electrophysiological properties of autonomic ganglion neurons. In Autonomic Ganglia, Ed. E.M. McLanchlan. Harwood Acad. Publishers, pp. 153-212.

Adams, D.J., Trequattrini, C., 1998. Opioid receptor-mediated inhibition of ω-conotoxin GVIA-sensitive calcium channel currents in rat intracardiac neurons. J. Neurophysiol. 79, 753-762.

Ardell, J.L., 2001. Neurohumoral control of cardiac function. In Sperelakis, N., ed. Heart Physiology and Pathophysiology, 4th Ed., San Diego, Academic Pres., pp. 45-59.

Ardell, J.L., Randall, W.C., 1986. Selective vagal innervation of sinoatrial and atrioventricular nodes in canine heart. Am. J. Physiol. 251, H764-H773.

Armour, J.A., 1999. Myocardial ischaemia and the cardiac nervous system. Cardiovasc. Res. 41, 41-54.

Armour, J.A., Smith, F.M., Losier, A.M., Ellenberger, H.H., Hopkins, D.A., 1995. Modulation of intrinsic cardiac neuronal activity by nitric oxide donors induces cardiodynamic changes. Am. J. Physiol. 268 (Regulatory Integrative Comp. Physiol. 37), R403-R413.

Billman, G.E., Hoskins, R.S., Randall, D.C., Randall, W.C., Hamlin, R.L., Lin, Y.C., 1989. Selective vagal postganglionic innervation of the sinoatrial and atrioventricular nodes in the non-human primate. J. Auton. Nerv. Syst. 26, 27-36.

Braas, K.M., May, V., 1999. Pituitary adenylate cyclase-activating polypeptides directly stimulate sympathetic neuron neuropeptide Y release through PAC_1 receptor isoform activation of specific intracellular signaling pathways. J. Biol. Chem. 274(39), 27702-27710.

Braas, K.M., May, V., Harakall, S.A., Hardwick, J.C., Parsons, R.L., 1998. Pituitary adenylate cyclase-activating polypeptide expression and modulation of neuronal excitability in guinea pig cardiac ganglia. J. Neurosci. 18(23), 9766-9779.

Calupca, M.A., Locknar, S.A., Zhang, L., Harrison, T.A., Hoover, D.B., Parsons, R.L., 2001. Distribution of cocaine- and amphetamine-regulated transcript peptide in the guinea pig intrinsic cardiac nervous system and co-localization with neuropeptides or transmitter synthetic enzymes. J. Comp. Neurol. 439, 73-86.

Calupca, M.A., Vizzard, M.A., Parsons, R.L., 2000a. Origin of pituitary adenylate cyclase-activating polypeptide (PACAP)- immunoreactive fibers innervating guinea pig parasympathetic cardiac ganglia. J. Comp. Neurol. 423, 26-39.

Calupca, M.A., Vizzard, M.A., Parsons, R.L., 2000b. Origin of neuronal nitric oxide synthase (NOS)-immunoreactive fibers in guinea pig parasympathetic cardiac ganglia. J. Comp. Neurol. 426, 493-504.

Chang, Y., Hoover, D.B., Hancock, J.C., 2000. Endogenous tachykinins cause bradycardia by stimulating cholinergic neurons in the isolated guinea pig heart. Am. J. Physiol. 278, R1483-R1489.

Cheng, Z., Powley, T.L., 2000. Nucleus ambiguus projections to cardiac ganglia of rat atria: an antegrade tracing study. J. Comp. Neurol. 424, 588-606.

Cheng, Z., Powley, T.L., Schwaber, J.S., Doyle, F.J. III., 1997. Vagal afferent innervation of the atria of the rat heart reconstructed with confocal microscopy. J. Comp. Neurol. 81, 1-17.

Cheng, Z., Powley, T.L., Schwaber, J.S., Doyle, F.J. III., 1999. Projections of the dorsal motor nucleus of the vagus to cardiac ganglia of rat atria: an antegrade tracing study. J. Comp. Neurol. 410, 320-341.

15. MAMMALIAN CARDIAC GANGLIA AS LOCAL INTEGRATION CENTERS: HISTOCHEMICAL AND ELECTROPHYSIOLOGICAL EVIDENCE

Conlon, K., Kidd, C., 1999. Neuronal nitric oxide facilitates vagal chronotropic and dromotropic actions on the heart. J. Auton. Nerv. Syst. 75, 136-146.
Cuevas, J., Adams, D.J., 1996. Vasoactive intestinal polypeptide modulation of nicotinic ACh receptor channels in rat intracardiac neurons. J. Physiol. 493.2, 503-515.
Cuevas, J., Adams, D.J., 1997. M4 muscarinic receptor activation modulates calcium channel currents in rat intracardiac neurons. J. Neurophysiol. 78, 1903-1912.
Cuevas, J., Adams, D.J., 2000. Substance P preferentially inhibits large conductance nicotinic ACh receptor channels in rat intracardiac ganglion neurons. J. Neurophysiol. 84, 1961-1970.
Cuevas, J., Harper, A.A., Trequattrini, C., Adams, D.J., 1997. Passive and active membrane properties of isolated rat intracardiac neurons: regulation by H- and M-currents. J. Neurophysiol. 78, 1890-1902.
Dalsgaard, C.J., Franco-Cereceda, A., Saria, A., Lundberg, J.M., Theodorsson-Norheim, E., Hokfelt, T., 1986. Distribution and origin of substance P- and neuropeptide Y-immunoreactive nerves in the guinea pig heart. Cell Tissue Res. 243, 477-485.
De Haven, W.I., Cuevas, J., 2002. PACAP modulates neuroexcitability in rat intracardiac neurons. Program No. 42.10 202 Abstract Viewer/Itinerary Planner. Washington, DC: Soc. for Neurosci. CD-ROM.
Edwards, F.R., Hirst, G.D.S., Klemm, M.F., Steele, P.A., 1995. Different types of ganglion cell in the cardiac plexus of guinea pigs. J. Physiol. 486.2, 453-471.
Fieber, L.A., Adams, D.J., 1991a. Acetylcholine-evoked currents in cultured neurons dissociated from rat parasympathetic cardiac ganglia. J. Physiol. 434, 215-237.
Fieber, L.A., Adams, D.J., 1991b. Adenosine triphosphate-evoked currents in cultured neurons dissociated from rat parasympathetic cardiac ganglia. J. Physiol. 434, 239-256.
Gatti, P.J., Johnson, T.A., McKenzie, J., Lauenstein J-M., Gray, A., Massari, V.J., 1997. Vagal control of left venricular contractility is selectively mediated by a cranioventricular intracardiac ganglion in the cat. J. Auton. Nerv. Syst. 66, 138-144.
Gatti, P.J., Johnson, T.A., Phan, P., Jordan III, I.K., Coleman, W., Massari, V.J., 1995. The physiological and anatomical demonstration of functionally selective parasympathetic ganglia located in discrete fat pads on the feline myocardium. J. Auton. Nerv. Sys. 51, 255-259.
Gerstheimer, F.P., Metz, J., 1986. Distribution of calcitonin gene-related peptide-like immunoreactivity in the guinea pig-heart. Anat. Embryol. 175, 255-260.
Gibbins, I.L., Furness, J.B., Costa, M., 1987. Pathway-specific patterns of the co-existence of substance P, calcitonin gene-related peptide, cholecystokinin and dynorphin in neurons of the dorsal root ganglia of the guinea-pig. Cell Tissue Res. 248, 417-437.
Hardwick, J.C., Mawe, G.M., Parsons, R.L., 1995. Evidence for afferent fiber innervation of parasympathetic neurons in the guinea-pig cardiac ganglion. J. Auton. Nerv. Syst. 53, 166-174.
Hardwick, J.C., Mawe, G.M., Parsons, R.L., 1997. Tachykinin-induced activation of non-specific cation conductance via NK3 neurokinin receptors in guinea-pig intracardiac neurones. J. Physiol. 504.1, 65-74.
Hassall, C.J.S., Saffrey, M.J., Belai, A., Hoyle, C.H.V., Moules, E.W., Moss, J., Schmidt, H.H.H.W., Murad, F., Forstermann, U., Burnstock, G., 1992. Nitric oxide synthase immunoreactivity and NADPH-diaphorase activity in a subpopulation of intrinsic neurones of the guinea pig heart. Neurosci. Lett. 143, 65-68.

Herring, N., Golding, S., Paterson, D.J., 2000. Pre-synaptic NO-cGMP pathway modulates vagal control of heart rate in isolated adult guinea pig atria. J. Mol. Cell. Cardiol. 32, 1795-1804.

Herring, N., Paterson, D.J., 2001. Nitric oxide-cGMP pathway facilitates acetylcholine release and bradycardia during vagal nerve stimulation in the guinea pig in vitro. J. Physiol. 535, 507-518.

Horackova, M., Armour, J.A., Byczko, Z., 1999. Distribution of intrinsic cardiac neurons in whole-mount guinea pig atria identified by multiple neurochemical coding: A confocal microscope study. Cell Tissue Res. 297, 409-421.

Inokuchi, H., McLachlan, E.M., 1995. Lack of evidence for P2X-purinoceptor involvement in fast synaptic responses in intact sympathetic ganglia isolated from guinea-pigs. Neurosci. 69, 651-659.

Jeong, S-W., Ikeda, S.R., Wurster, R.D., 1999. Activation of various G-protein coupled receptors modulates Ca^{2+} channel currents via PTX-sensitive and voltage-dependent pathways in rat intracardiac neurons. J. Auton. Nerv. Sys. 76, 68-74.

Jeong, S-W., Wurster, R.D., 1997a. Calcium channel currents in acutely dissociated intracardiac neurons from adult rats. J. Neurophysiol. 77, 1769-1778.

Jeong, S-W., Wurster, R.D., 1997b. Muscarinic receptor activation modulates Ca^{2+} channels in rat intracardiac neurons via a PTX- and voltage-sensitive pathway. J. Neurophysiol. 78, 1476-1490.

Kennedy, A.L., Harakall, S.A., Lynch, S.W., Braas, K.M., Hardwick, J.C., Mawe, G.M., Parsons, R.L., 1998. Expression and physiological actions of neuropeptide Y in guinea pig parasympathetic cardiac ganglia. J. Auton. Nerv. Syst. 71, 190-195.

Klimaschewski, L., Kummer, W., Mayer, B., Couraud, J.Y., Preissler, U., Philippin, B., Heym, C., 1992. Nitric oxide synthase in cardiac nerve fibers and neurons of rat and guinea pig heart. Circ. Res. 71, 1533-1537.

Konishi, S., Okamoto, T., Otsuka, M., 1985. Substance P as a neurotransmitter released from peripheral branches of primary afferent neurons producing slow synaptic excitation in autonomic ganglion cells. In: CC Jordan and P Oehme (eds) Substance P. Metabolism and Biological Action. Taylor& Francis Philadelphia PA, pp. 121-136.

Leger, J., Croll, R.P., Smith, F.M., 1999. Regional distribution and extrinsic innervation of intrinsic cardiac neurons in the guinea pig. J. Comp. Neurol. 407, 303-317.

Liu, D-M., Adams, D.J., 2001. Ionic selectivity of native ATP-activated (P2X) receptor channels in dissociated neurons from rat parasympathetic ganglia. J. Physiol. 534.2, 423-435.

Liu, D-M., Cuevas, J., Adams, D.J., 2000a. VIP and PACAP potentiation of nicotinic Ach-evoked currents in rat parasympathetic neurons is mediated by G-protein activation. Eur. J. Neurosci. 12, 2243-2251.

Liu, D-M., Katnik, C., Stafford, M., Adams, D.J., 2000b. P2Y purinoceptor activation mobilizes intracellular Ca^{2+} and induces a membrane current in rat intracardiac neurons. J. Physiol. 526.2, 287-298.

Lynch, S.W., Braas, K.M., Harakall, S.A., Kennedy, A.L., Mawe, G.M., Parsons, R.L., 1999. Neuropeptide Y (NPY) expression is increased in explanted guinea pig parasympathetic cardiac ganglia neurons. Brain Res. 827, 70-78.

Mawe, G.M., Talmage, E.K., Lee, K.P., Parsons, R.L., 1996. Expression of choline acetyltransferase immunoreactivity in guinea pig cardiac ganglia. Cell Tissue Res. 285, 281-286.

Merriam, L.A., Parsons, R.L., 2002. Pituitary adenylate cyclase-activating polypeptide (PACAP) enhances the hyperpolarization-activating nonselective catrionic current (I_H) in guinea pig parasympathetic neurons. Program No. 744.2. 2002 Abstract Viewer/Itinerary Planner. Washington, DC: Soc. for Neurosci. 2002. CD-ROM.

Pardini, B.J., Lund, D.D., Puk, D.E., 1992. Sites at which neuropeptide Y modulates parasympathetic control of heart rate in guinea pigs and rats. J. Auton. Nerv. Syst. 38, 139-146.

Parsons, R.L., Neel, D.S., McKeon, T.W., Carraway, R.E., 1987. Organization of a vertebrate cardiac ganglion: A correlated biochemical and histochemical study. J. Neurosci. 7(3), 837-846.

Parsons, R.L., Rossignol, T.M., Calupca, M.A., Hardwick, J.C., Braas, K.M., 2000. PACAP peptides modulate guinea pig cardiac neuron membrane excitability and neuropeptide expression. Annals of the New York Academy of Science. 921, 202-210.

Randall, W.C., Randall, D.C., Ardell, J.L., 1991. Autonomic regulation of myocardial contractility. In: Reflex Control of the Circulation, eds: I.H. Zucker and J.P. Gilmore, Boston MA, pp. 39-65.

Randall, W.C., Wurster, R.D., 1994. Peripheral innervation of the Heart. From Vagal Control of the Heart: Experimental Basis and Clinical Implications. Eds. M.N. Levy, P.J. Schwartz, Futura Publishing Co., Inc., Armonk, NY, pp. 21-32.

Seabrook, G.R., Fieber, L.A., Adams, D.J., 1990. Neurotransmission in neonatal rat cardiac ganglion in situ. Am. J. Physiol. 259(Heart Circ. Physiol. 28), H997-H1005.

Seebeck, J., Schmidt, W.E., Kilbinger, H., Neumann, J., Zimmerman, N., Herzig, S., 1996. PACAP induces bradycardia in guinea-pig heart by stimulation of atrial cholinergic neurons. Naunyn Schmiedebergs Arch. Pharmacol. 354, 424-430.

Selyanko, A.A., 1992. Membrane properties and firing characteristics of rat cardiac neurones in vitro. J. Auton. Nerv. Syst. 39, 181-190.

Selyanko, A.A., Skok, V.I., 1992a. Acetylcholine receptors in rat cardiac neurones. J. Auton. Nerv. Syst. 33, 33-48.

Selyanko, A.A., Skok, V.I., 1992b. Synaptic transmission in rat cardiac neurones. J. Auton. Nerv. Syst. 39, 191-200.

Smith, F.M., 1999. Extrinsic inputs to intrinsic neurons in the porcine heart in vitro. Am. J. Physiol. 276, R455-R467.

Smith, F.M., Hopkins, D.A., Armour, J.A., 1992. Electrophysiological properties of in vitro intrinsic cardiac neurons in the pig (Sus scrofa). Brain Res.Bull. 28, 715-725.

Sosunov, A.A., Hassall, C.J.S., Loesch, A., Turmaine, M., Feher, E., Burnstock, G., 1997. Neuropeptide Y-immunoreactive intracardiac neurones, granule containing cells and nerves associated with ganglia and blood vessels in the rat and guinea-pig heart. Cell Tissue Res. 289, 445-454.

Steele, P.A., Gibbins, I.L., Morris, J.L., Mayer, B., 1994. Multiple populations of neuropeptide-containing intrinsic neurons in the guinea pig heart. Neurosci. 62, 241-250.

Sternini, C., Brecha, N., 1985. Distribution and co-localization of neuropeptide Y- and tyrosine hydroxylase-like immunoreactivity in the guinea pig heart. Cell Tissue Res. 241, 93-102.

Tanaka, K., Chiba, T., 1998. The vagal origin of preganglionic fibers containing nitric oxide synthase in the guinea-pig heart. Neurosci. Lett. 252, 135-138.

Tanaka, K., Ohshima, H., Esumi, H., Chiba, T., 1993. Direct synaptic contacts of nitric oxide synthase-immunoreactive nerve terminals on the neurons of the intracardiac ganglia of the guinea pig. Neurosci. Lett. 158, 67-70.

Tompkins, J.D., Hoover, D.B., Hancock, J.C., 1999. Substance P evokes bradycardia by stimulation of postganglionic cholinergic neurons. Peptides 20, 623-628.

Urban, L., Papka, R.E., 1985. Origin of small primary afferent substance P-immunoreactive nerve fibers in the guinea-pig heart. J. Auton. Nerv. Syst. 12, 321-331.

Wharton, J., Polak, J.M., McGregor, J.P., Bishop, A.E., Bloom, S.R., 1981. The distribution of substance P-like immunoreactive nerves in the guinea pig heart. Neurosci. 6, 2193-2204.

Xi, X., Randall, W.C., Wurster, R.D., 1994. Electrophysiological properties of canine cardiac ganglion cell types. J. Auton. Nerv. Syst. 47, 69-74.

Xi-Moy, S.X., Dun, N.J., 1995. Potassium currents in adult rat intracardiac neurones. J. Physiol. 486.1, 15-31.

Xi-Moy, S.X., Randall, W.C., Wurster, R.D., 1993. Nicotinic and muscarinic synaptic transmission in canine intracardiac ganglion cells innervating the sinoatrial node. J. Auton. Nerv. Syst. 42, 201-214.

Xu, Z.-J., Adams, D.J., 1992a. Resting membrane potential and potassium currents in cultured parasympathetic neurones from rat intracardiac ganglia. J. Physiol. 456, 405-424.

Xu, Z.-J., Adams, D.J., 1992b. Voltage-dependent sodium and calcium currents in cultured parasympathetic neurones from rat intracardiac ganglia. J. Physiol. 456, 425-441.

Xu, Z.-J., Adams, D.J., 1993. α-Adrenergic modulation of ionic currents in cultured parasympathetic neurons from rat intracardiac ganglia. J. Neurophysiol. 69(4), 1060-1070.

Zhang, H., Cuevas, J., 2002. Sigma receptors inhibit high-voltage-activated calcium channels in rat sympathetic and parasympathetic neurons. J. Neurophysiol. 87, 2867-2879.

Zhang, L., Tompkins, J.D., Hancock, J.C., Hoover, D.B., 2001. Substance P modulates nicotinic responses of intracardiac neurons to acetylcholine in the guinea pig. Am. J. Physiol. Regul. Integr. Comp. Physiol. 281, R1792-R1800.

Chapter 16

PARASYMPATHETIC INFLUENCES ON CEREBRAL CIRCULATION: A LINK TO ARTERIAL BAROREFLEXES

William T. Talman
Laboratory of Neurobiology, Department of Neurology, Veterans Affairs Medical Center and University of Iowa, Iowa City, IA 52242

Abstract: It has long been known that arterial baroreflexes participate in regulating peripheral vascular tone, but there has been little evidence to link those reflexes to regulation of the cerebral circulation. In studies reviewed here we show that arterial baroreceptor reflexes modulate cerebrovascular tone through a pathway that connects the cardiovascular nucleus tractus solitarii with parasympathetic preganglionic neurons in the pons. Utilizing laser flowmetry in anesthetized rats we assessed autoregulation of cerebral blood flow while blood pressure was slowly increased in response to i.v. infusion of phenylephrine. Intact control animals demonstrated classic autoregulation and breakthrough, an increase of cerebral blood flow, when pressure exceeded the upper limit of autoregulation. However, breakthrough was attenuated or abolished in animals with lesions of baroreceptor nerves, the nucleus tractus solitarii where those nerves would terminate, the pontine preganglionic parasympathetic neurons, or the parasympathetic ganglionic nerves to the cerebral vessels. Similarly blocking synthesis of nitric oxide, which is released by the parasympathetic nerves from the pterygopalatine ganglia, also attenuated breakthrough of autoregulation during hypertension. While interruption of function of the pontine preganglionic neurons reduced cerebrovascular dilatation during hypertension, stimulation of those neurons caused cerebral vasodilatation. We performed anatomical studies to complement these physiological experiments and defined a pathway that could mediate baroreflex influences on cerebral blood vessels. Retrograde and anterograde tracing studies showed that neurons in the cardiovascular region of the nucleus tractus solitarii projected to and synapsed with preganglionic neurons of the superior salivatory nucleus. The synapses formed were of an asymmetric type and were thus consistent with excitatory synapses. These studies provide strong evidence that arterial baroreceptors may modulate cerebral blood flow through direct connections with pontine parasympathetic neurons. We conjecture that the pathway may participate in integrating the various functions mediated by those pontine parasympathetic preganglionic neurons.

Key words: baroreflex, cerebral circulation, hypertension, medulla, nitric oxide, parasympathetic, pons.

INTRODUCTION

There is considerable evidence that neural pathways may influence cerebrovascular resistance (CVR) and, thus, cerebral blood flow (CBF). A network of nerves investing cerebral arteries has been recognized since the classic work of Willis over 330 years ago (Edvinsson et al., 1993), but the functional implications of this innervation have begun to be appreciated only more recently. Nerves to cerebral vessels may be considered in four groups.

Sensory fibers of the trigeminal nerve comprise one physiologically relevant group. This group of nerves contribute to the pathogenesis of migraine headache (Moskowitz, 1993; Bolay et al., 2002) and may be responsible for post ischemic hyperemia (Macfarlane et al., 1991), a phenomenon related to release of calcitonin gene related peptide (CGRP) (Macfarlane et al., 1991). However, the afferents may contain not only CGRP but also nitric oxide synthase (NOS) (Nozaki et al., 1993). Therefore, NO· may also contribute to cerebral vasodilatation mediated by the trigeminal nerve.

Axons projecting from other central structures also innervate cerebral arteries and, upon stimulation, may profoundly affect CBF. Examples of such sources include the nucleus basalis, the substantia innominata, the centromedian nucleus of the thalamus, the fastigial nucleus of the cerebellum, the rostral ventrolateral medulla (RVLM), and local cortical neurons (Iadecola et al., 1987; Mraovitch and Seylaz, 1987; Adachi et al., 1990). Some of these sources of vascular innervation, e.g. the fastigial nucleus of the cerebellum, may not only modulate vasomotor tone but may also attenuate neuronal damage resulting from cerebral ischemia (Zhang and Iadecola, 1993).

Sympathetic nerves comprise a third source of cerebrovascular innervation. The sympathetic nerve fibers to cerebral arteries in the forebrain largely arise from the ipsilateral superior cervical ganglion while those to the posterior circulation arise largely from the stellate ganglion (Arbab et al., 1986). As with peripheral autonomics, norepinephrine and neuropeptide Y are both found in sympathetic nerves to cerebral vessels (Edvinsson et al., 1993). Although sympathetic activity may play little role in regulation of cerebral circulation during normotension, sympathetic activation may augment cerebral vasoconstriction during acute hypertension (Tamaki and

Heistad, 1986). Furthermore, sympathetic nerves may participate in militating against stroke in chronic hypertension (Sadoshima et al., 1981).

A *parasympathetic innervation* has also been described and found to arise from cell bodies in the pterygopalatine (previously called "sphenopalatine") ganglion (Hara et al., 1985). Preganglionic fibers project to the pterygopalatine ganglia from cell bodies in the superior salivatory nucleus (SSN), a nucleus that participates in control of lacrimation and salivation as well as cerebrovascular control (Loewy and Spyer, 1990). Compared with the innervations reviewed above, the functional role of parasympathetic nerves in cerebrovascular control has been less extensively studied (Blessing, 1997). However, those nerves have received recent attention because they provide the major nitroxidergic innervation to large (o.d.>400 µm) and to small (o.d.<100 µm) cerebral arteries in the forebrain (Nozaki et al., 1993; Toda et al., 1993; Yoshida et al., 1993; Suzuki et al., 1994; Ignacio et al., 1997; Kimura et al., 1997). Neurons of the pterygopalatine ganglion and their projections to cerebral vessels not only contain NOS but some also contain acetylcholine, vasoactive intestinal polypeptide (VIP), and CGRP (Hardebo et al., 1992; Nozaki et al., 1993; Kimura et al., 1997). Each of these putative transmitters could contribute to cerebral vasodilation seen with activation of postganglionic fibers (Goadsby, 1990), but data from several recent studies suggest that dilatation seen with stimulation of either preganglionic or ganglionic fibers is mediated by release of NO· (Toda et al., 2000a; Toda et al., 2000b). Acetylcholine, also colocalized with NOS in ganglionic fibers, may actually contribute to vasoconstriction through complex interactions with terminals that release NO· (Liu and Lee, 1999), but NO·-mediated vasodilatation is the prominent effect seen with activation of ganglionic fibers (Goadsby et al., 1996). Therefore, in contrast to the sympathetic system, which might augment vasoconstriction, the parasympathetic system may limit vasoconstriction and may do so in part through nitroxidergic mechanisms. It is likely that the parasympathetic input effects changes in CVR during hypertension but not under basal conditions (Tanaka et al., 1995). Our own studies, reviewed here, as well as those from other investigators support this possibility.

Initially we focused on the influence of the arterial baroreflex on cerebrovascular autoregulation. (Talman et al., 1994). The study was performed in adult male Sprague Dawley rats anesthetized with chloralose. Arterial blood pH, pCO_2 and O_2 and body temperature were carefully controlled. PaO_2 was maintained above 200 TORR by ventilating the rat with room air to which 100% O_2 was added. CBF was monitored by laser flowmetry or assessed by the radioactive microsphere technique as mean arterial pressure (MAP) was slowly raised by infusion of phenylephrine, angiotensin (AII), or vasopressin. CBF results were the same with either technique; therefore, further studies utilized laser flowmetry only. However, the microsphere studies were rewarding because they showed that changes in CBF were not confined to the parietal cortex from which CBF was recorded with laser flowmetry. Instead changes were seen in other cortical areas as well as in subcortical structures like the basal ganglia and thalamus.

In 5 rats with an intact baroreflex, CVR fell and CBF rose when MAP exceeded 155 ± 4 mmHg. From a basal MAP of 94 ± 8 mmHg to the point where CBF broke through, CVR and CBF were autoregulated. In 5 other rats acute bilateral transection (sinoaortic denervation; SAD) of both carotid sinus nerves (CSN) and aortic depressor nerves (ADN) did not affect CBF over the same pressure range (~100-155 mmHg). However, in contrast to intact animals, rats with SAD did not demonstrate breakthrough of CBF or CVR (Fig. 1) even though the MAP increased in these animals to even higher levels (172 ± 8 mmHg) than in the intact rat. Increases in CBF that did occur were modest and slow to develop. Our suggestion that breakthrough may, in part, be an active phenomenon under baroreflex control was met with considerable skepticism at first; but others now report that breakthrough may indeed be an active event rather than a passive loss of autoregulation (Paterno et al., 2000). However, there were other potential explanations for the effects that we saw with SAD.

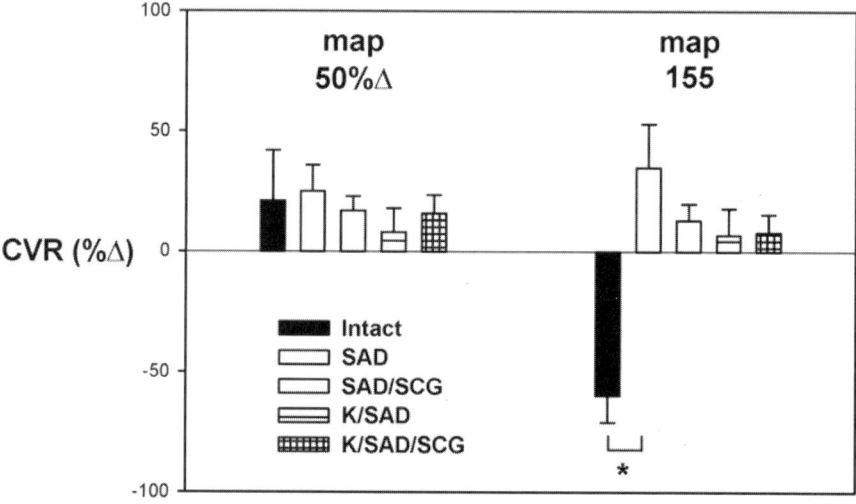

Figure 1. Autoregulation, an increase in cerebrovascular resistance (CVR) and relative maintenance of cerebral blood flow (CBF) as mean arterial pressure (MAP) is gradually increased, occurred in intact rats as well as in rats subjected to sinoaortic denervation (SAD), SAD plus removal of the superior cervical ganglia (SCG), SAD plus bilateral nephrectomy

(K), or SAD/K/SCG. However, as MAP was slowly increased to exceed 150 mmHg, CBF broke through autoregulation and CVR fell significantly (p<.05) only in the intact group.

For example, SAD acutely increases sympathetic nerve activity (Barres et al., 1992) that could in turn blunt increases in CBF associated with high MAP's (Tamaki and Heistad, 1986; Faraci et al., 1987). Therefore, we sought to determine if bilateral removal of the superior cervical ganglia, the source of sympathetic innervation to the cerebral vasculature (Alafaci et al., 1986; Arbab et al., 1986), eliminated effects of SAD on CBF during hypertension (Talman and Nitschke Dragon, 1995b) (Fig. 1). As in rats with SAD alone, CBF did not rise markedly during hypertension (167 ± 6 mmHg) in 8 rats that had also undergone complete superior cervical ganglionectomy. To control for other neuroendocrine influences of SAD that might affect CVR we also sought to determine if total denervation of the kidneys (to eliminate renin/angiotensin effects) alone or in combination with removal of the superior cervical ganglia affected the cerebrovascular changes seen in animals with SAD. Breakthrough was still significantly delayed in the animals with SAD. Thus, neither sympathetic nor renin/angiotensin influences were responsible for effects of SAD on autoregulation.

The differences in CBF (and CVR) responses to hypertension were not attributable to the rate of rise of MAP or to the change in pulse pressure produced by infusion of phenylephrine. Instead, they seemed related to interruption of the baroreflex. With one exception (see below) animals that did not manifest rapid breakthrough like that in intact controls treated with phenylephrine exhibited no baroreflex-mediated bradycardia in response to the elevation of MAP. In 6 rats the increased MAP elicited by i.v. infusion of angiotensin-II (AII) was not accompanied by baroreflex-mediated bradycardia. In these animals also rapid breakthrough did not occur. Therefore, AII infusion may be another model in which hypertension in the presence of downregulation (Casto and Phillips, 1986; Paton and Kasparov, 1999) of the baroreflex does not lead to breakthrough. It is important to contrast these experiments with exogenous AII as a pressor agent from those above. Again, plasma renin and, by inference, AII were shown not to contribute to the changes in autoregulation produced by SAD. In the latter, levels of plasma renin did not rise significantly after SAD and thus did not contribute either to a pressor effect or to a reduction of baroreflex activity (Talman and Nitschke Dragon, 1995b).

We found that breakthrough can be averted in some circumstances even if the arterial baroreflex is intact. For example (Talman and Nitschke Dragon, 1995a), 11 rats treated with L-nitroarginine (L-NA), an inhibitor of NOS, (Moncada et al., 1991) did not show rapid breakthrough of CBF and CVR yet manifested reflex bradycardia. The pattern of CBF and CVR changes in response to hypertension in these animals was similar to that seen in SAD treated animals. Others (Kelly et al., 1994) have found a similar influence of NOS on CBF during hypertension. In our studies MAP rose from a basal level of 123 ± 3 mmHg after L-NA to a maximum of 189 ± 2 mmHg after phenylephrine. CBF rose from 17.4 ± 1.2 laser units prior to L-NA by $57 \pm 9\%$ at maximal MAP; CVR, which was $7.5 \pm .6$ mmHg/laser unit at basal MAP rose by 2 ± 7 % at maximal MAP. Baroreflex-related changes in heart rate were present with heart rate falling from 356 ± 19 bpm at basal MAP to 309 ± 29 bpm at maximal MAP. Administration of L-arginine blocked the effects of L-NA but D-arginine did not. Another control consisted of animals pretreated with vasopressin to raise MAP to levels comparable to those produced by L-NA. Those animals demonstrated breakthrough of autoregulation like that seen in intact control animals. We have now found (unpublished observations) that breakthrough of autoregulation is also significantly attenuated when pial vessels are treated with a selective nNOS inhibitor delivered through a cranial window directly on those vessels prior to inducing hypertension.

Although these studies suggested that release of NO·, possibly under baroreflex control, participates in vasodilatation seen during the breakthrough phenomenon, the studies did not provide a direct link between the role of NO· and innervation to the cerebral vessels. Therefore, we investigated (Talman and Nitschke Dragon, 2000) whether interruption of axons from the pterygopalatine ganglion, the main source of NO· innervation to the cerebral vasculature, affected autoregulation as did SAD and NOS inhibition. When MAP was raised, as in other studies, CVR fell significantly ($p<0.05$) less in denervated animals than in intact animals

Thus, our data (Talman et al., 1994; Talman and Nitschke Dragon, 1995a; Talman and Nitschke Dragon, 1995b; Talman and Nitschke Dragon, 2000) suggest that arterial baroreflexes, and the parasympathetic nerves that they may influence, modulate CVR and CBF at high levels of arterial blood pressure. In all studies where the baroreflex was interrupted, the chemoreflex was likewise interrupted, but for the following reasons it seemed unlikely that chemoreceptor reflex interruption played a role in the responses. First, arterial pO_2 in all rats was maintained above 200 TORR and arterial pCO_2 was the same in all groups of animals studied. Peripheral chemoreceptors, being most responsive to decreases in arterial pO_2, (Korner, 1971) would have been significantly inhibited even in the intact group (without SAD). Thus, SAD would have had little added effect on chemoreflex activity. Second, AII, is known to inhibit the baroreflex but to have less predictable effects on the chemoreflex (Lee and Lumbers, 1981; Paton and Kasparov, 1999). AII had an effect on CBF like that of SAD.

16. PARASYMPATHETIC INFLUENCES ON CEREBRAL CIRCULATION: A LINK TO ARTERIAL BAROREFLEXES

Our suggestion that the baroreflex plays a role in autoregulation at high arterial pressures at first seemed to contrast with conclusions of some other studies. In fact, participation of arterial baroreceptors in regulation of CBF has been controversial (Rapela et al., 1967; Heistad and Marcus, 1976; Heistad et al., 1980), and anatomical pathways that might subserve any such influences have not been studied with respect to their participation in baroreceptor input to the cerebral vessels. Earlier studies had sought to identify an influence of the baroreflex on autoregulation. In contrast to our own studies, earlier published reports dealt with CVR and CBF over the range of arterial pressures where autoregulation remained intact. When MAP was varied within the autoregulatory range, CVR and CBF were autoregulated regardless of the activity of the baroreflex (Rapela et al., 1967; Heistad and Marcus, 1976). Our studies agree with these findings. However, arterial pressure was not raised above the level of autoregulation in the earlier studies; and in some cases in which effects of the baroreflex were studied in barodenervated animals, only the carotid sinuses, not the aortic mechanoreceptors, were denervated (Rapela et al., 1967).

It is interesting that our studies demonstrate baroreflex effects on autoregulation only at very high arterial pressures. Indeed, it is likely that this restricted zone of action accounts for our having demonstrated effects while earlier investigations did not. We conjecture that only a select group of mechanoreceptors participates in CBF regulation and that the group is comprised of those that transduce signals only when arterial pressure reaches high levels (Seagard et al., 1993; Seagard et al., 1995).

Results of some studies suggest that the baroreflex plays a role in regulation of CBF and some support our hypothesis that CBF is modulated by arterial baroreflexes during hypertension. For example, electrical stimulation of the nucleus tractus solitarii (cNTS), the site of termination of baroreflex (and other cardiovascular) afferents, led to marked increases of CBF that were independent of sympathetic and vagal effects of the stimulus (Nakai and Ogino, 1984). Opposite results, a significant fall of CBF, were obtained when cNTS was stimulated with high doses of glutamate to selectively stimulate neurons rather than fibers of passage (Maeda et al., 1990). However, despite efforts to limit the magnitude of the fall of MAP below basal values the influence on CBF of rapid changes in MAP cannot be excluded in that study. Rapid changes in MAP, even within the autoregulatory range, can cause concomitant changes in CBF in the same direction as the change in MAP (Williams et al., 1989). Therefore, it is difficult to interpret this study in the context of cNTS and baroreflex influences on CBF and CVR.

In another study, rats demonstrated normal levels of CBF even though their MAP was significantly elevated above the upper limit of autoregulation several hours after interruption of the baroreflex by bilateral lesions of the cNTS (Graham et al., 1982). Others have disputed that finding and have shown a loss of autoregulation and consequent marked increase in CBF during hypertension in rats similarly treated with bilateral cNTS lesions (Ishitsuka et al., 1986). The latter investigators suggested that the former study was flawed by the extended period of hypertension that the animals had sustained. However, some caution needs to be exercised in interpreting even the latter study because MAP was increased in rapid steps. As noted above, step changes in MAP may cause concomitant and proportional changes in CBF (Williams et al., 1989). In the same study the investigators also transected the baroreceptor nerves and studied effects on autoregulation. They found no effect, but again blood pressure was kept below 140 mmHg, thus within the range of autoregulation. We have now reconciled the differences in these studies and shown that breakthrough of autoregulation was attenuated and the baroreflex blocked after bilateral injection of 4% lidocaine into the cardiovascular region of NTS in 6 rats (Talman and Nitschke Dragon, 2002). Therefore, central, like peripheral, baroreflex interruption extends autoregulation during hypertension.

In addition we have established (Agassandian et al., 2002) that the cardiovascular region of NTS, in which arterial baroreceptor axons terminate, projects directly to preganglionic parasympathetic neurons of the superior salivatory nucleus, or SSN, from which axons project to the pterygopalatine ganglion. Ganglionic axons then project to cerebral arteries as well as to lacrimal and salivary glands (Loewy and Spyer, 1990). Our studies have amplified and added new findings to others that preceded them. One study that utilized tritiated amino acids to trace projections from cNTS in the rat showed tracer in the region of the SSN (Norgren, 1978). The SSN was not clearly defined in that paper, but the region to which cNTS projected seemed to correspond with the region where neuronal stimulation was shown to elicit cerebral vasodilatation in another study (Nakai et al., 1993). Even that physiological study did not fully define the effective sites in terms of SSN. Another study that utilized cats did provide some support for projections from NTS to SSN (Loewy and Burton, 1978). Rats were later studied by the same group (Spencer et al., 1990) who applied retrogradely transported virus to the pterygopalatine ganglion and found viral labeling in the NTS. Until our work, this study came closest to identifying NTS to SSN projections but, unlike our study, it did not establish monosynaptic connections between NTS and the regions of SSN involved in cerebrovascular control. Functional anatomical studies using *c-fos* expression as an index of baroreceptor projection pathways in the medulla focused attention on more caudal regions of medulla and did not comment on c-*fos* expression at the level of the SSN (Li and Dampney, 1992; McKitrick et al., 1992).

We conjecture it is within the region of SSN that baroreceptor influences on cerebral vasculature are effected. In a recent study (Agassandian et al., 2003) we found that bilateral interruption of function in the SSN led to attenuation of breakthrough of autoregulation at high arterial blood pressures while local stimulation of SSN neurons with the excitatory amino acid glutamate led to cerebral vasodilatation and increased CBF (Agassandian et al., 2002). These findings supported the conjectured role of the SSN in baroreceptor mediation of CBF responses to hypertension. Although the pathway from NTS to SSN may be important, we recognize the possibility that physiologically relevant projections from NTS may not only terminate on neurons of the SSN but may also project to sites, such as the lateral tegmental field (Ruggiero et al., 1982; Ruggiero et al., 1996). These sites could also influence CBF directly or via secondary projections to the SSN.

The SSN actually is nicely positioned to play an integrative role in central control of the cerebral circulation. For example, the SSN receives descending projections from the lateral hypothalamus (Hosoya et al., 1983; Sugiura et al., 1989) and neurons in the SSN respond to stimulation of orbital and anterior coronal gyri (Ishizuka and Murakami, 1989). It is not currently known, however, if these descending projections to SSN terminate at sites that may be involved in cerebrovascular control and, if so, how the integrative influences are reflected in normal physiology. We find it intriguing that the SSN, a functionally heterogeneous group of neurons that modulates not only CBF but also lacrimation, salivation and nasal mucosal blood flow could integrate all of these functions during specific behaviors. If so, what might be the common feature of these seemingly disparate functions. We conjecture that that feature is thermoregulation, which is, particularly in lower vertebrates, dependent on each of the individual functions.

Therefore, our studies not only have provided the first evidence that arterial baroreceptors may participate in cerebral vasodilatation during acute hypertension and shown the pathway (Fig. 2) that could carry the baroreceptor signals but also they have defined a potentially important central path for thermoregulatory control that may be critical for protection of the brain in health and disease. Indeed, some studies have suggested that the parasympathetic nerves to cerebral vessels may themselves protect against ischemic neuronal injury (Koketsu et al., 1992), which has also been shown to be ameliorated by lowering brain temperature.

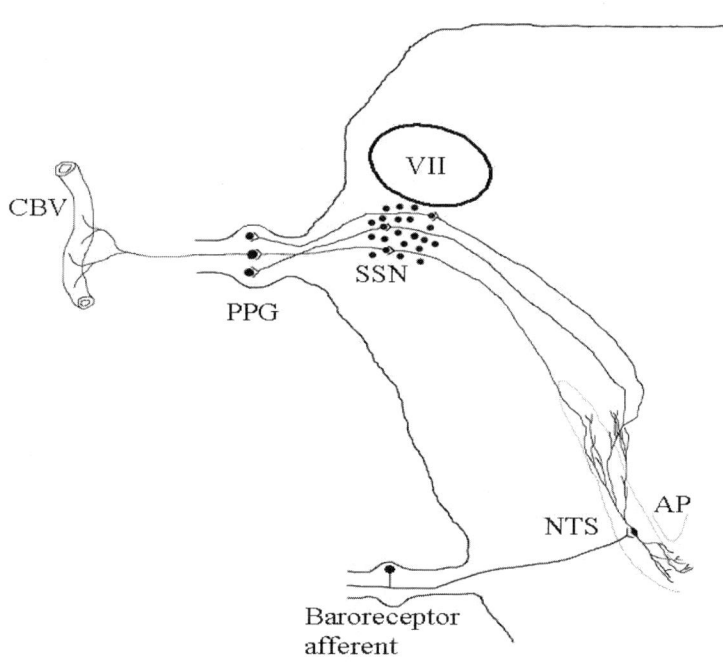

Figure 2. Diagrammatic representation of the proposed pathway from peripheral arterial baroreceptors to cerebral blood vessels. Baroreceptor nerves are known to terminate in the cardiovascular region of NTS at and around the level of the area postrema (AP). Our studies showed that NTS projects to SSN, which is located in the pons adjacent to the facial nerve (VII). There those projections synapse with preganglionic neurons of SSN, which, in turn, projects to the pterygopalatine ganglion (PPG), a primary source of nitroxidergic nerves to cerebral blood vessels (CBV).

REFERENCES

Adachi, T., Biesold, D., Inanami, O., Sato, A., 1990. Stimulation of the nucleus basalis of Meynert and substantia innominata produces widespread increases in cerebral blood flow in the frontal, parietal and occipital cortices. Brain Res. 514, 163-166.

Agassandian, K., Fazan, V.P.S., Adanina, V., Talman, W.T., 2002. Direct projections from the cardiovascular nucleus tractus solitarii to pontine preganglionic parasympathetic neurons: a link to cerebrovascular regulation. J. Comp. Neurol. 452, 242-254.

Agassandian, K., Fazan, V.P.S., Margaryan, N., Nitschke Dragon, D., Riley, J., Talman, W.T., 2003. A novel central pathway links arterial baroreceptors and pontine parasympathetic neurons in cerebrovascular control. Cell. Molec. Neurobiol.

Alafaci, C., Cowen, T., Crockard, H.A., Burnstock, G., 1986. Perivascular nerve types supplying cerebral blood vessels of the gerbil. Acta Physiol. Scand. 127 (Suppl. 552), 9-12.

Arbab, M.A.R., Wiklund, L., Svendgaard, N.A., 1986. Origin and distribution of cerebral vascular innervation from superior cervical, trigeminal and spinal ganglia investigated with retrograde and anterograde WGA-HRP tracing in the rat. Neurosci. 19, 695-708.

Barres, C., Lewis, S.J., Jacob, H.J., Brody, M.J., 1992. Arterial pressure lability and renal sympathetic nerve activity are dissociated in SAD rats. Am. J. Physiol. Regul. Integr. Comp. Physiol. 263, R639-R646.

Blessing, W.W., 1997. The lower brainstem and bodily homeostasis. Oxford: New York.

Bolay, H., Reuter, U., Dunn, A.K., Huang, Z., Boas, D.A., Moskowitz, M.A., 2002. Intrinsic brain activity triggers trigeminal meningeal afferents in a migraine model. Nature Med. 8, 136-142.

Casto, R., Phillips, M.I., 1986. Angiotensin II attenuates baroreflexes at nucleus tractus solitarius of rats. Am. J. Physiol. 250, R193-R198.

Edvinsson, L., MacKenzie, E.T., McCulloch, J., 1993. Cerebral blood flow and metabolism. Raven Press: New York.

Faraci, F.M., Mayhan, W.G., Werber, A.H., Heistad, D.D., 1987. Cerebral circulation: effects of sympathetic nerves and protective mechanisms during hypertension. Circ. Res. 61, II-102-II-106.

Goadsby, P.J., 1990. Sphenopalatine ganglion stimulation increases regional cerebral blood flow independent of glucose utilization in the cat. Brain Res. 506, 145-148.

Goadsby, P.J., Uddman, R., Edvinsson, L., 1996. Cerebral vasodilatation in the cat involves nitric oxide from parasympathetic nerves. Brain Res. 707, 110-118.

Graham, D.I., Grome, J.J., Kelly, P.A.T., MacKenzie, E.T., McCulloch, J., Reis, D.J., Talman, W.T., 1982. Cerebral circulatory effect of fulminating neurogenic hypertension. In: Heistad, D.D., Marcus, M.L., (Eds.), Cerebral blood flow: effects of nerves and neurotransmitters. Elsevier North Holland, New York, 493-502.

Hara, H., Hamill, G.S., Jacobowitz, D.M., 1985. Origin of cholinergic nerves to the rat major cerebral arteries: coexistence with vasoactive intestinal polypeptide. Brain Res. Bull. 14, 179-188.

Hardebo, J.E., Suzuki, N., Ekblad, E., Owman, C., 1992. Vasoactive intestinal polypeptide and acetylcholine coexist with neuropeptide Y, dopamine-β-hydroxylase, tyrosine hydroxylase, substance P or calcitonin gene-related peptide in neuronal subpopulations in cranial parasympathetic ganglia of rat. Cell Tissue Res. 267, 291-300.

Heistad, D.D., Gross, P.M., Busija, D.W., Marcus, M.L., 1980. Cerebral vascular response to loading and unloading of arterial baroreceptors. In: Sleight, P., (Ed.), Arterial baroreceptors and hypertension. Oxford University Press, Oxford, 210-217.

Heistad, D.D., Marcus, M.L., 1976. Total and regional cerebral blood flow during stimulation of carotid baroreceptors. Stroke 7, 239-243.

Hosoya, Y., Matsushita, M., Sugiura, Y., 1983. A direct hypothalamic projection to the superior salivatory nucleus neurons in the rat. A study using anterograde autoradiographic and retrograde HRP methods. Brain Res. 266, 329-333.

Iadecola, C., Arneric, S.P., Baker, H.D., Tucker, L.W., Reis, D.J., 1987. Role of local neurons in cerebrocortical vasodilation elicited from cerebellum. Am. J. Physiol. 252, R1082-R1091.

Ignacio, C.S., Curling, P.E., Childres, W.F., Bryan, R.M., Jr., 1997. Nitric oxide-synthesizing perivascular nerves in the rat middle cerebral artery. Am. J. Physiol. Regul. Integr. Comp. Physiol. 273, R661-R668.

Ishitsuka, T., Iadecola, C., Underwood, M.D., Reis, D.J., 1986. Lesions of nucleus tractus solitarii globally impair cerebrovascular autoregulation. Am. J. Physiol. 251, H269-H281.

Ishizuka, K., Murakami, T., 1989. Cortically evoked responses of superior salivary nucleus neurons in the cat. Proc. Finn. Dent. Soc. 85, 355-359.

Kelly, P.A.T., Thomas, C.L., Ritchie, I.M., Arbuthnott, G.W., 1994. Cerebrovascular autoregulation in response to hypertension induced by N^G-nitro-L-arginine methyl ester. Neurosci. 59, 13-20.

Kimura, T., Yu, J.G., Edvinsson, L., Lee, T.J., 1997. Cholinergic, nitric oxidergic innervation in cerebral arteries of the cat. Brain Res. 773, 117-124.

Koketsu, N., Moskowitz, M.A., Kontos, H.A., Yokota, M., Shimizu, T., 1992. Chronic parasympathetic sectioning decreases regional cerebral blood flow during hemorrhagic hypotension and increases infarct size after middle cerebral artery occlusion in spontaneously hypertensive rats. J. Cereb. Blood Flow Metab. 12, 613-620.

Korner, P.I., 1971. Integrative neural cardiovascular control. Physiol. Rev. 51, 312-367.

Lee, W.B., Lumbers, E.R., 1981. Angiotensin and the cardiac baroreflex response to phenylephrine. Clin. Exp. Pharmacol. Physiol. 8, 109-117.

Li, Y.-W., Dampney, R.A.L., 1992. Expression of c-fos protein in the medulla oblongata of conscious rabbits in response to baroreceptor activation. Neurosci. Lett. 144, 70-74.

Liu, J., Lee, T.J., 1999. Mechanism of prejunctional muscarinic receptor-mediated inhibition of neurogenic vasodilation in cerebral arteries. Am. J. Physiol. Heart Circ. Physiol. 276, 194-204.

Loewy, A.D., Burton, H., 1978. Nuclei of the solitary tract: efferent projections to the lower brain stem and spinal cord in the cat. J. Comp. Neurol. 181, 421-449.

Loewy, A.D., Spyer, K.M., 1990. Central regulation of autonomic functions. Oxford: New York.

Macfarlane, R., Moskowitz, M.A., Tasdemiroglu, E., Wei, E.P., Kontos, H.A., 1991. Postischemic cerebral blood flow and neuroeffector mechanisms. Blood Vessels 28, 46-51.

Maeda, M., Nakai, M., Krieger, A.J., Sapru, H.N., 1990. Chemical stimulation of the nucleus tractus solitarii decreases cerebral blood flow in anesthetized rats. Brain Res. 520, 255-261.

McKitrick, D.J., Krukoff, T.L., Calaresu, F.R., 1992. Expression of c-fos protein in rat brain after electrical stimulation of the aortic depressor nerve. Brain Res. 599, 215-222.

Moncada, S., Palmer, R.M.J., Higgs, E.A., 1991. Nitric oxide: Physiology, pathophysiology, and pharmacology. Pharmacol. Rev. 43, 109-142.

Moskowitz, M.A., 1993. Neurogenic inflammation in the pathophysiology and treatment of migraine. Neuro. 43 (Suppl 3), S16-S20.

Mraovitch, S., Seylaz, J., 1987. Metabolism-independent cerebral vasodilation elicited by electrical stimulation of the centromedian-parafascicular complex in rat. Neurosci. Lett. 83, 269-274.
Nakai, M., Ogino, K., 1984. The relevance of cardio-pulmonary-vascular reflex to regulation of the brain vessels. Jpn. J. Physiol. 34, 193-197.
Nakai, M., Tamaki, K., Ogata, J., Matsui, Y., Maeda, M., 1993. Parasympathetic cerebrovasodilator center of the facial nerve. Circ. Res. 72, 470-475.
Norgren, R., 1978. Projections from the nucleus of the solitary tract in the rat. Neurosci. 3, 207-218.
Nozaki, K., Moskowitz, M.A., Maynard, K.I., Koketsu, N., Dawson, T.M., Bredt, D.S., Snyder, S.H., 1993. Possible origins and distribution of immunoreactive nitric oxide synthase-containing nerve fibers in cerebral arteries. J. Cereb. Blood Flow Metab. 13, 70-79.
Paterno, R., Heistad, D.D., Faraci, F.M., 2000. Potassium channels modulate cerebral autoregulation during acute hypertension. Am. J. Physiol. Heart Circ. Physiol. 278, H2003-H2007.
Paton, J.F., Kasparov, S., 1999. Differential effects of angiotensin II on cardiorespiratory reflexes mediated by nucleus tractus solitarii - a microinjection study in the rat. J. Physiol. 521, 213-225.
Rapela, C.E., Green, H.D., Denison, A.B., 1967. Baroreceptor reflexes and autoregulation of cerebral blood flow in the dog. Circ. Res. 21, 559-568.
Ruggiero, D.A., Ross, C.A., Kumada, M., Reis, D.J., 1982. Reevaluation of projections from the mesencephalic trigeminal nucleus to the medulla and spinal cord: new projections. a combined retrograde and anterograde horseradish peroxidase study. J. Comp. Neurol. 206, 278-292.
Ruggiero, D.A., Tong, S., Anwar, M., Gootman, N., Gootman, P.M., 1996. Hypotension-induced expression of the c-fos gene in the medulla oblongata of piglets. Brain Res. 706, 199-209.
Sadoshima, S., Busija, D.W., Brody, M.J., Heistad, D.D., 1981. Sympathetic nerves protect against stroke in stroke-prone hypertensive rats. Hypertension 3 (Suppl. I), 124-127.
Seagard, J.L., Dean, C., Hopp, F.A., 1995. Discharge patterns of baroreceptor-modulated neurons in the nucleus tractus solitarius. Neurosci. Lett. 191, 13-18.
Seagard, J.L., Hopp, F.A., Drummond, H.A., Van Wynsberghe, D.M., 1993. Selective contribution of two types of carotid sinus baroreceptors to the control of blood pressure. Circ. Res. 72, 1011-1022.
Spencer, S.E., Sawyer, W.B., Wada, H., Platt, K.B., Loewy, A.D., 1990. CNS projections to the pterygopalatine parasympathetic preganglionic neurons in the rat: a retrograde transneuronal viral cell body labeling study. Brain Res. 534, 149-169.
Sugiura, Y., Terui, N., Hosoya, Y., 1989. Difference in distribution of central terminals between visceral and somatic unmyelinated (C) primary afferent fibers. J. Neurophysiol. 62, 834-840.
Suzuki, N., Fukuuchi, Y., Koto, A., Naganuma, Y., Isozumi, K., Konno, S., Gotoh, J., Shimizu, T., 1994. Distribution and origins of cerebrovascular NADPH-diaphorase-containing nerve fibers in the rat. J. Auton. Nerv. Syst. 49, S51-S54.
Talman, W.T., Nitschke Dragon, D., 1995a. Inhibition of nitric oxide synthesis extends cerebrovascular autoregulation during hypertension. Brain Res. 672, 48-54.
Talman, W.T., Nitschke Dragon, D., 1995b. Mechanisms for preserved cerebrovascular autoregulation during hypertension in rats after sinoaortic denervation. Clin. Exp. Pharmacol. Physiol. 22, S77-S79.

Talman, W.T., Nitschke Dragon, D., 2000. Parasympathetic nerves influence cerebral blood flow during hypertension in rat. Brain Res. 873, 145-148.

Talman, W.T., Nitschke Dragon, D., 2002. Inhibiting the nucleus tractus solitarii extends cerebrovascular autoregulation during hypertension. Brain Res. 931, 92-95.

Talman, W.T., Nitschke Dragon, D., Ohta, H., 1994. Baroreflexes influence autoregulation of cerebral blood flow during hypertension. Am. J. Physiol. Heart Circ. Physiol. 267, H1183-H1189.

Tamaki, K., Heistad, D.D., 1986. Response of cerebral arteries to sympathetic stimulation during acute hypertension. Hypertension 8, 911-917.

Tanaka, K., Fukuuchi, Y., Shirai, T., Nogawa, S., Nozaki, H., Nagata, E., Kondo, T., Suzuki, N., Shimizu, T., 1995. Chronic transection of post-ganglionic parasympathetic and nasociliary nerves does not affect local cerebral blood flow in the rat. J. Auton. Nerv. Syst. 53, 95-102.

Toda, N., Ayajiki, K., Tanaka, T., Okamura, T., 2000a. Preganglionic and postganglionic neurons responsible for cerebral vasodilation mediated by nitric oxide in anesthetized dogs. J. Cereb. Blood Flow Metab. 20, 700-708.

Toda, N., Ayajiki, K., Yoshida, K., Kimura, H., Okamura, T., 1993. Impairment by damage of the pterygopalatine ganglion of nitroxidergic vasodilator nerve function in canine cerebral and retinal arteries. Circ. Res. 72, 206-213.

Toda, N., Tanaka, T., Ayajiki, K., Okamura, T., 2000b. Cerebral vasodilatation induced by stimulation of the pterygopalatine ganglion and greater petrosal nerve in anesthetized monkeys. Neurosci. 96, 393-398.

Williams, J.L., Heistad, D.D., Siems, J.L., Talman, W.T., 1989. Effects of stimulation of fastigial nucleus on cerebral blood flow in cats. Am. J. Physiol. 257, H297-H304.

Yoshida, K., Okamura, T., Kimura, H., Bredt, D.S., Snyder, S.H., Toda, N., 1993. Nitric oxide synthase-immunoreactive nerve fibers in dog cerebral and peripheral arteries. Brain Res. 629, 67-72.

Zhang, F., Iadecola, C., 1993. Fastigial stimulation increases ischemic blood flow and reduces brain damage after focal ischemia. J. Cereb. Blood Flow Metab. 13, 1013-1019.

Chapter 17

BRAINSTEM PREMOTOR CARDIAC VAGAL NEURONS

David Mendelowitz
George Washington University, Department of Pharmacology, Ross 654, 2300 Eye Street, NW, Washington, DC 20037, USA

Abstract: This chapter focuses on the neurobiology of premotor cardiac vagal neurons that are located in the brainstem. These parasympathetic cardiac neurons play an essential role in the regulation of heart rate, and are responsible for, together with sympathetic activity, cardiovascular homeostasis. Since cardiac vagal neurons are intrinsically silent their activity is determined by the activity and transmitters released by the neurons that synapse upon these neurons. This chapter focuses on the synaptic activation of cardiac vagal neurons, their postsynaptic receptors and electrophysiological responses, and, in particular, how the cellular activity of these neurons is altered during different physiological and pathological conditions.

Key words: Ambiguus, parasympathetic, heart rate, heart, sudden infant death syndrome, bradycardia

INTRODUCTION

Heart rate in healthy individuals is determined primarily by the tonic and reflex control of parasympathetic activity that innervates the heart. In conscious and anesthetized animals there is a tonic level of parasympathetic nerve firing and little, if any, sympathetic activity at rest, as described in humans (Pickering et al., 1972), dogs (Scher and Young, 1970), cats (Kunze, 1972) and rats (Coleman, 1980; Stornetta et al., 1987). During increases in arterial pressure the initial reflex induced slowing of the heart is caused primarily, if not exclusively, by increases in cardiac vagal nerve activity (Scher and Young, 1970; Stornetta et al., 1987). During decreases in arterial pressure the baroreflex induced tachycardia is caused mostly by decreases in parasympathetic, in addition to increases in sympathetic nerve activity (Scher and Young, 1970; Spyer, 1981; Spyer and Gilbey, 1988).

When both parasympathetic and sympathetic activities are present, parasympathetic activity generally dominates the control of heart rate. Increases in parasympathetic activity evoke a bradycardia that is more pronounced when there is a high level of sympathetic firing (Levy and Zieske, 1969). During moderate or high level of parasympathetic activity changes in sympathetic firing elicit negligible changes in heart rate (Levy and Zieske, 1969).

However, cardiac vagal activity is diminished or unresponsive in many disease states, including hypertension, heart failure and sudden cardiac death (Eckberg et al., 1971; Waxman and Wald, 1977; Rardon and Bailey, 1983; La Rovere et al., 1988; Whitescarver et al., 1990; Vanoli et al., 1991). A delay in the inhibitory actions of this autonomic motor system after exercise is a powerful predictor of overall mortality (Cole et al., 1999). Restoration of normal cardiac vagal activity prevents ischemia and reperfusion induced arrhythmias, and decreases risk of sudden death after myocardial infarction, suggesting that increasing cardiac vagal activity could be an effective clinical target in some heart diseases (Eckberg et al., 1971; Waxman and Wald, 1977; Rardon and Bailey, 1983; La Rovere et al., 1988; Whitescarver et al., 1990; Vanoli et al., 1991; Townend and Littler, 1995).

A number of previously published reviews have extensively described the location of premotor cardiac vagal neurons, the anatomical pathways that project to premotor cardiac neurons and the efferent projections from these neurons to parasympathetic motorneurons in cardiac ganglia (Kalia, 1981; Loewy and Spyer, 1990; Taylor, 1994). This chapter will focus on the physiology of premotor cardiac vagal neurons, and in particular what cellular electrophysiological processes determine their activity during different physiological and pathophysiological conditions.

INTRINSIC ACTIVITY OF PREMOTOR CARDIAC VAGAL NEURONS

It is widely accepted that parasympathetic activity originates from the central nervous system rather than from peripheral ganglia. Preganglionic cardiac vagal fibers are tonically active with a firing pattern that is cardiac pulse synchronous and most active during post-inspiration and reduced during inspiration (Heymans and Neil, 1958; Kunze, 1972; Spyer, 1981; Gilbey et al., 1984; Loewy and Spyer, 1990). Section of the preganglionic fibers releases the heart from parasympathetic inhibition (Heymans and Neil, 1958), indicating that parasympathetic cardiac activity originating from the brainstem is necessary while parasympathetic cardiac cannot be sustained when only the activity of neurons in the cardiac ganglia are preserved.

A fundamental issue concerning the origin of parasympathetic cardiac activity is whether premotor cardiac vagal neurons in the brainstem possess spontaneous pacemaker activity or alternatively whether premotor cardiac vagal neurons are inherently silent and depend on synaptic activation to initiate and determine their firing. Extracellular recordings from premotor cardiac vagal neurons *in vivo* have indicated the great majority of these neurons (identified by antidromic stimulation) are silent (McAllen and Spyer, 1978; Jordan et al., 1982; Gilbey et al., 1984). Only two cardiac vagal neurons have been recorded intracellularly *in vivo*, and these neurons were silent (Gilbey et al., 1984). The lack of ongoing cardiac vagal activity in these anesthetized *in vivo* animals is, at first, somewhat unexpected since in conscious animals there is a high level of tonic cardiac vagal activity. However, in these *in vivo* experiments, excitatory pathways to cardiac vagal neurons were likely inhibited due to the trauma of the acute open-chest surgery, or anesthesia, which, in general, inhibits excitatory synaptic pathways. These results from *in vivo* experiments would suggest premotor cardiac vagal neurons do not possess spontaneous pacemaker activity but rather their activity is determined by synaptic activation of these neurons which is susceptible to anesthestics as well as other inhibitory influences present in these *in vivo* electrophysiological experiments.

The study of premotor cardiac vagal neurons in the central nervous system has been exceptionally difficult *in vivo*. This is due to many factors, including the relatively small population of cardiac vagal neurons, experimental difficulties antidromically activating these neurons from their cardiac fibers projecting to the heart, and the location of their soma deep within the brainstem in a poorly defined and heterogeneous nucleus ambiguus. Furthermore, these studies have been hampered by the deleterious effects of the required surgical procedures that include opening the chest, exposing the brainstem, anesthesia, and the difficulty controlling the experimental conditions, including blood pressure, in these *in vivo* preparations.

More recent attempts at examination of premotor cardiac vagal neurons, especially at the cellular level, have been greatly advanced with new methodological strategies including infrared visualization of neurons in *in vitro* brainstem slices coupled with the retention of physiological identity using retrograde fluorescent tracers. Using these techniques cardiac vagal neurons can be visually identified and electrophysiologically manipulated using patch clamp methodologies (Mendelowitz and Kunze, 1991; Mendelowitz, 1996; Mihalevich et al., 1996; Mendelowitz, 1998; Neff et al., 1998b; Neff et al., 1998a; Mendelowitz, 1999; Irnaten et al., 2001; Wang et al., 2001a, b; Wang et al., 2001c; Irnaten et al., 2002c; Irnaten et al., 2002a;

Irnaten et al., 2002b; Venkatesan et al., 2002a, b; Wang et al., 2002; Irnaten et al., 2003; Wang et al., 2003) without many of the confounding variables that are typically present in *in vivo* experiments.

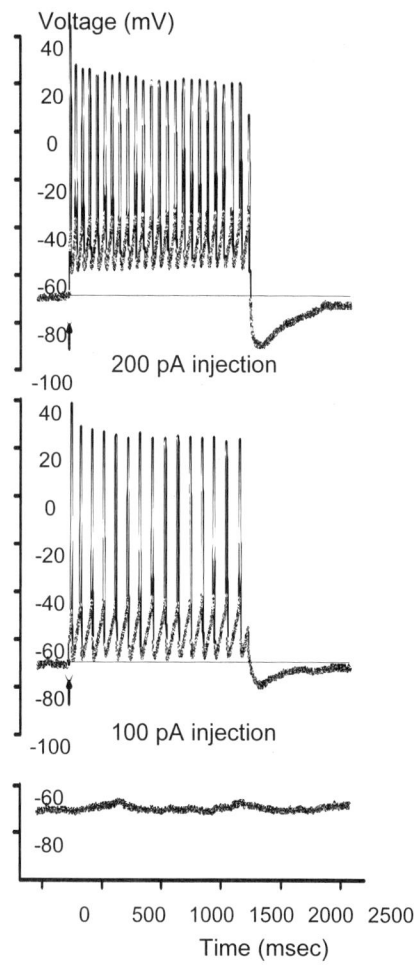

Figure 1. Bottom panel: parasympathetic cardiac neurons were inherently silent. Middle panel: injection of depolarizing current evoked repetitive firing with minimal delay (14.4 ± 1.0 ms, n=12) and little, if any, spike frequency adaptation (SFA; initial firing frequency 15.2 ± 1.5 Hz, final frequency 13.3 ± 1.1 Hz, not significant). Top panel: increasing the

depolarizing current (to 200pA) elicited a greater firing frequency, again with little SFA, and increased afterhyperpolarization that occurred after sustained activity.

Similar to the results obtained *in vivo*, cardiac vagal neurons recorded *in vitro* in the nucleus ambiguus are normally silent in the absence of synaptic activity (see Fig. 1). Cardiac vagal neurons do not display any pacemaker-like activity such as repetitive or phasic depolarizations or action potentials. However, depolarizing currents (as little as 100pA) are sufficient to evoke repetitive firing in cardiac vagal neurons. This activity occurs with little delay and minimal spike frequency adaptation during maintained depolarizing currents.

The voltage gated currents that have been characterized in these neurons and are responsible for these firing properties include a rapidly activating and inactivating Na^+ current (Mihalevich et al., 1996) and a voltage gated calcium current that is comprised nearly entirely of ω-agatoxin sensitive P/Q-type voltage gated calcium currents (Irnaten et al., 2003). Surprisingly the Na^+ current is relatively resistant to tetrodotoxin (TTX), requiring 10 mM TTX for complete block. Voltage gated potassium currents include a transient K^+ current that can be blocked by 4-aminopyridine (4AP), and a tetraethylammonium (TEA) sensitive delayed rectifyer K^+ channel. Also present, and responsible for the afterhyperpolarization that occurs after firing, is a calcium activated potassium channel that is blocked by apamin (Mendelowitz, 1996). These currents, among others, are responsible for the firing characteristics of cardiac vagal neurons and enable them to follow fast synaptic drive closely, without adaptation, as well as integrate long-lasting modulatory influences. As will be discussed later, many of these voltage gated currents are targets of neurotransmitters and anesthetics that modulate the activity of premotor cardiac vagal neurons.

Premotor cardiac vagal neurons possess follower-cell like properties such as the ability to follow synaptic activity closely without frequency adaptation. Also, unlike other brainstem neurons which may be involved in maintaining blood pressure or respiration in the absence of sensory information, such as those neurons in the nucleus tractus solitarius that fire spontaneously even when impinging sensory synapses are silenced, or neurons in the pre-Botzinger complex that may have intrinsic respiratory rhythmogenic activity, the firing of cardiac vagal neurons is fully dependent on critical synaptic input to these neurons. The synaptic inputs to cardiac vagal neurons are therefore important in maintaining normal heart rates and cardiac function, but seem to be easily silenced with stressful conditions, such as pain and discomfort, consistent with the high heart rate under these conditions (Porter et al., 1988; van Lieshout et al., 1991).

BAROREFLEX ACTIVATION OF PREMOTOR CARDIAC VAGAL NEURONS

Increases in activity from arterial baroreceptors and chemoreceptors is exclusively transmitted to neurons in the nucleus tractus solitarius (NTS) which constitutes the first synapse of the baroreflex pathway (Andresen and Kunze, 1994; Andresen and Mendelowitz, 1996). A major pathway to the nucleus ambiguus originates from the NTS (Andresen and Kunze, 1994; Andresen and Mendelowitz, 1996; Browning and Mendelowitz, 2003). Retrograde tracers (using either HRP or fluorescent tracers) injected into the nucleus ambiguus label neurons predominately ventral to the solitary tract in the NTS (Loewy and Spyer, 1990). Similar results have been obtained using viral retrograde tracing techniques that initially infect cardiac vagal neurons and then infect and identify neurons that project to premotor cardiac vagal neurons (Standish et al., 1994). These neurons which project to premotor cardiac vagal neurons, are mostly located in the ventral region of the NTS (Standish et al., 1994). Electrophysiological experiments, in which the NTS region is stimulated and post-synaptic responses are recorded in identified premotor cardiac vagal neurons (Neff et al., 1998a) demonstrate that there is an excitatory glutamatergic pathway from NTS to premotor cardiac vagal neurons in the nucleus ambiguus. Figure 2 illustrates the post-synaptic responses in premotor cardiac vagal neurons in response to stimulation of the NTS. Stimulation of this pathway activates post-synaptic excitatory amino acid receptors including a long lasting NMDA mediated current and a rapidly activating and inactivating non-NMDA receptor response with onset latencies of approximately 12 ms.

Stimulation of afferent fibers in the vagus nerve also consistently evoked glutamatergic post-synaptic currents in cardiac vagal neurons *in vitro*, and these polysynaptic responses occurred with a latency of approximately 44 ms (Evans et al., 2003). These results are very similar with the results from *in vivo* experiments. Kunze examined the latency of the entire baroreflex by stimulating the carotid sinus nerve (CSN) while recording the activity of cardiac vagal fibers. She found that the latencies of activity following CSN stimulation were 30-72 ms (Kunze, 1972). McAllen and Spyer, directly recording from the cell bodies of cardic vagal neurons, observed evoked activity with latencies of 20-50 ms upon stimulation of the CSN (McAllen and Spyer, 1978). Subtracting the time taken for the action potentials to be conducted through either the vagus nerve or CSN (approximately 2 ms, based on an estimate of 2 mm from nodose ganglia to NTS, and a conduction velocity of 1.0 m/s), the delay of the reflex response due to interactions within the NTS would be approximately 15 to 42 ms. This result, along with a highly variable response time, suggests that there is a complex polysynaptic circuitry within the NTS between receiving input from the afferent vagal fibers and relaying this signal to cardiac vagal neurons. However, regardless of the number of preceding synapses, the increase in excitatory amino acid neurotransmission from NTS neurons to cardiac vagal neurons activated in response to increases in baroreceptor activity is likely an essential pathway that acts to excite cardioinhibitory premotor vagal neurons, reduce heart rate, and restore blood pressure and heart rate to normal levels.

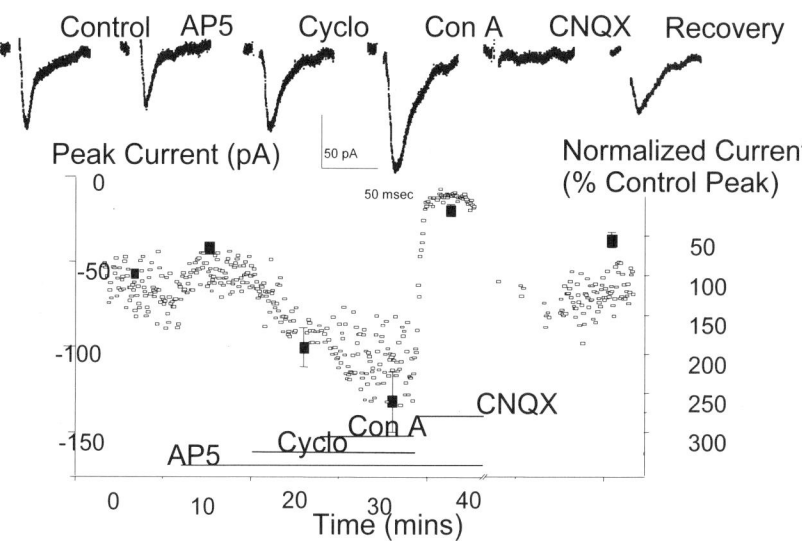

Figure 2. Stimulation of the NTS evokes rapidly activating and inactivating, and long lasting excitatory synaptic currents in premotor cardiac vagal neurons. The selective NMDA antagonist AP5 blocks the long lasting component. The non-NMDA currents are enhanced by both concanavalin A (Con A, 0.3 mg/ml), and cyclothiazide (Cyclo, 100 µM) which remove the desensitization of kainate and AMPA receptors, respectively. The non-NMDA antagonist CNQX blocks the AMPA and kainate synaptic currents.

Stimulation of the NTS also evokes an inhibitory GABAergic pathway to premotor cardiac vagal neurons (see Fig. 3) (Wang et al., 2001a). The physiological role of this GABAergic projection from the NTS to premotor cardiac vagal neurons has yet to be identified. Among the potential roles for this GABAergic pathway include a role in the inhibition of cardiac neurons that likely occurs during inspiration, and/or this inhibitory pathway from the NTS may be involved in synchronizing cardiac vagal activity to afferent input. For example, it is interesting to postulate that baroreceptor activity first excites premotor cardiac vagal activity via the glutamatergic pathway from the NTS, but then subsequently premotor cardiac vagal neurons may be inhibited by a second wave of inhibitory GABAergic neurotransmission. A delay between the initial excitatory and the following inhibitory neurotransmission to premotor cardiac vagal neurons would likely be a powerful mechanism to synchronize premotor cardiac vagal activity to NTS activity and baroreceptor discharge.

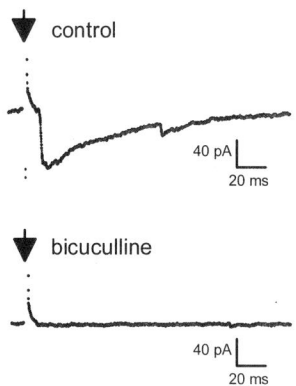

Figure 3. Stimulation of the NTS, illustrated by the arrows, evoked GABAergic currents in premotor cardiac vagal neurons. These evoked postsynaptic currents had a mean onset latency of 11.8 ± 1.1 ms (n=11). Bicuculline (10 μM), a $GABA_A$ antagonist, blocked these evoked synaptic currents.

RESPIRATORY SINUS ARRHYTHMIA: RESPIRATORY MODULATION OF PREMOTOR CARDIAC VAGAL ACTIVITY

While feedback from pulmonary stretch receptors and direct respiratory related changes in venous return and cardiac stretch can evoke respiratory related fluctuations in heart rate, the dominant source of respiratory sinus arrhythmia originates from the brainstem (Anrep et al., 1935). Respiratory sinus arrhythmia persists when the lungs are stationary (caused by muscle paralysis or constant flow ventilation), and the respiratory modulation of heart rate remains synchronized with brainstem respiratory rhythms even if artificial ventilation of the lungs and chemoreceptor activation occurs at different intervals (Spyer and Gilbey, 1988; Daly, 1991; Elghozi et al., 1991; Hrushesky, 1991; Shykoff et al., 1991). In both animals and humans, respiratory sinus arrhythmia is mediated via cardiac vagal activity. Respiratory sinus arrhythmia persists in experimental animals upon sectioning sympathetic pathways and in quadriplegic patients with spinal cord injury and sympathetic dysfunction (Inoue et al., 1990; Daly, 1991; Elghozi et al., 1991; Hrushesky, 1991; Shykoff et al., 1991). Blockade of parasympathetic cardiac activity abolishes respiratory sinus arrhythmia (Warner et al., 1986).

The respiratory system also influences heart rate by modulating the baroreceptor and chemoreceptor input to cardiac vagal neurons. In both animals and humans, the baroreceptor and chemoreceptor reflexes are inhibited during inspiration and are facilitated during post-inspiration and expiration or during a maintained phase of post-inspiration and apnea (Davidson et al., 1976; Eckberg and Orshan, 1977; McAllen and Spyer, 1978; Loewy and Spyer, 1990). This respiratory modulation of both reflexes persists after pulmonary denervation, as well as ventilatory paralysis, suggesting that this "gating" of the baroreceptor and chemoreceptor reflexes also occurs within the brainstem (Spyer and Gilbey, 1988; Loewy and Spyer, 1990; Daly, 1991).

The respiratory and cardiovascular systems are highly intertwined, both anatomically and physiologically. Respiratory and cardioinhibitory neurons are often co-localized in the same brainstem nuclei, and this is particularly evident in the nucleus ambiguus which is comprised of vagal cardioinhibitory neurons and the ventral respiratory group, which includes the pre-Botzinger complex. Anatomical studies of ventral respiratory group neurons demonstrate that these neurons have many axon collateral synapses within the nucleus ambiguus (Gauthier et al., 1980; Yamada et al., 1988; Duffin and Aweida, 1990; Zheng et al., 1991; Lipski et al., 1994; Pilowsky et al., 1994; Morillo et al., 1995).

Cardiac vagal activity (recorded from fibers in the cardiac nerve) has pronounced respiratory modulation. Cardiac vagal fibers fire most rapidly in post-inspiration and are often silent in inspiration (Kunze, 1972; McAllen and Spyer, 1978; Spyer, 1981). Unfortunately information concerning the transmitters and neurons responsible for this cardiorespiratory interaction *in vivo* is extremely difficult to obtain. In the only *in vivo* study in which intracellular recordings were successful (and in only 2 neurons) cardiac vagal neurons received inhibitory synaptic input during inspiration and a rapid depolarization during post-inspiration/expiration (Gilbey et al., 1984).

During inspiration input resistance decreased and injection of chloride reversed this hyperpolarization suggesting a chloride conductance was activated (Gilbey et al., 1984). This inhibitory chloride conductance would most likely be caused by activation of postsynaptic GABA or glycine receptors. Paradoxically, however, in a review by one of these investigators, it is stated "the inspiratory related inhibition of cardioinhibitory neurons is not antagonized by the iontophoretic application of either bicuculline or the glycine antagonist strychnine" (Loewy and Spyer, 1990). This seemingly conflicting data could be due to the small sample size, inability of microinjections to distinguish direct effects on cardiac vagal neurons from actions on presynaptic terminals or local interneurons, and anesthetic modulation of the respiratory activity and/or neurotransmission to cardiac vagal neurons. In summary, there is an extreme paucity of information concerning the transmitters, receptors and sites of action within the central nervous system responsible for cardio-respiratory interactions. Respiratory inputs do not seem to alter baroreceptor and chemoreceptor synapses at their first synapse in the nucleus tractus solitarius (NTS) (Spyer, 1981; Spyer and Gilbey, 1988). Rather, the little data that exists suggest the cardio-respiratory interactions occur within the nucleus ambiguus (Spyer, 1981; Gilbey et al., 1984).

Recent work *in vitro* has suggested nicotinic receptors may be involved in generating cardiorespiratory interactions. Activation of cholinergic receptors on cardiac vagal neurons is excitatory and this occurs by activation of both pre-synaptic and post-synaptic mechanisms (see Fig. 4). Nicotine, but not muscarinic agonists, activates post-synaptic receptors and a depolarizing inward current in vagal cardiac neurons (Neff et al., 1998b). In addition, nicotine acts at different presynaptic and postsynaptic sites to facilitate glutamatergic neurotransmission. Presynaptic nicotinic receptors increase the frequency of transmitter release and are sensitive to block by αBgtx, indicating these presynaptic receptors likely contain the α7 subunit of the nicotinic receptor (Neff et al., 1998b). Nicotine also augments post-synaptic non-NMDA currents via a αBgtx-insensitive receptor (Neff et al., 1998b).

The nicotinic facilitation of glutamatergic neurotransmission to premotor cardiac vagal neurons could either be due to the influx of calcium via the calcium permeable α7 nicotinic receptors at the presynaptic glutamatergic terminal, or alternatively, activation of nicotinic receptors may depolarize the presynaptic terminal sufficient to open voltage gated calcium channels which then facilitate neurotransmitter release. To test whether voltage dependent calcium channels were involved in the nicotinic excitation of cardiac vagal neurons the nicotine evoked inward current, increase in mini amplitude and increase in mini frequency were examined in the presence and absence of specific voltage gated calcium channel antagonists (Wang et al.,

2001b). All of the nicotine evoked responses were inhibited by the non-selective voltage dependent calcium channel blocker Cd (100μM). The P-type voltage dependent calcium channel blocker agatoxin IVA (100 nM) abolished the nicotine evoked responses (see Fig. 5). Nimodipine (2 μM), an antagonist of L-type calcium channels, inhibited the increase in mini amplitude and frequency, but did not block the ligand gated inward current. The N- and Q-type voltage dependent calcium channel antagonists conotoxin GVIA (1μM) and conotoxin MVIIC (5 μM) had no effect. The presynaptic and postsynaptic facilitation of glutamatergic neurotransmission to cardiac vagal neurons by nicotine therefore involves activation of agatoxin-IVA sensitive, and possibly L-type voltage dependent calcium channels (Wang et al., 2001b). The postsynaptic inward current elicited by nicotine is dependent upon activation of agatoxin-IVA sensitive voltage dependent calcium channels (Wang et al., 2001b).

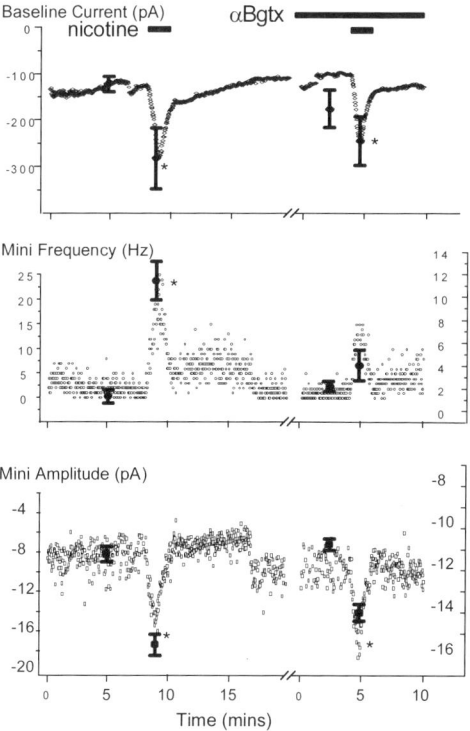

Figure 4. αBgtx prevented the increase in frequency but not amplitude of the nicotine elicited miniature synaptic events (minis). Inward currents were not strongly influenced by αBgtx,

top. However, αBgtx did selectively inhibit the nicotine induced increase in mini frequency. Mini amplitude was unaffected by αBgtx.

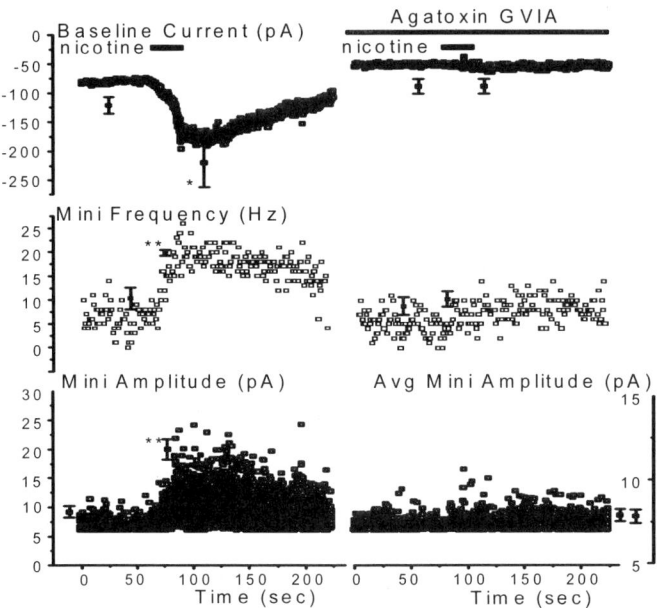

Figure 5. The specific voltage dependent calcium channel blocker ω-Agatoxin IVA (100 nM) completely abolished all presynaptic and postsynaptic nicotinic activation of cardiac vagal neurons. ω-Agatoxin IVA, blocked the nicotine induced inward current, increases in mini frequency and mini amplitude.

If activation of cholinergic receptors are responsible for the respiratory modulation of premotor cardiac vagal neurons, it is critical to identify the origin of the cholinergic pathway to premotor cardiac vagal neurons. One study has utilized a trans-synaptic viral tracer, pseudorabies virus, to identify cholinergic pathways to premotor cardiac vagal neurons (Irnaten et al., 2001). The Bartha strain of pseudorabies virus (PRV), an attenuated swine alpha herpes virus, can be used as a transsynaptic marker of neural circuits. Bartha PRV invades neuronal networks in the central nervous system through peripherally projecting axons, replicates in these parent neurons, and then travels transsynaptically to continue labeling the second and higher order neurons in a time dependent manner. Cholinergic neurons that project to cardiac vagal neurons include superior laryngeal neurons (Irnaten et al.,

2001). It is possible that these or other cholinergic neurons could influence cardiac vagal neurons via three independent mechanisms. One site of action would be via a direct activation of post-synaptic ligand gated nicotinic channels in cardiac vagal neurons, which would act to depolarize and excite cardiac vagal neurons during post-inspiration. An additional site of action would be presynaptic and evoke a nicotinic facilitation of presynaptic release of glutamate. A third action of the post-inspiratory neurons would be to augment glutamatergic neurotransmission by activating nicotinic receptors that facilitate post-synaptic non-NMDA receptors in cardiac vagal neurons. These latter two effects could constitute mechanisms by which respiratory inputs could gate or facilitate the baroreflex during post-inspiration. This does not rule out the possibility that nicotinic receptors may also be involved in other phases of respiration, such as enhancing inhibitory neurotransmission to premotor cardiac vagal neurons during inspiration.

OPIOID RECEPTOR MEDIATED ACTIVATION OF PREMOTOR CARDIAC VAGAL NEURONS

Hypotension typically evokes an increase in sympathetic and a decrease in parasympathetic activity which acts to restore blood pressure to normal levels (Heymans and Neil, 1958; Scher and Young, 1970; Spyer, 1981; Loewy and Spyer, 1990). In severe hemorrhagic and septic shock both the parasympathetic and sympathetic responses are initially compensatory but then a second paradoxical or "decompensatory" phase occurs in which the severe hypotension evokes a reflex bradycardia and a decrease in vascular resistance causing blood pressure to fall precipitously (Evans et al., 1989; Ohnishi et al., 1997). The stimulus for this second phase has been attributed to activation of vagal afferents from the heart while the reflex bradycardia is caused by an increase in efferent parasympathetic cardiac activity (Evans et al., 1989; Kwok and Dun, 1998). Application of low doses of opioid antagonists into the central nervous system (CNS), but not intravenously, block the decrease in heart rate during hemorrhagic shock indicating that activation of an endogenous opioid mechanism in the CNS is responsible for increasing cardiac vagal activity (Evans et al., 1989; Ohnishi et al., 1997; Ang et al., 1999).

Recent work (Irnaten et al., 2003) has demonstrated that opioids can alter the activity of premotor cardiac vagal neurons both by modulating voltage gated calcium currents in cardiac vagal neurons, as well as altering the synaptic inputs that impinge on premotor cardiac vagal neurons. The μ opioid receptor agonist endomorphin1 has been shown to inhibit the ω-agatoxin sensitive P/Q-type voltage gated calcium currents in premotor cardiac vagal neurons (see Fig. 6). This inhibition is mediated via a G-protein mediated pathway which can be blocked by pretreatment with pertusis toxin (Irnaten et al., 2003). It is possible that the inhibition of calcium currents may act to indirectly facilitate the activity of premotor cardiac vagal neurons by disinhibition, such as by a reduction in inhibitory calcium activated potassium currents.

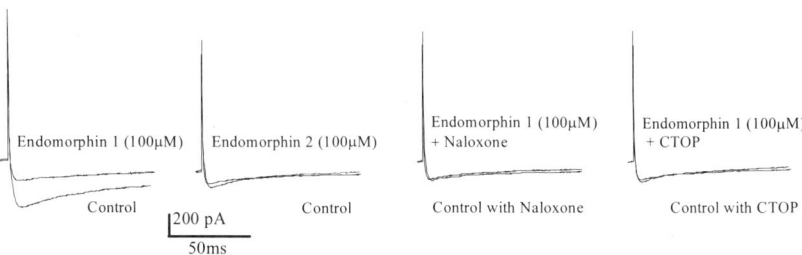

Figure 6. The effect of endomorphin 1 (100 μM) and endomorphin 2 (100 μM), on Agatoxin-IVA sensitive calcium channels is shown. Only Endomorphin-1 evoked an inhibition. This inhibition was prevented by prior application of naloxone as well as CTOP.

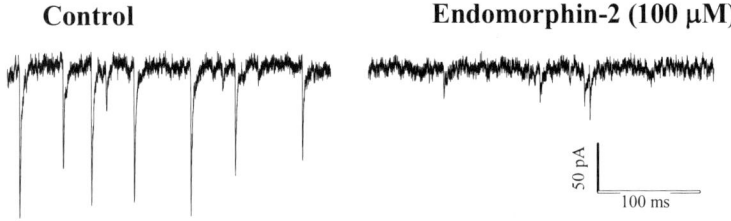

Figure 7. Endomorphin-2, but not endomorphin-1 or DAMGO inhibits both the amplitude and frequency of the GABAergic IPSCs to premotor cardiac vagal neurons.

To examine whether opioid receptor activation can alter the synaptic neurotransmission to premotor cardiac vagal neurons, a recent study examined the effects of endomorphin-1, endomorphin-2 and DAMGO (a synthetic, μ-selective agonist) on spontaneous GABAergic inhibitory postsynaptic currents (IPSCs) in premotor cardiac vagal neurons. Only endomorphin-2 (100 μM) produced a significant inhibition, of both the frequency and the amplitude of the spontaneous IPSCs in premotor cardiac vagal neurons (Fig. 7). The inhibitory effects of endomorphin-2 were blocked by naloxonazine (10 μM), a selective μ_1 receptor antagonist. Naloxonazine alone (10 μM) had a potentiating effect on the frequency of the GABAergic IPSCs (+161.43%) but not on the amplitude, indicating that GABA release to cardiac vagal neurons may be under tonic control of opioids acting at the μ_1 receptor. Endomorphin-2 did not reduce the responses evoked by exogenous application of GABA. These results indicate that endomorphin-2 acts on μ_1 receptors located on precedent neurons to decrease GABAergic input to cardiac vagal neurons located in the nucleus ambiguus. The subsequent increase in parasympathetic outflow to the heart may be one mechanism by which μ-selective opioids act to induce bradycardia.

Nociceptin, a newly discovered endogenous peptide, is the ligand for the opioid receptor-like receptor (ORL_1) (Henderson and McKnight, 1997; Meunier, 1997). Nociceptin acts to increase parasympathetic outflow to the heart, yet the mechanisms by which this is achieved are unknown. In a recent study the effects of nociceptin on spontaneous GABAergic input to cardiac parasympathetic neurons was examined. At 100μM nociceptin inhibited both the frequency and the amplitude of spontaneous GABAergic IPSCs in premotor cardiac vagal neurons (Fig. 8). Nociceptin also caused a novel postsynaptic inhibition of the responses evoked by exogenous application of GABA (Fig. 8). These results indicate nociceptin acts both on neurons precedent to cardiovascular neurons to decrease the activity of GABAergic neurons that synapse upon cardiovascular neurons, as well as directly inhibiting the postsynaptic currents evoked by GABA. This inhibition by nociceptin would increase parasympathetic outflow to the heart, thus providing a possible mechanism for nociceptin-induced bradycardia.

SENSITIVITY OF PREMOTOR CARDIAC VAGAL NEURONS TO ANESTHESICS

The intravenous anesthetic ketamine is associated with a hemodynamic profile that typically is characterized by increases in both blood pressure and heart rate. These cardiovascular actions of ketamine are commonly thought of as sympathoexcitatory and the increases in blood pressure and heart rate have been attributed to "enigmatic" mechanisms of activation of the central sympathetic nervous system. However, this rise in heart rate is paradoxical since increases in blood pressure normally elicit baroreflex induced decreases in heart rate. Few aspects of the underlying mechanisms responsible for these ketamine responses have been directly examined, particularly at central autonomic sites of action (McGrath et al., 1975; Blake and Korner, 1982). Although ketamine does not modify baroreceptors (Slogoff and Allen, 1974), the sensory limb of the arterial baroreflex, it may affect the central nervous system sites, likely autonomic regions below the pons to alter neurons within reflex pathway (McGrath et al., 1975; Blake and Korner, 1982). The heart rate responses could well reflect inhibition of central cardiac parasympathetic mechanisms.

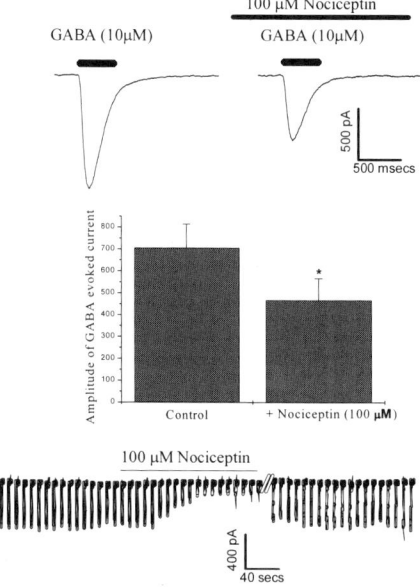

Figure 8. A typical control response to 10 μM GABA is significantly reduced by co-administration of nociceptin (100 μM). The averaged results from 6 neurons show a significant decrease (36.7 ± 9.8 %) in the amplitude of responses to exogenous GABA (10 μM) in premotor cardiac vagal neurons. The time course of a typical experiment shows responses to repetitive application of GABA (10 μM) are reduced with co-application of nociceptin (100 μM); this inhibition is reversed 10 minutes after drug washout.

Two studies have examined the cellular effects of ketamine on premotor cardiac vagal neurons. In the first study the effects of ketamine on the voltage gated currents in premotor cardiac vagal neurons was examined. Ketamine (10 µM – 1 mM) did not alter the current voltage relationships for the transient K^+ current and the delayed rectified K^+ current present in premotor cardiac vagal neurons. However, ketamine selectively depressed the total Na^+ current dose-dependently (10µM-1mM) (Fig. 9). In addition, ketamine shifted the Na^+ current inactivation curves to more negative potentials enhancing the Na^+ channel inactivation. One mechanism of action of ketamine is an inhibition of Na^+ but not K^+ channel function in brainstem premotor cardiac vagal neurons and these actions may contribute to the ketamine induced decrease in parasympathetic cardiac activity and increase in heart rate.

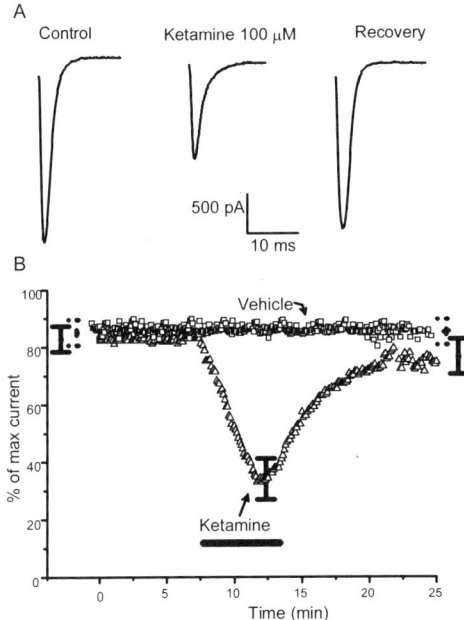

Figure 9. Ketamine (100 µM) reversibly inhibited the inward Na+ currents in premotor cardiac vagal neurons. To examine the time course and recovery of the ketamine evoked inhibition of Na+ current depolarizing steps from –100 mV to –40 mV were applied every 5 secs. Ketamine (100 µM) inhibited the peak Na+ current from control 85.5+3.6 to 31.3+10.4% of maximum (A and B) and recovery of the Na+ currents was nearly complete (B). Experiments performed with application of vehicle instead of ketamine indicated there was no time dependent run-down of the Na+ currents in the course of these experiments (B).

Ketamine has also been shown to inhibit components of the nicotine evoked excitation of premotor cardiac vagal neurons. At clinically relevant concentrations, ketamine inhibits two key nicotinic cholinergic excitatory mechanisms: reductions in postsynaptic nicotinic excitatory inward currents and elimination of the nicotinic enhancement of presynaptic release of glutamate (Fig. 10). The cholinergic modulation of cardiac parasympathetic activity can occur via nicotinic activation of direct post-synaptic inward currents as well as presynaptic nicotinic receptors that modulate the release of glutamate, and both of these responses are sensitive to ketamine. The third mechanism of action of nicotine, to increase glutamatergic non-NMDA synaptic responses, is insensitive to ketamine.

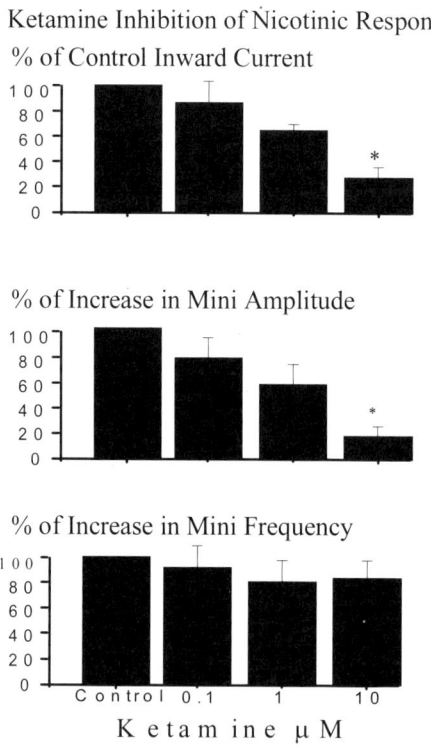

Figure 10. The nicotine-evoked inward current (normalized to control) was inhibited by increasing concentrations of ketamine and this inhibition was statistically significant at a ketamine concentration of 10 μM. Ketamine also significantly inhibited the nicotine evoked increase in mini amplitude and this inhibition was significant at a ketamine concentration of 10 μM. Ketamine, at all concentrations examined (0.1-10 μM) had no significant effect on mini amplitude.

The inhibitory GABAergic synaptic innervation of premotor cardiac vagal neurons is also a target of anesthetics. Barbiturates are among the most frequently used intravenous anesthetic agents for induction (Gilman et al., 1985). However barbiturates can cause central respiratory depression, an increase in heart rate and blunting of cardiovascular homeostatic reflexes (Murthy et al., 1982; Stornetta et al., 1987; Barringer and Bunag, 1990; Bedran-de-Castro et al., 1990; Watkins and Maixner, 1991). Results from animal studies have shown that pentobarbital decreases the gain of the baroreceptor reflex on the order of 50%, and this blunting of the baroreflex is caused nearly entirely by decreasing cardioinhibitory parasympathetic activity which dominates the control of heart rate (Stornetta et al., 1987; Bedran-de-Castro et al., 1990; Mendelowitz, 1999).

The most likely site of action of pentobarbital is the γ-aminobutyric acid type A ($GABA_A$) receptor (Whiting et al., 1999). Clinically relevant concentrations of barbiturates modulate $GABA_A$ receptors. The modulation includes potentiation of GABA responses, direct activation of $GABA_A$ receptors in the absence of GABA, and increased $GABA_A$ receptor desensitization (Ransom and Barker, 1975). At high concentrations barbiturates also inhibit $GABA_A$ receptor function (Robertson, 1989).

Activation of the ligand gated $GABA_A$ receptor by GABA increases Cl^- conductance causing hyperpolarization and neuronal inhibition (Whiting et al., 1999). The $GABA_A$ receptor complex is a pentamer which can be formed by a combination of subunits from at least 16 mammalian subunits, grouped in seven classes (α1-6, β1-3, γ1-3, δ, π, ε and θ) (Whiting et al., 1999). The minimal functional $GABA_A$ receptor comprises α and β subunits. The functional properties of $\alpha\beta$ receptors are modified by incorporation of γ, δ or ε subunits (Davies et al., 1997; Whiting et al., 1999). Modulation of $GABA_A$ function in sedation by benzodiazepines requires the presence of the γ subunit (Pritchett et al., 1989). By contrast modulation of the $GABA_A$ receptor by anesthetic agents is independent of this subunit (Pritchett et al., 1989; Jones et al., 1995). Recombinant $\alpha\beta\varepsilon$ and $\alpha\beta\delta$ receptors are both insensitive to benzodiazepines but can be distinguished from one another by virtue of the diminished anesthetic modulation seen upon incorporation of the ε subunit (Davies et al., 1997; Davies et al., 2001). However, the role of the ε subunit in $GABA_A$ receptor sensitivity to anesthetics is controversial (Pritchett et al., 1989; Jones et al., 1995; Davies et al., 1997; Neelands et al., 1999; Davies et al., 2001).

A recent study tested whether pentobarbital alters spontaneous inhibitory postsynaptic currents (IPSCs) in premotor cardiac vagal neurons in the brainstem nucleus ambiguus (Irnaten et al., 2002c). The hypothesis that expression of the GABA subunit into cardiac parasympathetic neurons using an adenovirus blocks the pentobarbital-evoked enhancement of spontaneous

GABAergic synaptic currents was also tested in this study (Irnaten et al., 2002c).

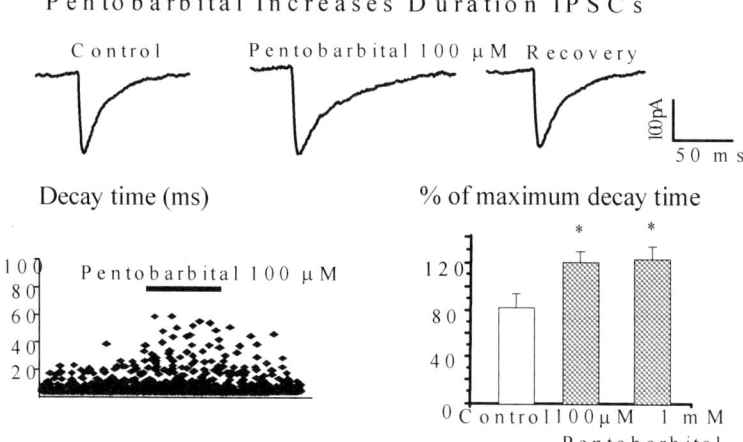

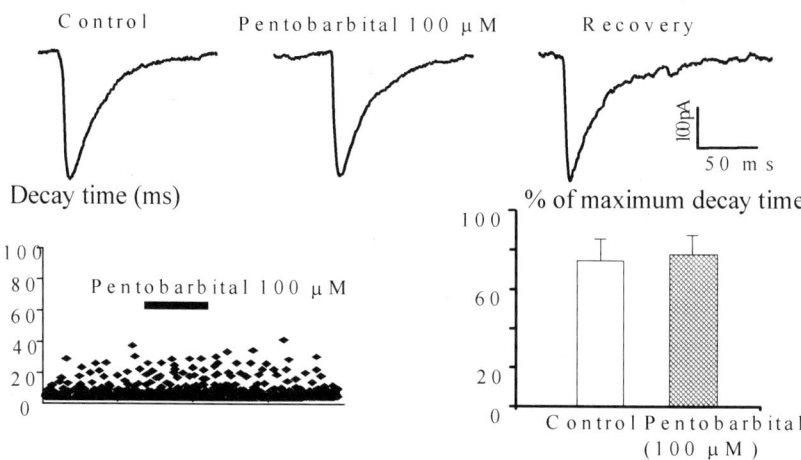

Figure 11. Pentobarbital increased the duration of spontaneous IPSCs as shown by comparing the average of 20 IPSCs during control (left), pentobarbital (100 μM) (middle) and recovery IPSCs (right). The time course of IPSC decay is shown for a representative experiment and the average from experiments in 9 cardiac parasympathetic neurons. Recombinant expression of the ε subunit in cardiac parasympathetic neurons renders their IPSCs insensitive to pentobarbital, bottom. Pentobarbital did not increase the duration of spontaneous IPSCs in cardiac parasympathetic neurons transfected with the adenovirus that expresses the ε subunit of the $GABA_A$ receptor as shown in a typical neuronal recording. The average of 20 IPSCs recorded under control conditions were indistinguishable from the

IPSCs recorded during application of pentobarbital (100 μM) and during washout in premotor cardiac vagal neurons transfected with the adenovirus that expresses the ε subunit.

As shown in figure 11, pentobarbital increased the decay time, but not the frequency or amplitude of the spontaneous IPSCs in premotor cardiac vagal neurons. These concentrations are clinically relevant as pentobarbital produces surgical anesthesia at concentrations of 200-300 μM (Richards, 1972). This action would augment the inhibition of premotor cardiac vagal neurons thereby reducing parasympathetic cardioinhibitory activity and increasing heart rate. An adenovirus was then used to selectively transfect premotor cardiac vagal neurons with the ε subunit of the GABA receptor. As shown in figure 11 (bottom) in the neurons transfected with the adenovirus that was engineered to co-express both GFP and the ε subunit of the GABA receptor, pentobarbital had no effect on the spontaneous IPSCs. The increase of decay-time induced by pentobarbital (100 μM) was prevented by the transfection of GABA ε-subunit receptor into premotor cardiac vagal neurons. This presumably occurred by expression and insertion of the ε subunit into postsynaptic $GABA_A$ receptors which renders the GABA receptors in cardiac parasympathetic neurons insensitive to pentobarbital (Irnaten et al., 2002c).

SUMMARY

In summary, there has been considerable progress during the past decade in our understanding the neurobiology of premotor cardiac vagal neurons. Many of the post-synaptic voltage and ligand gated channels in these neurons have been identified and characterized. A number of the essential synaptic connections to premotor cardiac vagal neurons, along with the transmitters and post-synaptic receptors that mediate these pathways, have been identified. However many of these synaptic pathways are likely very sensitive to peptides and other endogenous neuromodulatory agents and it is this modulation which may be responsible for many critical physiological responses in parasympathetic cardiac activity and ultimately heart rate. Future work will need to examine the modulation of the powerful glutamatergic, GABAergic and glycinergic pathways to premotor cardiac vagal neurons, and in particular, identify whether these pathways are activated and modulated during specific physiological conditions such as during the diving reflex, hemorrhage, hypoxia, and each phase of the respiratory cycle. It is hoped that a further understanding of the functional importance and pharmacological properties of the synaptic innervation of premotor cardiac vagal neurons may allow us to identify agents that can reduce cardiac vagal activity in pathological conditions with abnormally low heart rates and cardiac function such as Sudden Infant Death Syndrome, as well as increase vagal cardiac activity and reduce the fatality associated with cardiac hyperexcitability.

REFERENCES

Andresen, M.C., Kunze, D.L., 1994. Nucleus tractus solitarius--gateway to neural circulatory control. Annu. Rev. Physiol. 56,93-116.

Andresen, M.C., Mendelowitz, D., 1996. Sensory afferent neurotransmission in caudal nucleus tractus solitarius--common denominators. Chem. Senses 21, 387-395.

Ang, K.K., McRitchie, R.J., Minson, J.B., Llewellyn-Smith, I.J., Pilowsky, P.M., Chalmers, J.P., Arnolda, L.F., 1999. Activation of spinal opioid receptors contributes to hypotension after hemorrhage in conscious rats. Am. J. Physiol. 276, H1552-H1558.

Anrep, G.V., Pascual, F.W., Rossler, R., 1935. Respiratory variations of the heart rate. II The central mechanism of the respiratory arrhythmia and the interrelationships between the central and reflex mechanisms. Proc. Roy. Soc. 119, 218-231.

Barringer, D.L., Bunag, R.D., 1990. Differential anesthetic depression of chronotropic baroreflexes in rats. J. Cardiovasc. Pharmacol. 15, 10-15.

Bedran-de-Castro, M.T., Farah, V.M., Krieger, E.M., 1990. Influence of general anesthetics on baroreflex control of circulation. Braz. J. Med. Biol. Res. 23, 1185-1193.

Blake, D.W., Korner, P.I., 1982. Effects of ketamine and Althesin Anesthesia on baroreceptor--heart rate reflex and hemodynamics of intact and pontine rabbits. J. Auton. Nerv. Syst. 5, 145-154.

Browning, K.N., Mendelowitz, D., 2003. Musings on the wanderer: what's new in our understanding of vago-vagal reflexes?: II. Integration of afferent signaling from the viscera by the nodose ganglia. Am. J. Physiol. Gastrointest. Liver Physiol. 284, G8-G14.

Cole, C.R., Blackstone, E.H., Pashkow, F.J., Snader, C.E., Lauer, M.S., 1999. Heart-rate recovery immediately after exercise as a predictor of mortality. N. Engl. J. Med. 341, 1351-1357.

Coleman, T.G., 1980. Arterial baroreflex control of heart rate in the conscious rat. Am. J. Physiol. 238, H515-H520.

Daly, M.D., 1991. Some reflex cardioinhibitory responses in the cat and their modulation by central inspiratory neuronal activity. J. Physiol. 439, 559-577.

Davidson, N.S., Goldner, S., McCloskey, D.I., 1976. Respiratory modulation of barareceptor and chemoreceptor reflexes affecting heart rate and cardiac vagal efferent nerve activity. J. Physiol. 259, 523-530.

Davies, P.A., Kirkness, E.F., Hales, T.G., 2001. Evidence for the formation of functionally distinct α,β,γ, and ϵ GABA$_A$ receptors. J. Physiol. 537, 101-113.

Davies, P.A., Hanna, M.C., Hales, T.G., Kirkness, E.F., 1997. Insensitivity to anaesthetic agents conferred by a class of GABA(A) receptor subunit. Nature 385, 820-823.

Duffin, J., Aweida, D., 1990. The propriobulbar respiratory neurons in the cat. Exp. Brain Res. 81, 213-220.

Eckberg, D.L., Orshan, C.R., 1977. Respiratory and baroreceptor reflex interactions in man. J. Clin. Invest. 59, 780-785.

Eckberg, D.L., Drabinsky, M., Braunwald, E., 1971. Defective cardiac parasympathetic control in patients with heart disease. New Engl. J. Med. 285, 877-883.

Elghozi, J.L., Laude, D., Girard, A., 1991. Effects of respiration on blood pressure and heart rate variability in humans. Clin. Exp. Pharmacol. Physiol. 18, 735-742.

Evans, C., Baxi, S., Neff, R.A., Venkatesan, P., Mendelowitz, D., 2003. Synaptic activation of cardiac vagal neurons by capsaicin sensitive and insensitive sensory neurons. Brain Res. in press.

Evans, R.G., Ludbrook, J., Potocnik, S.J., 1989. Intracisternal naloxone and cardiac nerve blockade prevent vasodilatation during simulated haemorrhage in awake rabbits. J. Physiol. 409, 1-14.

Gauthier, P., Barillot, J.C., Dussardier, M., 1980. [Central interactions between laryngeal motoneurones. J. Physiol. (Paris) 76, 647-661.
Gilbey, M.P., Jordan, D., Richter, D.W., Spyer, K.M., 1984. Synaptic mechanisms involved in the inspiratory modulation of vagal cardio-inhibitory neurones in the cat. J. Physiol. 356, 65-78.
Gilman, A.G., Goodman L.S., Rall, T.W., Murad, F., (1985) The Pharmacological Basis of Therapeutics, Seventh Edition. New York, NY: Macmillan Publishing Company.
Henderson, G., McKnight, A.T., 1997. The orphan opioid receptor and its endogenous ligand--nociceptin/orphanin FQ. Trends Pharmacol. Sci. 18, 293-300.
Heymans, C., Neil, E., 1958. Reflexogenic Areas of the Cardiovascular System. London: Churchill.
Hrushesky, W.J., 1991. Quantitative respiratory sinus arrhythmia analysis. A simple noninvasive, reimbursable measure of cardiac wellness and dysfunction. Ann. NY Acad. Sci. 618, 67-101.
Inoue, K., Miyake, S., Kumashiro, M., Ogata, H., Yoshimura, O., 1990. Power spectral analysis of heart rate variability in traumatic quadriplegic humans. Am. J. Physiol. 258, H1722-H1726.
Irnaten, M., Wang, J., Chang, K.S., Andresen, M.C., Mendelowitz, D., 2002a. Ketamine inhibits sodium currents in identified cardiac parasympathetic neurons in nucleus ambiguus. Anesthesiology 96, 659-666.
Irnaten, M., Neff, R.A., Wang, J., Loewy, A.D., Mettenleiter, T.C., Mendelowitz, D., 2001. Activity of cardiorespiratory networks revealed by transsynaptic virus expressing GFP. J Neurophysiol. 85, 435-438.
Irnaten, M., Wang, J., Venkatesan, P., Evans, C., K Chang, K.S., Andresen, M.C., Mendelowitz, D., 2002b. Ketamine inhibits presynaptic and postsynaptic nicotinic excitation of identified cardiac parasympathetic neurons in nucleus ambiguus. Anesthesiology 96, 667-674.
Irnaten, M., Aicher, S.A., Wang, J., Venkatesan, P., Evans, C., Baxi, S., Mendelowitz, D., 2003. Mu-opioid receptors are located postsynaptically and endomorphin-1 inhibits voltage-gated calcium currents in premotor cardiac parasympathetic neurons in the rat nucleus ambiguus. Neurosci. 116, 573-582.
Irnaten, M., Walwyn, W.M., Wang, J., Venkatesan, P., Evans, C., Chang, K.S., Andresen, M.C., Hales, T.G., Mendelowitz, D., 2002c. Pentobarbital enhances GABAergic neurotransmission to cardiac parasympathetic neurons, which is prevented by expression of $GABA_A$ epsilon subunit. Anesthesiology 97, 717-724.
Jones, M.V., Harrison, N.L., Pritchett, D.B., Hales, T.G., 1995. Modulation of the $GABA_A$ receptor by propofol is independent of the gamma subunit. J. Pharmacol. Exp. Ther. 274, 962-968.
Jordan, D., Khalid, M.E., Schneiderman, N., Spyer, K.M., 1982. The location and properties of preganglionic vagal cardiomotor neurones in the rabbit. Pflugers Arch. 395, 244-250.
Kalia, M., 1981. Brain stem localization of vagal preganglionic neurons. J. Auton. Nerv. Syst. 3, 451-481.
Kunze, D.L., 1972. Reflex discharge patterns of cardiac vagal efferent fibres. J. Physiol. 222, 1-15.
Kwok, E.H., Dun, N.J., 1998. Endomorphins decrease heart rate and blood pressure possibly by activating vagal afferents in anesthetized rats. Brain Res. 803, 204-207.
La Rovere, M.T., Specchia, G., Mortara, A., Schwartz, P.J., 1988. Baroreflex sensitivity, clinical correlates, and cardiovascular mortality among patients with a first myocardial infarction. A prospective study. Circ. 78, 816-824.

Levy, M.N., Zieske, H., 1969. Autonomic control of cardiac pacemaker activity and atrioventricular transmission. J. Appl. Physiol. 27, 465-470.

Lipski, J., Zhang, X., Kruszewska, B., Kanjhan, R., 1994. Morphological study of long axonal projections of ventral medullary inspiratory neurons in the rat. Brain Res. 640, 171-184.

Loewy, A.D., Spyer, K.M. (1990) Central Regulation of Autonomic Functions: Oxford University Press.

McAllen, R.M., Spyer, K.M., 1978. The baroreceptor input to cardiac vagal motoneurones. J. Physiol. 282, 365-374.

McGrath, J.C., MacKenzie, J.E., Millar, R.A., 1975. Effects of ketamine on central sympathetic discharge and the baroreceptor reflex during mechanical ventilation. Br. J. Anaesth. 47, 1141-1147.

Mendelowitz, D., 1996. Firing properties of identified parasympathetic cardiac neurons in nucleus ambiguus. Am. J. Physiol. 271, H2609-H2614.

Mendelowitz, D., 1998. Nicotine excites cardiac vagal neurons via three sites of action. Clin. Exp. Pharmacol. Physiol. 25, 453-456.

Mendelowitz, D., 1999. Advances in Parasympathetic Control of Heart Rate and Cardiac Function. News Physiol. Sci. 14, 155-161.

Mendelowitz, D., Kunze, D.L., 1991. Identification and dissociation of cardiovascular neurons from the medulla for patch clamp analysis. Neurosci. Lett. 132, 217-221.

Meunier, J.C., 1997. Nociceptin/orphanin FQ and the opioid receptor-like ORL1 receptor. Eur. J. Pharmacol. 340, 1-15.

Mihalevich, M., Neff, R.A., Mendelowitz, D., 1996. Voltage-gated currents in identified parasympathetic cardiac neurons in the nucleus ambiguus. Brain Res. 739, 258-262.

Morillo, A.M., Nunez-Abades, P.A., Gaytan, S.P., Pasaro, R., 1995. Brain stem projections by axonal collaterals to the rostral and caudal ventral respiratory group in the rat. Brain Res. Bull. 37, 205-211.

Murthy, V.S., Zagar, M.E., Vollmer, R.R., Schmidt, D.H., 1982. Pentobarbital-induced changes in vagal tone and reflex vagal activity in rabbits. Eur. J. Pharmacol. 84, 41-50.

Neelands, T.R., Fisher, J.L., Bianchi, M., Macdonald, R.L., 1999. Spontaneous and gamma-aminobutyric acid (GABA)-activated GABA(A) receptor channels formed by epsilon subunit-containing isoforms. Mol. Pharmacol. 55, 168-178.

Neff, R.A., Mihalevich, M., Mendelowitz, D., 1998a. Stimulation of NTS activates NMDA and non-NMDA receptors in rat cardiac vagal neurons in the nucleus ambiguus. Brain Res. 792, 277-282.

Neff, R.A., Humphrey, J., Mihalevich, M., Mendelowitz, D., 1998b. Nicotine enhances presynaptic and postsynaptic glutamatergic neurotransmission to activate cardiac parasympathetic neurons. Circ. Res. 83, 1241-1247.

Ohnishi, M., Kirkman, E., Marshall, H.W., Little, R.A., 1997. Morphine blocks the bradycardia associated with severe hemorrhage in the anesthetized rat. Brain Res. 763, 39-46.

Pickering, T.G., Gribbin, B., Petersen, E.S., Cunningham, D.J., Sleight, P., 1972. Effects of autonomic blockade on the baroreflex in man at rest and during exercise. Circ. Res. 30, 177-185.

Pilowsky, P., Llewellyn-Smith, I.J., Lipski, J., Minson, J., Arnolda, L., Chalmers, J., 1994. Projections from inspiratory neurons of the ventral respiratory group to the subretrofacial nucleus of the cat. Brain Res. 633, 63-71.

Porter, F.L., Porges, S.W., Marshall, R.E., 1988. Newborn pain cries and vagal tone: parallel changes in response to circumcision. Child Dev. 59, 495-505.

Pritchett, D.B., Sontheimer, H., Shivers, B.D., Ymer, S., Kettenmann, H., Schofield, P.R., Seeburg, P.H., 1989. Importance of a novel GABAA receptor subunit for benzodiazepine pharmacology. Nature 338, 582-585.

Ransom, B.R., Barker, J.L., 1975. Pentobarbital modulates transmitter effects on mouse spinal neurones grown in tissue culture. Nature 254, 703-705.

Rardon, D.P., Bailey, J.C., 1983. Parasympathetic effects on electrophysiologic properties of cardiac ventricular tissue. J. Am. Coll. Cardiol. 2, 1200-1209.

Richards, C.D., 1972. On the mechanism of barbiturate anaesthesia. J. Physiol. 227, 749-767.

Robertson, B., 1989. Actions of anaesthetics and avermectin on GABAA chloride channels in mammalian dorsal root ganglion neurones. Br. J. Pharmacol. 98, 167-176.

Scher, A.M., Young, A.C., 1970. Reflex control of heart rate in the unanesthetized dog. Am. J. Physiol. 218, 780-789.

Shykoff, B.E., Naqvi, S.S., Menon, A.S., Slutsky, A.S., 1991. Respiratory sinus arrhythmia in dogs. Effects of phasic afferents and chemostimulation. J. Clin. Invest. 87, 1621-1627.

Slogoff, S., Allen, G.W., 1974. The role of baroreceptors in the cardiovascular response to ketamine. Anesth. Analg. 53, 704-707.

Spyer, K.M., 1981. Neural organisation and control of the baroreceptor reflex. Rev. Physiol. Biochem. Pharmacol. 88, 24-124.

Spyer, K.M., Gilbey, M.P., 1988. Cardiorespiratory interactions in heart-rate control. Ann. NY Acad. Sci. 533, 350-357.

Standish, A., Enquist, L.W., Schwaber, J.S., 1994. Innervation of the heart and its central medullary origin defined by viral tracing. Sci. 263, 232-234.

Stornetta, R.L., Guyenet, P.G., McCarty, R.C., 1987. Autonomic nervous system control of heart rate during baroreceptor activation in conscious and anesthetized rats. J. Auton. Nerv. Syst. 20, 121-127.

Taylor, E.W., 1994. The evolution of efferent vagal control of the heart in vertebrates. Cardiosci. 5, 173-182.

Townend, J.N., Littler, W.A., 1995. Cardiac vagal activity: a target for intervention in heart disease. Lancet 345, 937-938.

van Lieshout, J.J., Wieling, W., Karemaker, J.M., Eckberg, D.L., 1991. The vasovagal response. Clin. Sci. (Lond) 81, 575-586.

Vanoli, E., De Ferrari, G.M., Stramba-Badiale, M., Hull, S.S., Jr., Foreman, R.D., Schwartz, P.J., 1991. Vagal stimulation and prevention of sudden death in conscious dogs with a healed myocardial infarction. Circ. Res. 68, 1471-1481.

Venkatesan, P., Wang, J., Evans, C., Irnaten, M., Mendelowitz, D., 2002a. Endomorphin -2 inhibits GABAergic inputs to cardiac parasympathetic neurons in the nucleus ambiguus. Neurosci. 111, 699-705.

Venkatesan, P., Wang, J., Evans, C., Irnaten, M., Mendelowitz, D., 2002b. Nociceptin inhibits gamma-aminobutyric acidergic inputs to cardiac parasympathetic neurons in the nucleus ambiguus. J. Pharmacol. Exp. Ther. 300, 78-82.

Wang, J., Irnaten, M., Mendelowitz, D., 2001a. Characteristics of spontaneous and evoked GABAergic synaptic currents in cardiac vagal neurons in rats. Brain Res. 889, 78-83.

Wang, J., Irnaten, M., Mendelowitz, D., 2001b. Agatoxin-IVA-sensitive calcium channels mediate the presynaptic and postsynaptic nicotinic activation of cardiac vagal neurons. J. Neurophysiol. 85, 164-168.

Wang, J., Irnaten, M., Venkatesan, P., Evans, C., Mendelowitz, D., 2002. Arginine vasopressin enhances GABAergic inhibition of cardiac parasympathetic neurons in the nucleus ambiguus. Neurosci. 111(3), 699-705.

Wang, J., Wang, X., Irnaten, M., Venkatesan, P., Evans, C., Baxi, S., Mendelowitz, D., 2003. Endogenous Acetylcholine and Nicotine Activation Enhances GABAergic and Glycinergic Inputs to Cardiac Vagal Neurons. J. Neurophysiol. 89(5), 2473-2481.

Wang, J., Irnaten, M., Neff, R.A., Venkatesan, P., Evans, C., Loewy, A.D., Mettenleiter, T.C., Mendelowitz, D., 2001c. Synaptic and neurotransmitter activation of cardiac vagal neurons in the nucleus ambiguus. Ann. NY Acad. Sci. 940, 237-246.

Warner, M.R., deTarnowsky, J.M., Whitson, C.C., Loeb, J.M., 1986. Beat-by-beat modulation of AV conduction. II. Autonomic neural mechanisms. Am. J. Physiol. 251, H1134-H1142.

Watkins, L., Maixner, W., 1991. The effect of pentobarbital anesthesia on the autonomic nervous system control of heart rate during baroreceptor activation. J. Auton. Nerv. Syst. 36, 107-114.

Waxman, M.B., Wald, R.W., 1977. Termination of ventricular tachycardia by an increase in cardiac vagal drive. Circ. 56, 385-391.

Whitescarver, S.A., Ott, C.E., Kotchen, T.A., 1990. Parasympathetic impairment of baroreflex control of heart rate in Dahl S rats. Am. J. Physiol. 259, R76-83.

Whiting, P.J., Bonnert, T.P., McKernan, R.M., Farrar, S., Le Bourdelles, B., Heavens, R.P., Smith, D.W., Hewson, L., Rigby, M.R., Sirinathsinghji, D.J., Thompson, S.A., Wafford, K.A., 1999. Molecular and functional diversity of the expanding GABA-A receptor gene family. Ann. NY Acad. Sci. 868, 645-653.

Yamada, H., Ezure, K., Manabe, M., 1988. Efferent projections of inspiratory neurons of the ventral respiratory group. A dual labeling study in the rat. Brain Res. 455, 283-294.

Zheng, Y., Barillot, J.C., Bianchi, A.L., 1991. Are the post-inspiratory neurons in the decerebrate rat cranial motoneurons or interneurons? Brain Res. 551, 256-266.

Chapter 18

GENES REGULATING CARDIOVASCULAR FUNCTION AS REVEALED USING VIRAL VECTORS

Julian F.R. Paton, Hidefumi Waki, Mohan Raizada* and Sergey Kasparov
Department of Physiology, School of Medical Sciences, University of Bristol, Bristol, BS8 1TD, UK; *Department of Physiology, University of Florida, College of Medicine, PO Box 100274, Gainesville, FL 32610, USA

Abstract: This chapter illustrates briefly the usefulness of viral vectors in studying genes regulating the cardiovascular system in both health and disease. Some advantages of using viral vectors are discussed and recent examples illustrated. A discussion follows on possible refinements for viral vectors as tools for systems physiology. This includes the use of cell-specific promoters, longer term expression with lentivirus and 'vigilent' vectors.

Key words: adenovirus, hypertension, baroreceptor reflex, rat, somatic gene transfer

INTRODUCTION

Sequencing the genome of a number of species including man has now been accomplished. In parallel there has been an explosion in genetic engineering that now provides us with a wide range of new molecular tools for gene delivery and manipulation. By tethering genomic information with these molecular tools we can begin to apply this strategy to fundamental questions relating to physiology. But, it is true that "new" does not automatically mean "better". For an integrative physiologist who is interested in understanding the relationship between genome and function such as control of heart rate or blood pressure, the new opportunities presented are quite overwhelming. However, before these opportunities can be fully exploited, there needs to be a good appreciation of their potential strengths and weaknesses. Some skeptics may reason that it is always wise to wait for others to first "discover" the artifacts associated with the new methods such as somatic gene transfer. This chapter hopes to motivate the hesitant by demonstrating the use of molecular tools for understanding systems level physiology. Here, we focus on the application of viral vectors to understanding mechanisms that control the cardiovascular system in health and disease.

WHY THE CARDIOVASCULAR SYSTEM?

To quote from Sydney Brenner (1994): *"Neurobiology still has a long way to go to contend with consciousness. Perhaps we should be content to work instead on unconsciousness for a while, and find out about all those processes going on in our brains that we don't know about directly and need science to tell us the answers"*.

One of those processes that *"we don't know about directly"* is the autonomic nervous system and integral to this is cardiovascular homeostasis. The cardiovascular system lends itself to study since it offers the ability to both manipulate and monitor numerous variables on the sensory as well as the motor side. For example, arterial pressure can be manipulated to stimulate baroreceptors precisely while measuring central neuronal responses, sympathetic nerve discharge and cardiac vagal motor activity simultaneously. Moreover, the system can be studied at all levels (sub-cellular to systems) both *in vitro* and *in vivo*. As diseases related to cardiovascular homeostasis are quite wide-spread these studies have obvious medical relevance. The choice of species is an important step. Despite the vast number of transgenic mice available (see below), the rat remains the species of choice for cardiovascular studies. Indeed, much of our understanding of central control of this system is based on the rat. Not only is the larger size an advantage but there are also numerous animal models of cardiovascular disease such as those prone to hypertension (spontaneously hypertensive rat), sensitive to salt loading (Dahl rat) and high cholesterol diets (Zucker rat). With the use of radio telemetry the way is open for non-invasive studies from which definitive cardiovascular data regarding arterial pressure, heart rate and baroreceptor reflex sensitivity can be obtained.

CHOOSING AN APPROPRIATE MOLECULAR STRATEGY

In round figures there are 30,000 genes in man. 20,000 are expressed in the brain of which 15,000 have unknown function (see http://expression.gnf.org/cgi-bin/index.cgi or http://www.hugeindex.org or http://bioinformatics.weizmann.ac.il/cards/). This raises the question as to the function of these genes. Certainly this is a massive challenge for the future. Alberts et al. (1994) stated: *"The ultimate test of the function of a gene or altered gene is to insert it into an organism and see what effect it has."*

This quotation is helpful in that it provides a steer towards a strategy. Two ways in which gene transfer can come about are germline transgenesis and somatic gene transfer. The former involves the injection of cloned DNA sequences into germ cells whereas a vector is required for the transfer of foreign nucleic acid into differentiated cells. For a number of reasons we adopted somatic gene transfer because it:

1. avails developmental compensation caused by the altered gene as is the case with transgenics
2. allows control data to be obtained in the same animal prior to gene manipulation
3. is not limited to mice and to a particular strain of mice chosen as a progenitor
4. is relatively fast and cheap: no animal colonies need be maintained
5. allows site specific gene manipulation i.e. gene delivery can be made to a selected organ or its sub-part
6. can be controlled temporally using inducible promoters
7. can be directed to a particular cell type by using a cell-specific promoter

Somatic Gene Transfer: Over-Expression and Dominant Negative Proteins

It is a relatively easy task to enhance activity of a certain intracellular pathway: this can be achieved by expressing excessive amounts of relevant proteins or introducing their constitutively active mutants. However, suppression of a specific signalling pathway has proven to be a much more difficult task. One of the best established ways of doing this is to express a mutated version of the protein of interest. The latter behaves in a dominant negative fashion by interfering with the endogenous proteins.

Examples follow where somatic gene transfer was used in two different ways to either over-express a transgene or to block an endogenous intracellular signalling cascade.

Can Over Expression of Protein Kinase B (PKB) Protect Against Myocardial Ischaemia?

Miao et al. (2000) used a replication defective adenovirus to constitutively express protein kinase B (PKB; also known as Akt) in the heart (Fig 1). They hypothesised that over-expression of PKB would protect against myocardial ischaemia induced surgically by temporary ligation of the left anterior descending coronary artery. This was based on evidence that PKB (a) phosphorylates endothelial nitric oxide synthase (eNOS) and hence increases the nitric oxide to cause vasodilatation (Luo et al., 2000); (b) stimulates angiogenesis (Kureishi et al., 2000); (c) inhibits apoptotic death by inhibiting pre-apoptotic molecules (Datta et al., 1999). The virus was given by intra-coronary injection. Fig. 1 shows that over expression of PKB significantly reduced the volume of infarcted tissue within the heart following an ischaemic insult. Importantly, viral control experiments were performed in which an adenovirus expressing β-galactosidase failed to protect the heart.

CONTROL

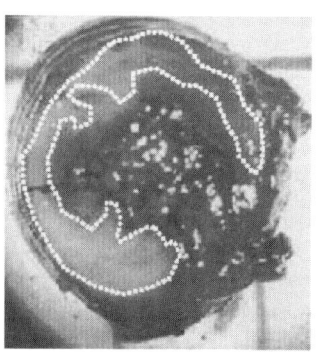

Heart transfected with
AdV-β-galactosidase

EXPERIMENT

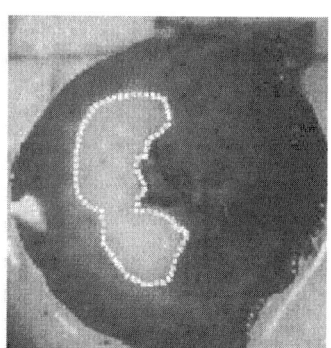

Heart transfected
with AdV-Akt

Figure 1. Overexpression of protein kinase B (Akt) in the heart protects against ischaemia. The dotted lines encompass the dead tissue (stained white), which was greatly reduced in adenovirus (AdV) transfected hearts expressing Akt (right) compared to those in which the reporter gene β-galactosidase was expressed (left). Data reproduced from Miao et al. (2000); permission pending.

Whilst this is a good example of how effective somatic gene expression can be, it should be emphasised that over expression of protein is not "physiological" and the specificity of the effects may not be guaranteed. First, the expressed protein may become so concentrated that it is toxic to the cell. Second, the expressed gene is not restricted solely to the cells in which PKB is endogenously produced. In other words, ectopic expression results in very high PKB activity in all cell types including those where it is normally low or even absent. This can hardly be seen as physiological. Nevertheless, this does not diminish the potential therapeutic benefits that might be achieved.

Expressing a Dominant Negative Protein to Interrupt an Endogenous Process Chronically

In our recent studies we have shown that the cardiac component of the baroreceptor reflex can be attenuated centrally by angiotensin II acting within the nucleus of the solitary tract (NTS) and that this requires activation of eNOS and subsequent release of nitric oxide (Paton et al., 2001; 2002; Wong et al., 2002). In these experiments a dominant negative form of eNOS (TeNOS) (Kantor et al., 1996) was expressed using an adenovirus with a non-cell specific human cytomegalovirus promoter. eNOS is a homodimer and TeNOS combines with endogenous eNOS protein. Because TeNOS is expressed constitutively it "mops up" all the endogenous eNOS rendering it inactive. Based on this, we hypothesised that chronic blockade of eNOS in the NTS should enhance baroreceptor reflex sensitivity. In rats fitted with radio transmitter devices signalling arterial blood pressure, we recorded arterial pressure and calculated both heart rate and the spontaneous baroreceptor reflex sensitivity using sequence analysis of the arterial pressure signal (Waki et al., 2003). Following a control period, we microinjected an adenovirus that expressed a dominant negative TeNOS protein in the caudal part of the NTS; the approach was both site-specific and presumably only affected cells in which eNOS was endogenously active. Figure 2 shows that the cardiac baroreceptor reflex was potentiated following manipulation of a single gene within the NTS. This effect persisted as long as the transgene was present in the NTS. Figure 2 shows immunocytochemical evidence for the expression of the transgene. In our study, adenoviral transfection to express a reporter gene (enhanced green fluorescent protein) was without effect on baroreceptor reflex sensitivity (Waki et al., 2003).

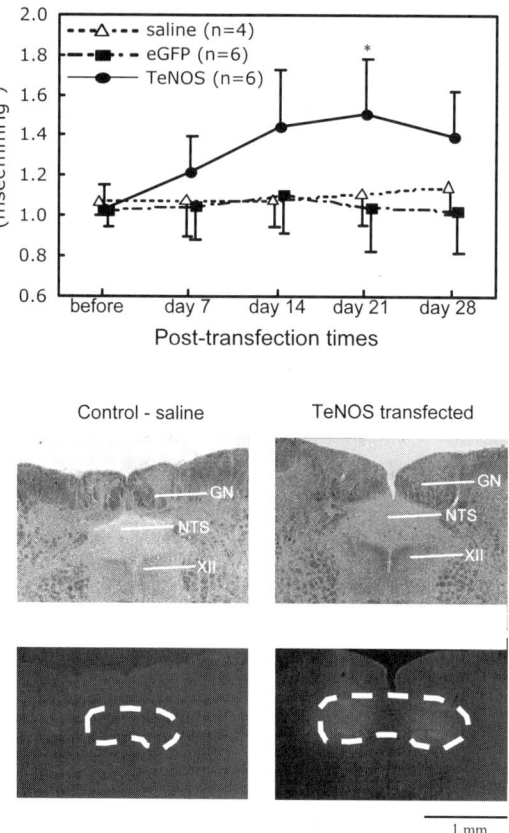

Figure 2. Adenoviral mediated transfection of the nucleus of the solitary tract (NTS) to induce expression of a dominant negative protein (TeNOS; see Kantor et al., 1996) was used to block endothelial nitric oxide synthase activity chronically. This resulted in an enhancement of the spontaneous baroreceptor reflex gain (sBRG, top panel), which remained significant for three weeks. Below on the right shows a photomicrograph of structures in the dorsomedial medulla and beneath a neighbouring section showing enhanced TeNOS immunofluorescence in TeNOS transfected animals. These rats showed increased baroreceptor reflex sensitivity. On the left are two sections from a control rat (light field above and immuofluorescent section below) showing no TeNOS immunoreactivity. Abbreviations: GN, gracile nucleus; XII, hypoglossal motonucleus. Data from Waki et al. (2003) with permission.

It should be emphasised that chronic pharmacological manipulation of a localised region of the NTS would be technically impossible. Our recent data indicate that eNOS is constitutively active within the NTS and that its physiological function is to depress the sensitivity of the baroreceptor reflex around its operating point. The question now arises as to whether eNOS activity in NTS increases in hypertension and, if so, does this contribute to the high blood pressure.

VIRAL VECTORS AS TOOLS: PRO'S AND CON'S WITH THEIR USE

No experimental technique or preparation is perfect. The adenovirus is useful but has its limitations. These include problems related to inflammation although this appears to be of minor significance when used in the brain. Some other vectors, such as adeno-associated virus or lentivirus may be used as they produce a minimal immune reaction. Unlike with drugs, concentration of a transgene is difficult to control and measure and therefore studying a dose-response relationship is not possible. Nevertheless, viral vectors provide an efficient means of gene delivery and may achieve what no drug can offer, i.e., selective targeting of a particular cell type. Moreover, in some cases genetic approaches may circumnavigate problems of poor drug specificity. Finally, viral gene delivery is ideal for studying chronic effects and one injection is sufficient to obtain a lasting effect (Geddes et al., 1999; Kishi et al., 2003; Waki et al., 2003).

Making Viral Vectors Smarter – The Way Forward ?

Longer Term Expression

From our experience, and that of others, adenoviral gene delivery leads to expression lasting 2-3 weeks (Paton et al., 2001; Kishi et al., 2003; Waki et al., 2003). Others have seen transgene after a month using adenoviral mediated gene delivery (Harding et al., 1998). Recently, we have demonstrated that lentiviral gene delivery results in much more lasting gene expression *in vivo* (Coleman et al., 2003). Using a lentivirus we have seen strong expression within the paraventricular nucleus upwards of 2 months (Fig. 3).

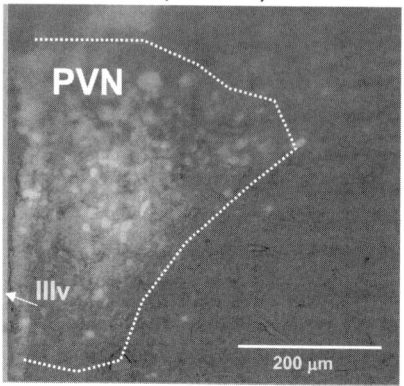 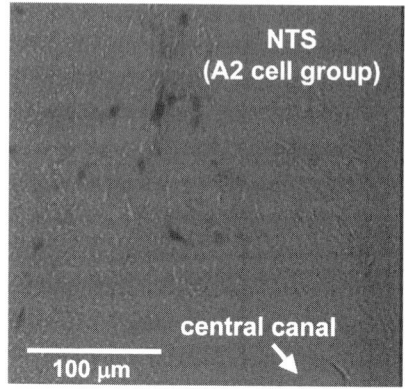

Figure 3. Left hand side: Lentiviruses allow longer transgene expression than adenoviruses. This example shows robust expression of enhanced green fluorescent protein within the paraventricular nucleus after two months. Abbreviation: IIIv, IIIrd ventricle. Reproduced from Coleman et al. (2003) with permission. Right hand side: Cell specific expression of β-galactosidase in the A2 cell group located within the nucleus of the solitary tract (NTS) after one week. Transgene expression was induced by the virtue that A2 neurones contain a specific transcription factor for synthesizing catecholamines. The viral construct used contained a promoter region that was responsive to this transcription factor. Data are unpublished (Kasparov, S. et al.).

Cell Specific Promoters

Any organ is comprised of a range of cellular phenotypes. Within the brain, for example, there are neurons, glia and blood vessels. Each of these again comes in a number of flavours. In order to understand their specific contributions to physiological and pathological processes, it may be essential to selectively manipulate these cells using viral constructs. For example, catecholamine containing neurons are known to play a role in arterial pressure regulation and exist within the brain. We have recently targeted them selectively using a cell-specific promoter (Kasparov et al., unpublished data). The A2 cell group expressing β-galactosidase was driven using an artificial promoter (Hwang et al., 2002) based on dopamine- β-hydroxylase – an enzyme present in these neurons which catalyses the conversion of dopamine to noradrenaline. Clearly, this now gives the ability to manipulate genes within these neurones without affecting their neighbours.

Vigilant Vectors

Phillips et al. (2002) have engineered a most ingenious construct, which they term a *"vigilant vector"*. A vigilant vector is described as one of potential therapeutic importance for life threatening diseases that is safe, stable, incorporates a gene switch/biosensor, has a cell-specific promoter and drives both a transgene and reporter gene. The latter allows monitoring of the activity of the transgene. The scenario is that myocardial ischaemia is a progressive asymptomatic condition that is often noticed too late following a cardiac arrest. The notion presented by Phillips' group is that early detection of such a disease should be used as the trigger to induce beneficial gene expression in the appropriate cell type. The construct designed for ischaemic heart disease contains as a promoter hypoxia inducible factor (HIF1) upstream of the cell specific promoter - myosin light chain kinase, which is found only in myocytes, that drives the transgene of interest and a reported (enhanced green fluorescent protein). The transgene suggested was antisense for angiotensin II type 1 receptors. In heart disease this receptor subtype is over expressed and via actions on the myocardium and arterioles exacerbates the problems of myocardial ischaemia. Thus, the construct is automatically driven when hypoxia builds up within the heart. This triggers protein expression confined to the heart muscle cells.

CONCLUSIONS

It is now clear that temporal, spatial and cellular control of gene expression is a powerful approach when applied to an integrating system such as the central control of the circulation. Effort now needs to be given in the development of more sophisticated constructs to allow genetic manipulation of specific cell types. In any organ this is essential since phenotypes are intermingled. With the power of molecular biology new experimental tools can be made to determine the physiological role of different genes in specific cell phenotypes. This is exciting and essential for the further development of novel therapeutic strategies to life threatening diseases such as hypertension and myocardial ischaemia.

ACKNOWLEDGEMENTS

Julian F.R. Paton is funded by the British Heart Foundation (BS 93003). Hidefumi Waki was funded by a grant from the Japanese Space Forum.

REFERENCES

Brenner, S., 1994. Loose ends. Current Biology 4, 864.
Coleman, J.E., Huentelman, M., Kasparov, S., Metcalfe, B.L., Paton, J.F.R., Katovich, M.J., Semple-Rowland, S.L., Raizada, M.K., 2003. Efficient large-scale production and concentration of HIV-1-based lentiviral vectors for use in vivo. Functional Genomics 12, 221-228.
Datta, S.R., Brunet, A., Greenberg, M.E., 1999. Cellular survival: a play in three Akts. Genes and Development 13, 2905-2927.
Geddes B.J., Harding, T.C., Lightman S.L., Uney JB. (1999). Assessing viral gene therapy in neuroendocrine models. Frontiers in Neuroendocrinology 20, 296-316.
Harding, T.C., Geddes, B.J., Murphy, D., Knight, D., Uney, J.B., 1998. Switching transgene expression in the brain using an adenoviral tetracyclineregulatable system. Nature Biotechnology 16, 553-555.
Kantor, D. B., Lanzrein, M., Stary, S. J., Sandoval, G. M., Smith, W. B., Sullivan, B. M., Davidson, N., Schuman, E.M., 1996. A role for endothelial NO synthase in LTP revealed by adenovirus-mediated inhibition and rescue. Science 274, 1744-1748.
Kishi, T. Hirooka, Y., Kimura, Y., Sakai, K., Ito, K., Shimokawa, H., Takeshita, A., 2003. Overexpression of eNOS in RVLM improves impaired baroreflex control of heart rate in SHRSP. Hypertension 41, 255-260
Kureishi, Y., Luo, Z., Shiojima, I., Bialik, A., Fulton, D., Lefer, D.J., Sessa, W.C., Walsh, K., 2000. The HMG-CoA reductase inhibitor simvastatin activates the protein kinase Akt and promotes angiogenesis in normocholesterolemic animals. Nature Med. 9, 1004-1010.
Luo, Z., Fujio, Y., Kureishi, Y., Rudic, R.D., Daumerie, G., Fulton, D., Sessa, W.C., Walsh, K., 2000. Acute modulation of endothelial Akt/PKB activity alters nitric oxide-dependent vasomotor activity in vivo. J. Clin. Inv. 106, 439-499.
Miao, W., Luo, Z., Kitsis, R.N., Walsh, K., 2000. Intracoronary, adenovirus-mediated Akt gene transfer in heart limits infarct size following ischaemia-reperfusion injury in vivo. J. Molec. Cell. Cardio. 32, 2397-2402.
Paton J.F.R., Deuchars, J., Ahmad, Z., Wong, L.-F., Murphy, D., Kasparov, S., 2001. Adeno viral vector demonstrates that angiotensin II induced depression of the cardiac baroreflex is mediated by endothelial nitric oxide synthase in the nucleus tractus solitarii. J. Physiol. 531.2, 445-458.
Paton, J.F.R., Kasparov, S., Paterson, D.J., 2002. Site-Specific Differential Modulation of Cardiac Autonomic Control by Nitric Oxide. TINS 25, 626-631.
Phillips, M.I., Tang, Y., Schmidt-Ott, K., Quian, K., Kagiyama, S., 2002. Vigilent vector: Heart specific promoter in an adeno associated virus vector for cardioprotection. Hypertension 39, 651-655.
Waki, H., Kasparov, S., Wong, L.-F., Murphy, D., Shimizu, T., Paton, J.F.R., 2003. Chronic inhibition of eNOS activity in NTS enhances baroreceptor reflex in conscious rats. J. Physiol. 546.1, 233-242.
Wong, L.-F. Polson, J., Murphy D., Paton. J.F.R., Kasparov, S., 2002. Molecular and pharmacological dissection of the intracellular pathways mediating the angiotensin II induced baroreceptor reflex depression in the nucleus of the solitary tract. FASEB J. 16, 1595-1601.

Index

A

Acetylcholine
  in rat brain, 84
Acetylcholine esterase
  in rat brain, 84
Acid Sensing Ion Channel 2 (ASIC2)
  functions of, 10
Action potential responses
  to mechanical stimulation, 10f
Adenocortico tropic hormone (ACTH)
  release of, 121
Adenosine
  ganglionic neurotransmission effected by, 311
  as neurotransmitter, 43
Adenoviruses
  transgene expression of, 406f
Adrenergic cells
  of medulla oblongata, 188
Afferent integration
  initiation of, 64
Afferent nerves
  chemical stimulation of, 108
  electrical stimulation of, 108
Afferent synapses
  strong, 63–64
Agatoxin-IVA
  effect of endomorphin 1 on, 384f
  effect of endomorphin 2 on, 384f
Amino acids
  neurochemistry of, 269–270
Anesthetics
  chemoreflex and, 33–34
Angiotensin-1 receptors (AT1R)
  in RVLM, 202
Angiotensin II
  effect of, on presympathetic neurons, 202
  effect of, on sympathetic ganglia, 317
  role of, 174
ANP
  in hypothalamus, 175–176
Anterior hypothalamus (AHN). *See also* Hypothalamus
  angiotensin II injections into, 174
  atrial ANP in, 175–176
  Chronic clonidine microinfusions into, 168–169
  sympathoinhibitory function of, 167
Aortic depressor nerve (ADN)
  baroreceptor activity in, 5f
  electrical stimulation of, 108
AP-7
  effect of, on chemoreflex, 33
Apolipoprotein E (apoE)
  effect of, on baroreceptor sensitivity, 15
Arcuate nucleus (Arc)
  efferent projections from, 127f, 128
Arginine vasopressin (AVP)
  release of, 120
ATP
  as neurotransmitter, 43
Atrial natriuretic peptide
  effect of, on sympathetic ganglia, 318–319
Autonomic cardiovascular control
  basic organization of, 60f

B

Baroreceptor sensory transduction, 8–13
Baroreceptors
  activation mechanisms of, 9f
  arterial pathway, 366f
  in cerebral vasodilation, 365
  effect of apolipoprotein E (apoE) on, 15
  effect of oxidative stress on, 14–15
  effect of platelet activation on, 15–16
  isolated, 7, 8f
  locations of, 1
  myelinated A-type, 69
  neural pathways activated by, 104
  response discharge of, 69
Baroreflex
  adaptation, 11–12
  afferent class issues and, 69
  aortic, 88–89
  assessing, 3
  carotid sinus, 90
  chemical sensitivity of, 12–13
  clinical importance of, 19
  effect of nitric oxide on, 12–13

gene transfer and, 6–7
in "knockout" mice, 5, 6t
mechanical components of, 4
negative feedback regulation of, 2
neural components of, 4
neurohumoral activation and, 17–18
nitric oxide release under, 362
overview of, 85–88
paracrine modulation of, 13–16
premotor cardiac vagal neurons activated by, 376–378
reactive oxygen species effecting, 12–13
resetting, 11–12
role of, 19
spontaneous sensitivity of, 3–4
systemic, 88
in transgenic animals, 5, 6t
Baroreflex BP buffereing capacity
assessing, 3
Baroreflex pathways
afferent, 2f
central, 2f
efferent, 2f
Barosensitive and Bulbospinal (BSBS) neurons
action potentials from, 195–196
axonal conduction velocities of, 191
CCK-8 inhibition of, 193
electrophysical characterization of, 190–192
excitatory drives to, 195–197
inhibitory drives to, 194–195
inputs to, 194–198
role of, 208
sympathetic reflexes of, 197–198
Bilateral carotid artery occlusion (BCO)
measurement of, 6
Bilateral electrolytic lesions
of PVN, 47
Blood pressure (BP). *See also* Mean arterial pressure (MAP), Pulsatile arterial pressure (PAP)
bulbospinal C1 cells and, 188–190
circadian rhythms of, 179f
dietary supplementation effecting, 180f
increased, in brain stem, 104–105
increased, in CVLM, 107
increased, in forebrain, 104–105
pharmacological reduction of, 100–102
sensing changes in, 1
of spontaneously hypertensive rats, 165f
sympathetic ganglia influencing, 320
of Wistar Kyoto rats, 165f

BNC1. *See* Acid Sensing Ion Channel 2 (ASIC2)
Bradycardic response, 88
Brain
neonate, 200
sympathetic preganglionic neuron input from, 267–268
Brain derived neurotrophic factor (BDNF)
as autocrine factor, 16
Brainstem
angiotensin II in, 174
chemical stimulation in, 108–109
electrical stimulation in, 108–109
hemorrhage in, 103
increased blood pressure in, 104–105
serotonin-containing neurons in, 226–227
Brown adipose tissue (BAT)
stimulation of, 252f
Bulbospinal C1 cells
blood pressure control and, 188–190
effect of serotonin on, 201
mechanically dissociated, 199
vesicular glutamate transporter 2 (VGLUT2) in, 191–192
Bulbospinal raphe
firing probability in, 229f
Bulbospinal serotonin neurons
in parapyramidal region, 226

C
Calbindin
in sympathetic preganglionic neurons, 270
Capsaicin (CAP)
depolarization of, 347
NTS neurons responding to, 72f
VR1 binding with, 70
Cardiac ganglia
chemical phenotype of neurons in, 339–341
histochemical organization of, 338–344
neurochemical expression in, 341t
Cardiac neurons
chemical phenotype of, 339–341
cholinergic, 340f
effect of PACAP peptides on, 348f
electrophysiological properties of, 336–338
fibers surrounding, 342f
functional properties of, 336
neurochemical modulation of, 344
voltage dependent calcium channels of, 338

Index 413

Cardiac parasympathetic pathway, 62
Cardiac vagal activity
  respiratory modulation of, 379–380
Cardiac vagal neurons
  paraventricular nucleus (PVN) projection to, 123
Cardio-pulmonary reflex, 90–91
Cardiovascular sympathetic pathway, 62–63
Cardiovascular system
  molecular strategies for studying, 400–401
Carotid chemoreflex
  neural pathways involved in, 91–92
Carotid sinus nerve (CSN)
  electrical simulation of, 108
CART
  projections associated with, 128
CARTp-IR fibers, 343
Catecholamine neurotoxin
  effect of, on blood pressure, 227f
  effect of, on heart rate, 227f
  effect of, on renal nerve activity, 227f
Catecholaminergic neurons
  activation of, 101
Catecholamines
  effect of, on presympathetic neurons, 201–202
Caudal medullary raphe neurons
  excitation of, 249f
  response to hemorrhage of, 248
Caudal medullary raphe nuclei
  role of, in gastric function, 254
Caudal midline raphe nuclei (CMR)
  role of, 220
Caudal raphe magnus
  activation of, 251
CCK-8
  BSBS neurons inhibited by, 193
Cellular substrates
  for cardiovascular response, 247
Central nucleus of amygdala (CNA)
  stimulation experiments on, 109–110
Cerebral blood flow (CBF)
  neural pathway influence on, 358
Cerebral vasodilation
  baroreceptor participation in, 365–366
Cerebrovascular resistance (CVR)
  autoregulation and, 360
  neural pathway influence on, 358
Chemical stimulation
  of afferent nerves, 108
  in brainstem, 108–109
  in forebrain, 109–110
Chemoreceptors
  integration of, 34
Chemoreflex
  in anesthetized animals, 33–34
  historical aspects of, 32–33
  neural pathways of, 34–36
  neuromodulation of, 45–46
  neurotransmission of, 36–44
  pathophysiological implications of, 51–52
  symapthoexcitatory component of, 39
Chemosensation
  role of serotonin in, 232
Choline acetyltransferase (ChAT)
  co-localization of, 342f
  immunohistochemistry, 288
  in rat brain, 84
Cholinergic mechanisms
  of ganglionic neurotransmission, 312–313
Cholinergic nTS mechanisms, 84–85
Circardian rhythms
  of blood pressure, 179f
Clonidine
  microinfusions of, in anterior hypothalamus (AHN), 168–169
Colocalisation studies
  of rostral ventromedial medulla (RVMM), 223
Corticotrophin releasing factor (CRF)
  in parvocellular neurones, 121
  on presympathetic neurons, 205
Cricoarytenoid (CT)
  reconstruction of, 234f

D

DNQX
  effect of, on chemoreflex, 33
  effect of, on heart rate, 40f
  effect of, on mean arterial pressure, 40f
  effect of, on pulsatile arterial pressure, 40f
Dominant negative proteins
  expression of, 403
Dopamine
  blocking antagonists of, 126
Dorsal motor nucleus of vagus (dmnX)
  descending projections to, 150
Dorsal root ganglia (DRG)
  sensory neurons of, 64
Dorsal vagal complex (DVC)
  neurons innervating, 152
  neurons projecting, 156f
  stimulation of, 253
Dorsomedial hypothalamic nucleus (DMH)
  efferent projections from, 127f, 128–133

preoptic area of, 130
DPCPX
  effect of, on chemoreflex, 43

E

Efferent-motor pathways
  transversing sympathetic ganglia, 319
Electrical stimulation
  of afferent nerves, 108
  in brainstem, 108–109
  in forebrain, 109–110
Endomorphin 1
  effect of on Agatoxin-IVA, 384f
Endomorphin 2
  effect of on Agatoxin-IVA, 384f
Enkephalin
  distribution patterns of, 308f
  ganglia reaction to, 307f
  IML innervation by, 278f
  immunoreactive axons of, 275
  in rostral ventrolateral medulla, 204
ENOS
  gene transfer of, 14f
Estrogen
  on hypertension, 177
Excitatory postsynaptic currents (EPSCs)
  in NTS neurons, 70f
Excitatory postsynaptic potentials (EPSPs)
  pharmacology, 293f
  of sympathetic preganglionic neurons, 292–294
  vagal trunk stimulation and, 337

F

5-HT immunoreactive nerve fibers, 224f
Forebrain
  chemical stimulation in, 109–110
  electrical stimulation in, 109–110
  hemorrhage in, 104
  increased blood pressure in, 105–106
  NP infusion in, 102
Fos
  brain regions expressing, 106f
  expression of, 100
  hemorrhage activating, 103
  in PVN, 105
4-aminopyridine (4-AP)
  inactivating potassium current blocked by, 72

G

GABA
  control responses to, 386f
  inhibition from, 65
  paraventricular nucleus neurons associated with, 133
  role of, in chemoreflex, 46
GABAergic nTS mechanisms, 83–84
Ganglionic transmission
  postsynaptic effects of, 312–315
  presynaptic modulation, 310–312
Gastric function
  effect of caudal raphe pallidus, 256f
  role of caudal medullary raphe nuclei in, 254
Gene transfer
  baroreceptor resetting by, 14f
  baroreflex function and, 6–7
Glucagon-like peptide
  effect of, on presympathetic neurons, 205–206
Glutamate
  role of, in presympathetic neuron discharge, 196
Glutamate receptors
  antagonists of, 68
Glutamatergic nTS mechanisms, 83

H

Heart failure
  presympathetic neurons and, 207–208
  sympathetic ganglia and, 324–325
Heart rate (HR)
  effect of catecholamine neurotoxin on, 227f
  effect of DNQX on, 40f
  effect of hypoxic-hypoxia on, 37f
  effect of KCN on, 39f
  effect of kynurenic acid, 42f
  effect of lidocaine on, 50f
  effect of muscimol on, 45f
  effect of potassium cyanide on, 38f
  effect of serotonin neurotoxin on, 227f
  effect of Sham-lesion on, 48f
Hemorrhage
  in brainstem, 103
  caudal medullary raphe neuron response to, 248
  in forebrain, 104
  increases in plasma osmolality, 131
Hippocampus
  long-term potentation of, 316

Index

Histamine
  ganglionic neurotransmission effected by, 311
Hyperpolarization-activated inward current, 11
Hypertension
  cerebrovascular resistance response to, 361
  effect of estrogen on, 177
  pathology of, 110–111
  presympathetic neurons and, 207–208
Hypotension
  increases in plasma osmolality, 131
  NP induced, 102
  platelet activation triggering, 16f
Hypothalamus. *See also* Anterior hypothalamus
  atrial ANP in, 175–176
  cardiovascular connecting regions of, 119–120
  functional role of paraventricular nucleus of, 130–133
  lateral, 126–128
  posterior areas of, 128–129

I

Immunohistochemical studies
  of paraventricular nucleus neurons, 125
In situ hybridisation studies
  of paraventricular nucleus neurons, 125
Inactivating potassium current (IKA)
  4-AP blocking, 72
Inhibitory postsynaptic responses (IPSPs)
  of sympathetic preganglionic neurons, 294
Integrated renal activity (INA)
  effect of catecholamine neurotoxin on, 227f
  effect of serotonin neurotoxin on, 227f
Interneurons
  controlling sympathetic preganglionic neurons, 269
  electrophysical characterization of, 271
  injury induced changes in, 273
  neurochemistry of, 269–271
  neuropeptide-immunoreactive inputs provided by, 274–275
  Spinal autonomic, 270
Intracardiac ganglia
  neural inputs to, 341–344
Intracardiac neurons
  effect of norepinephrine on, 344–345
  effect of NPY on, 344–345

K

KCN
  effect of, on chemoreflex, 36
  effect of, on heart rate, 39f
  effect of, on mean arterial pressure, 39f
  effect of, on pulsatile arterial pressure, 39f
Ketamine
  nicotinic responses inhibited by, 388f
  responses to, 387f
"Knockout" mice
  baroreflex in, 5, 6t
Kynurenic acid
  effect of, on blood pressure, 196
  effect of, on heart rate, 42f
  effect of, on mean arterial pressure, 42f
  effect of, on pulsatile arterial pressure, 42f

L

L-glutamate (L-Glu)
  in RVLM, 108–109
  uses of, 82
Lateral hypothalamus (LH), 126–128
  depressor effects on, 127–128
  efferent projections from, 127f
  pressor effects on, 127–128
Lateral parabrachial nucleus (LPBN)
  paraventricular nucleus (PVN) projection to, 122
Lentiviruses
  transgene expression of, 406f
Lidocaine
  effect of, on heart rate, 50f
  effect of, on mean arterial pressure, 50f
  effect of, on pulsatile arterial pressure, 50f
Long-term potentiation (LTP)
  of sympathetic ganglia, 315–316

M

Magnocellular neuroendocrine neurons
  of paraventricular nucleus (PVN), 148, 153
Mean arterial pressure (MAP). *See also* Blood pressure (BP), Pulsatile arterial pressure (PAP)
  changes in, after ANP injection, 177f
  chemoreceptor activation and, 33
  effect of catecholamine neurotoxin on, 227f
  effect of DNQX on, 40f
  effect of hypoxic-hypoxia on, 37f
  effect of KCN on, 39f

effect of kynurenic acid, 42f
effect of lidocaine on, 50f
effect of muscimol on, 45f
effect of potassium cyanide on, 38f
effect of serotonin neurotoxin on, 227f
effect of Sham-lesion on, 48f
Mechanosensitive ion channels
  implications of, 9
Medulla oblongata
  adrenergic cells of, 188
Medullary raphe neurons
  influence of, on pelvic organ functions, 255–258
  parasympathetic vagal outflow control and, 253–255
Medullary raphe nuclei
  definition of, 245
Medullo-apinal cardiovascular areas, 87f
Membrane depolarization
  mechanically-induced, 11
Membrane potential responses
  to mechanical stimulation, 10f
Midline raphe obscurus
  role of neurons in, 228
Motor control
  role of serotonin in, 233
Muscarinic
  M2 receptors, 312–313
  in rat brain, 84
Muscimol
  effect of, on heart rate, 45f
  effect of, on mean arterial pressure, 45f
  effect of, on pulsatile arterial pressure, 45f
Myocardial Ischaemia, 401–402

N

NADPH-diaphorase
  in RVLM, 206
Neural control systems
  development of, 59
Neurochemical expression
  plasticity of, 341t
Neurohumoral activation
  baroreflex sensitivity and, 17–18
Neuronal circuits
  controlling sympathetic preganglionic neurons, 269
Neuronal NOS (nNOS)
  HR variability and, 18
  location of, 16–17
Neuropeptide Y (NPY)
  effect of, on ganglionic transmission, 345f

effect of, on intracardiac neurons, 344–345
IML innervation by, 277f
Neuropeptides
  cholinergic, 306
  immunoreactive inputs of, 274–275
  neurochemistry of, 270
Neurotensin
  postsynaptic actions of, 313–314
Ni^{2+}-sensitive, low threshold Ca^{2+} conductance (IT)
  properties of, 152–153
Nicotinic transmission, 200, 381f, 382f
  ketamine inhibition of, 388f
Nitric oxide (NO)
  effect of, on baroreceptor sensitivity, 12–14
  effect of, on presympathetic neurons, 206–207
  postsynaptic actions of, 314–315
  synaptic modulation and, 295–296
Nitric oxide synthase (NOS)
  immunoreactivity for, 340f
  in sympathetic preganglionic neurons, 270
NK1 receptors
  RT-PCR analysis of, 324f
NMDA receptors
  in phrenic motor nucleus, 92
  role of, in chemoreflex processing, 38
  role of, in NTS, 39
Non-cholinergic transmitters
  postsynaptic actions of, 313–315
Non-NMDA receptors
  in nTS, 89
Norepinephrine (NE)
  effect of, on ganglionic transmission, 345f
  effect of, on intracardiac neurons, 344
Nucleui tracti solitarii (NTS)
  adenoviral mediated transfection of, 404f
  afferent information in, 61–62
  capsaicin-sensitive, 72f
  cardiovascular region of, 364
  caudal commissural, 34
  chemoreflex afferents at, 32
  chemoreflex neuromodulation in, 45–46
  excitatory postsynaptic currents produced in, 70f
  heterogeneity of, 73
  lateral, 34
  paraventricular nucleus (PVN) projection to, 123
  respiratory functions mediated by, 44
  role of, in cardiovascular regulation, 82
  role of NMDA receptors in, 39

Index

stimulation of, 377f
synapse formation in, 67f
Nucleus ambiguus (NA)
  descending projections to, 150

O

l-glutamate
  effect of, on chemoreflex, 35
Opioid peptides
  effect of, on presympathetic neurons, 203–204
  postsynaptic actions of, 314
Opioid receptors
  expression of, 203
  premotor cardiac vagal neuron activation by, 383–385
  subtypes of, 85
Opioidergic nTS mechanisms, 85
Orexins
  effect of, on presympathetic neurons, 205
Organum vasculosum of lamina terminalis (OVLT)
  role of, in plasma sodium regulation, 172
Oxidative stress
  effect of, on baroreceptor sensitivity, 14–15
Oxytocin (OT)
  effect of, on presympathetic neurons, 205
  release of, 120

P

P2 receptors
  suramin blockade of, 44f
PACAP peptides
  depolarizing effects of, 347–350
  effect of, on cardiac neurons, 348f
  effect of phospholipase (PLC) inhibitor on, 349f
Pain pathways
  in parapyramidal region, 231
Pancreas
  vagal parasympathetic input to, 253
Parabrachial nucleus (PBN)
  in cardiovascular control, 49–51
  role of, 32
Paracrine mediators
  ganglionic transmission effected by, 311
Paracrine modulation
  of baroreceptor sensitivity, 13–16
Paragigantocellular reticular nucleus (PGi)
  bilateral lesion of, 258f
Parapyramidal region
  bulbospinal serotonin neurons in, 226
  neuron activation in, 225
  pain pathways in, 231
  significance of, 220
Parasympathetic innervation, 359
Paraventricular nucleus (PVN)
  anatomical subdivisions of, 149f
  autonomic related neurons of, 150–152
  bilateral electrolytic lesions of, 47
  chemical stimulation of, 109
  cytoarchitectural organization of, 148–149
  divisions of, 120
  DVC projecting neurons of, 156f
  electrical stimulation of, 109
  electrophysical properties of pre-autonomic, 152–154
  functional role of, 130–133
  functional subdivisions of, 149f
  GABAergic, 133
  immunohistochemical studies of, 125
  low-threshold spike (LTS) of neurons of, 153f
  neuron firing patterns of, 154f
  neuron heterogeneity in, 156–157
  neuron morphological patterns of, 154–156
  plasma volume regulatory role of, 132
  preautonomic neurons of, 151f
  principal autonomic efferent projections from, 121f
  projection to cardiac vagal neurons, 123
  projection to lateral parabrachial nucleus (LPBN), 122
  projection to nuclei tracti solitarii (NTS), 123
  projection to periaqueductal grey (PAG), 122
  projection to spinal sympathetic neurons, 124–126
  projection to ventral medulla, 124
  role of, 32
  role of, in central autonomic control, 149–150
  as sympathoexcitatory component, 46–49
Parvocellular neuroendocrine neurons
  of paraventricular nucleus (PVN), 148
Parvocellular neurones
  corticotrophin releasing factor in, 121
Parvocellular pre-autonomic neuroendocrine neurons
  of paraventricular nucleus (PVN), 148
Pentobarbital

responses to, 390f
Periaqueductal grey (PAG)
   paraventricular nucleus projection to, 122
Phenylephrine (PHE)
   effect of, on VLM, 105
   uses for, 104
Phenylethanolamine-N-methyl transferase (PNMT)
   upregulation of, 207
   uses of, 188
Phospholipase (PLC) inhibitor
   effect of, on PACAP peptide, 349f
Phrenic motor nucleus (PMN)
   NMDA receptors in, 92
Plasma sodium regulation
   detection of, 171–'73
   diets effecting, 169–171
   pathways of, 173f
   role of organum vasculosum in, 172
Platelet activation
   effect of, on baroreceptor sensitivity, 15–16
   hypotension triggered by, 16f
Posterior cricoarytenoid (PCA)
   reconstruction of, 234f
Posterior hypothalamic region (post. Hyp)
   efferent projections from, 127f, 128–129
Postganglionic sympathetic neurons
   influencing cardiac function, 320
Postganglionic transmitters, 308–310
Postsynaptic differentiation, 71–74
Preganglionic sympathetic neurons
   cholinergic, 306
Preganglionic transmitters, 306–308
Premotor cardiac vagal neurons
   baroreflex activation of, 376–378
   follower-cell properties of, 375
   intrinsic activity of, 372–375
   opioid receptor activation of, 383–385
   respiratory modulation of, 378–383
   sensitivity of, to anesthetics, 385–391
Preoptic area
   of dorsomedial hypothalamic nucleus (DMH), 130
Presympathetic cardiac neurons, 374f
Presympathetic neurons
   effect of angiotensin II on, 202–203
   effect of catecholamines on, 201–202
   effect of corticotrophin releasing factor on, 205
   effect of glucagon-like peptide on, 205–206
   effect of nitric oxide on, 206
   effect of opioid peptides on, 203–204
   effect of orexins on, 205
   effect of oxytocin on, 205
   effect of peptides on, 202–207
   effect of serotonin on, 201–202
   effect of substance P, 204
   glutamatergic inputs of, 200
   heart failure and, 207–208
   hypertension and, 207–208
   immunoreactivity to, 204
   intrinsic properties of, 198–199
   organization of, 192–194
   role of glutamate in discharge of, 196
   synaptic properties of, 199–200
Presynaptic differentiation, 65–71
Presynaptic modulation
   in ganglionic transmission, 310–312
Primary afferents
   injury induced changes in, 272
Prostacyclin
   effect of, on baroreceptor sensitivity, 13–14
Protein kinase B (PKB)
   in heart, 402f
   over expression of, 401–402
Pseudorabies
   Bartha strain of, 382
   viral retrograde tracing, 255
Pulsatile arterial pressure (PAP). *See also* Blood pressure (BP), Mean arterial pressure (MAP)
   effect of DNQX on, 40f
   effect of hypoxic-hypoxia on, 37f
   effect of KCN on, 39f
   effect of kynurenic acid, 42f
   effect of lidocaine on, 50f
   effect of muscimol on, 45f
   effect of potassium cyanide on, 38f
   effect of Sham-lesion on, 48f
Putative autocrine factors
   effect of, on baroreceptor sensitivity, 16

R

Raphe neurons
   role of, vasoconstriction, 246–248
Raphe pallidus
   role of neurons in, 228
Reactive oxygen species (ROS)
   baroreceptor activity effected by, 12–13
Renin-angiotensin-aldosterone system
   activation of, 17–18
Respiratory sinus arrhythmia, 378–383

Retrochiasmatic area
  efferent projections from, 128
Rostral medullary raphe
  activation of pallidus of, 251
  role of, in vasoconstriction, 251
  thermoregulation and, 250–253
Rostral ventrolateral medulla (RVLM)
  angiotensin I receptors in, 202
  electrophysical characterization of, 190–192
  enkephalin in, 204
  GABA-containing neurons in, 83
  intrinsic properties of neurons of, 198–199
  L-glutamate injections in, 108–109
  neuron activity in, 220
  neuron organization of, 192–194
  neurons of, 34
  nitric oxide in, 206–207
  nuclei locations in, 221f
  presympathetic neurons of, 190
  regions surrounding, 220
  synaptic inputs from, 289
  synaptic properties of neurons of, 199–200
Rostral ventromedial medulla (RVMM)
  anatomy of, 222–223
  cardiovascular functions of, 225–231
  chemosensation functions of, 232
  definition of, 222–223
  functions of, in pain, 231
  microinjection sites in, 226
  motor control functions of, 233
  neurochemistry of, 223–225
  neuron variation in, 222
  sexual functions of, 234–235
  sleep-wake cycle functions of, 233–234
  synaptic inputs from, 289
  thermoregulatory functions of, 233
RT-PCR analysis
  of NK1 receptors, 324f

S

Sensory transduction
  baroreflex involvement with, 4
Serotonin
  effect of, on presympathetic neurons, 201–202
  role of, in chemosensation, 232
  role of, in motor control, 233
  role of, in sexual function, 234–235
  role of, in sleep/wake cycle, 233–234
  role of, in thermoregulation, 232–233
Serotonin efflux, 228f
Serotonin neurotoxin
  effect of, on blood pressure, 227f
  effect of, on heart rate, 227f
  effect of, on renal nerve activity, 227f
Sexual function
  effect of bilateral lesion on, 258f
  role of medullary raphe neurons in, 257
  role of serotonin in, 234–235
Sham-lesion (SHAM)
  effect of, on heart rate, 48f
  effect of, on mean arterial pressure, 48f
  effect of, on pulsatile arterial pressure, 48f
Sinoaortic denervation
  effect of, in brainstem, 104–105
  neuron activation reduced by, 107
Sleep/wake cycle
  role of serotonin in, 233–234
Sodium nitroprusside (NP)
  baroreflex activity after administering, 5
  decreasing blood pressure with, 100–101
  hypotension induced by, 102
Solitary tract (ST)
  electrical activation of, 70f
Somatic gene transfer, 401–404
Somatic motor neurons
  of thoracic spinal cord, 287
Spike initiating zone (SIZ)
  action potential at, 8
Spinal sympathetic neurons
  paraventricular nucleus (PVN) projection to, 124–126
Spontaneously hypertensive rats (SHRs)
  blood pressure of, 165f
  effect of substance P on, 321f
  hypertension in, 110–111
  plasma sodium concentrations in, 171f
  plasma sodium rhythm in, 170f
  salt-sensitive hypertension in, 166f
  tachykinergic system of, 323
Stellate ganglion
  reactions of, 307f, 309f
Substance P
  depolarization effects of, 346–347
  effect of, on presympathetic neurons, 204
  immunoreactive axons of, 274–275
  immunoreactivity for, 340f
  recorder tracings of, 321f
  role of, in neurotransmission, 42
Superior cervical ganglion (SCG)
  dopaminergic neurons in, 308
  model system of, 310
Sympathetic ganglia
  anatomical complexity of, 305

chemical coding in, 306–310
early concepts of, 304–305
effect of Angiotensin II on, 317–318
effect of atrial natriuretic peptide on, 318–319
efferent-motor pathways transversing, 319
heart failure and, 324–325
in hypertension, 321–324
influencing blood pressure, 320
input integration of, 305–306
long-term potentiation of, 315–316
sensory innervation of, 319
Sympathetic nerve activity (SNA)
excessive, 17
excitatory responses in, 230f
Sympathetic nerves
role of, in cerebrovascular innervation, 358–359
Sympathetic preganglionic neurons (SPN)
afferent information to, 268
central circuits including, 267f
excitatory postsynaptic potentials of, 292–294
IML innervation by, 276f
immunoreactive inputs for, 274–275
inhibitory postsynaptic responses of, 294
injury induced changes in, 272
input from brain to, 267–268
input to, 219–220
interneuronal circuits contributing to, 266–268
interneurons controlling, 269
location of, 286–287
membrane currents of, 291
membrane properties of, 290
morphology of, 288
phenotypes of, 289
in rats, 287f
role of, 266
short range local circuits involving, 268
synaptic inputs of, 289
synaptic transmission modulation of, 294–295
synaptic transmission of, 292
in thoracolumbar spinal cord, 101–102
Sympathetic reflexes
role of BSBS neurons in, 197
Sympathoexcitatory component
pathophysiological implications of, 51–52
Sympathoinhibition
spinally-projecting neuron mediation of, 248

T

Tachykinergic system
upregulation of, 323
Tachykinin receptors
activation of, 346–347
Tachykinins
depolarization of, 347f
Thermogenesis
in brown adipose tissue, 252
Thermoregulation
role of serotonin in, 232–233
rostral medullary raphe and, 250–253
Thoracic spinal cord
somatic motor neurons of, 287
3-hydroxy-4-methoxyphenylglycol (MOPEG)
levels of, 167f
Transgenic animals
baroreflex in, 5, 6t
Transmural arterial BP
changes in, 3
TRH
role of, in gastric function, 256
synthesizing, 254
Trigeminal nerve
sensory fibers of, 358
Tyrosine hydroxylase (TH)
immunoreactivity to, 308, 340f

V

Vagal efferent control
molecular determinants of, 18
Vasoactive hormones
effect of, on baroreceptor activity, 12
Vasoconstriction
role of raphe neurons in, 246–248
role of rostral medullary raphe neurons in, 251
Vasomotor tone
control of, 194–195
origin of, 195–197
Ventrolateral medulla (VLM)
effect of phenylephrine on, 105
paraventricular nucleus (PVN) projection to, 124
Vesicular glutamate transporter 2 (VGLUT2)
in bulbospinal C1 cells, 191–192
Vestibulo-sympathetic reflexes
of cats, 193–194
Viral vectors
cell specific promoters for, 406
long-term expression of, 405

uses of, 405
vigilant, 407
Visceral afferents
  central axons of, 67f
Voltage-gated ion channels, 11–13
VR1
  capsaicin binding with, 70

W
Wistar-Kyoto rats (WKY)
  blood pressure of, 165f
  effect of substance P on, 321f
  hypertension in, 110–111
  plasma sodium concentrations in, 171f
  plasma sodium rhythm in, 170f
  salt-sensitive hypertension in, 166f